Trends in Control
A European Perspective

Springer
Berlin
Heidelberg
New York
Barcelona
Budapest
Hong Kong
London
Milan
Paris
Tokyo

**Books are to be returned on or before
the last date below.**

Alberto Isidori (Ed.)

Trends in Control
A European Perspective

With 101 Figures

 Springer

Professor Alberto Isidori
Dipartimento di Informatica e Sistemistica,
Università degli Studi di Roma "La Sapienza",
Via Eudossiana 18, 00184 Roma, Italy

ISBN 3-540-19967-5 Springer-Verlag Berlin Heidelberg New York

British Library Cataloguing in Publication Data
Trends in Control: European Perspective
  I. Isidori, Alberto
  629.89
ISBN 3-540-19967-5

Library of Congress Cataloging-in-Publication Data
Trends in control : a European perspective / Alberto Isidori, ed.
      p. cm.
  Papers prepared for the European Control Conference held in Rome, Italy,
Sept. 5-8, 1995.
  Includes bibliographical references.
  ISBN 3-540-19967-5 (alk. paper)
  1. Automatic control - Congresses. I. Isidori, Alberto.
TJ212.2.T73 1995   95-20217
629.8 - dc20   CIP

Apart from any fair dealing for the purposes of research or private study, or criticism or review, as permitted under the Copyright, Designs and Patents Act 1988, this publication may only be reproduced, stored or transmitted, in any form or by any means, with the prior permission in writing of the publishers, or in the case of reprographic reproduction in accordance with the terms of licences issued by the Copyright Licensing Agency. Enquiries concerning reproduction outside those terms should be sent to the publishers.

© Springer-Verlag London Limited 1995
Printed in Great Britain

The publisher makes no representation, express or implied, with regard to the accuracy of the information contained in this book and cannot accept any legal responsibility or liability for any errors or omissions that may be made.

Typesetting: Camera ready by authors
Printed and bound by the Athenæum Press Ltd., Gateshead
69/3830-543210 Printed on acid-free paper

Preface

This book contains the text of the plenary lectures and the mini-courses of the European Control Conference (ECC 95) held in Rome, Italy, September 5–September 8, 1995. In particular, the book includes nine essays in which a selected number of prominent authorities present their views on some of the most recent developments in the theory and practice of control systems design and three self-contained sets of lecture notes.

Some of the essays are focused on the topic of robust control. The article by J. Ackermann describes how to robustly control the rotational motions of a vehicle, to the purpose of simplifying the driver's task. The contribution by H. Kwakernaak presents a detailed discussion of the requirements that performance and robustness impose on control systems design and of the symmetric roles of sensitivity and complementary sensitivity functions. The article by P. Boulet, B.A. Francis, P.C .Hughes and T. Hong describes an experimental testbed facility, called Daisy, whose dynamics emulate those of a real large flexible space structure and whose purpose is to test advanced identification and control design methods. The article of K. Glover discusses recent advances in uncertain system modeling, analysis and design, with reference to a flight control case study that has been test flown.

The other essays describe advances in fundamental problems of control theory. The article by V.A. Yakubovich is a survey of certain new infinite-horizon linear-quadratic optimization problems. The contribution by A.S. Morse is a detailed presentation of the major problems of analysis and design occurring in the largely unexplored area of logic-based switching control systems. The article of H. Kimura and F.Okunishi proposes a new design framework of control systems, based on chain scattering representation of the plant. The contribution by C. Scherer is a thorough presentation of the solution to the mixed H_2/H_∞ control problem with reduced order controllers for time-varying systems, in terms of solvability of differential linear matrix inequalities. The paper by B. De Moor and P. Van Overschee presents the basic notions on subspace identification algorithms for linear systems.

In addition to these lectures, the "special events" of ECC 95 included three mini-courses. The first mini-course, organized by C.G. Cassandras, S. Lafortune and G.J. Olsder, is an introduction to modeling, control and optimization of discrete event systems. The second mini-course, organized by

J.-M. Coron, L. Praly and A. Teel is devoted to the survey of a number of recent advances in the problem of feedback stabilization of nonlinear systems. The third mini-course, organized by R. Frezza, P. Perona, G. Picci and S. Soatto, discusses how the introduction of computer vision in a control loop raises exciting and yet unexplored problems in system theory.

The Organizing Committee of ECC 95 would like to take this occasion to thank all of those who have contributed to this book for their outstanding works and for their timely response.

Rome, March 30, 1995

Alberto Isidori, *ECC 95 Chairman*

Table of Contents

Safe and Comfortable Travel by Robust Control
  J. Ackermann 1

Symmetries in Control System Design
  H. Kwakernaak 17

Universal Regulators in Linear-Quadratic Optimization Problems
  V. A. Yakubovich 53

Control Using Logic-Based Switching
  A. S. Morse 69

Daisy: A Large Flexible Space Structure Testbed for Advanced Control Experiments
  B. Boulet, B.A. Francis, P.C. Hughes, T. Hong 115

Progress in Applied Robust Control
  K. Glover 141

Chain-Scattering Approach to Control System Design
  H. Kimura, F. Okunishi 151

Mixed H_2/H_∞ Control
  C. Scherer 173

Introduction to the Modelling, Control and Optimization of Discrete Event Systems
  C.G. Cassandras, S. Lafortune, G.J. Olsder 217

Feedback Stabilization of Nonlinear Systems: Sufficient Conditions and Lyapunov and Input-output Techniques
  J.M. Coron, L. Praly, A. Teel 293

System-Theoretic Aspects of Dynamic Vision
  R. Frezza, P. Perona, G. Picci, S. Soatto 349

Numerical Algorithms for Subspace State Space System Identification
  B. De Moor, P. Van Overschee 385

Safe and Comfortable Travel by Robust Control

Juergen Ackermann

Institute for Robotics and System Dynamics, DLR, German Aerospace Research Establishment, Oberpfaffenhofen,
Wessling 82230, Germany

Abstract. Travellers do not want to be rotated more than required by the travel path, but most drivers are not good at controlling rotations. A generic controller for generic vehicles is introduced by which the driver is concerned only with lateral motions to keep a point mass on the travel path. disturbance torques are attenuated by feedback control.

1 On Travelling

People like to travel. Fig. 1 shows a traveller and his vehicle.

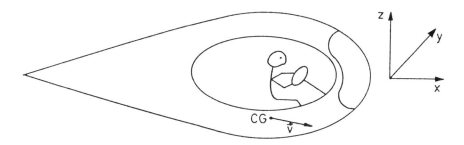

Fig. 1. Traveller in a vehicle with coordinates x, y, z and velocity $\vec{v}$

The vehicle is optimized for motion in the x-direction. Therefore we travel such that the velocity vector $\vec{v}$ at the center of gravity (CG) is aligned with the x-axis, practically it forms a small slip angle with the x-axis and the component v_x is dominating compared to v_y and v_z, where v_x, v_y and v_z are the velocity components in the respective directions.

Travelling means following a path from A to B in an inertial coordinate system x_0, y_0, z_0; $\vec{v}$ is tangential to the path. The velocity component $v = v_x$ satisfies $v \geq 0$ and we distinguish the phases
$\dot{v} > 0$ acceleration
$\dot{v} = 0$ cruising
$\dot{v} < 0$ braking.
In this paper only the cruising phase with a constant $v > 0$ will be analyzed.

The vehicle is equipped with actuators like rudders, engines, gas or water jets, wheels that can be steered and accelerated by motors and brakes. Commands by the driver (pilot, helmsman) are converted by such devices into forces f_x, f_y, f_z in the respective directions and torques m_x, m_y, m_z around the respective axes. Usually it is a difficult task to model the nonlinear dynamics between driver command inputs and force and torque outputs. The driver learns to master these nonlinear dynamics and to adapt to changing operating conditions, for example for driving a car on a dry or a slippery road, or flying an airplane in different altitudes.

The largest force is f_x, it provides the acceleration $\dot{v}$ and balances the sum of all drag forces when travelling in water or air. The forces f_y and f_z are used to keep the CG on the planned path. There are different preferences of travellers; the pre-planners like to select only the destination B and the departure time at A, the ad-hoc planners enjoy the freedom of changing the path at any time or making an unplanned stop at a nice place. The latter driver is primarily interested in controlling f_y when travelling on the surface of the earth. Guidance by commanding f_y (and possibly f_z) is called "point mass steering", because the vehicle is considered only as a point mass at its CG, that must be kept on the path.

The actual vehicle is not a point mass, it has a moment of inertia and torques that rotate it. Only small rotations are necessary in order to follow the path curvature, other rotations are not needed, they are even dangerous. Safe and comfortable travel is for example provided by railroads that constrain the rotations around the y and z axis mechanically to very small deviations from those that are required by the purpose of following the track. For rotations around the x axis active control can improve the comfort as shown by the pendolino train. Also in other vehicles the traveller does not want to be rotated unnecessarily. So, why do vehicles rotate?

The first reason is that typical actuators do not only produce forces in the y and z directions. In an aircraft f_z is normally produced via a change of the angle of attack by the elevator (unless the aircraft is equipped with direct lift control by a pair of rudders). The same applies to f_y of a ship or car. Even in cars with four-wheel stering a lane change maneuver requires some yaw rotation. A helicopter must rotate around the x or y axis for acceleration in the y or x direction. Therefore rotations are unavoidable. Torques in m_x may also be produced intentionally for passenger comfort in curves (e.g. in an aircraft).

A second reason for rotations are disturbance torques resulting for example from wind and water flow. Vehicles are designed essentially symmetric with respect to the (x, z)-plane but there are asymmetries for example for an aircraft engine failure, flat tire or one-sided ice or aquaplaning on the road.

For the driver it is a difficult task to control the rotations. This applies in particular to unexpected disturbance torques. It takes the driver between 0.5 and 1 seconds to react, and then he/she may overreact and produce driver-induced oscillations by too high gain.

In the present paper a concept for automatic feedback control of the rotational motions is proposed. It is based on feedback of angular velocities and accelerations. The primary goal is to achieve a triangular decoupling such that the driver is concerned only with a subsystem for the generation of the force f_y (and/or f_z) in order to keep the vehicle on the planned path. All rotations that deviate from those nominally associated with steering maneuvers should be automatically corrected by the feedback system. Since the gyro and accelerometer feedback works without the reaction time of the driver, it counteracts immediately to disturbances. Thereby large rotations, that may bring the vehicle to its physical limits, are avoided.

An essential feature of the control concept is, that it is robust with respect to uncertain vehicle mass and velocity and it is robust with respect to the uncertain nonlinear steering dynamics producing the forces f_y and f_z.

2 The Vehicle Model

Consider the vehicle motion in the (x,y)-plane, i.e. the horizontal plane. The model and control concepts apply as well to the (x,z)-plane, i.e. the vertical plane. Couplings between these two rotations are neglected because the control system will keep the angular rates small.

Assume that the lateral force f_y and the torque m_z are produced by a pair of forces f_f and f_r as shown in Fig. 2.

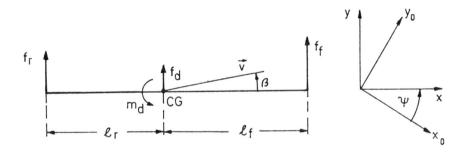

Fig. 2. Steering forces f_f and f_r and disturbance force f_d and torque m_d, (x_0, y_0) is an inertially fixed coordinate system and ψ is the heading angle.

The velocity vector $\vec{v}$ forms a sideslip angle β with the x-axis. (For the vertical plane of aircraft motion β is replaced by the angle of attack α.)

The distances ℓ_f and ℓ_r from the CG are chosen such that one of them corresponds to the location where the steering command of the driver acts, e.g. at the front axle of a car or at the rudder of a ship or airplane. The other distance then describes the position where the reactive forces may be lumped, e.g. the lift of an airplane or the center of lateral forces on the underwater part of a ship or the rear axle of a car. For uniqueness assume that f_f is controlled by the driver and f_r is a reactive force.

Assume that also a disturbance force f_d and a disturbance torque m_d act on the vehicle as indicated in Fig. 2. The total force and torque are then

$$\begin{bmatrix} f_y \\ m_z \end{bmatrix} = \begin{bmatrix} 1 & 1 \\ \ell_f & -\ell_r \end{bmatrix} \begin{bmatrix} f_f \\ f_r \end{bmatrix} + \begin{bmatrix} f_d \\ m_d \end{bmatrix} \qquad (1)$$

The forces are unknown, also the mass and moment of inertia of the vehicle is uncertain. We can, however, measure the effects of the forces by a rate gyro, that measures the yaw rate $r = \dot{\psi}$, and by two accelerometers for the lateral accelerations a_f and a_r at the positions of the forces f_f and f_r respectively. If they cannot be mounted in these positions, then a_f and a_r may also be calculated by linear interpolation between the measurements by two accelerometers in different positions, e.g. at the front and rear ends of the vehicle.

For the mass m and moment of inertia J_z w.r.t. the z-axis we assume that both are uncertain, but tied together by a known radius of inertia i, i.e.

$$J_z = mi^2 \qquad (2)$$

with $m \in [m^-; m^+]$. This mass distribution may also be represented by two rigidly connected point masses as shown in Fig. 3.

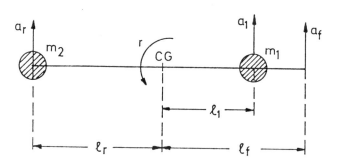

Fig. 3. Representation of the mass distribution by two rigidly connected point masses m_1 and m_2

The quantities m and J_z are related with m_1 and m_2 by

$$m = m_1 + m_2 \qquad (3)$$

$$J_z = m_1 \ell_1^2 + m_2 \ell_r^2 \qquad (4)$$

$$m_1 \ell_1 = m_2 \ell_r \qquad (5)$$

Equation (4) may be expressed using (5) as $J_z = m_2 \ell_r \ell_1 + m_1 \ell_1 \ell_r$ and with (2) and (3)

$$J_z = m \ell_1 \ell_r = mi^2 \qquad (6)$$

2 The Vehicle Model

i.e. the distance ℓ_1 of the mass m_1 from the CG is determined by

$$\ell_1 = \frac{i^2}{\ell_r} \tag{7}$$

and it does not involve the uncertain vehicle mass m. The above calculation only serves the purpose to define ℓ_1, in our model we use m and $\ell_1 = J_z/m\ell_r$ rather than m_1 and m_2. The lateral acceleration of the mass m_1 is by interpolation between a_f and a_r

$$a_1 = \lambda a_f + (1-\lambda)a_r \ , \ \lambda = (\ell_r + \ell_1)/(\ell_r + \ell_f) \tag{8}$$

(For vehicles steered by f_r the mass m_1 is assumed in a distance ℓ_f from the CG, the mass m_2 then has a distance $\ell_2 = i^2/\ell_f$ from the CG.)

Fig. 2 shows that the velocity vector $\vec{v}$ forms an angle $\beta + \psi$ with the inertial coordinate system (x_0, y_0), thus the equations of motion for $\dot{v} = 0$ and small β are

$$\begin{bmatrix} mv(\dot{\beta} + \dot{\psi}) \\ m\ell_r\ell_1\ddot{\psi} \end{bmatrix} = \begin{bmatrix} f_y \\ m_z \end{bmatrix}$$

and with $\dot{\psi} = r$ and (1)

$$\begin{bmatrix} mv(\dot{\beta} + r) \\ m\ell_r\ell_1\dot{r} \end{bmatrix} = \begin{bmatrix} 1 & 1 \\ \ell_f & -\ell_r \end{bmatrix} \begin{bmatrix} f_f \\ f_r \end{bmatrix} + \begin{bmatrix} f_d \\ m_d \end{bmatrix} \tag{9}$$

Solving for $\dot{\beta}$ and $\dot{r}$ yields

$$\begin{bmatrix} \dot{\beta} \\ \dot{r} \end{bmatrix} = \begin{bmatrix} 1/mv & 1/mv \\ \ell_f/m\ell_r\ell_1 & -1/m\ell_1 \end{bmatrix} \begin{bmatrix} f_f \\ f_r \end{bmatrix} - \begin{bmatrix} 1 \\ 0 \end{bmatrix} r + \begin{bmatrix} f_d/mv \\ m_d/m\ell_r\ell_1 \end{bmatrix} \tag{10}$$

In order to complete the state equations, we must establish, how the forces f_f and f_r depend on β and r. Also an input in form of a deflection angle δ_f of a rudder or wheel enters into the actuated force, here f_f. Fig. 4 shows the deflection angle and the orientation of the local velocity vector $\vec{v}_f$.

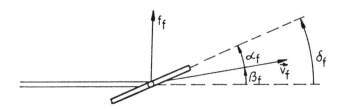

Fig. 4. Angles at a rudder or wheel, f_f is the force acting on the vehicle.

If the velocity vector $\vec{v}_f$ is aligned with the rudder or wheel, then it produces no steering force. The steering force f_f depends on the angle

$$\alpha_f = \delta_f - \beta_f \tag{11}$$

where δ_f is the steering angle commanded by the driver and/or the controller; β_f is the sideslip angle in a distance ℓ_f in front of the CG. The kinematic relationship between different sideslip angles is illustrated by Fig. 5.

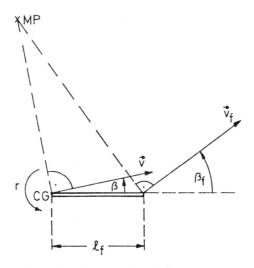

Fig. 5. Kinematic relationship between β and β_f

For a rotation of the vehicle around a momentary pole MP the velocity vectors $\vec{v}$ and $\vec{v}_f$ must be perpendicular to the radius from MP and they must have the same projection on the x-axis

$$v_f \cos \beta_f = v \cos \beta$$

The velocity components in the y-direction are

$$v_f \sin \beta_f = v \sin \beta + \ell_f r$$

Dividing both sides of the second equation by the respective sides of the first one yields

$$\tan \beta_f = \tan \beta + \frac{\ell_f r}{v \cos \beta}$$

and for small slip angles

$$\beta_f = \beta + \ell_f r / v \tag{12}$$

2 The Vehicle Model

We assume that the dependence of f_f on α_f is described by an unknown nonlinear differential equation

$$\dot{\mathbf{x}}_f = g_f(\mathbf{x}_f, \alpha_f)$$

$$f_f = h_f(\mathbf{x}_f, \alpha_f) \tag{13}$$

A similar equation also applies to the rear lateral force f_r, i.e.

$$\dot{\mathbf{x}}_r = g_r(\mathbf{x}_r, \alpha_r)$$

$$f_r = h_r(\mathbf{x}_r, \alpha_r) \tag{14}$$

where $\delta_r = 0$, i.e. $\alpha_r = -\beta_r$ and $\beta_r = \beta - \ell_r r/v$. Now combine (13) and (14) with (10) to the state space model

$$\begin{bmatrix} \dot{\beta} \\ \dot{\mathbf{x}}_f \\ \dot{r} \\ \dot{\mathbf{x}}_r \end{bmatrix} = \begin{bmatrix} (h_f + h_r)/mv - r \\ g_f \\ (\ell_f h_f - \ell_r h_r)/m\ell_r \ell_1 \\ g_r \end{bmatrix} + \begin{bmatrix} f_d/mv \\ 0 \\ m_d/m\ell_r \ell_1 \\ 0 \end{bmatrix} \tag{15}$$

The arguments of h_f, h_r, g_f and g_r are $\mathbf{x}_f, \mathbf{x}_r, \alpha_f = \delta_f - \beta - \ell_f r/v$, and $\alpha_r = -\beta + \ell_r r/v$. Actuated input is δ_f, disturbance inputs are f_d and m_d. Measured quantities are the velocity v (for gain scheduling), the yaw rate r and the lateral accelerations a_1 and a_r of the masses m_1 and m_2 respectively. They are related with the quantities in (15) by

$$a_1 = a_{CG} + \ell_1 \dot{r}$$

$$= (f_f + f_r + f_d)/m + (\ell_f f_f - \ell_r f_r + m_d)/m\ell_r$$

$$= [(\ell_r + \ell_f)h_f + \ell_r f_d + m_d]/m\ell_r \tag{16}$$

Note that the force f_r does not enter into a_1.

$$a_r = a_{CG} - \ell_r \dot{r}$$

$$= (f_f + f_r + f_d)/m - (\ell_f f_f - \ell_r f_r + m_d)/m\ell_1$$

$$= [(\ell_1 - \ell_f)h_f + (\ell_1 + \ell_r)h_r + f_d \ell_1 - m_d]/m\ell_1 \tag{17}$$

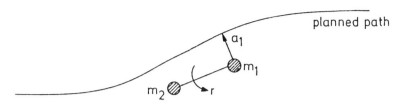

Fig. 6. Illustration of decoupling

3 Robust Decoupling of Lateral and Yaw Motions

In this section we present a decoupling concept that provides the driver with a situation as illustrated by Fig. 6.

The driver should be concerned only with the lateral acceleration a_1 of the front mass m_1 in order to bring m_1 on top of his planned path. The yaw motion involving r should be robustly decoupled from the task of the driver. This purpose is achieved by the main result of this paper:

Theorem 1. *The control law*

$$\dot{\delta}_c = -r + \frac{\ell_f - \ell_1}{v(\ell_r + \ell_1)}(a_1 - a_r) + k_L(v)w \tag{18}$$

$$\delta_f = \delta_c + w$$

robustly decouples a_1 from the influence of the yaw rate r. The steering input commanded by the driver is w. □

Before the proof of the theorem, the control law and its decoupling effect is first illustrated by Figs. 7 and 8.

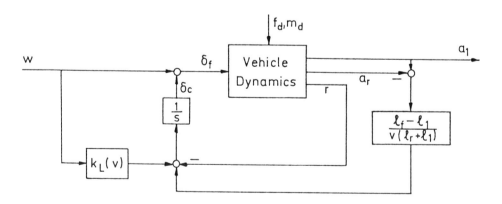

Fig. 7. Block diagram of robust decoupling control law

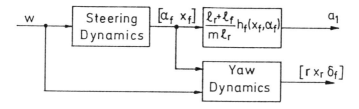

Fig. 8. Structure of the decoupled control system

Remark Note that the control law (18) is simplified for an "ideal mass distribution" $\ell_1 = \ell_f$; then no accelerometers for a_1 and a_r are needed and the gain scheduling by the velocity applies only to the prefilter $k_L(v)$. In our experiments with a car (BMW 735i) it turned out that it has the ideal mass distribution, so $\ell_1 = \ell_f$ is not an exotic case. Note also that the term $(a_1 - a_r)/(\ell_r + \ell_1)$ is equal to $\dot{r}$ and may alternatively be generated by differentiation of r.

Proof of the Theorem Substitute (16) and (17) into the control law (18)

$$\dot{\delta}_c = -r + \frac{\ell_f - \ell_1}{mv\ell_r\ell_1}[\ell_f h_f - \ell_r h_r + m_d] + k_L(v)w \tag{19}$$

Equation (15) is now augmented by (19)

$$\begin{bmatrix} \dot{\beta} \\ \dot{\mathbf{x}}_f \\ \dot{r} \\ \dot{\mathbf{x}}_r \\ \dot{\delta}_c \end{bmatrix} = \begin{bmatrix} (h_f + h_r + f_d)/mv - r \\ g_f \\ (\ell_f h_f - \ell_r h_r + m_d)/m\ell_r\ell_1 \\ g_r \\ (\ell_f h_f - \ell_r h_r + m_d)(\ell_f - \ell_1)/mv\ell_r\ell_1 - r + k_L(v)w \end{bmatrix} \tag{20}$$

and the argument α_f of h_f and g_f is introduced by (11), (12), and (18) as $\alpha_f = \delta_f - \beta_f = \delta_c + w - \beta - \ell_f r/v$. Then for $\dot{v} = 0$ $\dot{\alpha}_f = \dot{\delta}_f + \dot{w} - \dot{\beta} - \ell_f \dot{r}/v$ and α_f is introduced as a model variable instead of β, i.e.

$$\begin{bmatrix} \dot{\alpha}_f \\ \dot{\mathbf{x}}_f \\ \dot{r} \\ \dot{\mathbf{x}}_r \\ \dot{\delta}_c \end{bmatrix} = \begin{bmatrix} -h_f(\mathbf{x}_f,\alpha_f)(\ell_f + \ell_r)/mv\ell_r \\ g_f(\mathbf{x}_f,\alpha_f) \\ [\ell_f h_f(\mathbf{x}_f,\alpha_f) - \ell_r h_r(\mathbf{x}_r,\alpha_r)]/m\ell_r\ell_1 \\ g_r(\mathbf{x}_r,\alpha_r) \\ [\ell_f h_f(\mathbf{x}_f,\alpha_f) - \ell_r h_r(\mathbf{x}_r,\alpha_r)](\ell_f - \ell_1)/mv\ell_r\ell_1 - r \end{bmatrix} +$$

$$+ \begin{bmatrix} k_L(v)w + \dot{w} - (m_d + f_d\ell_r)/mv\ell_r \\ 0 \\ m_d/m\ell_r\ell_1 \\ 0 \\ k_L(v)w + m_d(\ell_f - \ell_1)/mv\ell_r\ell_1 \end{bmatrix} \quad (21)$$

The "steering dynamics" of Fig. 8 are then given by the first two rows of (21) i.e.

$$\begin{bmatrix} \dot{\alpha}_f \\ \dot{\mathbf{x}}_f \end{bmatrix} = \begin{bmatrix} -h_f(\mathbf{x}_f, \alpha_f)(\ell_f + \ell_r)/mv\ell_r \\ g_f(\mathbf{x}_f, \alpha_f) \end{bmatrix} +$$

$$+ \begin{bmatrix} k_L(v)w + \dot{w} - (m_d + f_d\ell_r)/mv\ell_r \\ 0 \end{bmatrix} \quad (22)$$

There is no coupling from the states $r, \mathbf{x}_r, \delta_c$ of the "yaw dynamics". The yaw dynamic states are unobservable from $[\alpha_f, \mathbf{x}_f]$ and thereby also from

$$a_1 = [(\ell_f + \ell_r)h_f(\mathbf{x}_f, \alpha_f) + \ell_r f_d + m_d]/m\ell_r \quad (23)$$

This completes the proof of the theorem.

Note that decoupling does not imply stability. Stability depends on the nonlinear differential equations (13), (14). For nondynamic nonlinear characteristics with positive slope, however, stability is guaranteed [9]. In other cases the two decoupled subsystems can be stabilized separately, if there is a further actuator for the yaw subsystem (e.g. four-wheel steering of cars).

4 Linearized Steering Dynamics

The decoupled steering dynamics become particularly simple for linearized rudder or tire characteristics without dynamics, i.e.
$g_f(\mathbf{x}_f, \alpha_f) = 0$, $h_f(\mathbf{x}_f, \alpha_f) = c_f \alpha_f$
with a stiffness coefficient $c_f > 0$. Then (22) and (23) reduce to

$$\dot{\alpha}_f = \frac{-c_f(\ell_f + \ell_r)}{mv\ell_r}\alpha_f + k_L(v)w + \dot{w} - \frac{1}{mv\ell_r}(m_d + f_d\ell_r)$$

$$a_1 = [c_f(\ell_f + \ell_r)\alpha_f + \ell_r f_d + m_d]/m\ell_r \quad (24)$$

and by Laplace transformation

$$a_1(s) = \frac{v[k_L(v) + s]}{1 + \frac{mv\ell_r}{c_f(\ell_f + \ell_r)}s} w(s) + \frac{s}{m\ell_r(s + \frac{c_f(\ell_f + \ell_r)}{mv\ell_r})}[m_d(s) + \ell_r f_d(s)] \quad (25)$$

4 Linearized Steering Dynamics

The steering subsystem is stable with one negative real pole at

$$s_f = -\frac{c_f(\ell_f + \ell_r)}{mv\ell_r} \tag{26}$$

If the driver commands a step $w(s) = 1/s$, he feels an immediate reaction

$$a_{10} = \frac{c_f(\ell_f + \ell_r)}{m\ell_r} \tag{27}$$

and the exponential response with time constant $T = -1/s_f$ approaches the steady-state value

$$a_{1st} = vk_L(v) \tag{28}$$

Fig. 9 shows the response a_1 in a solid line (also the case $a_{10} > a_{1st}$ occurs at low velocities v).

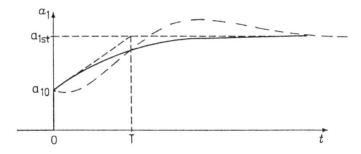

Fig. 9. Lateral acceleration a_1 of mass m_1 for decoupled vehicle (solid line) and conventional vehicle (dashed line).

For comparison we also calculate the response for the conventional vehicle without decoupling control. Its linearized steering dynamics without rudder or tire dynamics are obtained from (15) with
$g_f(\mathbf{x}_f, \alpha_f) = 0$, $h_f(\mathbf{x}_f, \alpha_f) = c_f\alpha_f$, $g_r(\mathbf{x}_r, \alpha_r) = 0$, $h_r(\mathbf{x}_r, \alpha_r) = c_r\alpha_r$, $c_r > 0$

$$\begin{bmatrix} \dot{\beta} \\ \dot{r} \end{bmatrix} = \begin{bmatrix} (c_f\alpha_f + c_r\alpha_r)/mv - r \\ (c_f\ell_f\alpha_f - c_r\ell_r\alpha_r)/m\ell_r\ell_1 \end{bmatrix} + \begin{bmatrix} f_d/mv \\ m_d/m\ell_r\ell_1 \end{bmatrix}$$

The output a_1 is by (16) $a_1 = [(\ell_r + \ell_f)c_f\alpha_f + \ell_r f_d + m_d]/m\ell_r$ where, by analogy with (11), $\alpha_r = \delta_r - \beta_r$, $\delta_r = 0$ and β_r is, by analogy with (12), $\beta_r = \beta - \ell_r r/v$, thus $\alpha_r = -\beta + \ell_r r/v$ and $\alpha_f = \delta_f - \beta_f = \delta_f - \beta - \ell_f r/v$ and the linear state space model of the uncontrolled car is

$$\begin{bmatrix} \dot{\beta} \\ \dot{r} \end{bmatrix} = \begin{bmatrix} \dfrac{-(c_f + c_r)}{mv} & \dfrac{(\ell_r c_r - \ell_f c_f)}{mv^2} - 1 \\ \dfrac{c_r \ell_r - c_f \ell_f}{m\ell_r \ell_1} & \dfrac{-(c_f \ell_f^2 + c_r \ell_r^2)}{mv\ell_r \ell_1} \end{bmatrix} \begin{bmatrix} \beta \\ r \end{bmatrix} +$$

$$+ \begin{bmatrix} \dfrac{c_f}{mv} \\ \dfrac{c_f \ell_f}{m\ell_r \ell_1} \end{bmatrix} \delta_f + \begin{bmatrix} \dfrac{f_d}{mv} \\ \dfrac{m_d}{m\ell_r \ell_1} \end{bmatrix} \tag{29}$$

$$a_1 = \begin{bmatrix} \dfrac{-c_f(\ell_r + \ell_f)}{m\ell_r} & \dfrac{-c_f \ell_f(\ell_r + \ell_f)}{mv\ell_r} \end{bmatrix} \begin{bmatrix} \beta \\ r \end{bmatrix} + \dfrac{c_f(\ell_r + \ell_f)}{m\ell_r}\delta_f + \dfrac{f_d}{m} + \dfrac{m_d}{m\ell_r}$$

By Laplace transformation we obtain the input-output relation for a steering input $\delta_f(s) = w(s)$

$$a_1(s) = \dfrac{c_f(\ell_r + \ell_f)(b_0 + b_1 s + b_2 s^2)}{(c_0 + c_1 s + s^2)} w(s) \tag{30}$$

$$c_0 = \dfrac{c_f + c_r(\ell_f + \ell_r)^2}{m^2 v^2 \ell_r \ell_1} + \dfrac{c_r \ell_r - c_f \ell_f}{m\ell_r \ell_1}$$

$$c_1 = \dfrac{c_r + c_f}{mv} + \dfrac{c_r \ell_r^2 + c_f \ell_f^2}{mv\ell_r \ell_1}$$

$$b_0 = \dfrac{c_r}{m^2 \ell_r \ell_1}$$

$$b_1 = \dfrac{c_r(\ell_1 + \ell_r)}{m^2 v \ell_r \ell_1}$$

$$b_2 = \dfrac{1}{m\ell_r} \tag{31}$$

For a step input $w(s) = 1/s$ commanded by the driver the immediate reaction at the output a_1 is

$$a_{10} = c_f(\ell_r + \ell_f)b_2 = \dfrac{c_f(\ell_r + \ell_f)}{m\ell_r} \tag{32}$$

A comparison with (27) shows that the immediate reaction of the vehicle has not been changed by feedback, which is also obvious from Fig. 7 where the immediate response of δ_c is equal to zero.

The steady-state response is

$$a_{1st} = \frac{c_f(\ell_r + \ell_f)b_0}{c_0}$$

$$= \frac{(\ell_r + \ell_f)v^2}{(\ell_r + \ell_f)^2 + mv^2(\frac{\ell_r}{c_f} - \frac{\ell_f}{c_r})} \tag{33}$$

A comparison with (28) shows how $k_L(v)$ must be chosen in order to have the same steady-state response:

$$k_L(v) = \frac{(\ell_r + \ell_f)v}{(\ell_r + \ell_f)^2 + mv^2(\frac{\ell_r}{c_f} - \frac{\ell_f}{c_r})} \tag{34}$$

This prefilter, shown in Fig. 7, is scheduled by the measured velocity. For m, c_f and c_r, however, only nominal values are substituted such that the steady-state response of the decoupled vehicle becomes identical to the steady-state response of a vehicle with nominal values of m, c_f and c_r as they are substituted in $k_L(v)$.

The main difference between the uncontrolled and the controlled vehicle in terms of the lateral acceleration a_1 at the front mass m_1 is that the uncontrolled vehicle has a second order response with oscillations whose frequency and damping depend on all uncertain parameters m, v, c_f and c_r. An example is shown in Fig. 9 in a dashed line. For the controlled vehicle the transient still depends on the uncertain parameters via the uncertain time constant $T = mv\ell_r/c_f(\ell_f + \ell_r)$ as shown in Fig. 9, but it is an exponential, i.e. monotonous response that does not surprise or unnecessarily shake the traveller. In fact the time constant T can be made arbitrarily small by high-gain feedback of a_1 to $\dot{\delta}_c$.

5 Disturbance Attenuation

Practically it is easy for the driver to compensate the effect of a lateral force f_d in the course of normal steering. Therefore we consider only disturbance torques m_d as they occur for example by wind or water flow or failures that destroy the symmetry of the vehicle (e.g. engine failure in an aircraft, flat tire of a car).

The disturbance attenuation properties of the decoupling control law (18) are best seen in responses to step inputs at m_d. Such step responses require, however, simulations with specific numeric parameter values. The safety advantages for a car under crosswind and μ-split braking are quantified for example in [7].

Here we pursue analytical solutions for generic vehicles. We will determine the frequency band for which sinusoidal disturbances m_d are attenuated by the controller. By Bode's theorem [11], however, this means that disturbances at other frequencies are amplified. We will explicitely determine the limit frequency that separates these bands. This analysis is performed for the yaw rate r as an output because the yaw rate r is the main concern after we have tight control over a_1, see Fig. 6.

The transfer function from $m_d(s)$ to $r(s)$ is calculated for the conventional (uncontrolled) vehicle from (29) as

$$r(s) = \frac{d_0 + d_1 s}{c_0 + c_1 s + s^2} m_d(s) \qquad (35)$$

with c_0, c_1 as in (32) and

$$d_0 = \frac{c_f + c_r}{m^2 v \ell_r \ell_1}$$

$$d_1 = \frac{1}{m \ell_r \ell_1}$$

For the decoupled system the linearized version of (21) with $w = 0$, $\dot{w} = 0$, $f_d = 0$, $g_f = g_r = 0$, $h_f = c_f \alpha_f$, $h_r = c_r \alpha_r$ yields

$$r_{dec}(s) = \frac{s(d_0 + d_1 s)}{(e_0 + e_1 s + s^2)(e_2 + s)} \cdot \frac{\ell_f c_f}{v} m_d(s) \qquad (36)$$

$$e_0 = \frac{c_r}{\ell_1 m}$$

$$e_1 = \frac{c_r(\ell_1 + \ell_r)}{\ell_1 m v}$$

$$e_2 = \frac{c_f(\ell_r + \ell_f)}{\ell_r m v}$$

Note that $r_{dec}(s)$ has a factor s in the numerator, i.e. the steady-state response to a step torque m_d becomes zero by decoupling. Also note that $e_0 > 0$, $e_1 > 0$, $e_2 > 0$, i.e. the yaw subsystem is stable.

For comparison of the disturbance influences on the open and closed loop we define

$$\rho_r(s) = \frac{r_{dec}(s)}{r(s)} \qquad (37)$$

as disturbance attenuation ratio. The influence of a yaw disturbance on r is reduced by decoupling for all frequencies for which $|\rho_r(j\omega)| < 1$.

$$\rho_r(s) = \frac{\ell_f c_f s (c_0 + c_1 s + s^2)}{v(e_0 + e_1 s + s^2)(e_2 + s)} \qquad (38)$$

Now $|\rho_r(j\omega)|^2 = \rho_r(j\omega)\rho_r(-j\omega) = B(\omega^2)/A(\omega^2)$ and $|\rho_r(j\omega)|^2 < 1$ if and only if $A(\omega^2) - B(\omega^2) > 0$. The resulting quadratic equation in ω^2 is

$$A(\omega^2) - B(\omega^2) = a\omega^4 + b\omega^2 + c = 0 \qquad (39)$$

where

$$a = -(\ell_f - \ell_1)[2c_r\ell_r^2(\ell_f - \ell_1) + c_f\ell_f^2(\ell_f + \ell_1) + 2c_f\ell_f\ell_r\ell_1]m^2v^2$$
$$\quad - 2\ell_f\ell_r\ell_1 m^3 v^4$$

$$b = (2c_r\ell_r - c_f\ell_f)\ell_f m^2 v^4 + 2c_r(\ell_f + \ell_r)^2[c_f(\ell_f - \ell_1) - c_r\ell_r]mv^2$$
$$\quad - c_f c_r^2(\ell_f + \ell_r)^2(\ell_f - \ell_1)(\ell_f + 2\ell_r + \ell_1)$$

$$c = c_f c_r^2 (\ell_f + \ell_r)^2 v^2$$

Equation (39) can be explicitly solved for the limit frequency ω_ℓ

$$\omega_\ell^2 = \frac{-b + \sqrt{b^2 - 4ac}}{2a} \qquad (40)$$

The yaw disturbances are attenuated for frequencies $\omega < \omega_\ell$ and amplified for frequencies $\omega > \omega_\ell$.

6 Conclusions

A generic decoupling controller for generic vehicles has been presented. For its implementation a gyro for the yaw rate r, two accelerometers for the lateral accelerations a_1 and a_r for feedback, and the velocity v for gain scheduling must be measured. The accelerometers can be avoided by differentiations of r. In the case of an ideal mass distribution $\ell_1 = \ell_f$ the accelerations a_1 and a_r are not needed at all. The properties of the decoupled system are summarized as follows:

The yaw rate r is not observable from the lateral acceleration a_1 of a front point mass m_1. Therefore the influences of uncertain parameters (mass, velocity, tire-road contact) that enter into the yaw dynamics are largely kept away from a_1. The only task of the driver is to keep the front mass m_1 on top of his planned path by commanding a lateral acceleration a_1. This command is executed via a first order steering transfer function with one zero. The immediate reaction of the vehicle to a steering step input is the same as in the uncontrolled case. Then an exponential transient follows (rather than a largely uncertain second order resoponse of the conventional vehicle). The control of the position of m_1 can be made even faster by additional feedback of a_1 if there are no essential actuator delays. The steady-state step response can be made equal to that of a vehicle with nominal parameter values by a velocity-scheduled prefilter $k_L(v)$.

The decoupled vehicle has significant safety advantages in situations when unexpected yaw disturbance torques occur. In the uncontrolled vehicle the driver typically reacts only after 0.5 to 1 second when the vehicle has gained already a significant yaw rate and sideslip angle and may approach physical limits. The driver may even make things worse by overreactions leading to driver-induced oscillations. In contrast, the feedback system reacts immediately and keeps the yaw rate and sideslip angle small. The disturbance attenuation properties of robust decoupling are analyzed in the frequency domain, where the limit frequency

ω_ℓ is explicitely determined up to which disturbance attenuation occurs. The decoupling property is not only robust with respect to large uncertainties in the vehicle mass m and velocity v, but also with respect to an unknown nonlinear dynamic relationship between the command and the lateral force at a rudder or steered wheel. Such additional dynamics may, however, cause stability problems.

7 Bibliographical Notes

There is a vast literature on control of specific vehicles. The reader is referred to books on specific classes of vehicles, e.g.

References

1. B. Etkin: Dynamics of Atmospheric Flight. Wiley, New York, 1972.
2. A.E. Bryson: Control of Spacecraft and Aircraft. Princeton University Press, Princeton, N.J., 1994.
3. R. Brockhaus: Flugregelung. Springer, Berlin, 1994.
4. M. Mitschke: Dynamik der Kraftfahrzeuge. Springer, Berlin, 1990.
5. W. Kortüm, P. Lugner: Systemdynamik und Regelung von Fahrzeugen. Springer, Berlin, 1994.

The concept of robust decoupling was first formulated for automobiles with ideal mass distribution in:

6. J. Ackermann: Verfahren zum Lenken von Straßenfahrzeugen mit Vorder- und Hinterradlenkung. (Method of steering road vehicles having front-wheel and rear-wheel steering). German Patent No. 4028 320, European Patent No. 0474 130, US Patent No. 537057 (Priority Sept. 6, 1990)

Significant safety advantages of this concept were shown in simulations in:

7. J. Ackermann, W. Sienel: Robust yaw damping of cars with front and rear wheel steering. IEEE Trans. on Control Systems Technology, vol. 1, no. 1, pp. 15-20, 1993.

The generalization to non-ideal mass distribution was made in:

8. J. Ackermann: Robust decoupling of car steering dynamics with arbitrary mass distribution. Proc. 1994 American Control Conference, Baltimore, vol. 2, pp. 1964-1968.

The generalization to nonlinear tire characteristics is in

9. J. Ackermann: Robust decoupling, ideal steering dynamics, and yaw stabilization of 4WS cars. Automatica, 1994, vol. 30, pp. 1761-1768.

The generalization to nonlinear dynamics of the tire is in

10. W. Sienel: Robust decoupling for active car steering holds for arbitrary dynamic tire characteristics. Submitted to European Control Conference, Rome, 1995.

The present paper introduces the generalization to generic vehicles, a modification of the feedforward path of the control law for immediate reaction of the vehicle to a steering step command, and an analytical result on disturbance attenuation for frequencies below a limit frequency ω_ℓ.
The existence of such a limit follows from

11. H. Bode: Network analysis and feedback amplifier design. Van Nostrand, Princeton, N.J. 1945.

Symmetries in Control System Design

Huibert Kwakernaak

Systems and Control Group, University of Twente
P. O. Box 217, 7500 AE Enschede, The Netherlands
Telephone +31-53-893457, fax +31-53-340733
e-mail h.kwakernaak@math.utwente.nl

Abstract. When the robustness of linear feedback control systems with respect to perturbations of the loop gain and the inverse of the loop gain is analyzed then the sensitivity and complementary sensitivity functions turn out to have symmetric roles. This leads to interesting conclusions about the requirements that performance and robustness impose on control system design. In many respects performance and robustness support each other.

1 Introduction

Classical control is going through a revival. *Feedback* is the key to control. After having been lost from view during the "optimal control" era in the sixties feedback theory came — weakly – back in the picture in the "LQG" era during the seventies.

LQG lacks equipment to handle plant and parameter uncertainty. It focuses on *performance* – in particular, disturbance attenuation and measurement noise abatement.

Early in the eighties the term robustness was coined. It emphatically expresses the ability to cope with plant and parameter uncertainty. The new word certainly helped to bring an old and important subject back into the limelight.

During the optimal control period the control community acquired an acquaintance with relatively sophisticated mathematical techniques. This was not unprecedented: Classical control — think of the work of Bode and Nyquist — relies on complex function theory, part of which is comparatively deep. This familiarity with various powerful mathematical tools came in useful in the new attack on feedback control.

Doyle's work ([Doy79] is an early reference) on modelling and assessing the effects of uncertainty relies on the small gain theorem and bounds on system norms. It initially focused almost entirely on the notion of *stability robustness*. Minimizing suitable norms allows to optimize or at least improve robustness. As a result, $\mathcal{H}_\infty$ optimization was invented as a feedback system design tool.

$\mathcal{H}_\infty$ optimization theory gave a group of gifted researchers the opportunity to indulge in some lovely mathematics — and to avoid the design issues. By the early nineties the mathematical problem was more or less solved, and practical algorithms — the "two Riccati solution" — were implemented on popular platforms such as MATLAB.

The time has now arrived to explore the application of these techniques to control system design. At this point in history it is natural to take another look at some of the basic issues of control system design, in particular linear control system design. This is what is done in this paper. We review the fundamental design targets. We discuss the various essential open- and closed-loop system functions such as the loop gain and the sensitivity and complementary sensitivity functions. We translate the design targets into requirements on these functions.

Their names imply that the sensitivity and complementary sensitivity functions play complementary roles — hence, exhibit a symmetry. Another symmetry occurs in the roles played by high frequencies and low frequencies. The symmetry extends to the significance of poles and zeros.

In robustness analysis a further symmetry may be discerned in the effects of loop gain perturbations and perturbations of the inverse loop gain.

Looking over the modern robustness literature easily the mistaken impression arises that only high-frequency uncertainties are important, or at least, that only such uncertainty can be handled. Low-frequency plant perturbations often are at least as significant. They arise from load variations, set point changes and environmental effects. By studying inverse loop gain perturbations rather than the loop gain perturbations themselves these low-frequency perturbations come into focus.

It is perhaps not always understood that well-designed feedback control systems from the point of view of performance have a natural robustness with respect to both low- and high-frequency perturbations and uncertainty. It is the *crossover region* that is critical, both for performance and robustness. Performance and robustness are companions in the low- and high-frequency regions but often also in the intermediate region.

This is our list of symmetric notions:

low frequencies – high frequencies
sensitivity function – complementary sensitivity function
loop gain perturbations – inverse loop gain perturbations
performance – robustness
zeros – poles

In Section 2 we discuss the performance of two-degree-of-freedom control systems and introduce the various open- and closed-loop system functions. This is where we recognize the symmetric roles of the sensitivity function and its complement, and of low and high frequencies.

Section 3 is devoted to a review of stability and robustness questions. Here we encounter the symmetric roles of loop gain perturbations and inverse loop gain perturbations.

Section 4 deals with the limits of performance imposed by right-half plane open-loop poles and zeros. The effects of right-half plane zeros on the one hand and right-half plane poles on the other exhibit a marked symmetry.

Section 5 concerns other perturbations than loop and inverse loop perturbations. It is shown how the two types of perturbations may be linked by what we call fractional perturbations. This is where it becomes clear how crucial the crossover region is, both for performance and for robustness.

In Section 6 we review what the various control system design techniques that are available — classical, QFT, LQG, $\mathcal{H}_2$ and $\mathcal{H}_\infty$-optimization, μ-synthesis — have to offer in the light of the "symmetry" analysis.

Some of the insight that has been gained is applied in Section 7 to a sampled date case study that is currently circulating. In Section 8 the conclusions are summarized.

Many of the results reviewed in this paper are not new. Portions of this paper were adapted from course and lecture notes [BK94] [Kwa94]. This explains the prevailing didactic tone.

2 Performance and the closed-loop system functions

We study the classical two-degree-of-freedom configuration of Fig. 1. The design targets are

- closed-loop stability,
- adequate disturbance attenuation,
- adequate command input response, and
- robustness with respect to plant uncertainty and parameter variations,

within the limitations imposed by

- plant capacity, and
- measurement noise.

In this section we translate the performance design targets for the linear time-invariant two-degree-of-freedom feedback system of Fig. 1 into requirements on various closed-loop frequency response functions.

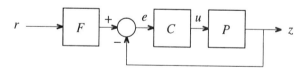

Fig. 1. Two-degree-of-freedom feedback system

Closed-loop stability. Suppose that the feedback system of Fig. 1 is open-loop stable. By the Nyquist stability criterion, for closed-loop stability the loop gain should be shaped such that the Nyquist plot of the loop gain $L = PC$ does not encircle the point -1.

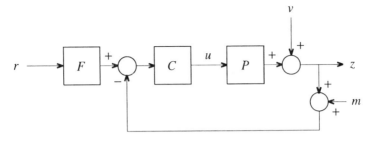

Fig. 2. Two-degree-of-freedom feedback loop with disturbances and measurement noise

Disturbance attenuation and bandwidth — the sensitivity function.
To study disturbance attenuation, consider the block diagram of Fig. 2. The signal v represents the equivalent disturbance at the output of the plant and w the measurement noise. Ignoring the reference input r and the measurement noise m, the signal balance equation is $z = v - Lz$. Solution for z results in

$$z = \underbrace{\frac{1}{1+L}}_{S} v = Sv, \tag{1}$$

where S is the sensitivity function of the closed-loop system. The smaller $|S(j\omega)|$ is, with $\omega \in \mathbb{R}$, the more the disturbances are attenuated at the angular frequency ω. $|S|$ is small if the magnitude of the loop gain L is large. Hence, for disturbance attenuation it is necessary to shape the loop gain such that it is large over those frequencies where disturbance attenuation is needed.

Making the loop gain L large over a large frequency band easily results in error signals e and resulting plant inputs u that are larger than the plant can absorb. Therefore, L can only be made large over a limited frequency band. Figure 3(a) shows an "ideal" shape of the sensitivity function.

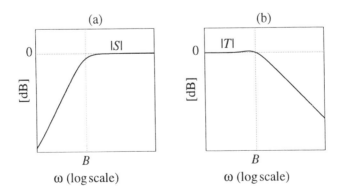

Fig. 3. (a) "Ideal" sensitivity function. (b) A corresponding complementary sensitivity function.

2 Performance and the closed-loop system functions

The larger the "capacity" of the plant is, that is, the larger the inputs are the plant can handle before it saturates or otherwise fails, the larger the maximally achievable bandwidth B usually is. For plants whose transfer functions have zeros with nonnegative real parts, however, the maximally achievable bandwidth is limited by the location of the right-half plane zero closest to the origin. This is discussed in Section 4.

It may be necessary to impose further requirements on the shape of the sensitivity function if the disturbances have a distinct frequency profile. Consider for instance the situation that the actual disturbances enter the plant internally, or even at the plant input, and that the plant is highly oscillatory. Then the equivalent disturbance at the output is also oscillatory. To attenuate these disturbances effectively the sensitivity function should be small at and near the resonance frequency. For this reason it sometimes is useful to replace the requirement that S be small with the requirement that $|S(j\omega)V(j\omega)|$ be small over a suitable low frequency range. The shape of the weighting function V reflects the frequency content of the disturbances. If the actual disturbances enter the system at the plant input a possible choice is to let $V = P$. This leads to considering the closed-loop system function [Lun91]

$$R = PS = \frac{P}{1+PC}. \tag{2}$$

R could be called the *input disturbance sensitivity function*. Besides characterizing the closed-loop response to disturbances at the plant input it serves to assess the closed-loop response to nonzero initial conditions of the plant. Design requirements on R clearly may be translated into requirements on S, and vice-versa.

Command response—the complementary sensitivity function. The response of the two-degree-of-freedom configuration of Fig. 2 to the command signal r follows from the signal balance equation $z = PC(-z + Fr)$. Solution for z results in

$$z = \underbrace{\frac{PC}{1+PC}}_{H} F \; r. \tag{3}$$

The *closed-loop transfer function* H may be expressed as

$$H = \underbrace{\frac{L}{1+L}}_{T} F = TF, \tag{4}$$

with $L = PC$ the loop gain and $T = 1 - S$ the complementary sensitivity function.

Adequate loop shaping ideally results in a complementary sensitivity function T that is close to 1 up to the bandwidth, and transits smoothly to zero above this frequency. The desired shape for the sensitivity function S of Fig. 3(a) implies a matching shape for the complementary sensitivity function $T = 1 - S$. Figure 3(b) shows a possible shape[1] for the complementary sensitivity T corresponding to the sensivity function of Fig. 3(a). Thus, without a prefilter F (that is, with $F = 1$), the closed-loop transfer function H ideally is low-pass with the same bandwidth as the frequency band for disturbance attenuation.

Like for the sensitivity function, the plant dynamics impose limitations on the shape that T may assume. In particular, right-half plane plant poles constrain the frequency above which T may be made to roll off. This is discussed in Section 4.

Plant capacity—the input sensitivity function. Any physical plant has limited "capacity," that is, can absorb inputs of limited magnitude only. In terms of Laplace transforms we have from Fig. 2 the signal balance $u = C(Fr - m - v - Pu)$. This may be solved for u as

$$u = \underbrace{\frac{C}{I + CP}}_{U} (Fr - m - v). \tag{5}$$

The function U determines the sensitivity of the plant input to disturbances and the command signal. It is sometimes known as the *input sensitivity function*.

If the loop gain $L = CP$ is large then the input sensitivity U approximately equals the *inverse* $1/P$ of the plant transfer function. If the open-loop plant has zeros in the right-half complex plane then $1/P$ is unstable. This is why the right-half plane open-loop plant poles limit the closed-loop bandwidth. The input sensitivity function U may only be made equal to $1/P$ up to the frequency which equals the magnitude of the right-half plane plant zero with the smallest magnitude.

The input sensitivity function U is connected to the complementary sensitivity function T by the relation

$$T = UP. \tag{6}$$

By this connection requirements on the input sensitivity function U may be translated into corresponding requirements on the complementary sensitivity T, and vice-versa.

Measurement noise. To study the effect of measurement noise on the closed-loop output we again consider the configuration of Fig. 2. By solving the signal balance $z = v + PC(Fr - m - z)$ for the output z we find

$$z = \underbrace{\frac{1}{1 + PC}}_{S} v + \underbrace{\frac{PC}{1 + PC}}_{T} Fr - \underbrace{\frac{PC}{1 + PC}}_{T} m. \tag{7}$$

---

[1] Note that S and T are complementary, not $|S|$ and $|T|$.

2 Performance and the closed-loop system functions

This shows that the influence of the measurement noise m on the control system output is determined by the complementary sensitivity function T. For low frequencies, where by the other design requirements T is close to 1, the measurement noise fully affects the output. This emphasizes the need for good, low-noise sensors.

The four system functions. In this section four closed-loop system functions are introduced:
$$\begin{bmatrix} \frac{1}{1+PC} & \frac{P}{1+PC} \\ \frac{C}{1+PC} & \frac{PC}{1+PC} \end{bmatrix} = \begin{bmatrix} S & R \\ U & T \end{bmatrix}. \tag{8}$$

The functions constitute the closed-loop transfer matrix from the input (v_1, v_2) to the output (z_1, z_2) in Fig. 4.

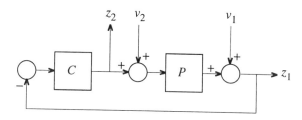

Fig. 4. The four closed-loop system functions

Review of the design requirements. We summarize the conclusions of this section as follows:

- The sensitivity S should be small at low frequencies to achieve disturbance attenuation and good command response.
- The complementary sensitivity T should be small at high frequencies to prevent exceeding the plant capacity and adverse effects of measurement noise.
- In the intermediate frequency region peaking of both S and T should be avoided to prevent overly large sensitivity to disturbances and excessive influence of the measurement noise.

The sensitivity function S and the complementary sensitivity function T are related to the loop gain L as
$$S = \frac{1}{1+L}, \quad T = \frac{L}{1+L}. \tag{9}$$

The sensitivity S may be made small at low frequencies by making
$$|L(j\omega)| \gg 1 \text{ at low frequencies.} \tag{10}$$

Likewise, the complementary sensitivity function may be made small at high frequencies by designing such that

$$|L(j\omega)| \ll 1 \text{ at high frequencies.} \tag{11}$$

In the intermediate region the loop gain L changes from large to small — that is, $|L(j\omega)|$ crosses the zero dB line. In the intermediate frequency region, also known as the *crossover region*, the loop gain should avoid the point -1.

3 Robustness

In this section we consider the *stability* and *performance robustness* of SISO feedback systems with the configuration of Fig. 1. Stability robustness is the property that the closed-loop system remains stable under changes of the plant and the compensator. Performance robustness means that even if the closed-loop system remains stable under perturbation certain specified closed-loop system functions or properties are not overly affected by the perturbations.

For simplicity we assume that the system is *open-loop stable*, that is, both P and C represent the transfer function of a stable system.

We also assume the existence of a *nominal* feedback loop with loop gain L_o, which is the loop gain that is supposed to be valid under nominal circumstances.

Stability margins. The closed-loop system of Fig. 1 remains stable under perturbations of the loop gain L as long as the Nyquist plot of the perturbed loop gain does not encircle the point -1 Intuitively, this may be accomplished by "keeping the Nyquist plot of the nominal feedback system away from the point -1."

The classic *gain margin* and *phase margin* are well-known indicators that show how closely the Nyquist plot approaches the point -1.

The gain and phase margin do not necessarily adequately characterize robustness. It is easy to think of loop gains with excellent gain and phase margins but where a relatively small *joint* perturbation of gain and phase suffices to destabilize the system.

For this reason Landau (see for instance [LRCV94]) introduced another margin. The *modulus margin*[2] s_m is smallest distance of any point on the Nyquist plot to the point -1. Hence, $s_m = \min_{\omega \in \mathbb{R}} |1 + L(j\omega)|$, so that

$$\frac{1}{s_m} = \max_{\omega \in \mathbb{R}} \frac{1}{|1 + L(j\omega)|} = \max_{\omega \in \mathbb{R}} |S(j\omega)|. \tag{12}$$

Thus, the modulus margin is the inverse of the peak value of the sensitivity function.

If the loop gain L approaches the point -1 closely then so does the *inverse loop gain* $1/L$. Therefore we may define the smallest distance of any point on

---
[2] French: *marge de module*.

the inverse Nyquist plot (that is, the locus of $1/L$) to the point -1 as another robustness margin, denoted r_m. We have

$$r_m = \min_{\omega \in \mathbb{R}} \left|1 + \frac{1}{L(j\omega)}\right| = \min_{\omega \in \mathbb{R}} \left|\frac{1+L(j\omega)}{L(j\omega)}\right|, \qquad (13)$$

so that

$$\frac{1}{r_m} = \max_{\omega \in \mathbb{R}} |T(j\omega)|. \qquad (14)$$

Thus, r_m is the inverse of the peak value of the complementary sensitivity function T.

Adequate margins of these types are not only needed for robustness, but also to achieve a satisfactory time response of the closed-loop system. If the margins are small, the Nyquist plot approaches the point -1 closely. This means that the stability boundary is approached closely, manifesting itself by closed-loop poles that are very near the imaginary axis. These closed-loop poles may cause an oscillatory response.

Robustness for loop gain perturbations. The robustness specifications discussed so far are all rather qualitative. They break down when the system is not open-loop stable, and, even more spectacularly, for MIMO systems. We introduce a more refined measure of stability robustness by considering the effect of plant perturbations on the Nyquist plot more in detail. For the time being the assumptions that the feedback system is SISO and open-loop stable are upheld. Both may be relaxed.

Naturally, we suppose the nominal feedback system to be well-designed so that it is stable. We investigate whether the feedback system *remains* stable when the loop gain is perturbed from the nominal loop gain L_o to the actual loop gain L.

By the Nyquist criterion, the Nyquist plot of the nominal loop gain does not encircle the point -1, as shown in Fig. 5. The actual closed-loop system is stable if also the Nyquist plot of the actual loop gain L does not encircle -1.

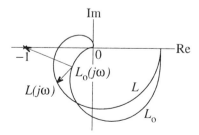

Fig. 5. Nominal and perturbed Nyquist plots

It is easy to see by inspection of Fig. 5 that the Nyquist plot of L definitely does not encircle the point -1 if for all $\omega \in \mathsf{R}$ the distance $|L(j\omega) - L_o(j\omega)|$ between any point $L(j\omega)$ and the corresponding point $L_o(j\omega)$ is *less* than the distance $|L_o(j\omega) + 1|$ of the point $L_o(j\omega)$ and the point -1, that is, if

$$|L(j\omega) - L_o(j\omega)| < |L_o(j\omega) + 1| \quad \text{for all } \omega \in \mathsf{R}. \tag{15}$$

This is equivalent to

$$\frac{|L(j\omega) - L_o(j\omega)|}{|L_o(j\omega)|} \cdot \frac{|L_o(j\omega)|}{|L_o(j\omega) + 1|} < 1 \quad \text{for all } \omega \in \mathsf{R}. \tag{16}$$

It follows that the system is not destabilized if

$$\frac{|L(j\omega) - L_o(j\omega)|}{|L_o(j\omega)|} < \frac{1}{|T_o(j\omega)|} \quad \text{for all } \omega \in \mathsf{R}. \tag{17}$$

The factor $|L(j\omega) - L_o(j\omega)|/|L_o(j\omega)|$ in this expression is the *relative* size of the perturbation of the loop gain L from its nominal value L_o. T_o is the nominal complementary sensitivity function. The larger the magnitude of the complementary sensitivity function, the smaller is the allowable perturbation.

This result is the SISO version of the celebrated result of Doyle [Doy79]. It is a *sufficient* condition. This means that there may well exist perturbations that do not satisfy (17) but nevertheless do not destabilize the closed-loop system. With a suitable modification the condition is also necessary. Suppose that the relative perturbations are known to bounded in the form

$$\frac{|L(j\omega) - L_o(j\omega)|}{|L_o(j\omega)|} \leq |W(j\omega)| \quad \text{for all } \omega \in \mathsf{R}, \tag{18}$$

with W a given function. Then the condition (17) is implied by the inequality

$$|T_o(j\omega)| < \frac{1}{|W(j\omega)|} \quad \text{for all } \omega \in \mathsf{R}. \tag{19}$$

Thus, if the latter condition holds, robust stability is guaranteed for all perturbations satisfying (18). Moreover, (19) is not only sufficient but also *necessary* to guarantee stability for *all* perturbations satisfying (18) [Vid85]. Such perturbations are said to "fill the uncertainty envelope."

The stability robustness condition has been obtained under the assumption that the open-loop system is stable. In fact, it also holds for open-loop unstable systems, *provided* the number of right-half plane poles remains invariant under perturbation.

Inverse loop gain perturbations. According to the Nyquist criterion, the closed-loop system remains stable under perturbation as long as under perturbation the Nyquist plot of the loop gain does not cross the point -1. Equivalently, the closed-loop system remains stable under perturbation as long as the *inverse*

3 Robustness

$1/L$ of the loop gain does not cross the point -1. Thus, the sufficient condition (15) may be replaced with the sufficient condition

$$\left| \frac{1}{L(j\omega)} - \frac{1}{L_o(j\omega)} \right| < \left| \frac{1}{L_o(j\omega)} + 1 \right| \quad \text{for all } \omega \in \mathsf{R}. \tag{20}$$

Dividing by the inverse $1/L_o$ of the nominal loop gain we find that a sufficient condition for robust stability is that

$$\left| \frac{\frac{1}{L(j\omega)} - \frac{1}{L_o(j\omega)}}{\frac{1}{L_o(j\omega)}} \right| < \left| \frac{\frac{1}{L_o(j\omega)} + 1}{\frac{1}{L_o(j\omega)}} \right| = |1 + L_o(j\omega)| = \frac{1}{|S_o(j\omega)|} \tag{21}$$

for all $\omega \in \mathsf{R}$.

Again the result may be generalized to a sufficient and necessary condition. The closed-loop system remains stable under all relative perturbations of the inverse loop gain that satisfy the bound

$$\left| \frac{\frac{1}{L(j\omega)} - \frac{1}{L_o(j\omega)}}{\frac{1}{L_o(j\omega)}} \right| \le |W(j\omega)| \quad \text{for all } \omega \in \mathsf{R}, \tag{22}$$

with W a given function, if and only if

$$|S_o(j\omega)| < \frac{1}{|W(j\omega)|} \quad \text{for all } \omega \in \mathsf{R}. \tag{23}$$

These results remain true as long as no open-loop zero crosses the imaginary axis. Note that the role of the right-half plane poles now is taken by the right-half plane zeros.

Performance robustness. Feedback system performance is determined by the sensitivity function S, the complementary sensitivity function T, the input sensitivity function M, and the closed-loop transfer function H, successively given by

$$S = \frac{1}{1+L}, \quad T = \frac{L}{1+L}, \tag{24}$$

$$M = \frac{C}{1+L} = SC, \quad H = \frac{L}{1+L}F = TF. \tag{25}$$

We consider the extent to which each of these functions is affected by plant variations. For simplicity we suppose that the system environment is sufficiently controlled so that the compensator transfer function C and the prefilter transfer function F are not subject to perturbation. Inspection of (24–25) shows that under this assumption we only need to study the effect of perturbations on S and T: The variations in M are proportional to those in S, and the variations in H are proportional to those in T.

It is not difficult to establish that if the loop gain changes from its nominal value L_o to its actual value L then the corresponding *relative change* of the sensitivity function S may be expressed as

$$\frac{S - S_o}{S} = -T_o \frac{L - L_o}{L_o}. \qquad (26)$$

Similarly, the relative change of the complementary sensitivity function may be written as

$$\frac{T - T_o}{T} = S_o \frac{L - L_o}{L}. \qquad (27)$$

These relations show that for the sensitivity function S to be robust with respect to changes in the loop gain we desire the nominal complementary sensitivity function T_o to be small. On the other hand, for the complementary sensitivity function T to be robust we wish the nominal sensitivity function S_o to be small. These requirements are conflicting, because S_o and T_o add up to 1 and therefore cannot simultaneously be small.

The solution is to have each small in a different frequency range. Normal control system design specifications require S_o to be small at low frequencies (below the bandwidth). This causes T to be robust at low frequencies, which is precisely the region where its values are significant. Complementarily, T_o is required to be small at high frequencies, causing S to be robust in the high frequency range.

4 Limits of performance

Causality on the one hand, and the pole-zero pattern of the plant on the other, limit the performance[3] of the closed-loop system. In particular, right-half plane poles (which cause the open-loop system to be unstable) and right-half plane zeros (which make the inverse plant unstable) set bounds to what may be achieved. The poles and zeros play symmetric roles.

Bode's sensitivity integral. A well-known result of Bode's pioneering work [Bod45] is *Bode's sensitivity integral.* If L has at least two more poles than zeros then the sensitivity function S must satisfy

$$\int_0^\infty \log |S(j\omega)| \, d\omega = \pi \sum_i \mathrm{Re} \ p_i. \qquad (28)$$

The right-hand side is formed from the open right-half plane poles p_i of L, included according to their multiplicity. If the pole-zero excess is one then the integral on the left-hand side of (28) is finite but may be negative.

Suppose for the time being that L has no open right-half plane poles, so that the sensitivity integral vanishes. Then the integral over all frequencies of

---

[3] The title of this section has been taken from [BB91] and [Eng88].

4 Limits of performance

$\log |S|$ is zero. This means that $\log |S|$ both assumes negative and positive values, or, equivalently, that $|S|$ both assumes values less than 1 and values greater than 1. Figure 6 illustrates that low frequency disturbance attenuation may only be achieved at the cost of high frequency disturbance amplification. If the open-loop system has right-half plane poles then this statement holds even more emphatically, especially if the right-half plane poles are far into the right-half plane.

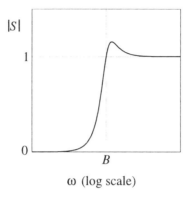

Fig. 6. Low frequency disturbance attenuation may only be achieved at the cost of high frequency disturbance amplification.

If the pole-zero excess of the plant is zero or one then disturbance attenuation is possible over all frequencies.

We next consider the equivalent of the Bode sensitivity integral for the *complementary sensitivity*. Consider a closed-loop system with integrating action. Then

$$S(j\omega) \xrightarrow{\omega \to 0} 0, \qquad T(j\omega) \xrightarrow{\omega \to 0} 1, \tag{29}$$

$$S(j\omega) \xrightarrow{\omega \to \infty} 1, \qquad T(j\omega) \xrightarrow{\omega \to \infty} 0. \tag{30}$$

Clearly, S and T exhibit a symmetry with respect to frequency inversion. In fact, it may be proved that if L is proper and the loop has integrating action of at least order 2 then the complementary sensitivity T satisfies

$$\int_0^\infty \log |T(1/j\omega)| \, d\omega = \pi \sum_i \text{Re } \frac{1}{z_i}, \tag{31}$$

where the z_i are the right-half plane *zeros* of the loop gain L. The role of the poles is now taken by the zeros. Small values of T at high frequencies may only be achieved at the expense of a peak value that is greater than 1. The closer the right-half plane zeros are to the imaginary axis, the more disadvantageous the situation is.

The natural limitations on stability, performance and robustness are aggravated by the presence of right-half plane plant poles and zeros. We elaborate upon this in the remainder of this section.

Tradeoffs for the sensitivity function More extensive discussions on the limits of performance caused by right-half plane zeros and poles may be found in [Eng88] for the SISO case and [FL88] for both the SISO and the MIMO case.

A central result is the *Zames-Francis equality*, which is based on the Poisson integral formula from complex function theory. The result that follows was originally obtained by [FZ84] and [ZF83]. Suppose that the closed-loop system of Fig. 1 is stable, and that the loop gain has a right-half plane zero z. Then the sensitivity function $S = 1/(1+L)$ satisfies

$$\int_{-\infty}^{\infty} \log(|S(j\omega)|) \underbrace{\frac{\text{Re } z}{(\text{Re } z)^2 + (\omega - \text{Im } z)^2}}_{w_z(\omega)} d\omega = \pi \log |B_{\text{poles}}^{-1}(z)|. \qquad (32)$$

B_{poles} is the *Blaschke product*

$$B_{\text{poles}}(s) = \prod_i \frac{p_i - s}{\bar{p}_i + s}, \qquad (33)$$

formed from the open right-half plane poles p_i of the loop gain $L = PC$. The overbar denotes the complex conjugate. To analyze the consequences of this relationship we rewrite it in the form

$$\int_0^\infty \log(|S(j\omega)|) \, dW_z(\omega) = \log |B_{\text{poles}}^{-1}(z)|, \qquad (34)$$

with W_z the function

$$W_z(\omega) = \int_0^\omega w_z(\eta) \, d\eta = \frac{1}{\pi} \arctan \frac{\omega - \text{Im } z}{\text{Re } z} + \frac{1}{\pi} \arctan \frac{\omega + \text{Im } z}{\text{Re } z}. \qquad (35)$$

The function W_z increases monotonically from 0 to 1. Its steepest increase is about the frequency $\omega = |z|$. The expression $\log |B_{\text{poles}}^{-1}(z)|$ is positive.

The Zames-Francis equality strengthens the Bode integral because of the weighting function W_z included in the integrand. The quantity $dW_z(j\omega) = w_z(j\omega)d\omega$ may be viewed as a weighted length of the frequency interval. The larger the weight w_z is, the more the interval contributes.

The weighting function determines to what extent small values of $|S|$ at low frequencies need to be compensated by large values at high frequencies. An important consequence of the Zames-Francis equality is that if $|S|$ is required to be small in a certain frequency band—in particular, a low-frequency band—then it necessarily peaks in another band.

We summarize the qualitative effects of right-half plane zeros of the plant on the shape of the sensitivity function S [Eng88].

(a) The upper limit of the frequency band over which effective disturbance attenuation is possible is constrained from above by the magnitude of the smallest right-half plane zero.
(b) If the plant has unstable poles then the achievable disturbance attenuation is further impaired. This effect is especially pronounced when one or several right-half plane pole-zero pairs are close (because this makes $|B^{-1}_{\text{poles}}(z)|$ large).
(c) If the plant has no right-half plane zeros then the maximally achievable bandwidth is solely constrained by the plant capacity. As seen in the next subsection the right-half plane pole with largest magnitude constrains the *smallest* bandwidth that is required.

Trade-offs for the complementary sensitivity function. Symmetrically to the results for the sensitivity function well-defined trade-offs hold for the complementary sensitivity function. The role of the right-half plane zeros is now taken by the right-half plant open-loop *poles*, and vice-versa. This is seen by writing the complementary sensitivity function as

$$T = \frac{L}{1+L} = \frac{1}{1+\frac{1}{L}}. \tag{36}$$

Comparison with the Zames-Francis equality (32) leads to the (correct) conjecture that for any right-half plane open-loop pole p we have

$$\int_{-\infty}^{\infty} \log(|T(j\omega)|) \frac{\operatorname{Re} p}{(\operatorname{Re} p)^2 + (\omega - \operatorname{Im} p)^2} \, d\omega = \pi \, \log |B^{-1}_{\text{zeros}}(p)|, \tag{37}$$

with B_{zeros} the Blaschke product

$$B_{\text{zeros}}(s) = \prod_i \frac{z_i - s}{\bar{z}_i + s} \tag{38}$$

formed from the open right-half plane *zeros* z_i of L.

We consider the implications of the equality (37) for the shape of T. Whereas the sensitivity S is required to be small at *low* frequencies, T needs to be small at *high* frequencies. By an argument that is symmetric to that for the sensitivity function it follows that if excessive peaking of the complementary sensitivity function at low and intermediate frequencies is to be avoided then $|T|$ may only be made small at frequencies that *exceed* the magnitude of the open-loop right-half plane *pole* with *largest* magnitude. Again, close right-half plane pole-zero pairs make things worse.

We summarize the qualitative effects of right-half plane zeros of the plant on the shape achievable for the complementary sensitivity function T [Eng88].

(a) The lower limit of the band over which the complementary sensitivity function T may be made small is constrained from below by the magnitude of the largest right-half plane open-loop pole. Practically, the lower limit is always greater than this magnitude.

(b) If the plant has right-half plane zeros then the achievable reduction of T is further impaired. This effect is especially pronounced when one or several right-half plane pole-zero pairs are very close.

Figure 7 shows the difficulties caused by right-half plane zeros and poles of the loop gain L and, hence, the plant transfer function P. The sensitivity S can only be small up to the magnitude of the smallest right-half plane zero. The complementary sensitivity T can only start to roll off to zero at frequencies greater than the magnitude of the largest right-half plane pole. The crossover region, where S and T assume their peak values, extends over the intermediate frequency range.

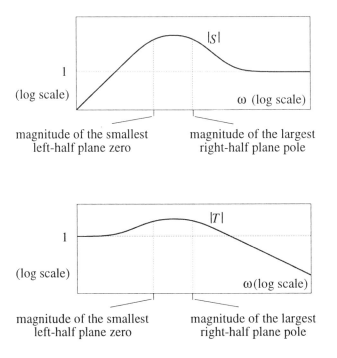

Fig. 7. Right-half plane zeros and poles constrain S and T

5 More general perturbations

More general perturbations may be considered with Doyle's basic perturbation model [Doy79] of Fig. 8. The block marked H represents the feedback system. H is sometimes called the *interconnection matrix*. The block marked Δ_H models the perturbation. It is nominally 0. Both H and Δ_H are transfer matrices of stable

5 More general perturbations

systems. By the small gain theorem, a sufficient condition for the perturbed system to be stable is that

$$\|\Delta_H\| < \frac{1}{\|H\|}. \tag{39}$$

$\|\cdot\|$ is a suitable operator norm. A very useful norm is the ∞-norm, defined as

$$\|H\|_\infty = \sup_{\omega \in \mathbb{R}} \sigma_{\max}(H(j\omega)). \tag{40}$$

If M is a complex-valued matrix then $\sigma_{\max}(M)$ is the largest singular value of M, that is, the largest eigenvalue of $M^H M$. This is also the largest eigenvalue of MM^H. The superscript H denotes the complex conjugate transpose.

If the perturbations Δ_H have been scaled such that $\|\Delta_H\|_\infty \leq 1$ then the perturbed system is stable for all such perturbations if and only if

$$\|H\|_\infty < 1. \tag{41}$$

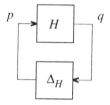

Fig. 8. Basic perturbation model

Loop gain and inverse loop gain perturbations. The loop gain and inverse loop gain perturbations considered before are special cases of the basic model. Figure 9 shows how they are modeled. The blocks V and W represent weighting filters that scale the perturbations such that $\|\delta_L\|_\infty \leq 1$ and $\|\delta_{L^{-1}}\|_\infty \leq 1$.

The loop gain perturbation model of Fig. 9(a) represents the perturbed loop gain as $(I+\Delta_L)L$, with $\Delta_L = V\delta_L W$. It is easy to check that the interconnection matrix is by $H = -WTV$, with $T = L(I + L)^{-1}$ the complementary sensitivity matrix. Hence, in the SISO case the closed-loop system is stable for all loop gain perturbations Δ_L such that $|\Delta_L(j\omega)| \leq |V(j\omega)W(j\omega)|$ for all $\omega \in \mathbb{R}$ if and only if

$$|T(j\omega)| < \frac{1}{|V(j\omega)W(j\omega)|} \quad \text{for all } \omega \in \mathbb{R}. \tag{42}$$

The inverse loop gain perturbation model of Fig. 9(b) represents the perturbed inverse loop gain as $(I + \Delta_{L^{-1}})L^{-1}$, with $\Delta_{L^{-1}} = V\delta_{L^{-1}}W$. In this case the

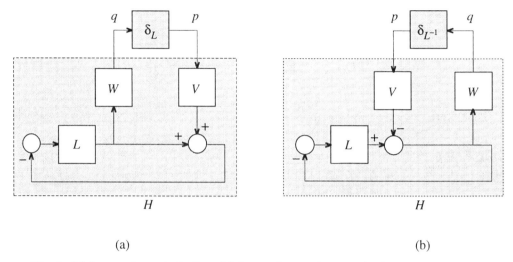

Fig. 9. (a) Loop gain perturbation. (b) Inverse loop gain perturbation

interconnection matrix is $H = -WSV$, with $S = (I + L)^{-1}$ the sensitivity matrix. In the SISO case the perturbed system is stable for all inverse loop gain perturbations $\Delta_{L^{-1}}$ such that $|\Delta_{L^{-1}}(j\omega)| \leq |V(j\omega)W(j\omega)|$ for all $\omega \in \mathsf{R}$ if and only if

$$|S(j\omega)| < \frac{1}{|V(j\omega)W(j\omega)|} \quad \text{for all } \omega \in \mathsf{R}. \tag{43}$$

Fractional perturbations. Stability robustness analysis of feedback systems based on perturbations of the loop gain or its inverse is simple, but often overly conservative.

Another model encountered in the literature[4] relies on what we term here *fractional perturbations*. It combines, in a way, loop gain and inverse loop gain perturbations. In this analysis, the loop gain L is represented as

$$L = ND^{-1}, \tag{44}$$

where the *denominator* D is a square nonsingular rational or polynomial matrix, and N a rational or polynomial matrix. Any rational transfer matrix L may be represented like this in many ways. If D and N are polynomial then the representation is known as a (right) *polynomial matrix fraction representation*. If D and N are rational and proper with all their poles in the open left-half complex plane then the representation is known as a (right) *rational matrix fraction representation*.

Figure 10 shows the fractional perturbation model. Inspection shows that the

---

[4] The idea originates from Vidyasagar [VSF82] [Vid85]. It is elaborated in [MG90].

5 More general perturbations

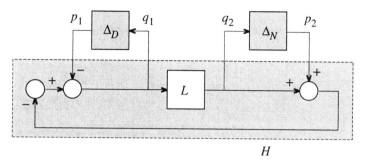

Fig. 10. Fractional perturbation model

perturbation is given by

$$L \longrightarrow (I+\Delta_N)L(I+\Delta_D)^{-1}, \tag{45}$$

or

$$ND^{-1} \longrightarrow (I+\Delta_N)ND^{-1}(I+\Delta_D)^{-1}. \tag{46}$$

Hence, the denominator and numerator are perturbed as $D^{-1} \longrightarrow (I+\Delta_D)^{-1}D^{-1}$ and $N \longrightarrow N(I+\Delta_N)$, respectively, or

$$D \longrightarrow D(I+\Delta_D), \qquad N \longrightarrow N(I+\Delta_N). \tag{47}$$

Thus, Δ_D and Δ_N represent *proportional perturbations of the denominator and of the numerator*, respectively.

By block diagram substitution it is easily seen that the configuration of Fig. 10 is equivalent to that of Fig. 11, where

$$p = -\Delta_D q_1 + \Delta_N q_2 = \underbrace{[-\Delta_D \quad \Delta_N]}_{\Delta_L} \begin{bmatrix} q_1 \\ q_2 \end{bmatrix}. \tag{48}$$

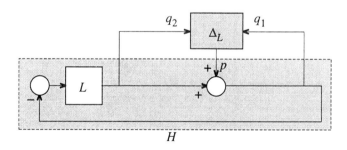

Fig. 11. Equivalent configuration

The interconnection matrix is

$$H = \begin{bmatrix} S \\ -T \end{bmatrix}. \tag{49}$$

Investigation of the frequency dependence of the largest singular value of $H(j\omega)$ yields information about the largest possible perturbations Δ_L that leave the loop stable.

It is useful to allow for scaling by representing the perturbations as $\Delta_D = V\delta_D W_1$ and $\Delta_N = V\delta_N W_2$, where V, W_1, and W_2 are suitably chosen (stable) rational matrices such that $\|\delta_L\|_\infty \leq 1$, with $\delta_L = [-\delta_D \quad \delta_N]$. Accordingly, the interconnection matrix changes to

$$H = \begin{bmatrix} W_1 SV \\ -W_2 TV \end{bmatrix}. \tag{50}$$

The largest singular value of $H(j\omega)$ equals the square root of the largest eigenvalue of

$$H^T(-j\omega)H(j\omega) = V^T(-j\omega)[S^T(-j\omega)W_1^T(-j\omega)W_1(j\omega)S(j\omega)$$
$$+ T^T(-j\omega)W_2^T(-j\omega)W_2(j\omega)T(j\omega)]V(j\omega). \tag{51}$$

For SISO systems this is the scalar function

$$|H(j\omega)|^2 = |V(j\omega)|^2[|S(j\omega)|^2|W_1(j\omega)|^2 + |T(j\omega)|^2|W_2(j\omega)|^2]. \tag{52}$$

Application of the fractional perturbation model. We consider how to arrange the fractional perturbation model. In the SISO case, without loss of generality we may take the scaling function V equal to 1. Then W_1 represents the scaling factor for the denominator perturbations and W_2 that for the numerator perturbations. Accordingly we have

$$\|H\|_\infty^2 = \sup_{\omega \in \mathbb{R}} \left(|S(j\omega)|^2|W_1(j\omega)|^2 + |T(j\omega)|^2|W_2(j\omega)|^2 \right). \tag{53}$$

For well-designed control systems the sensitivity function S is small at low frequencies while the complementary sensitivity function T is small at high frequencies. Hence, W_1 may be large at low frequencies and W_2 large at high frequencies without violating the robustness condition $\|H\|_\infty < 1$. This means that at low frequencies we may allow large denominator perturbations, and at high frequencies large numerator perturbations.

The most extreme point of view is to structure the perturbation model such that all low-frequency perturbations are denominator perturbations, and all high-frequency perturbations are numerator perturbations. Since we may trivially write

$$L = \frac{1}{\frac{1}{L}}, \tag{54}$$

5 More general perturbations

modelling low-frequency perturbations as pure denominator perturbations implies modelling low-frequency perturbations as *inverse* loop gain perturbations. Likewise, modelling high-frequency perturbations as pure numerator perturbations implies modelling high-frequency perturbations as loop gain perturbations. This amounts to taking

$$|W_1(j\omega)| = \begin{cases} W_{L^{-1}}(\omega) & \text{at low frequencies,} \\ 0 & \text{at high frequencies,} \end{cases} \tag{55}$$

$$|W_2(j\omega)| = \begin{cases} 0 & \text{at low frequencies,} \\ W_L(\omega) & \text{at high frequencies,} \end{cases} \tag{56}$$

with $W_{L^{-1}}$ a bound on the size of the inverse loop perturbations, and W_L a bound on the size of the loop perturbations. Obviously, the boundary between "low" and "high" frequencies lies in the crossover region.

Another way of dealing with this perturbation model is to modify the stability robustness test to checking whether for each $\omega \in \mathbb{R}$

$$|\Delta_{L^{-1}}(j\omega)| < \frac{1}{|S(j\omega)|} \quad \text{or} \quad |\Delta_L(j\omega)| < \frac{1}{|T(j\omega)|}. \tag{57}$$

This test amounts to verifying whether either the proportional loop gain perturbation test succeeds or the proportional inverse loop gain test. Figure 12 illustrates this. Obviously, its results are less conservative than the individual tests.

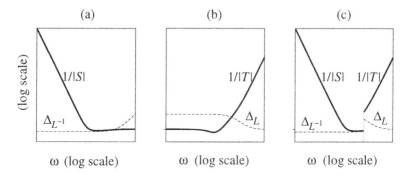

Fig. 12. (a), (b) Separate stability tests. (c) Combined test.

Feedback systems have good stability robustness in the frequency regions where either the sensitivity is small (at low frequencies) or the complementary sensitivity is small (at high frequencies). In the crossover region neither sensitivity is small. Hence, the feedback system is not robust for perturbations that strongly affect the crossover region.

In the crossover region the uncertainty therefore should be limited. On the one hand this limitation restricts *structured* uncertainty—caused by load variations and environmental changes—that the system can handle. On the other hand, *unstructured* uncertainty—deriving from neglected dynamics and parasitic effects—should be kept within bounds by adequate modelling.

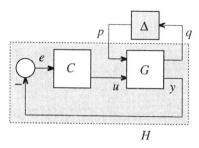

Fig. 13. Other perturbations

Other perturbations. To study more general plant perturbations we may consider the configuration of Fig. 13. Δ models the uncertainty in the plant. The interconnection matrix of the plant by itself is assumed to be given by

$$\begin{bmatrix} q \\ y \end{bmatrix} = \begin{bmatrix} G_{11} & G_{12} \\ G_{21} & G_{22} \end{bmatrix} \begin{bmatrix} p \\ u \end{bmatrix}. \tag{58}$$

Interconnecting the uncertainty block with the open-loop plant we have the loop balance equation $p = \Delta q = \Delta(G_{11}p + G_{12}u)$, so that $p = (I - \Delta G_{11})^{-1}\Delta G_{12}u$. Hence, the perturbed plant is described by

$$y = \underbrace{[G_{22} + G_{21}(I - \Delta G_{11})^{-1}\Delta G_{12}]}_{P} u. \tag{59}$$

To find the interconnection matrix H of the closed-loop system we solve the loop balance equation $u = -C(G_{22}u + G_{21}p)$, to find $p = -(I + CG_{22})^{-1}CG_{21}u$. It follows that

$$q = \underbrace{[G_{11} - G_{12}(I + CG_{22})^{-1}CG_{21}]}_{H} p. \tag{60}$$

The interconnection matrix H may be rewritten as

$$H = G_{11} - G_{12}(I + L)^{-1}LG_{22}^{-1}G_{21} = G_{11} - G_{12}TG_{22}^{-1}G_{21}, \tag{61}$$

where $L = CG_{22}$ is the loop gain and $T = (I + L)^{-1}L$ the complementary sensitivity matrix.

(a) If $G_{11} = 0$ then H reduces to

$$H = -G_{12}TG_{22}^{-1}G_{21}, \tag{62}$$

so that the perturbation is of the type of a *loop gain perturbation*.

(b) If $G_{11} - G_{12}G_{22}^{-1}G_{21} = 0$ then H reduces to

$$H = G_{11} - G_{12}TG_{22}^{-1}G_{21} = G_{11} - G_{12}(I - S)G_{22}^{-1}G_{21}$$
$$= G_{12}SG_{22}^{-1}G_{21}, \qquad (63)$$

so that the perturbation is of the type of an *inverse loop gain perturbation*.

Assuming that the necessary inverses exist we may write

$$H = G_{12}SG_{12}^{-1}G_{11} + G_{12}T(G_{12}^{-1}G_{11} - G_{22}^{-1}G_{21}). \qquad (64)$$

This shows that in general the effect of the perturbation involves both S and T. In the scalar case (64) takes the form

$$H = G_{11}S + \frac{G_{11}G_{22} - G_{12}G_{21}}{G_{22}}T. \qquad (65)$$

In the crossover region, where neither S nor T is small, this function determines how critical the design is.

6 Design

In this section we very briefly review what the various best known design techniques can do in the light of the analysis.

Design targets. The design targets for the low, high and intermediate frequency regions are clear:

- LF: Make the gain L large, and, consequenctly, S small and $T \approx 1$.
- HF: Make the gain L small, and consequently, T small and $S \approx 1$.
- IF: Make L avoid the point -1 as widely as possible. As a result, reduce undesirable peaking of both S and T.

Classical design. Classical design methods achieve the LF target by using lag compensation, in particular, by introducing integrating action. The HF target is achieved by low-pass filter action in the compensator (with a high cut-off frequency). In the IF region the loop gain is shaped by lead compensation and, if needed, the use of a notch filter. The classical techniques work well for simple problems but fail for more complex problems, in particular, for multivariable systems.

Design by QFT. The starting point for design by quantitative feedback theory — QFT — is to specify upper and lower bounds for the magnitude of the complementary sensitivity function T and uncertainty regions in the complex plane for the plant frequency response. Typically the bounds on $|T|$ are tight at low frequencies and much looser at high frequencies beyond the desired bandwidth for the closed-loop transfer function. These specifications are satisfied by a semi-graphical method of shaping the loop gain in a Nichols chart. In a way the problems in the crossover region are resolved by shifting it to a frequency region where the stability margin is less critical. The design method is potentially powerful but unfortunately no or little support — which it certainly needs — seems available in the standard control system design environments. Extensions to multivariable problems are available in the literature but here the lack of tools is even more poignant.

LQG and $\mathcal{H}_2$ design. The linear-quadratic-Gaussian (LQG) paradigm centers around optimization of the time domain response. Design methods that are based on it shape the closed-loop system functions S and T only indirectly. There is no explicit provision for controlling the loop gain in the crossover region. Time responses for LQG systems typically are well-damped, however. This implies good stability margins. There is no way to design for *specific* perturbations, such as parameter variations. Multivariable problems are as easily handled as single-variable systems.

Rephrasing the LQG problem as an $\mathcal{H}_2$ optimization problem naturally leads to the introduction of frequency dependent weighting functions. This provides more flexibility in the design but still does not allow to incorporate information about the plant uncertainty directly.

$\mathcal{H}_\infty$-optimization. Design by $\mathcal{H}_\infty$ optimization relies on the minimization of the ∞-norm of a suitably chosen closed-loop transfer matrix. A well-known version is the *mixed sensitivity problem,* which amounts to the minimization of the ∞-norm of

$$H = \begin{bmatrix} W_1 SV \\ -W_2 UV \end{bmatrix}. \tag{66}$$

S is the sensitivity matrix and U the input sensitivity matrix. V is a shaping filter and W_1 and W_2 are weighting filters. In the SISO case we have

$$\begin{aligned} \|H\|_\infty^2 &= \sup_{\omega \in \mathbb{R}} |H(j\omega)|^2 \\ &= \sup_{\omega \in \mathbb{R}} |V(j\omega)|^2 \left(|W_1(j\omega)|^2 |S(j\omega)|^2 + |W_2(j\omega)|^2 |U(j\omega)|^2 \right). \end{aligned} \tag{67}$$

By choosing the weighting functions V, W_1 and W_2 suitably — within the limitations imposed by the pole-zero configuration of the plant — the design targets that S be small at low frequencies and T small at high frequencies may be achieved very well. Integral control of any desired order and high-frequency compensator roll-off of any order may easily be accomplished. Since the input sensitivity U and the complementary sensitivity T are related by $T = PU$ shaping U is equivalent to shaping T.

By a technique known as *partial pole placement* [Kwa93] the dominant poles may be suitably placed. This serves to control the bandwidth and the dominant time response. Though of course multivariable systems are inherently more complex than single-variable systems they may easily be handled in the mixed sensitivity approach.

The mixed sensitivity approach is less suitable for design problems where robustness is critical because of large perturbations in the crossover region. The reason is that as we saw for such problems the sensitivity and complementary sensitivity functions separately are not always good predictors for robustness.

μ-Synthesis. Doyle [Doy84] has developed an attractive theory that allows to consider block structured perturbations. The theory admits both dynamical perturbations and real parameter variations, but the associated design method — known as μ-synthesis — only allows dynamical perturbations. This method may be of great help for critical problems where the uncertainty in the crossover region is adequately modeled as dynamical uncertainties. For *parameter uncertainty* — often the critical factor — the method may be of no avail.

7 Design study of a sampled data control system

In this section we present a case study. The problem is to design a sampled-data controller for a flexible transmission system [LRCV94].

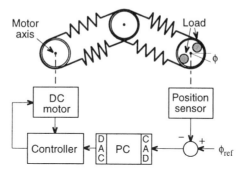

Fig. 14. Schematic representation of a flexible transmission system

System specifications. Figure 14 shows a schematic diagram of the transmission. This laboratory setup has been built at the Grenoble Automatic Control Laboratory. The system consists of three horizontal pulleys connected by two flexible belts. The first pulley is driven by a DC motor, whose position is controlled by local feedback. The third pulley may be loaded with small disks of different weights. Its position is measured by a position sensor, sampled and digitized. A PC is used to control the system.

Table 1. DATA FOR THE FLEXIBLE TRANSMISSION

Load	$A(z)$	$B(z)$
0%	$z^4 - 1.41833z^3 + 1.58939z^2 - 1.31608z + 0.88642$	$0.28261z + 0.50666$
50%	$z^4 - 1.99185z^3 + 2.20265z^2 - 1.84083z + 0.89413$	$0.10276z + 0.18123$
100%	$z^4 - 2.09679z^3 + 2.31962z^2 - 1.93353z + 0.87129$	$0.06408z + 0.10407$

The plant is described as a discrete-time system. In terms of z-transforms it has a transfer function of the form

$$P(z) = \frac{B(z)}{A(z)}. \tag{68}$$

The polynomials A and B depend on the load and are specified in Table 1. The sampling period of the digital controller is $T_s = 50$ ms. Figure 15 gives the Bode plots of the open loop frequency response functions. It shows the two vibration modes of the transmissions and their variations.

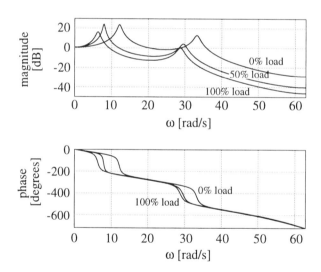

Fig. 15. Frequency responses of the flexible transmission

A one- or two-degree-of-freedom feedback controller is to be designed that achieves the following specifications:

1. A peak value of less than 6 dB of the sensitivity function for all loads.
2. A delay margin [LRCV94] of at least 40 ms, that is, an extra delay in the feedback loop of 40 ms or less does not destabilize the system.
3. Perfect rejection of constant disturbances (by integrating action).

7 Design study of a sampled data control system

4. Disturbance attenuation (that is, a sensitivity function less than 1) in the low-frequency band from 0 to $0.01/T_s = 0.2$ Hz for all loads.
5. Overshoot less than 10% for a step change in the reference input for all loads.
6. A rise time (to within 90% of the final value) of less than 1 s for a step change in the reference input for all loads.
7. Attenuation of step disturbances filtered by $1/A$ to less than 10% of the measured peak value within 1.2 s.
8. A peak value of less than 10 dB of the input sensitivity function in the frequency range $0.4/T_s$ to $0.5/T_s$ (8 to 10 Hz). In this frequency region the plant has very low gain.

Table 2. OPEN-LOOP POLES AND ZEROS OF THE CONTINUOUS-TIME SYSTEM

Load	Poles	Zeros
0%	$-1.5861 \pm j43.9697$	140.91
	$-0.5356 \pm j12.5534$	40, 40, 40
50%	$-1.5370 \pm j36.5572$	144.76
	$-0.2929 \pm j7.9750$	40, 40, 40
100%	$-1.54040 \pm j35.1616$	168.19
	$-0.5428 \pm j6.3307$	40, 40, 40

Transformation to a continuous-time problem. We transform the problem to a continuous-time problem by the Tustin transformation

$$z = \frac{1 + \frac{sT}{2}}{1 - \frac{sT}{2}}. \tag{69}$$

We choose T equal to the sampling period T_s. Table 2 lists the continuous-time open-loop poles and zeros. After tranfsormation, the continuous-time radial frequency ω_c is related to the discrete-time angular frequency ω_d as

$$\frac{\omega_c T}{2} = \tan \frac{\omega_d T}{2}. \tag{70}$$

We consider the implications of the design specifications (after transformation to the continuous-time domain) on the sensitivity function and the complementary sensitivity function.

1. The design requirements 1, 3, 4, 6 and 7 impose specifications on the sensitivity function S.
   (a) Requirement 1 implies that S needs to have a peak value of less than 6 dB.
   (b) Requirement 3 means that S has to zero at zero frequency,

(c) Requirement 4 implies that $|S|$ needs to be smaller than 0 dB up to the frequency 0.2 Hz (1.26 rad/s).
(d) Requirement 6 implies that the bandwidth should be sufficiently large. If it is not, it may be corrected by the prefilter.
(e) Requirement 7 implies that the function S/A should be sufficiently small. Equivalently, the input disturbance sensitivity function $SB/A = SP = R$ needs to be sufficiently small. This is especially important in the frequency region about 10 rad/s where the plant frequency response peaks.
2. The design requirements 2 and 8 have implications for the complementary sensitivity function T. To satisfy requirement 2 the complementary sensitivity T should be small at high frequencies. This is also what is needed to satisfy design requirement 8.
3. Design requirements 5, 6 and 7 forbid excessive peaking of both S and T.

We aim to achieve these targets by introducing integrating action, a high frequency roll-off of the compensator of 0 or 20 dB/decade, and a bandwidth of about 10 rad/s.

Note that the plant has a triple open-loop zero at 40 rad/s. This means that robustness cannot be achieved by using high gain and thus making the bandwidth large. The three open-loop zeros constitute an essential limitation on the bandwidth. The zeros are a consequence of the inherent delay of three sampling periods of the discrete-time system.

The perturbations of the flexible transmission system are caused by changes in a single physical parameter—the load. We assess the resulting proportional and proportional inverse perturbations.

Proportional and propoportional inverse perturbations. We tentatively choose the 50% load case as the nominal plant. Figures 16(a) and (b) show the magnitudes of the proportional plant perturbations and the proportional inverse plant perturbations, given by

$$\frac{P - P_o}{P_o} \quad \text{and} \quad \frac{\frac{1}{P} - \frac{1}{P_o}}{\frac{1}{P_o}}, \tag{71}$$

respectively. P_o is the nominal plant transfer function and P the perturbed transfer function. The plots show that there are important perturbations in what is likely to be the crossover region.

Parametrization of the plant transfer function. The plant perturbations are caused by a single scalar parameter, the moment of inertia of the third pulley. Inspection of Fig. 14 suggests that the equations of motion may be represented as

$$J_1 \ddot{\theta} = -k_1(\theta - u) + k_2(\phi - \theta) - \rho_1 \dot{\theta}, \tag{72}$$
$$J_2 \ddot{\phi} = -k_2(\phi - \theta) - \rho_2 \dot{\phi}. \tag{73}$$

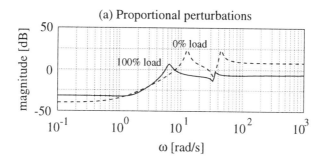
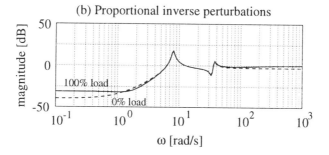

Fig. 16. Proportional and proportional inverse perturbations of the flexible transmission system.

J_1 is the moment of inertia of the second pulley and J_2 that of the third. The angle of rotation of the first pulley is the input u and θ is the angle of the second pulley. Finally, k_1 and k_2 are the spring constants of the first and second belt, respectively, and ρ_1 and ρ_2 are the friction constants of the second and third pulley, respectively. Laplace transformation and solution for ϕ results in

$$\phi = \underbrace{\frac{k_1 k_2}{(J_1 s^2 + \rho_1 s + k_1 + k_2)(J_2 s^2 + \rho_2 s + k_2) - k_1 k_2}}_{P(s)} u. \tag{74}$$

We rewrite the variable parameter J_2 as

$$J_2 = \frac{J_{2,\,\max} + J_{2,\,\min}}{2} + p \frac{J_{2,\,\max} - J_{2,\,\min}}{2}, \tag{75}$$

where the real parameter p varies between -1 and 1. Substitution into the plant transfer function P defined by (74) shows that P may be represented as

$$P = \frac{N_o}{D_o + pE}, \tag{76}$$

with N_o, D_o and E polynomials. It is not very difficult to estimate the coefficients of these polynomials from the data in Table 1. Table 3 lists the polynomials N_o,

Table 3. Parametrization $P = N_o/(D_o + pE)$ of the plant transfer function

| $N_o(s) = 0.013209s^4 - 3.5699s^3 + 301.58s^2 - 10373s + 127029$ |
| $D_o(s) = 1.3913s^4 + 5.5122s^3 + 1947.8s^2 + 1957.9s + 119577$ |
| $E(s) = s^4 + 4.0619s^3 + 1122.88s^2 + 1245.8s$ |

Load	Parameter value
0%	$p_1 = -1$
50%	$p_2 = 0.11179$
100%	$p_3 = 1$

D_o and E and the values p_1, p_2 and p_3 corresponding to the three loads. The three models represent the three perturbed transfer functions quite accurately.

We establish the perturbation model of Fig. 13. From (59) we have

$$P = G_{22} + G_{21}(I - \Delta G_{11})^{-1}\Delta G_{12} = \frac{G_{22} + \Delta(G_{12}G_{21} - G_{11}G_{22})}{1 - \Delta G_{11}}. \tag{77}$$

Rewriting (76) as

$$P = \frac{\frac{N_o}{D_o}}{1 + p\frac{E}{D_o}} \tag{78}$$

we identify $\Delta = -p$ and

$$G_{11} = \frac{E}{D_o}, \qquad G_{22} = \frac{N_o}{D_o}, \qquad G_{12}G_{21} - G_{11}G_{22} = 0. \tag{79}$$

As (78) shows, and (79) confirms, the perturbation is of the inverse loop gain type. In fact, the function E/D_o is the relative inverse loop gain perturbation whose magnitude is plotted in Fig. 16(b).

$\mathcal{H}_\infty$ design methods necessitate considering $p = -\Delta$ as a *dynamical* perturbation with ∞-norm less than 1. In the case at hand the perturbation is real. Clearly it is overly conservative to represent Δ as a dynamical perturbation. By inspection of Fig. 16 $\|E/N_o\|_\infty$ is seen to be greater than 1. Hence, there exists a dynamical perturbation Δ that destabilizes the open-loop system. There is no real perturbation bounded by $|p| \leq 1$ that destabilizes the open-loop system, however.

From (65) it follows that the interconnection function of the closed-loop system for the plant perturbations is

$$H = G_{11}S + \frac{G_{11}G_{22} - G_{12}G_{21}}{G_{22}}T = \frac{E}{D_o}S. \tag{80}$$

As expected for a denominator or inverse loop gain perturbation the sensitivity function is the relevant function.

7 Design study of a sampled data control system

Mixed sensitivity design. The mixed sensitivity approach is ideally suited for frequency shaping. Before selecting the weighting functions we need to determine the requirements on the sensitivity function S and its complement T.

We consider the proportional and inverse proportional perturbations of Fig. 16 and their implications on S and T.

In the low-frequency region robustness is achieved because the sensitivity S is small. The performance specifications point to a closed-loop bandwidth of at least 1 rad/s. Moreover S has to be sufficiently small in the region around 10 rad/s, where $|P|$ peaks. The triple right-half plane open-loop zero at 40 rad/s precludes a bandwidth that is larger than, say, about 10 rad/s. Figure 16 shows that there are important perturbations, both of the inverse loop gain and the loop gain, in the resulting crossover region. It seems best to assign the crossover region in the frequency region around, say, 2 to 3 rad/s, because this is where the inverse perturbations are relatively small.

We attempt to achieve the design specifications by a mixed sensitivity design according to the following strategy:

1. Choose V to achieve robust damping at the frequency peaks. To this end, V should be large near these peaks. This serves to make PS small.
2. Use the integrator-in-the-loop method to obtain integrating action and low-frequency disturbance attenuation.
3. Use W_2 to secure adequate high-frequency roll-off.

As the nominal plant we choose the plant $P = N_o/D_o$ that corresponds to setting $p = 0$ in (76). In the standard mixed sensitivity approach we have

$$V = \frac{M}{D}, \qquad (81)$$

where D is the (nominal) denominator of the plant, and M a polynomial of the same degree as D. The roots of M reappear as closed-loop poles. Choosing M amounts to re-assigning the open-loop plant poles to new locations. This is partial pole placement. For the case at hand D_o has as its roots the open-loop poles $\pi_{1,2} = -0.4541 \pm j8.0176$ and $\pi_{3,4} = -1.5269 \pm j36.4756$. To damp the former pair we reassign it to a double pole at $-|\pi_{1,2}| = -8.0304$. The eigenfrequency of the second pair is well into the high-frequency region so we leave the pair in place.

There are several ways to obtain integral action in the mixed sensitivity approach [BK94]. The *constant disturbance model* method is to include a factor $1/s$ in the shaping filter V. To allow compensation of constant disturbances it is necessary to include at the same time a factor s in the weighting function W_2. The *constant error suppression* method consists of including a factor $1/s$ in the weighting function W_2. Both methods make the overall system non-stabilizable, which precludes the use of the standard two-Riccati equation algorithm. In both cases the offending factor $1/s$ may be brought "inside the loop" by block diagram substitutions. This leads to the *integrator-in-the loop* method, which amounts to modifying the nominal plant transfer function from $P = N/D$ to

$$P(s) = \frac{N(s)}{sD(s)}. \tag{82}$$

Correspondingly, the shaping filter $V = M/D$ is changed to

$$V(s) = \frac{(s+\alpha)M(s)}{sD(s)}, \tag{83}$$

with α to be chosen. Once the modified problem has been solved the extra factor $1/s$ in the plant transfer function is absorbed into the compensator.

It remains to select the weighting functions W_1 and W_2. Without loss W_1 may be taken equal to 1. W_2 may simply be chosen as $W_2(s) = c$, with the constant c to be determined. This results in $\mathcal{H}_\infty$ optimal compensators that are proper but not necessarily strictly proper. Because of the integrating factor $1/s$ that is included in the eventual compensator this actually has a high frequency roll-off of at least 20 dB/decade.

We choose $\alpha = 20$ to achieve the planned bandwidth cut-off frequency. After a litle experimenting we arrive at the value $c = 1$. This leads to the results shown in Fig. 17. Figure 18 shows the perturbed sensitivity functions for the three loads.

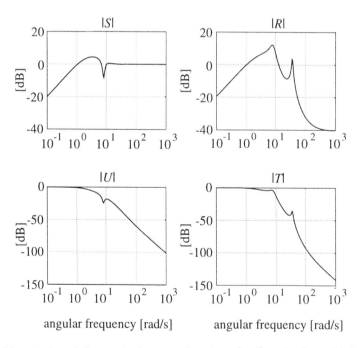

Fig. 17. Magnitudes of the nominal system functions for the mixed sensitivity design

7 Design study of a sampled data control system

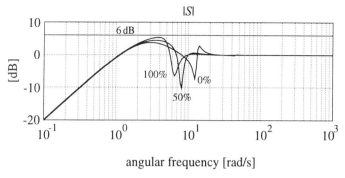

Fig. 18. Perturbed sensitivity functions for the mixed sensitivity design

The system is stabilized for all loads, but not all the design specifications are met. The sensitivity function crosses the 0 dB line at 1 rad/s rather than at 1.26 rad/s. The response to step disturbances at the plant input for the 0% load case (not shown here) is too oscillatory and does not quite settle down in the required 1.2 s.

The results may be marginally altered and improved by experimenting with the weights but it does not look as if they can be adjusted to satisfy the specifications on all points. The reason is that the low- and high-frequency regions may be shaped very well by using the mixed sensivity problem and other $\mathcal{H}_\infty$ methods. These methods are not suited, however, for problems with real parameter perturbations that strongly affect the crossover region.

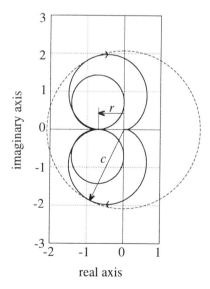

Fig. 19. Nyquist plot of SE/D_o

Figure 19 shows the Nyquist plot of the interconnection function SE/D_o for the parameter p. The system becomes unstable if for some real $p \in [-1,1]$ the Nyquist plot of $-pSE/D_o$ encircles the point -1. For real parameter perturbations the distance r to 0 of the point on the plot and the real axis that is furthest away from the origin is the inverse of the magnitude of the smallest parameter value that destabilizes the system. For complex (dynamical) perturbations the distance c to 0 of the point on the plot furthest away from the origin is the inverse of the magnitude of the smallest perturbation that destabilizes the system.

Clearly, c is much larger than r, which means that the system may be destabilized by dynamical perturbations that are considerably smaller than the smallest real perturbation that destabilizes the system. This explains why there is no reason to expect that μ-synthesis will be of much help for this problem.

The problem at hand is *critical* in the sense that the perturbations are large and the specifications can barely be met, if at all. From an integrated control engineering point of view such problems should be avoided by making sure that the plant has enough capacity to handle its tasks comfortably. In the present case the difficulties may for instance be resolved by increasing the sampling rate. Consider quadrupling the sampling rate, for instance, This leaves the variable poles and zeros of Table 2 more or less unchanged but the right-half plane zeros at 40 rad/s move to 160 rad/s. This allows to increase the bandwidth — provided the motor delivers the power.

8 Conclusions

Low frequencies and high frequencies play symmetric roles in control system design. The sensitivity function and complementary sensitivity function manifest this symmetry. The symmetry appears when robustness is analyzed not only for loop gain perturbations but also for perturbations of the inverse of the loop gain.

Performance and robustness often go hand in hand. Low sensitivity at low frequencies is needed for disturbance reduction, but also provides protection against low-frequency plant variations. Small complementary sensitivity at high frequencies is not only required for robustness against parasitic effects and neglected dynamics but also to prevent exceeding the plant capacity and to abate the effect of measurement noise.

For various reasons the crossover region is the most critical frequency region for control system design. Peaking of the sensitivity and the complementary sensitivity function in the crossover region disadvantages both performance and robustness.

It follows that any design method that does well for performance generally also achieves robustness [LA95]. This robustness may not be satisfactory for perturbations caused by real parameter variations that strongly affect the crossover region. The well-known analytical design methods — LQG, $\mathcal{H}_2$ and $\mathcal{H}_\infty$-optimization, μ-synthesis — fail to handle such critical perturbations.

References

[BB91] S. P. Boyd and C. H. Barratt. *Linear Controller Design: Limits of Performance.* Prentice Hall, Englewood Cliffs, NJ, 1991.

[BK94] O. Bosgra and H. Kwakernaak. Design methods for control systems. Dutch Graduate Network on Systems and Control, Spring 1994. Lecture notes.

[Bod45] H. W. Bode. *Network Analysis and Feedback Amplifier Design.* Van Nostrand, New York, 1945.

[Doy79] J. C. Doyle. Robustness of multiloop linear feedback systems. In *Proc. 17th IEEE Conf. Decision & Control,* pages 12–18, 1979.

[Doy84] J. C. Doyle. Lecture Notes, ONR/Honeywell Workshop on Advances in Multivariable Control, Minneapolis, Minn., 1984.

[Eng88] S. Engell. *Optimale lineare Regelung: Grenzen der erreichbaren Regelgüte in linearen zeitinvarianten Regelkreisen,* volume 18 of *Fachberichte Messen–Steuern–Regeln.* Springer-Verlag, Berlin, etc., 1988.

[FL88] J. S. Freudenberg and D. P. Looze. *Frequency Domain Properties of Scalar and Multivariable Feedback Systems,* volume 104 of *Lecture Notes in Control and Information Sciences.* Springer-Verlag, Berlin, etc., 1988.

[FZ84] B. A. Francis and G. Zames. On H^∞-optimal sensitivity theory for SISO feedback systems. *IEEE Trans. Aut. Control,* 29:9–16, 1984.

[Kwa93] H. Kwakernaak. Robust control and H^∞-optimization. *Automatica,* 29:255–273, 1993.

[Kwa94] H. Kwakernaak. La commande robuste: Optimisation à sensibilité mixte. In A. Oustaloup, editor, *La robustesse — analyse et synthèse de commandes robustes.* Hermès, Paris, 1994.

[LA95] L. Lublin and M. Athans. An experimental comparison of $\mathcal{H}_2$ and $\mathcal{H}_\infty$ designs for an interferometer testbed. In B. A. Francis and A. R. Tannenbaum, editors, *Feedback Control, Nonlinear Systems, and Complexity,* volume 202 of *Lecture Notes in Control and Information Sciences.* Springer-Verlag, Heidelberg, 1995.

[LRCV94] I. D. Landau, F. Rolland, C. Cyrot, and A. Voda. Régulation numérique robuste: Le placement de pôles avec calibrage de la fonction de sensibilité. In A. Oustaloup, editor, *La robustesse — analyse et synthèse de commandes robustes.* Hermès, Paris, 1994.

[Lun91] M. Lundh. *Robust Adaptive Control.* PhD thesis, Department of Automatic Control, Lund, Sweden, 1991.

[MG90] D. C. McFarlane and K. Glover. *Robust Controller Design Using Normalized Coprime Factor Plant Descriptions,* volume 146. Springer-Verlag, Berlin, etc., 1990.

[Vid85] M. Vidysagar. *Control Systems Synthesis—A Factorization Approach.* MIT Press, Cambridge, MA, 1985.

[VSF82] M. Vidyasagar, H. Schneider, and B. A. Francis. Algebraic and topological aspects of feedback stabilization. *IEEE Trans. Aut. Control,* 27:880–894, 1982.

[ZF83] G. Zames and B. A. Francis. Feedback, minimax sensitivity, and optimal robustness. *IEEE Trans. Aut. Control,* 28:585–601, 1983.

Universal Regulators in Linear-Quadratic Optimization Problems

V. A. Yakubovich

St.Petersburg University,
St.Petersburg 198904, Russia

1. Introduction

In this report we present a short survey and analysis of certain new infinite-horizon linear-quadratic optimization problems. They differ from well-known ones (see [1]–[5] and others) mainly in the assumption that there is no full information on the parameters of external disturbance and the output reference signal. It turns out that in certain such problems the optimal regulator is not unique and a special optimal regulator can be found that does not depend on unknown parameters. Therefore it can be used in this uncertain situation. We shall say that this regulator is universal in a given class of unknown parameters. The cases of existence of a universal regulator might seem to be very exceptional because such regulator solves a family of optimization problems simultaneously. But at least one such example is known very well in the control theory: in the linear-quadratic optimization problem without external disturbances the optimal regulator $u = Kx$ does not depend on the initial state value $x(0) = a$. We will consider other similar and more complicated cases.

2. Preliminaries

Let us give the formal definition of the optimal and the ε-optimal universal regulator.

Let $X = \{x\}$, $U = \{u\}$ be some sets, $D \subset X \times U$ be a given subset, $Y = \{y\}$ be a linear space and $F : X \times U \to Y$, $\Phi : X \times U \to \mathbb{R}^1$ be given mappings. We call elements $x \in X$ "states", $u \in U$ "controls", $(x, u) \in X \times U$ "processes" and $(x, u) \in D$ "admissible processes". Consider an optimization problem

$$\text{to minimize } \Phi(x, u) \text{ subject to } (x, u) \in D, \quad F(x, u) = 0_Y. \qquad (*)$$

Let (x^0, u^0) be its solution. Suppose we have constructed a mapping $G : X \times U \to Z$ into a certain linear space Z such that the set $\{(x^0, u^0)\}$ of all optimal processes (x^0, u^0) coincides with the set of all solutions of the system

$$F(x, u) = 0_Y, \quad G(x, u) = 0_Z. \qquad (2.1)$$

Then the equation $G(x, u) = 0_Z$ is called *an optimal regulator* in the problem $(*)$. It is clear that this definition agrees with the generally accepted notion.

Let $\Phi_{\min} = \text{Inf}[\Phi(x,u) : (x,u) \in D, F(x,u) = 0_Y]$. If $\Phi(x,u) \leq \Phi_{\min} + \varepsilon$ for all solutions of (2.1) then the equation $G(x,u) = 0_Z$ is called ε-*optimal regulator*. Now let us assume that $F = F_c$, $\Phi = \Phi_c$ depend on an abstract parameter $c \in \Xi$ where Ξ is a certain set. (The parameter c includes all unknown values in our optimization problem; further we call Ξ "the class of uncertainties".) In general the optimal process $(x^0(c), u^0(c))$ depends also on c. If G does not depend on c and a regulator $G(x,u) = 0_Z$ is optimal (ε-optimal) for every $c \in \Xi$ in the problem (*) with $F = F_c$, $\Phi = \Phi_c$ then we will call it *a universal optimal (ε-optimal) regulator* in the class Ξ. If an universal optimal regulator exists (or an universal ε-optimal regulator exists for every $\varepsilon > 0$) then we say that the common minimum (common infimum) situation takes place.

Often optimization problems are considered in which the regulator is a variable to be found. Let $\boldsymbol{D}_{\text{reg}} = \{G\}$ be a set of mappings $G = G(x,u)$ (the set of "admissible regulators"). We suppose that for any G in $\boldsymbol{D}_{\text{reg}}$ the corresponding regulator $G(x,u) = 0_Z$ is admissible in the following sense: (2.1) implies $(x,u) \in D$. We do not assume that every $(x,u) \in D$ is generated by a certain regulator $G = 0$, $G \in \boldsymbol{D}_{\text{reg}}$. Let us suppose that for (x,u) satisfying (2.1) $\Phi(x,u) = \Phi[G]$ in fact depends only on G, that is $\Phi(x,u) = \text{const}$ on each set (2.1), $G \in \boldsymbol{D}_{\text{reg}}$. Consider the regulator optimization problem

$$\text{to minimize } \Phi[G] \text{ subject to } G \in \boldsymbol{D}_{\text{reg}}. \qquad (**)$$

Obviously the infimum in the problem (**) is greater or equal to the infimum in the problem (*). If these two infima coincide, then an optimal (ε-optimal) regulator for the problem (**) works out an optimal (ε-optimal) process for the initial problem (*), and in this sense it solves this problem. Examples of such situation is well known in the control theory; other examples will be encountered below.

Let us suppose now again that $F = F_c$, $\Phi = \Phi_c$ depend on parameter $c \in \Xi$ and that this parameter is unknown. Consider for simplicity the case of the existence of an optimal process $[x^0(c), u^0(c)]$ and of an optimal regulator G_c^0 for every $c \in \Xi$. If they really depend on c then we cannot use these solutions in practice. In this case we have to change the settlement of problem. For example we can bear in mind "the worst" case and minimize $M(G) = \max_{c \in \Xi} \Phi_c(G)$ similarly to the H^∞-approach. Fig. 1 illustrates this case; G_j^0 denotes optimal regulator in problem (*) for $c = c_j$ and G_*^0 denotes the solution of the new minimax problem. This situation is unpleasant because the value of the cost functional $\Phi_{c_1}(G_*^0)$ which we obtain can exceed much the optimal value $\Phi_{c_1}(G_1^0)$ for a parameter c_1 that will fall in reality (see Fig. 1). It is important to note that this disadvantage is absent in the common minimum situation (see Fig. 2). So it is important to find out cases when the common minimum (common infimum) situation takes place.

One can expect that such cases occur very seldom. We will see that, in spite of expectations, the common minimum (infimum) situation takes place in a number of practically important optimization problems.

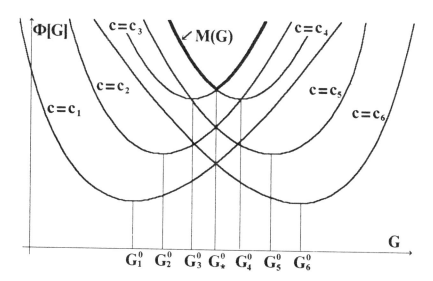

Fig. 2.1.

So we are going to find optimization problems where the common minimum (common infimum) situation holds and to solve them. Note that in the common minimum situation the optimal process $[x^0(c), y^0(c)]$ is not a satisfactory solution (unlike universal optimal or ε-optimal regulator) because $[x^0(c), u^0(c)]$ depends on c, as a rule.

Let us explain why in some cases one can expect the common minimum situation. Let X, U, Y be Banach spaces, $D = X \times U$ and $L(x, u) = \Phi(x, u) + l^* F(x, u)$ be the Lagrangian function of problem (*), $l^* \in Y^*$. Under some assumptions the Lagrange principle contends that the optimal process satisfies the equation

$$L'_x(x, u) = 0, \qquad L'_u(x, u) = 0 \qquad (2.2)$$

(see for example [6, Ch. 4]). If bounded inverse operator $(F'_x)^{-1}$ exists then (2.2) transforms into the equation $G^0(x, u) = 0$, where $G^0(x, u) = \Phi'_u - \Phi'_x (F'_x)^{-1} F'_u$. Suppose now that (2.2) is not only necessary but also sufficient condition for optimality. Then $G^0(x, u) = 0$ is an optimal regulator. We see that $G^0(x, u) = 0$ is an universal optimal regulator in the class $\Xi = \{c\}$ if L'_x, L'_u (or $\Phi'_x, \Phi'_u,$ F'_x, F'_u) do not depend on $c \in \Xi$.

Consider for example the standard optimization problem: to minimize the functional

$$\Phi[x(\cdot), u(\cdot)] = \int_0^T \mathfrak{G}[x(t), u(t), t]\, dt \qquad (2.3)$$

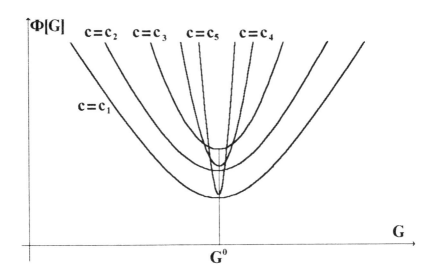

Fig. 2.2.

subject to constraints

$$\dot{x}(t) = f[x(t), u(t), t], \qquad x(0) = a. \tag{2.4}$$

(Here $x(t) \in \mathbb{R}^n$, $u(t) \in \mathbb{R}^m$, $\mathfrak{G}$, f, $\mathfrak{G}'_x$, f'_x are continuous functions). We take for $\boldsymbol{X}$, $\boldsymbol{Y}$, $\boldsymbol{U}$ the spaces of continuous functions $\boldsymbol{X} = \boldsymbol{Y} = C\{[0,T] \to \mathbb{R}^n\}$, $\boldsymbol{U} = C\{[0,T] \to \mathbb{R}^m\}$, and define the mapping F by

$$y(\cdot) = F[x(\cdot), u(\cdot)] \iff y(t) = x(t) - a - \int_0^t f[x(s), u(s), s]\, ds.$$

Then we obtain a special case of the optimization problem $(*)$.

It is simple to verify that $(F'_x)^{-1}$ exists. Since Φ'_x, Φ'_u, F'_x, F'_u do not depend on a the optimal regulator $G^0(x, u) = 0$ is universal in the class $\Xi = \{a\}$.

Obviously, $G^0(x, u) = 0$ is an optimal universal regulator also in the class of uncertainties $\Xi_1 = \{a, \varphi(\cdot)\}$ if f has the form $f(x, u, t) = f_0(x, u, t) + \varphi(t)$.

We did not discuss here if these universal optimal regulators are realizable and robust. These problems must be considered. It is well known from the control theory that for the class $\Xi = \{a\}$ and for special f, $\mathfrak{G}$ (linear f and quadratic $\mathfrak{G}$) the optimal regulator $u = K(t)x$ is realizable. It can be shown that for the class $\Xi = \{a, \varphi(\cdot)\}$ it is not realizable in general because $u(t)$ depends on $x(s)$, $t \leq s \leq T$.

We will consider below some common minimum (common infimum) situation problems in which the universal optimal (ε-optimal) regulators are realizable and robust. A crucial property of these problems is that they are special infinite–horizon problems and the disturbances are supposed to belong to more narrow classes.

3. Optimal damping of forced oscillations under unknown harmonic external disturbances

The problem of damping of forced oscillations often arises in the modern engineering. There are many results in this field (see [7]–[10] and others). We will consider the following mathematical version of this problem. Let $0 \le t < \infty$, $x(t) \in \mathbb{R}^n$, $u(t) \in \mathbb{R}^m$, $y(t) \in \mathbb{R}^k$, $\varphi(t) \in \mathbb{R}^l$ and

$$\frac{dx(t)}{dt} = Ax(t) + Bu(t) + f^0\varphi(t), \qquad y(t) = Cx(t) \qquad (3.1)$$

be a time–invariant plant. Here

$$\varphi(t) = \varphi_1 e^{i\omega_1 t} + \ldots + \varphi_N e^{i\omega_N t} \qquad (3.2)$$

is an external disturbance with known frequencies $\omega_1 < \ldots < \omega_N$ and with unknown complex vector amplitudes $\varphi_1, \ldots, \varphi_N$. So in this case $\Xi = \{a, \varphi_1, \ldots, \varphi_N\}$ is the class of uncertainties, where $a = x(0)$.

Let the cost functional be

$$\Phi = \limsup_{T \to \infty} \frac{1}{T} \int_0^T \mathfrak{G}[x(t), u(t)]\,dt, \qquad (3.3)$$

where $\mathfrak{G}$ is a real quadratic form

$$\mathfrak{G}(x, u) = \begin{bmatrix} x \\ u \end{bmatrix}^* \begin{bmatrix} G & g \\ g^* & \Gamma \end{bmatrix} \begin{bmatrix} x \\ u \end{bmatrix}. \qquad (3.4)$$

Consider two classes $\mathfrak{A}$ and $\mathfrak{A}_{\text{reg}}$ of admissible processes. The class $\mathfrak{A} = \{x(\cdot), u(\cdot)\}$ is defined only by the "weak" stability condition

$$\frac{1}{\sqrt{t}}|x(t)| \to 0 \text{ as } t \to \infty, \qquad (3.5)$$

by the natural condition $|x|+|\dot{x}|+|u| \in L_2(0,T), \forall T > 0$ and by equation (3.1). It is a very wide class, in particular it contains the processes produced by nonlinear and by nonrealizable regulators. The class $\mathfrak{A}_{\text{reg}} = \{(x(\cdot), u(\cdot))\}$ consists of processes which are implemented by a certain stabilizing regulator

$$\alpha(p)u = \beta(p)y, \qquad \left(p = \frac{d}{dt}\right) \qquad (3.6)$$

and satisfy (3.1). Here $\alpha(\lambda)$, $\beta(\lambda)$ are $m \times m$ and $m \times k$ matrix polynomials and $\det \alpha(\lambda) \neq 0$. We remind that regulator (3.6) is called stabilizing if the system (3.1), (3.6) is stable and $\Psi_x(\infty) = 0$, $\Psi_u(\infty) = 0$ where Ψ_x, Ψ_u are the transform matrices from $f = f^0 \varphi$ to x and u:

$$\begin{bmatrix} \Psi_x \\ \Psi_u \end{bmatrix} = \Xi(\lambda)^{-1} \begin{bmatrix} I_n \\ 0 \end{bmatrix}, \quad \Xi(\lambda) = \begin{bmatrix} \lambda I_n - A & (-B) \\ \beta C & (-\alpha) \end{bmatrix}. \tag{3.7}$$

The system (3.1), (3.6) is stable iff $\det \Xi(\lambda) \neq 0$ as $\mathrm{Re}\,\lambda \geq 0$. Obviously the regulator (3.6) is realizable.

Let us suppose first that φ_j are known and consider two optimization problems: to minimize the functional (3.3) in each of the classes $\mathfrak{A}$ and $\mathfrak{A}_{\mathrm{reg}}$. For $(x(\cdot), u(\cdot)) \in \mathfrak{A}_{\mathrm{reg}}$ the functional Φ depends only on regulator (3.6). Therefore these problems are of a type (*) and (**), Section 1, respectively. (In our case the equation $F[x(\cdot), u(\cdot)] = 0_{\mathbf{Y}}$ is the first equation (3.1) together with $x(0) = a$). Of course we wish to find an universal optimal regulator (3.6) in the class Ξ, but it is not clear a priori whether it exists.

Let us make one remark. We can consider (3.1) together with

$$\frac{d\tilde{\varphi}}{dt} = \Omega \tilde{\varphi} \qquad (\Omega = \mathrm{diag}(i\omega_1 I_l, \ldots, i\omega_N I_l), \quad \tilde{\varphi} = \mathrm{col}[\varphi_j e^{i\omega_j t}]). \tag{3.8}$$

In the system (3.1), (3.8) a new state–vector is $[x, \tilde{\varphi}]$. It might seem at first that in our optimization problem a standard linear-quadratic control theory result can be applied to the system (3.1), (3.8), which says that there exists an optimal universal regulator of the form

$$u = K_1 x + K_2 \tilde{\varphi} \qquad (K_1, K_2 = \mathrm{const}). \tag{3.9}$$

Then x, $\tilde{\varphi}$ should be estimated through y, and we would obtain an universal regulator in the class Ξ. However this approach fails because our optimization problem differs from the problems considered in this theory: the admissible process and the functional are different, the system (3.1), (3.8) is neither controllable nor observable and by other reasons. Nevertheless this gives us a hope that the universal regulator exists in our problem.

Let us assume that the system (3.1) is stabilizable by output y (there exists a matrix E such that $A + BEC$ is a Hurwitz matrix) and that the following frequency-domain condition (FDC) holds:

$$\mathfrak{G}(\tilde{x}, \tilde{u}) \geq \delta(|\tilde{x}|^2 + |\tilde{u}|^2) \tag{3.10}$$

for some $\delta > 0$ and for all $\tilde{x} \in \mathbb{C}^n$, $\tilde{u} \in \mathbb{C}^m$, $\omega \in \mathbb{R}^1$, satisfying the equation $i\omega \tilde{x} = A\tilde{x} + B\tilde{u}$. This is a natural condition because if it fails in a strong way, i.e. there are $\tilde{x}_0$, $\tilde{u}_0$, ω_0 such that $i\omega_0 \tilde{x}_0 = A\tilde{x}_0 + B\tilde{u}_0$, $\mathfrak{G}(\tilde{x}_0, \tilde{u}_0) < 0$, then $\inf \Phi = -\infty$ for some φ of the kind (3.2).

In the optimal damping problem the form $\mathfrak{G}$ is usually of the type $\mathfrak{G} = |C_0 x + D_0 u|^2 + u^* \Gamma_0 u$, $\Gamma_0 = \Gamma_0^* \geq 0$, $\Gamma = D_0 D_0^* + \Gamma_0 > 0$. In this case if $\det[i\omega I_n - A] \neq 0$ for all $\omega \in \mathbb{R}^1$, then the FDC holds.

By the KYP–lemma the FDC is equivalent to the existence of (necessarily unique) $n \times n$ and $n \times m$ matrices $H = H^*$, K which satisfy the identity

$$\mathfrak{G}(x, u) + 2x^*H(Ax + Bu) = |\Gamma^{\frac{1}{2}}(u - Kx)|^2 \tag{3.11}$$

and such that the $A^{(K)} = A + BK$ is a Hurwitz matrix. (See for example [11, 12]). Note that identity (3.11) can be transformed into an equation for the matrix H (the algebraic matrix Ricatti equation or, historically more correctly, the Lur'e equation). There are different methods to compute the matrices H, K and we will take next that they are known.

Consider first the optimization problem in the class $\mathfrak{A}$ under the assumptions that φ_j are known. Let $\mathcal{E} = \{\varepsilon(\cdot)\}$ be the class of all functions ε satisfying the condition $T^{-1} \int_0^T |\varepsilon(t)|^2 dt \to 0$ as $T \to \infty$. Put

$$\kappa_j = \Gamma^{-1} B^* \left(i\omega_j I_n + \left[A^{(K)} \right]^* \right)^{-1} H f^0 \varphi_j, \quad \kappa(t) = \sum_{j=1}^{N} \kappa_j e^{i\omega_j t}. \tag{3.12}$$

The following result is known in its essential part (see for example [13, 14]).

Theorem 3.1. *For any $x(0) = a$ and any $\varepsilon \in \mathcal{E}$ the regulator*

$$u(t) = Kx(t) + \kappa(t) + \varepsilon(t) \tag{3.13}$$

gives a process, which is optimal in the class $\mathfrak{A}$, and any process, optimal in $\mathfrak{A}$ is obtained in this way.

The regulator (3.13) is unsatisfactory in many respects. The main is the following. It can be shown that any arbitrary small mistake in the estimates of the frequencies ω_j, which are used in the regulator (3.13), produces a finite jump in the value of the cost functional Φ. So this regulator is not optimal from the practical point of view. Moreover it depends on complex amplitudes φ_j, which we would like to be unknown.

It turns out, however, that we can achieve the goal to construct a universal regulator in the class $\Xi = \{a, \omega_1, \ldots, \omega_N\}$. It has the form (3.6). Put

$$A_\lambda = \lambda I_n - A, \qquad \delta(\lambda) = \det A_\lambda, \qquad Q_\lambda = A_\lambda^{-1} \delta(\lambda), \tag{3.14}$$

$$V_j = \Gamma^{-1} B^* (i\omega_j I_n - A^{(K)})^{-1} H f^0, \qquad W_j = \delta(i\omega_j) C A_{i\omega_j}^{-1} f^0, \tag{3.15}$$

$$U_j = \left[I_m + K(i\omega_j I_n - A^{(K)})^{-1} B \right] V_j + K(i\omega_j I_n - A^{(K)})^{-1} f^0. \tag{3.16}$$

Suppose that $\det W_j^* W_j > 0$ and (without the loss of generality) that A is a Hurwitz matrix. Let $\Psi_v(\lambda) = \Psi_u - K\Psi_x$ be the transform matrix in the system (3.1), (3.6) from $f = f^0 \varphi$ to $v = u - Kx$. A solution of the optimization problem in the class $\mathfrak{A}_{\text{reg}}$ is given by the following theorem [15]:

Theorem 3.2. 1° *Any stabilizing regulator (3.6) for which the interpolation relations*

$$\Psi_v(i\omega_j)f^0 = V_j, \qquad j = 1, \ldots, N \tag{3.17}$$

hold is a universal (in Ξ) optimal regulator. In particular, we can take

$$\alpha(\lambda) = \gamma(\lambda)CQ_\lambda B + \rho(\lambda)I_m, \qquad \beta(\lambda) = \gamma(\lambda)\delta(\lambda) \tag{3.18}$$

where ρ is an arbitrary scalar Hurwitz polynomial, $\deg \rho > \deg(\gamma CQ_\lambda)$ and γ is an arbitrary $m \times k$ matrix polynomial such that

$$\gamma(i\omega_j) = \rho(i\omega_j)U_j(W_j^*W_j)^{-1}W_j^* + \Delta\gamma_j, \qquad \Delta\gamma_j W_j = 0, \quad j = 1, \ldots, N. \tag{3.19}$$

2° *The infima of Φ in the classes $\mathfrak{A}$ and $\mathfrak{A}_{\text{reg}}$ coincide.*

The last statement shows that no nonlinear or nonrealizable regulator will yield a smaller value of the cost functional in comparison with the mentioned linear realizable regulator. Note that (3.17) is equivalent to relations $\Psi_u(i\omega_j)f^0 = U_j$, $j = 1, \ldots, N$ and that for the choice (3.18) $\Psi_u = \gamma CQ_\lambda \rho^{-1}$ and the characteristic polynomial of (3.1), (3.6) is $\delta(\lambda)\rho(\lambda)^m$.

It is easy to see that the regulator given by this theorem is robust with respect to a small uncertainty in the knowledge of frequencies $\omega_1, \ldots, \omega_N$ and so has no disadvantages mentioned above. Since the close–loop system (3.1), (3.6), (3.18) is stable regulator (3.6) is also robust with respect to small additive white-noise perturbations of the external disturbance.

It is easy to see that Φ has a form $\Phi = \Phi_1|\varphi_1|^2 + \ldots + \Phi_N|\varphi_N|^2$ for any stabilizing regulator (3.6). Here Φ_j depends on (α, β) for the class $\mathfrak{A}_{\text{reg}}$. Since (3.18) is optimal universal regulator $\Phi_1, \ldots, \Phi_N$ achieve simultaneously its minimum on the regulator (3.18). Theorem 3.2 gives a full solution in the following sense. Two regulators (3.6) $\alpha u = \beta y$ and $\alpha' u = \beta' y$ are called $\mathcal{H}$–equivalent (Hurwitz equivalent) if there exist matrix polynomials α_0, β_0 such that $\alpha = \mu\alpha_0$, $\beta = \mu\beta_0$, $\alpha' = \nu\alpha_0$, $\beta' = \nu\beta_0$ for some stable polynomials μ, ν (det $\mu \neq 0$, det $\nu \neq 0$ as Re$\lambda \geq 0$). It is clear that Ψ_x, Ψ_u, Ψ_v are the same for $\mathcal{H}$-equivalent regulators and if one such regulator is stabilizing, then so is the other, with the same Φ.

Theorem 3.3. *Any optimal universal (in Ξ) regulator is obtained in the way given by Theorem 3.2, modulo $\mathcal{H}$-equivalence.*

It is obvious that an optimal universal (in Ξ) regulator is adaptive (in Ξ) as well. If the optimal forced regime has established and after that the values of φ_j change then a new optimal regime establishes after some time-period and so on.

Example. Consider the scalar equation

$$\frac{d^2y}{dt^2} + \mu\frac{dy}{dt} + \nu y = u + \varphi_0 + \text{Re}(\varphi_1 e^{i\omega_1 t}) \tag{3.20}$$

with unknown φ_0 and φ_1. Here $\mu > 0$, $\nu > 0$, $\omega_1 > 0$, φ_0 is real, φ_1 is complex. We wish to find universal regulator

3 Optimal damping under unknown disturbances

$$\alpha(p)u = \beta(p)y \qquad \left(p = \frac{d}{dt}\right) \qquad (3.21)$$

that does not depend on unknown φ_0, φ_1 and minimizes the value

$$\Phi = (y^2 + \Gamma u^2)\text{mean} = \limsup_{T\to\infty} \frac{1}{T}\int_0^T \left(y(t)^2 + \Gamma u(t)^2\right) dt$$

in the class $\mathfrak{A}$. $\Gamma > 0$ is given. The regulator (3.11) must be stabilizing.

Remind that the class $\mathfrak{A}$ contains all the processes satisfying the condition

$$\frac{1}{t}\left(|y(t)|^2 + \left|\frac{dy}{dt}\right|^2\right) \to 0 \text{ as } t \to \infty.$$

These processes can be produced by not only stabilizing regulators (3.21) but also by nonrealizable or realizable nonlinear regulators an so on.

Using Theorem 3.2 we obtain the following result[1]. Let

$$\delta(\lambda) = \lambda^2 + \mu\lambda + \nu, \qquad \psi(\lambda) = \lambda^2 + \psi_2\lambda + \psi_1,$$

$$\psi_1 = \sqrt{\nu^2 + \frac{1}{\Gamma}}, \qquad \psi_2 = \sqrt{\mu^2 - 2\nu + 2\psi_1}.$$

The solution $\begin{bmatrix} h_1 \\ h_2 \end{bmatrix}$ of Lur'e equation is $h_1 = \nu - \psi_1$, $h_2 = \mu - \psi_2$. (It can be obtained simply by using the frequency method of the solution of Lur'e equations, see, for example, [12, 18].)

The coefficients α, β of the universal optimal regulator (3.21) are defined by

$$\alpha(\lambda) = \gamma(\lambda) + \rho(\lambda), \qquad \beta(\lambda) = \gamma(\lambda)\delta(\lambda), \qquad (3.22)$$

where

$$\rho(\lambda) = \lambda^4 + \rho_1\lambda^3 + \rho_2\lambda^2 + \rho_3\lambda + \rho_4,$$

is an arbitrary Hurwitz polynomial,

$$\gamma(\lambda) = \gamma_0 + \frac{\text{Im}\gamma_1}{\omega_1}\lambda + \frac{\gamma_0 + \text{Re}\gamma_1}{\omega_1^2}\lambda^2,$$

$$\gamma_0 = \rho_0 U_0, \qquad \gamma_1 = \rho(i\omega_1)U_1,$$

$$U_0 = -\frac{1}{1+\Gamma\nu}, \qquad U_1 = -\frac{1}{1+\Gamma|\delta(i\omega_1)|^2}.$$

(Note that these formulas can be obtained also by the following way. We just have to rewrite Φ as $\Phi = |\varphi_0|^2\Phi_0 + |\varphi_1|^2\Phi_1$ and to find ρ and γ by minimizing Φ_0 and Φ_1 simultaneously. By Theorem 3.2 such ρ, γ exsist.) The characteristic polynomial of the closed–loop optimal system (3.20), (3.21), (3.22) is $\rho(\lambda)\delta(\lambda)$.

By Theorem 3.2 any regulator (3.21), (3.22) will be optimal universal regulator if ρ and γ are polynomials with real coefficients, ρ is Hurwitz polynomial, $\deg \gamma < \deg \rho$ and the interpolation relations $\gamma(0) = \rho(0)U_0$, $\gamma(i\omega_1) = \rho(i\omega_1)U_1$ hold.

---

[1] The calculations were fulfilled by St. Petersburg University student A. Shirjaev.

4. Optimal damping. Stochastic case

Infinite–horizon optimization problems for time–invariant system (3.1) with stoch-astic stationary external disturbances $\varphi(t)$ of different kinds are well studied (see [2, 3] and others). Usually the spectral density $S_\varphi(\theta)$ of the disturbance is supposed to be a rational function, and the Wiener–Hopf technique or the state space methods (the separation theorem, Kalman filter, the Lur'e equations) are applied. It is well known that in these cases the optimal regulator depends on $S_\varphi(\theta)$ as a rule. Therefore if $c = S_\varphi(\cdot)$ is taken as an abstract parameter in Section 1, we obtain the situation of absence of common infimum (Fig. 1).

We will consider a similar problem under other assumptions. It will be supposed that only an upper bound $s(\theta)$ of $S_\varphi(\theta)$ is known:

$$\|S_\varphi(\theta)\| \leq s(\theta), \tag{4.1}$$

and that $s(\theta)$ tends to zero quickly as $|\theta| \to \infty$ (exponentially for example), therefore S_φ cannot be rational. It turns out that under these assumptions the common infimum situation takes place (Fig. 2). In fact, Φ does not achieve its minimum in this problem, so we can find only ε-optimal regulators. Now let us pass to exact formulations.

Consider the plant (3.1) where the pair (A, b) is stabilizable and $\varphi(t)$ is a stochastic process defined by the stochastic integral

$$\varphi(t) = \int_{-\infty}^{+\infty} e^{i\theta t} W_\varphi(i\theta)\, dz_\theta. \tag{4.2}$$

Here $W_\varphi(i\theta) \in L_2(-\infty, \infty)$ and $z_\theta \in \mathbb{R}^l$, z_θ is a process with uncorrelated increments:

$$E(dz_\theta) = 0, \qquad E(dz_{\theta_1} dz_{\theta_2}^*) = \delta(\theta_1 - \theta_2) I_l\, d\theta_1\, d\theta_2. \tag{4.3}$$

The spectral density $S_\varphi(\theta) = W_\varphi(i\theta) W_\varphi(i\theta)^*$ is supposed to be unknown, we know only $s(\theta)$ in the estimate (4.1).

Consider two classes $\mathfrak{A}$ and $\mathfrak{A}_{\text{reg}}$ of admissible processes $[x(\cdot), u(\cdot)]$. The class $\mathfrak{A}$ is defined by the conditions: the pair $[x(\cdot), u(\cdot)]$ is a stochastic process with $E|x(0)|^2 < \infty$, $E|u(t)|^2 < \infty$, $E|u(t)|^2$ is locally integrable, equation (3.1) is fulfilled and

$$\frac{1}{t} E|x(t)|^2 \to 0 \quad \text{as } t \to \infty.$$

The class $\mathfrak{A}_{\text{reg}}$ consists of processes which satisfy equations (3.1), (3.6) where (3.6) is a stabilizing regulator. Obviously $\mathfrak{A}_{\text{reg}} \subset \mathfrak{A}$.

Let $\mathfrak{G}$ be the form (3.4), and suppose that FDC (3.10) holds. The problems under consideration are to minimize the functional

$$\Phi = \limsup_{T \to \infty} \frac{1}{T} \int_0^T E\mathfrak{G}[x(t), u(t)]\, dt \tag{4.4}$$

in the class $\mathfrak{A}$ and in the class $\mathfrak{A}_{\mathrm{reg}}$. If $[x(\cdot), u(\cdot)] \in \mathfrak{A}_{\mathrm{reg}}$ then Φ depends only on regulator (3.6). Therefore these problems are of type (*) and (**), Section 1.

Define the process $\kappa(t)$ by

$$\kappa(t) = \int_{-\infty}^{+\infty} e^{i\theta t} W_\kappa \, dz_\theta, \qquad (4.5)$$

$$W_\kappa = \Psi_\kappa H f^0 W_\varphi(i\theta), \qquad \Psi_\kappa = \Gamma^{-1} B^* \left(i\theta I_n + \left[A^{(K)}\right]^*\right)^{-1}.$$

Let $\mathcal{E}$ be the set of all stochastic processes satisfying conditions

$$E|\varepsilon(t)|^2 < \infty, \qquad \frac{1}{T} E \int_0^T |\varepsilon(t)|^2 dt \to 0 \text{ as } T \to \infty. \qquad (4.6)$$

The next theorem follows for example from Theorem 1 [16].

Theorem 4.1. *For any $x(0)$ and any $\varepsilon(\cdot) \in \mathcal{E}$ the regulator $u(t) = Kx(t) + \kappa(t) + \varepsilon(t)$ gives a process, which is optimal in the class $\mathfrak{A}$ and any process, optimal in $\mathfrak{A}$ is obtained in this way.*

However, like the situation of Theorem 3.1, the regulator $u(t) = Kx(t) + \kappa(t) + \varepsilon(t)$ is not realizable even if $\varphi(t)$ is an observable process. This follows from formulas (4.5) and from the fact that $A^{(K)}$ is a Hurwitz matrix.

Consider now the class $\mathfrak{A}_{\mathrm{reg}}$. Let $\Psi_v(\lambda)$ be the transform matrix which was introduced in Section 2. It is defined in a usual way through α, β. We put $\Gamma = I_m$ without the loss of generality.

Lemma 4.1. *Let $[x(\cdot), u(\cdot)] \in \mathfrak{A}_{\mathrm{reg}}$, and let $\Psi_\kappa(i\theta)$ be defined by (4.5). Then*

$$\Phi[(\cdot), u(\cdot)] = \int_{-\infty}^{+\infty} |(\Psi_v - \Psi_\kappa) f^0 W_\varphi|^2 d\theta + \Phi_{\min}, \qquad (4.7)$$

where $\Phi_{\min}$ is the infimum of Φ in the class $\mathfrak{A}$.

Suppose for simplicity that $y \equiv x$. Then α, β are defined (nonuniquely) by Ψ_v so that (3.6) is stabilizing regulator. Note that Ψ_v is stable and Ψ_κ is unstable rational functions. Consider first the case of a rational W_φ. Using Wiener–Hopf techniques it is easy to find Ψ_v which minimizes the integral in (4.7). This Ψ_v will determine the optimal regulator (3.6) in the class $\mathfrak{A}_{\mathrm{reg}}$. (May be this way of computations is more convenient than the usual one). We see that as a rule this regulator depends on W_φ and $\mathrm{Inf}_{\mathfrak{A}_{\mathrm{reg}}} \Phi > \Phi_{\min} = \mathrm{Inf}_{\mathfrak{A}} \Phi$.

Consider now the case of quickly decreasing W_φ. Namely let Ξ_s be the class of processes (4.2) such that their spectral density $S_\varphi(\theta) = W_\varphi(i\theta)W_\varphi(i\theta)^*$ satisfies the estimate (4.1) with a given function $s(\theta)$ such that

$$\int_{-\infty}^{+\infty} \frac{\ln s(\theta)}{1+\theta^2} d\theta = -\infty. \qquad (4.8)$$

(The condition (4.8) holds if for example $s(\theta) \sim e^{-\gamma|\theta|}$ as $|\theta| \to \infty$, $\gamma > 0$). Let $\varphi \in \Xi_s$. Using some results of complex analysis it can be shown that in this case we can choose a stable Ψ_v to make the integral in (4.7) arbitrary small. This Ψ_v will depend on s but not on W_φ (or S_φ). It means that $\text{Inf}_{\mathfrak{A}_{\text{reg}}} \Phi = \Phi_{\min} = \text{Inf}_{\mathfrak{A}} \Phi$, and that for any $\varepsilon > 0$ we can design an ε-optimal regulator, universal in the class Ξ_s. So we can construct ε-optimal regulator in the class $\mathfrak{A}$ (and consequently in $\mathfrak{A}_{\text{reg}}$) without the knowledge of the spectral density $S_\varphi(\theta)$. It is necessary to know only the scalar function $s(\theta)$.

Explicit formulas for α, β in this case are given in [16]. They are analogous to (3.18) and give a full solution.

5. Tracking problems

There are different settlements of these problems and many results in this field (see [1]–[5] and others). We consider certain special cases, in which universal regulators exist.

Consider the time–invariant plant described by the equations

$$\frac{dx}{dt} = Ax + Bu + f(t), \qquad y = Cx + Du, \tag{5.1}$$

where $x(t) \in \mathbb{R}^n$, $u(t) \in \mathbb{R}^m$, $y(t) \in \mathbb{R}^k$, the pair (A, B) is stabilizable, the pair (A, C) is observable. The external disturbance $f(\cdot)$, as well as a given output reference signal $r(\cdot)$ (the desired $y(\cdot)$) are deterministic functions bounded on $(0, \infty)$.

Define admissible regulators as stabilizing regulators of the type

$$\alpha(p)u = \beta(p)y + \gamma(p)r \qquad \left(p = \frac{d}{dt}\right) \tag{5.2}$$

where $\alpha(p), \beta(p), \gamma(p)$ are $m\times m, m\times k, m\times k$ matrix polynomials, $\det \alpha \neq 0$. The problem under consideration is to find an admissible regulator which minimizes the cost functional

$$\Phi_0 = \limsup_{T\to\infty} \frac{1}{T} \int_0^T (y-r)^* G(y-r)\, dt \tag{5.3}$$

(with $G = G^* > 0$) for any $f(\cdot)$, $r(\cdot)$ in a certain class Ξ. Since (5.2) is a stabilizing regulator it is clear that for fixed $f(\cdot), r(\cdot)$ the functional Φ_0 depends only on the regulator but not on the solution of (5.1), (5.2). Therefore it is a problem of the type (∗∗), Section 1.

It is naturally to study first the question whether a regulator (5.2) exists such that

$$|y(t) - r(t)| \to 0 \qquad \text{as } t \to \infty \tag{5.4}$$

for every reference signal $r(\cdot)$ and for any solution of (5.1), (5.2). Obviously this regulator minimizes the functional Φ_0 and therefore it is universal in the class $\Xi = \{r(\cdot)\}$ of all reference signals. It is easy to see that the mentioned property is possible only if $f(t) \equiv 0$.

Define A_λ, $\delta(\lambda)$, Q_λ by formulas (3.14). Let

$$W(\lambda) = D + CA_\lambda^{-1}B, \qquad V(\lambda) = \delta W = D\delta + CQ_\lambda B. \qquad (5.5)$$

Assume for simplicity of formulations that $m = k$ and V is a nondegenerate matrix polynomial, that is $\det V(\lambda) \neq 0$.

Theorem 5.1. *Let $f(t) \equiv 0$ and A be a Hurwitz matrix.*

$1°$ Assume that all poles of $V(\lambda)^{-1}$ lie in the half-plane $\operatorname{Re}\lambda < 0$. Let $\xi(\lambda)$ be an arbitrary $m \times m$ matrix polynomial, $\rho(\lambda)$ be an arbitrary scalar Hurwitz polynomial such that $\deg \rho > \deg[\xi(\lambda)CQ_\lambda]$ and $\delta\rho V^{-1}$ is a polynomial. Put

$$\alpha(\lambda) = \rho I_m + \xi V, \qquad \beta(\lambda) = \xi \delta, \qquad \gamma(\lambda) = \delta\rho V^{-1}. \qquad (5.6)$$

Then (5.2) is a stabilizing regulator and (5.4) holds for every reference signal $r(\cdot)$ and for any solution of (5.1), (5.2).

$2°$ If a stabilizing regulator (5.2) with the property (5.4) exists then all poles of V^{-1} (all zeros of $\det V(\lambda)$) lie in the open left half-plane and modulo $\mathcal{H}$-equivalence all such regulators are formed in mentioned above way.

The part $2°$ can be proved by using Lemma 1 [15], which gives the general formulas for the stabilizing regulators. Let us prove the part $1°$.

Compute the characteristic determinant $\Delta(\lambda)$ of the system (5.1), (5.2). Using the Shur formula we obtain

$$\Delta(\lambda) = \det \begin{bmatrix} A_\lambda & -B \\ -\beta C & (\alpha - \beta D) \end{bmatrix} = \delta \cdot \det[\alpha - \beta D - \beta C A_\lambda^{-1} B] = \delta \rho^m.$$

Since δ and ρ are Hurwitz polynomials, the system (5.1), (5.2) is stable.

It is easy to see that the transform matrices $\Psi_u(\lambda)$, $\Psi_x(\lambda)$ from f to u and from f to x in the system (5.1), (5.2) are $\Psi_u = \frac{\xi C Q_\lambda}{\rho}$, $\Psi_x = A_\lambda^{-1}(B\Psi_u + I_n)$. Therefore $\Psi_u(\infty) = 0$, $\Psi_x(\infty) = 0$, that is regulator (5.2) is stabilizing.

Let us show that there exist functions u^0, ε such that

$$\left.\begin{array}{rcl} \delta(p)r &=& V(p)u^0 \\ \alpha(p)u^0 &=& (\beta(p) + \gamma(p))r + \varepsilon \\ \delta(p)\varepsilon &=& 0 \end{array}\right\} \qquad (5.7)$$

Let u^0 be an arbitrary solution of the first equation in (5.7), it exists because of $\det V(\lambda) \neq 0$. It follows from (5.6) that $(\beta + \gamma)V = \delta\alpha$, therefore $\delta(\beta + \gamma)r = (\beta + \gamma)\delta r = (\beta + \gamma)Vu^0 = \delta\alpha u^0$. Hence we obtain two last equations in (5.7) by putting $\varepsilon = \alpha u^0 - (\beta + \gamma)r$.

Let x, y, u be an arbitrary solution of (5.1), (5.2) with $f \equiv 0$. Since $Q_\lambda A_\lambda = A_\lambda Q_\lambda = \delta I_n$, we have $\delta(p)x = Q_p A_p x = Q_p Bu$. Hence

$$\delta(p)y = V(p)u, \qquad \alpha(p)u = \beta(p)y + \gamma(p)r. \qquad (5.8)$$

Put $\Delta y = y - r$, $\Delta u = u - u^0$. It follows from (5.7), (5.8) that

$$\left.\begin{array}{r}\delta(p)\Delta y - V(p)\Delta u = 0 \\ \beta(p)\Delta y - \alpha(p)\Delta u + \varepsilon = 0 \\ \delta(p)\varepsilon = 0\end{array}\right\} \quad (5.9)$$

The characteristic determinant of this system equals $(-1)^m \delta \det(\alpha - \beta W) = (-1)^m \delta \rho^m$ and is a Hurwitz polynomial. Therefore $|\Delta y| \to 0$, $|\Delta u| \to 0$ as $t \to \infty$.

Remark. Without the condition $\deg \rho > \deg(\xi C Q_\lambda)$ the system (5.1), (5.2) remains stable and (5.4) holds.

In the case $m = k = 1$ of scalar control and output the universal regulator given by Theorem 5.1 has the form

$$V(\eta + \xi)u = \delta(\xi y + \eta r), \quad (5.10)$$

where ξ and η are arbitrary polynomials with η Hurwitz.

Next we consider briefly the case when the assumptions of Theorem 5.1 are not fulfilled. Instead of (5.3) consider the following more general functional to be minimized

$$\Phi = \limsup_{T \to \infty} \frac{1}{T} \int_0^T \left[(y - r)^* G(y - r) + u^* \Gamma u\right] dt. \quad (5.11)$$

Here $G \geq 0$, $\Gamma \geq 0$ and we suppose that the frequency–domain condition holds for the form $\mathfrak{G} = y^* G y + u^* \Gamma u$. It can be shown that the common infimum situation (Fig. 2) does not take place for the class $\Xi = \{f(\cdot), r(\cdot)\}$ of all $f(\cdot)$, $r(\cdot)$. Consider the class Ξ_h of all harmonic $f(\cdot)$, $r(\cdot)$

$$f(t) = f^0 \varphi(t), \qquad \varphi(t) = \varphi_1 e^{i\omega_1 t} + \ldots + \varphi_N e^{i\omega_N t}, \quad (5.12)$$
$$r(t) = r_1 e^{i\omega_1 t} + \ldots + r_2 e^{i\omega_N t} \quad (5.13)$$

with fixed (known) frequencies $\omega_1 < \ldots < \omega_N$ and unknown φ_j, r_j. Here f^0 is a given (known) $n \times l$ matrix and $\varphi_j \in \mathbb{C}^l$, $r_j \in \mathbb{C}^k$. For this class the results analogous that described in Section 2 can be obtained. Under some assumptions the universal optimal regulator (5.2) exists. For the case $y \equiv x$ explicit formulas can be found in [17].

Consider now the case of stochastic $f(\cdot)$, $r(\cdot)$ with the cost functional

$$\Phi = \limsup_{T \to \infty} \frac{1}{T} E \int_0^T \left[(y - r)^* G(y - r) + u^* \Gamma u\right] dt \quad (5.14)$$

under the same assumptions on G, Γ. We suppose $f(\cdot)$, $r(\cdot)$ to be stationary processes whose spectral density matrices $S_f(\theta)$, $S_r(\theta)$ satisfy the previous conditions

$$\|S_f(\theta)\| \leq s(\theta), \qquad \|S_r(\theta)\| \leq s(\theta), \quad (5.15)$$

$$\int_{-\infty}^{+\infty} \frac{\ln s(\theta)}{1 + \theta^2} d\theta = -\infty. \quad (5.16)$$

Let the class of uncertainties Ξ_s be the class of all $f(\cdot)$, $r(\cdot)$ satisfying these conditions (with a fixed $s(\cdot)$). Under some not restrictive assumptions, there exists an ε-universal optimal regulator in this class for any $\varepsilon > 0$. So, in the contrast to the well investigated case of rational S_f, S_r, we do not need to know S_f, S_r, it is enough to know only their upper bound $s(\theta)$.

6. Some other cases of the existence of a universal regulator

Many results described above can be extended to discrete systems. A universal regulators for optimal damping of forced harmonic oscillations in such systems has been designed in [20]. A case of nonstationary stochastic external disturbances is also studied there. By an idea of A. Lindquist, this specific stochastic optimization problem is decomposed into a countable number of deterministic problems, which differ only in amplitudes and phases of external disturbances. Since a universal regulator solves simultaneously all these problems, it also solves the initial stochastic problem.

Some results, analogous to that described above can be obtained for periodic continuous systems, for hybrid sampled–data systems of the type considered in [19] for time–invariant systems with delays and for some others.

Characteristic properties of all such problems are that they are infinite–horizon, the cost functional is of the type (3.3) or (4.4) and the optimal process is not unique.

A. S. Matveev considered some linear–quadratic optimal control problems similar to [18, 21] without external disturbances but with quadratic constraints. He proved the existence of an optimal universal regulator in the class $\Xi = \{x(0)\}$ of all the initial states. It is interesting that this universal regulator is nonlinear.

References

1. Anderson, B., Moore, J. : Optimal Control. Linear quadratic methods. Prentice–Hall International. Inc. Englewood Cliffs. NJ(1989)
2. Åström, K. : Introduction to Stochastic Control Theory. Academic Press. New York and London (1970)
3. Caines, P. : Linear Stochastic Systems. John Wiley & Sons. New York (1988)
4. Wonham, W. : Linear Multivariable Control: A Geometric Approach. Springer-Verlag. New York (1979)
5. Pervozvanskii, A. A. : The course of automatic control theory. Nauka. Moscow (1986) (Russian)
6. Matveev, A. S., Yakubovich, V. A. : Abstract Theory of Optimal Control. Ed. of St. Petersburg University. St. Petersburg (1994)
7. Frolov, K. V., Furman, F. A. : Applied theory of vibration protected systems. Mashinostroenie (1980) (Russian)
8. Frolov, K. V. : Vibration in Engineering. Mashinostroenie (1981) (Russian)

9. Genkin, M. D., Elezov, V. G., Yablonski, V. D. : Methods of Controlled Vibration Protection of Engines. Nauka. Moscow (1985) (Russian)
10. Guicking, D. : Active Noise and Vibration Control. Reference bibliography. Drittes Physical Institute Univ. of Goettingen (Jan., 1990)
11. Popov, V. M. : Hyperstability of Control Systems. Springer–Verlag (1973)
12. Yakubovich, V. A. : A frequency theorem in control theory. Sibirskij Mat. Zh. **4** (1973) 386–419 (Russian) English transl. in Siberian Math. J.
13. Andreev, V. A., Kazarinov, Yu. F., Yakubovich, V. A. : Doklady Acad. Nauk USSR **202, 6** (1972) 1247–1250 (Russian) English transl. in Soviet Math. Dokl.
14. Yakubovich, V. A. : Linear-quadratic problem of optimal damping of forced oscillations under unknown harmonic external disturbance. Doklady Acad. Nauk **333, 2** (1993) 170–172 (Russian)
15. Yakubovich, V. A. : Optimal damping of forced oscillations by the system output. Doklady Acad. Nauk **337, 3** (1994) 323–327 (Russian)
16. Yakubovich, V. A. : Universal regulator for optimal damping of forced stochastic oscillations in a linear system. Doklady Akad. Nauk **338, 1** (1994) 19–24 (Russian)
17. Yakubovich, V. A. : The problem of optimal tracking of deterministic harmonic signals with a known spectrum. Doklady Acad. Nauk **337, 4** (1994) 463–466 (Russian)
18. Yakubovich, V. A. : Nonconvex optimization problem: The infinite–horizon linear–quadratic control problem with quadratic constraints. System & Control Letters **19** (1992) 13–21
19. Rosenvasser, Ye. N. : Linear theory of digital control in continuous time. Nauka. Moscow (1994) (Russian)
20. Lindquist, A., Yakubovich, V. A. : Optimal damping of forced oscillations in discrete–time systems.(to appear)
21. Matveev, A. S., Yakubovich, V. A. : Nonconvex problems of global optimization. St. Petersburg Math. J. **4, 6** (1993) 1217–1243

Control Using Logic-Based Switching

*A. S. Morse**

Yale University, New Haven, CT 06520 USA

1 Introduction

Between the well-studied areas of discontinuous control [1], [2] on the one hand and sampled data control [3] on the other lies the largely unexplored area of logic-based switching control systems. By a logic-based switching controller is meant a controller whose subsystems include not only familiar dynamical components {integrators, summers, gains, etc.} but logic-driven elements as well {e.g., [4]}. More often than not the predominately logical component within such a system is called a supervisor [5], a mode changer [6], a gain scheduler, or something similar. Within the last decade a number of analytical studies of such systems have emerged, mainly in the area of self-adjusting control [7, 8, 9, 10, 11, 12, 13, 14, 15, 16]. These studies and others have shown that much can be gained by using logic-based switching together with more familiar techniques in the synthesis of feedback controls. The overall models of systems composed of such logics together with the processes they are intended to control are concrete examples of *hybrid dynamical systems* [17, 18, 19]. The aim of this paper is to give a brief tutorial review of four different classes of hybrid systems of this type - each consists of a continuous-time process to be controlled, a parameterized family of candidate controllers, and an event driven switching logic. Three of the logics, called *prerouted switching, hysteresis switching* and *dwell-time switching* respectively, are simple strategies capable of determining in real time which candidate controller should be put in feedback with a process in order to achieve desired closed-loop performance. The fourth, called *cyclic switching*, has been devised to solve the long-standing stabilizability problem which arises in the synthesis of identifier-based adaptive controllers because of the existence of points in parameter space where the estimated model upon which certainty equivalence synthesis is based, loses stabilizability.

In section 2, we discuss several basic issues common to supervised control systems of all types. In most cases of interest, the job of a supervisor is to

---

* The author's research was supported by NSF Grant n. ECS-9206021, AFOSR Grant n. F49620-94-1-0181, and ARO Grant n. DAAH04-95-1-0114

orchestrate the switching of a sequence of candidate controllers into feedback or series with a process, so as to achieve some prescribed goal. No matter what the goal might be, the underlying architecture of the supervised control system is pretty much the same - at least in concept. In section 2 we make the point that such "multi-controller" architectures can usually be implemented most efficiently as "state-shared" parameter-dependent controllers.

In section 3 we briefly discuss two examples of logic-based switching controllers which arise in nonadaptive applications. The first is an 'intelligent' control strategy devised to maximize system performance while at the same guaranteeing that hard-bound saturation constraints are satisfied [20]. The second is a simple, time-invariant, chatter-free, switching logic with one state variable, which is capable of asymptotically stabilizing a particular bilinear system of current interest called the "nonholonomic integrator" [21].

The aim of §4 is to explain the concepts of prerouted, hysteresis, dwell-time, and cyclic switching. Although each of these strategies is applicable to a variety of systems [11, 12, 13, 14, 22, 16, 23], for the sake of uniformity all are reviewed within the context of a single prototype problem - the set-point control of a siso linear system with large-scale parametric uncertainty [24]. The problem is formulated in §4.1.

The concept of "prerouted switching" is closely allied with the idea of a "nonestimator based supervisor"; both topics are discussed in §4.2.

Section 4.3 focuses on the idea of an estimator-based supervisor. It is within this context that the concepts of hysteresis switching and dwell-time switching are explained. The idea of cyclic switching is then reviewed in section 4.4

The logics discussed in §4 are conceptually straight forward. What's interesting about them theoretically is the set of technical questions they generate. Most of the questions have to do with dynamical systems in which switching is non-terminating, non-chattering and asynchronous. Many unanswered questions exist. Some are briefly discussed in §5.

2 Multi-Controllers

Perhaps the simplest architecture one can think of for a feedback system employing a family of controllers is that depicted in Figure 1. That is, the measured output y of a process to be controlled drives a bank of controllers, each controller generating a candidate {possibly vector-valued} feedback signal u_i. The control signal applied to the process at each instant of time is then

$$u \stackrel{\Delta}{=} u_\eta$$

where $\eta : [0, \infty) \to \mathcal{I}$ is a piecewise-constant switching signal taking values in the family's index set $\mathcal{I}$. The generation of such a switching signal is typically carried out by some type of hybrid dynamical system which depending on the situation might be called a tuner, a supervisor, a mode-changer, or something similar. In the sequel we shall refer to such architectures informally as *multi-controllers*.

2 Multi-Controllers

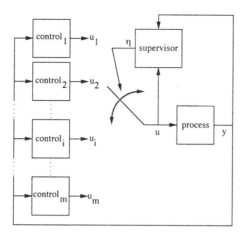

Fig. 1. Multi-Control

Many multi-controller configurations can be implemented using a much simpler architecture than Figure 1 would suggest. The key factor which makes this possible is simply that at any instant of time only *one* of the constituent controller is to be applied to the process. Because of this, at each time t it is only necessary to generate one candidate control signal. Often this means significant simplification can be achieved if all control signals are generated by a single system. In other word, rather than implementing each of the controllers in the family as a separate dynamical system, one can often achieve the same end using a single controller with adjustable parameters. The idea is quite straight forward and is called *state sharing*.

For example suppose that it is desired to implement a finite {or even countable} family of siso linear controllers with reduced transfer functions

$$\kappa_i(s) = \frac{\alpha_i(s)}{\beta_i(s)}, \quad i \in \mathcal{I}$$

where each $\beta_i(s)$ is a monic polynomial. Assuming a fixed upper bound n for the McMillan Degrees of the $\kappa_i(s)$, it is always possible to "cover" this family with a parameter-dependent transfer function $h_q(s)$ whose denominator is of degree n and whose parameter vector q takes values in a linear space of dimension not exceeding $2n+1$. In fact, for any positive integer $\bar{n} \leq 2n+1$, it is always possible to pick a subset $\mathcal{Q} \subset \mathbb{R}^{\bar{n}}$ with the same cardinality as $\mathcal{I}$, and a parameter-dependent transfer function $h_q(s)$ so that for each $i \in \mathcal{I}$ there is a $q_i \in \mathcal{Q}$ such that $\kappa_i(s) = h_{q_i}(s)$ after cancellation of common poles and zeros. Moreover it is always possible to choose $h_q(s)$ in such a way that whenever such pole-zero cancellations occur, they occur at prescribed stable locations.

Having constructed such an $h_q(s)$, the above multi-controller can be implemented as a parameter dependent system $\Sigma_C(\sigma)$ of the form

$$\dot{x}_C = A_\sigma x_C + b_\sigma y \tag{1}$$

$$u_\sigma = f_\sigma x_C + g_\sigma y \qquad (2)$$

where $\{A_q, b_q, f_q, g_q\}$ is a n-dimensional realization of $h_q(s)$ and σ is a piecewise constant switching signal taking values in $\mathcal{Q}$. The resulting multi-control system would then appear as in Figure 2.

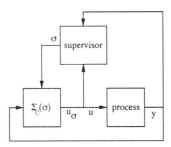

Fig. 2. State-Shared Multi-Controller Implementation

If the supervisor is allowed to re-initialize Σ_C's state at switching times, then this implementation can generate exactly the same feedback control signal as would have been generated had the original architecture been employed.

For multi-controller families consisting of more than just a few controllers, this state-shared implementation is clearly a lot less complicated than a direct implementation of the original multi-controller. Moreover state-sharing frees one from having to be concerned about the boundedness of the out-of-loop control signals which would be present in a direct implementation of the original multi-controller architecture.

There are of course a great many different ways to realize $h_q(s)$. The only essential requirement of any such realization is that it be a "globally detectable, globally stabilizable" system; i.e. for each fixed value of $q \in \mathcal{Q}$, the linear system $\{A_q, b_q, f_q, g_q\})$ should be stabilizable and detectable[2]. For without stabilizability, closed-loop boundedness of u and y cannot be assured and without detectability boundedness of Σ_C's state cannot be assured even if u and y are. One familiar structure which is globally detectable is of the form

$$\left\{ \begin{bmatrix} A & 0 \\ 0 & A \end{bmatrix} + \begin{bmatrix} b \\ 0 \end{bmatrix} f_q, \begin{bmatrix} 0 \\ b \end{bmatrix}, f_q, d_q \right\}$$

where (A, b) is a parameter-independent, n-dimensional siso, controllable pair with A stable. Another is $\{A + k_q f, b_q, f, d_q\}$ where (f, A) is an n-dimensional, parameter-independent observable pair. This particular realization is actually observable for all $q \in \mathbb{R}^{\bar{n}}$; moreover in the event that d_q is constant on $\mathcal{Q}$, this

---

[2] The reader should recognize that any such parameter-dependent system will *always* have points in $\mathcal{Q}$ at which it is not controllable and observable if the transfer functions being realized are not all of the same McMillan Degree.

realization guarantees that there will be a "bumpless" transfer between control signals when σ switches; i.e., $u \triangleq u_\sigma$ is continuous, even at those times at which σ changes values. Of course bumpless transfer can also be achieved with state-reinitialization, whether d_q is constant on $\mathcal{Q}$ or not.

It is fairly clear that the preceding ideas apply to multi-controller families of mimo finite dimensional controllers configured in almost any way imaginable. It is also clear that the number of (fixed-parameter) controllers one might contemplate implementing in a particular multi-controller application need not be finite nor even countable. In other words the complexity of a multi-controller is not so much a function of a number of controllers in a family as it is of the number of algebraically independent gains needed to parameterize the family.

3 Examples

In the sequel are several examples of {nonadaptive} logic-based switching controllers.

3.1 Smart Governors

An important problem of continuing interest is that of developing feedback controllers for linearly modeled processes whose associated inputs and outputs are required to satisfy hard-bound magnitiude constraints. Remarkable advances have recently been made in the development of implementable algorithms for the stabilization of such systems [25, 26]. At the same time there has also been a growing interest in the development of "smart controllers" employing logic aimed not only at maintaining loop stability, but at enhancing system performance as well [27, 28, 29, 20]. One configuration characteristic of this line of research is as follows.

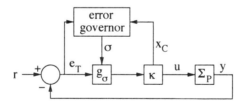

Here Σ_P represents a linear process with an input saturation constraint, g_σ is an adjustable gain, and κ is a linear controller. The idea is to design κ to meet performance specifications in the absence of saturation constraints; this is done for the case $g_\sigma = 1$. The error governor is designed to adjust g's value to give the best performance possible subject to the requirement that the saturation constraints are satisfied. This is accomplished, roughly speaking, by leaving g set at 1 whenever r is 'small' and by reducing g's value when r is 'large' by as much as is required to insure that there is no saturation. The error governor which accomplishes this is a logical circuit which carries out the required computations in real time. A generalized {discrete-time} version of the preceding with greatly reduced computational requirements has been proposed in [27].

An even more elaborate multi-controller architecture, aimed at a similar problem has been suggested in [20]. The problem addressed is to bring to zero from an admissible start, the state $x \triangleq \{x_P, x_C\}$ of the system $\Sigma(\sigma)$ depicted in the following figure while not violating a set of prespecified state constraints along the way.

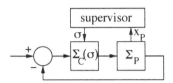

Associated with each fixed control index $q \in \mathcal{Q}$ is a *maximal admissible set* $\mathcal{S}_q$. A state x_0 is in $\mathcal{S}_q$ just in case each point on the closed-loop trajectory of $\Sigma(q)$ emanating from x_0, satisfies the aforementioned state constraints. According to [20], it is possible to use the theory of maximal output admissible sets [30] to design controllers so that $\mathcal{S}_q \subset \mathcal{S}_{q+1}$ and in addition so that controller q achieves better performance than controller $q+1$ when the system is initialized at a state in $\mathcal{S}_q \cap \mathcal{S}_{q+1}$. In [20] it is then explain how to construct a supervisor which successively switches σ to smaller and smaller values to as to achieve better and better performance while satisfying state constraints.

3.2 Nonholonomic Integrators

For more than a decade it has been known that there are nonlinear systems which are locally null controllable but which nevertheless cannot be locally asymptotically stabilized with any smooth, time-invariant controller [31]. A prototypical example of this is the bilinear system

$$\dot{x} = u$$
$$\dot{y} = v$$
$$\dot{z} = xv - yu$$

which is sometimes called the "nonholonomic integrator" [32]. Nonholonomic systems such as this have evoked considerable interest in recent years [33]. This has been especially true of the nonholonomic integrator itself. For example, a number of time-varying, periodic controllers have been devised which asymptotically stabilize the above system {cf. [32]}. In addition, by appealing to the theory of sliding modes [1], it has been recently shown that the simple discontinuous control $u = -x + y(\text{sign}(z))$ $v = -y - x(\text{sign}(z))$ will drive $x, y,$ and z to zero provided one admits generalized solution in the sense of [34]. It turns out

3 Examples

to be possible to achieve asymptotic stability without chattering using a time-invariant logic-based switching controller. One strategy which accomplishes this uses a multi-controller of the form

$$\begin{bmatrix} u \\ v \end{bmatrix} = g_\sigma(x, y, z)$$

where

$$g_1 = \begin{bmatrix} 1 \\ 1 \end{bmatrix} \qquad g_2 = \begin{bmatrix} x + yz \\ y - xz \end{bmatrix} \qquad g_3 = \begin{bmatrix} -x + yz \\ -y - xz \end{bmatrix} \qquad g_4 = \begin{bmatrix} 0 \\ 0 \end{bmatrix}$$

and σ is a piece-wise constant switching signal taking values in $\mathcal{I} \triangleq \{1, 2, 3, 4\}$. σ is generated by a supervisor of the form

$$\{x, y, z\} \longrightarrow \boxed{w = \begin{bmatrix} x^2 + y^2 \\ z^2 \end{bmatrix}} \xrightarrow{w} \boxed{\Sigma_S} \longrightarrow \sigma$$

where Σ_S is a switching logic whose input is w and whose state and output are both σ. Σ_S's definition requires one to pick four appropriately structured overlapping regions $\mathcal{R}_q$, $q \in \mathcal{I}$ which together cover the closed positive quadrant $\Omega \triangleq \{(r_1, r_2) : r_1 \geq 0, r_2 \geq 0\}$ in $\mathbb{R}^2$. One possible set of regions is

$$\mathcal{R}_1 \triangleq \{(r_1, r_2) : r_2 < 2\pi(r_2), \ (r_1, r_2) \in \Omega\}$$
$$\mathcal{R}_2 \triangleq \{(r_1, r_2) : \pi(r_1) < r_2 < 4\pi(r_1), \ (r_1, r_2) \in \Omega\}$$
$$\mathcal{R}_3 \triangleq \{(r_1, r_2) : r_2 > 3\pi(r_1), \ (r_1, r_2) \in \Omega\}$$
$$\mathcal{R}_4 \triangleq \{(0, 0)\}$$

where $\pi(r_1) \triangleq (1 - e^{-r_1})$. Σ_S's internal logic is then defined by the computer diagram

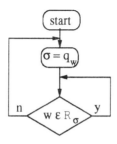

where

$$q_w \triangleq \min_q \{q : w \in \mathcal{R}_q, \ q \in \mathcal{I}\}$$

In interpreting this diagram it is to be understood that σ's value at each of its switching times $\bar{t}$ is its limit from above as $t \downarrow \bar{t}$. Thus if $\bar{t}_i$ and $\bar{t}_{i+1}$ are any two successive switching times, then σ is constant on $[\bar{t}_i, \bar{t}_{i+1})$.

It can be shown that with this switching logic, chattering cannot occur and that x, y and z must tend to zero no matter how they and σ are initialized [21]. It can also be shown that the origin $x = y = z = 0$ is 'Lyapunov stable' in an appropriately defined sense. We refer the reader to [35] for a different application of a switching logic similar to the one we've been discussing.

4 Self-Adjusting Control

The aim of this section is to give a brief tutorial overview of four different classes of logic-based switching control systems - each consists of a continuous-time process to be controlled, a parameterized family of linear controllers, and an event driven switching logic. Three of the logics, called *prerouted switching*, *hysteresis switching* and *dwell-time switching* respectively, are simple strategies capable of determining in real time which controller from a family of candidates should be put in feedback with a process in order to achieve desired closed-loop performance. The fourth, called *cyclic switching*, has been devised to solve the long-standing stabilizability problem which arises in the synthesis of identifier-based adaptive controllers because of the existence of points in parameter space where the estimated model upon which certainty equivalence synthesis is based, loses stabilizability. Although each of these strategies is applicable to a variety of problems, the sake of uniformity all are explained within the context of a single prototype problem - the set-point control of a siso linear system with large-scale parametric uncertainty §4.1. The concept of prerouted switching is closely allied with the idea of a "nonestimator based supervisor"; both topics are discussed in §4.2. Hysteresis switching and dwell-time switching are explained in §4.3 in connection with the concept of an estimator-based supervisor. Cyclic switching is discussed in §4.4

4.1 The Problem

The prototype problem we want to consider is basic: to construct a control system capable of driving to and holding at a prescribed set-point, the output of a process modeled by a dynamical system with large scale parametric uncertainty. Assume the process admits the model of a siso controllable, observable linear system Σ_P with control input u and measured output y. Further assume that Σ_P's transfer function from u to y is a member of a known class of admissible strictly proper transfer functions $\mathcal{C}_P$. In view of the requirements of set-point control, assume that the numerator of each transfer in $\mathcal{C}_P$ is nonzero at $s = 0$.

The specific design goal is to construct a positioning or set-point control system capable of causing y to approach and closely track any constant reference input r. Towards this end we introduce a *tracking error*

$$\mathbf{e_T} \stackrel{\Delta}{=} r - y \tag{3}$$

4 Self-Adjusting Control

and an integrating subsystem to generate u; i.e.,

$$\dot{u} = v \qquad (4)$$

Here v is a control signal which will be defined in the sequel.

As our concern is mainly with supervisory control, we are going to take as given, a parameterized family of proper, reduced controller transfer functions $\mathcal{K} \triangleq \{\kappa_q : q \in \mathcal{Q}\}$ which has the property that for each transfer function $\tau \in \mathcal{C}_P$, there is at least one controller transfer function $\kappa \in \mathcal{K}$ which internally stabilize feedback interconnection shown in Figure 3.

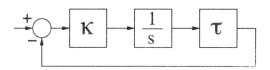

Fig. 3. Feedback Interconnection

The sub-system to be supervised is thus of the form

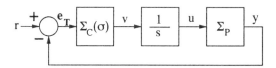

Fig. 4. Supervised Sub-System $\Sigma(\sigma)$

where $\Sigma_C(q)$ is a parameter-dependent, globally detectable/stabilizable realization of κ_q with state x_C. In the sequel we shall describe various types of supervisors capable of generating σ so as to at least achieve set-point regulation {i.e., $e_T \to 0$} and global boundedness.

4.2 Nonestimator-Based Supervisor

A 'nonestimator-based' supervisor is a hybrid dynamical system whose input is a suitably defined "tuning error" e and whose output is σ.

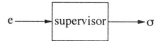

A tuning error is a linear {possibly parameter-dependent} function of the measurable signals in the sub-system $\Sigma(\sigma)$ shown in Figure 4. The key requirements governing the selection of e are as follows.

Tuning Error Requirements:

1. For each fixed $q \in \mathcal{Q}$, $\Sigma(q)$ must be detectable through e.
2. For each constant r and each $q \in \mathcal{Q}$, e must vanish on $\Sigma(q)$'s equilibrium state.

The global detectablilty requirement is fundamental. Its significance has been discussed in a broader context in [36].

One definition of e which satisfies both requirements for the problem under consideration is

$$e \triangleq \begin{bmatrix} e_T \\ v \end{bmatrix}$$

There are many other acceptable choices as well.

Assume that e has been defined so that the preceding requirements are satisfied. The sub-system depicted in Figure 4 then admits a state space model of the form

$$e = C_\sigma \bar{x}$$
$$\dot{\bar{x}} = A_\sigma \bar{x} + b_\sigma r \qquad (5)$$

where $\bar{x}$ is the composite state

$$\bar{x} \triangleq \begin{bmatrix} x_P \\ u \\ x_C \end{bmatrix}$$

and A_q, b_q and C_q are parameter-dependent matrices determined by the definition of e and the coefficient matrices of Σ_P and Σ_C. The position of the integrator in Figure 4 is important [24]: its location guarantees that for each fixed r, the equilibrium state of (5), namely

$$\bar{x}_0 \triangleq -A_q^{-1} b_q r,$$

is independent of $q \in \mathcal{Q}$. Because of this and the assumption that e satisfies Tuning Error Requirement 2, it is possible to write

$$e = C_\sigma x$$
$$\dot{x} = A_\sigma x \qquad (6)$$

where $x \triangleq \bar{x} - \bar{x}_0$.

What we want to do is to explain how to construct a supervisor whose output σ causes $x \to 0$ as $t \to \infty$. The *only* properties of (6) which we will exploit are the following:

Properties of (C_q, A_q) :

1. There exists a parameter value $q^* \in \mathcal{Q}$ for which A_{q^*} is a stability matrix.
2. (C_q, A_q) is detectable for each $q \in \mathcal{Q}$.

4 Self-Adjusting Control

The first property is a consequence of the assumption that for each transfer function $\tau \in \mathcal{C}_P$ there is a transfer function $\kappa \in \mathcal{K}$ which stabilizes the system shown in Figure 3. The second property follows from Tuning Error Requirement 1.

Apart from the preceding, nothing is assumed about (6) other than that e can be measured. In particular, neither the parameter-dependent pair (C_q, A_q) nor q^* are presumed to be known. Of course with so little known, one should not expect to come up with a supervisor worthy of actual implementation unless perhaps $\mathcal{Q}$ is a finite set with a small number of elements.

Within the parameter-adaptive framework proposed in [36], a nonestimator-based supervisor would be called a "prerouted" parameter tuner. All parameter tuners, be they prerouted or not, are based on the same underlying strategy which roughly speaking is to keep adjusting σ until e is "small" in some suitably defined sense. Although there are a great many different methods for accomplishing this, in most instances tuning is carried out in one of two fundamentally different ways depending on whether the 'path' σ takes in $\mathcal{Q}$ is 'prerouted' or not. For the prerouted case, tuning is achieved by moving σ through $\mathcal{Q}$ along a prespecified path or route, using on-line {i.e., real-time} data to decide only if and when or how fast to change σ from one value along the path to the next. In contrast, for the non-prerouted case, the path in $\mathcal{Q}$ along which σ is adjusted is not prespecified off-line but instead is determined in real time from the values of various measured signals.

The basic idea of prerouted tuning was devised by Mårtensson with the expressed purpose of delineating the theoretical limits of what might be achieved with any adaptive algorithm [7]. Over the past decade many refinements and modifications of the concept have appeared [8, 9, 10, 37, 38]. Although these modified algorithms differ from each other in many ways, all share certain underlying features in common. In most cases prerouted tuners consist of the cascade connection of two subsystems, one a *scheduling logic* Σ_S and the other a memoryless map $h : \{1, 2, \ldots, \infty\} \to \mathcal{Q}$ called a *routing function*.

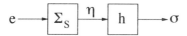

h is invariably required to have the *revisitation property*: That is, for any $q \in \mathcal{Q}$ and any positive integer i there must exist an integer $j \geq i$ at which $h(j) = q$. In other words, h must have the property that the prerouted path $h(1), h(2), \ldots$ *revisits* {i.e., passes through} each point in $\mathcal{Q}$ infinitely often. For this to be possible, $\mathcal{Q}$ must clearly be at least a countable set[3]. Assuming this to be the case, it is always possible to define a routing function with the revisitation property. One way to do this is as follows.

---

[3] Actually in Martenson's original work $\mathcal{Q}$ is a continuum and the elements of the sequence $h(1), h(2), \ldots$ are only required to get close to {rather than equal} previously visited ones [39].

1. If $\mathcal{Q} = \{q_1, q_2, \ldots, q_m\}$ is a finite set, define h to be the m periodic function whose first m values are $q_1, q_2, \ldots, q_{m-1}$ and q_m respectively.
2. If $\mathcal{Q} = \{q_1, q_2, \ldots, \}$ is not a finite set, define h be the function whose sequence of values $h(1), h(2), \ldots$ are the elements of the sequence $q_1, q_1, q_2, q_1, q_2, q_3, q_1, q_2, q_3, q_4, q_1, \ldots$

There are many possible ways to define Σ_S, depending what one is trying to accomplish. For illustrative purposes, we shall take Σ_S to be a hybrid dynamical system whose input is e and whose output is a piecewise-constant switching signal η taking values in the set of positive integers. Σ_S's state consists of four variables - η, a timing signal τ, a piecewise-continuous 'performance signal' π and a piecewise constant 'sampled performance signal' $\bar{\pi}$. Both π and $\bar{\pi}$ take values in $[0, \infty)$. Timing signal τ takes values in the closed interval $[0, \tau_D]$, where τ_D is a preselected positive number called a *dwell time*. Σ_S's dynamics are defined by the following computer diagram.

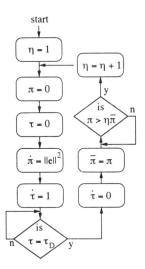

The functioning of Σ_S is as follows. During the first τ_D time units after the algorithm is initiated, τ is increased linearly from 0 to τ_D using a reset integrator and π is increased from 0 according to the rule

$$\dot{\pi} = ||e||^2 \qquad (7)$$

Just at the end of this period, τ is reset to zero, the reset integrator is turned off, and $\bar{\pi}$ is set equal to the present value of π. So long as π remains less than or equal to $\sigma\bar{\pi}$, the updating of π continues according to (7). If and when π becomes greater than $\eta\bar{\pi}$, η is incremented by 1, π and $\bar{\pi}$ are reset to zero, and the entire process is repeated. Note that the time between any two successive switchings of η can never be smaller than τ_D. Said differently, η "dwells" at each of its values for at least τ_D time units. Because of this, infinitely fast switching cannot occur

4 Self-Adjusting Control

so existence and uniqueness of solutions to the differential equations involved is not an issue.

Analysis: There is a fairly straight forward way to go about analyzing the type of supervisory control system we've just described. The key step is to prove that switching stops in finite time. In particular, for the problem at hand the trick is to show that there must be a finite integer $\bar{\eta}$, depending only on the pair (C_q, A_q) and not on the initial value of x, which η cannot exceed in value. Before we address this claim, let us consider its consequences. What the claim implies is that η can switch at most a finite number of times and therefore that there must be a time $\bar{t}$ beyond which η is constant and $\pi \leq \bar{\eta}\bar{\pi}$. The latter assures that

$$\int_{\bar{t}}^{t} ||e(s)||^2 ds \leq \bar{\eta}\bar{\pi}, \quad t \geq \bar{t}$$

and thus that e has a finite $\mathcal{L}^2[0, \infty)$ norm. Moreover since (C_q, A_q) is detectable on $\mathcal{Q}$ and η is fixed at some value $\bar{q} \in \mathcal{Q}$, $(C_{\bar{q}}, A_{\bar{q}})$ is detectable[4]. As a consequence, for $t \geq \bar{t}$ it is possible to rewrite (6) as

$$\dot{x} = (A_{\bar{q}} + KC_{\bar{q}})x - Ke$$

where K is any matrix which stabilizes $A_{\bar{q}} + KC_{\bar{q}}$. Therefore $x \to 0$ since e has a finite $\mathcal{L}^2[0, \infty)$ norm. In fact, because $\dot{\bar{x}} = A_{\bar{q}}\bar{x}$ is a time invariant linear system, x must go to zero as fast at $e^{-\lambda t}$, $-\lambda$ being the largest of the real parts of $A_{\bar{q}}$'s stable eigenvalues. In other words, to prove that $x \to 0$ {and consequently that $\bar{x}$ has a finite limit and that $e_T \to 0$} it is enough to show that there is an integer $\bar{\eta}$ which η cannot exceed in value.

Here briefly is the idea. Let p be any point in $\mathcal{Q}$ at which A_p is a stability matrix. If $C_p = 0$, let $\bar{\eta}$ be the first positive integer such that $h(\bar{\eta}) = p$. Then either there is an interval $[t_0, t_1)$ of maximal length on which $\eta = \bar{\eta}$ or η never gets as large as $\bar{\eta}$. If the latter is true, then we are done. On the other hand, if the former is true, then $\pi = 0$ on $[t_0, t_1)$ so for such t, $\pi \leq \bar{\eta}\bar{\pi}$. Because of Σ_S's definition, this means that no more switching can occur, that $t_1 = \infty$ and thus that η can grow no larger than $\bar{\eta}$.

Now suppose $C_p \neq 0$. Reduce (C_p, A_p) to an observable pair $(\bar{C}, \bar{A})$ by picking any full rank matrix R whose kernel is the unobservable space of (C_p, A_p) and solving the linear equations $C_p = \bar{C}R$, $RA_p = \bar{A}R$ for $\bar{C}$ and $\bar{A}$ respectively. Note that $\bar{A}$ must be a stability matrix because A_p is.

Let $G(t)$ denote the observability Gramian

$$G(t) \triangleq \int_0^t e^{\bar{A}'s} \bar{C}' \bar{C} e^{\bar{A}s} ds$$

Note that $G(\infty)$ must exist because of $\bar{A}$'s stability. Moreover, $G(\tau_D)$ must be positive definite because of the observability of $(\bar{C}, \bar{A})$. This implies that

$$\mu \triangleq \sup_x \frac{x'G(\infty)x}{x'G(\tau_D)x} < \infty$$

---

[4] There is of course no reason to assume that $A_{\bar{q}}$ is a stability matrix.

and that
$$G(\infty) \leq \mu G(\tau_D) \tag{8}$$

Now let η_p be the least integer no smaller than μ for which $h(\eta_p) = p$. Because of the revisitation property, η_p must necessarily exist. In view of (8)
$$G(\infty) \leq \eta_p G(\tau_D) \tag{9}$$

We claim that $\bar{\eta} \triangleq \eta_p$ has the desired property. To prove that this is so, we may as well assume that there is an interval on which $\eta = \bar{\eta}$. For if this were not the case then σ could not exceed $\bar{\eta}$ and we would be done.

Let $[t_0, t_1)$ denote the largest interval on which $\eta = \bar{\eta}$. For $t \in [t_0, t_1)$, $\sigma = h(\eta) = p$ and
$$\pi(t) = \int_{t_0}^{t} ||C_p e^{A_p(w-t_0)} x(t_0)||^2 dw \leq \int_0^{\infty} ||\bar{C} e^{\bar{A}s} Rx(t_0)||^2 ds = ||\sqrt{G(\infty)} Rx(t_0)||^2$$

From this, (9) and the definitions of $\bar{\eta}$ and $\bar{\pi}$ it follows that for $t \in [t_0, t_1)$,
$$\pi(t) \leq \eta_q ||\sqrt{G(\tau_D)} Rx(t_0)||^2 = \bar{\eta} \int_0^{\tau_D} ||\bar{C} e^{\bar{A}s} Rx(t_0)||^2 ds = \bar{\eta}\bar{\pi}$$

Thus because of Σ_S's definition, no more switching can occur, $t_1 = \infty$ and thus η can grow no larger than $\bar{\eta}$. ∎

There are many provably correct versions of the algorithm we've just analyzed [7, 8, 9, 10, 37, 38]. All employ a tuning error satisfying the aforementioned requirements, a performance signal, a routing function and a switching logic similar to the one we've described. Usually τ_D is an increasing function of η rather than a constant. In most cases, the proof technique employed relies on the cessation of switching in finite time. The selection of π and the definition of Σ_S are made to insure that this is so.

Although nonestimator based supervisors are prerouted tuners, the converse is not necessarily true. For example, it is quite possible for a supervisor employing "estimators" to use prerouted tuning to generate σ. Supervisors admitting this structure have in fact been studied in [10]. This reference actually examines the convergence properties of a variety of estimator-based switching logics.

The findings of [40] and earlier work clearly suggest that some of the concepts we've covered here have a universal character and may well be extendable to significantly broader classes of problems than have been considered so far. In the sequel we briefly summarize some preliminary thoughts along these lines.

Generalization: Let Q be a countable set. Suppose that for each $q \in Q$, $A_q : \mathbb{R}^n \to \mathbb{R}^n$ is a smooth, possibly nonlinear function and that for some $q^* \in Q$, the zero state of
$$\dot{x} = A_{q^*}(x)$$
is a globally asymptotically stable equilibrium. Assume that for each piecewise constant switching signal $\sigma : [0, \infty) \to Q$, all solutions to the differential equation
$$\dot{x} = A_{\sigma}(x) \tag{10}$$

… Our aim is to briefly outline how one might go about constructing a nonestimator based supervisor, not depending on q^* or precise knowledge of the A_q, which cause all "supervised" (i.e., closed-loop) solutions to (10) to tend to zero as $t \to \infty$.

Suppose it is possible to construct a smooth function $b : \mathbb{R}^n \mapsto \mathbb{R}$ such that

$$||A_q(x)|| \leq ||b(x)||, \quad \forall x \in \mathbb{R}^n, \quad q \in \mathcal{Q} \tag{11}$$

and for some $q \in \mathcal{Q}$

$$\sup_{z \in \mathbb{R}^n} \frac{\int_0^\infty (||\phi(t,z)||^2 + ||b(\phi(t,z))||^2) dt}{\int_0^{\tau_D} (||\phi(t,z)||^2 + ||b(\phi(t,z))||^2) dt} = \mu < \infty \tag{12}$$

where $\phi : [0, \infty) \times \mathbb{R}^n \to \mathbb{R}^n$ is the flow of

$$\dot{x} = A_q(x) \tag{13}$$

initialized at z. Requirement (11) is relatively mild and can typically be satisfied without precise knowledge of the A_q. Implicit in (12) is the requirement that the zero state of (13) is {at least} an asymptotically stable equilibrium; in fact, for the requirement to make sense as it stands, all solutions to (13) would have to have finite $\mathcal{L}^2[0, \infty)$ norms.

We claim that the supervisor we've already described will accomplish the prescribed task provided

$$e \triangleq \begin{bmatrix} x \\ b(x) \end{bmatrix} \tag{14}$$

The reasoning upon which this claim is based is as follows.

First of all note that satisfaction of (12) guarantees that η cannot exceed the least integer $\bar{\eta}$ no smaller than μ for which $h(\bar{\eta}) = q$. The argument which justifies this assertion exploits the inequality

$$\int_{t_0}^\infty \left\| \begin{bmatrix} \phi(t-t_0, z) \\ b(\phi(t-t_0, z)) \end{bmatrix} \right\|^2 dt \leq \bar{\eta} \int_{t_0}^{t_0+\tau_D} \left\| \begin{bmatrix} \phi(t-t_0, z) \\ b(\phi(t-t_0, z)) \end{bmatrix} \right\|^2 dt, \quad t_0 \geq 0, \; z \in \mathbb{R}^n$$

and is essentially the same as before. The inequality is a consequence of (12).

At this point we need a good working definition of detectability for nonlinear systems. Suppose we agree to call a smooth dynamical system of the form

$$\dot{x} = A(x)$$
$$e = C(x) \tag{15}$$

detectable if there exists a positive definite, radially unbounded, continuously differentiable function $V : \mathbb{R}^n \to \mathbb{R}$ such that

$$\frac{\partial V}{\partial x} A(x) - ||C(x)||^2 < 0, \quad x \in \mathbb{R}^n, \; x \neq 0 \tag{16}$$

- The definition characterizes detectability more as a generalization of stability than of observability; note for example, that if $C(x) = x$, (15) may not satisfy the definition, even though for this example (15) would certainly have to be considered an observable system.
- In the linear case when $C(x) = Cx$ and $A(x) = Ax$, the standard definition of detectability is known to be equivalent to the existence of a positive definite matrix P which satisfies the matrix inequality $PA + A'P - C'C < 0$ [41]; since $V \triangleq x'Px$ satisfies (16), the definition of detectability proposed here thus has the virtue of reducing to the standard one in the linear case.
- It can be easily shown that if (15) is a detectable nonlinear system and e has a finite $\mathcal{L}^2[0, \infty)$ norm along some solution x, then x must tend to zero. Thus the proposed definition fulfills the intuitively appealing requirement that smallness of the output of a detectable system ought to imply smallness of the system's state.

Returning to our problem we point out that (11) implies that for each fixed $q \in \mathcal{Q}$, the dynamical system

$$\dot{x} = A_q(x)$$
$$e = \begin{bmatrix} x \\ b(x) \end{bmatrix} \tag{17}$$

is detectable through e. This can be verified using the function $V \triangleq \frac{1}{2}||x||^2$.

The steps involved in showing that $x \to 0$ are clear. Since switching stops, π is bounded which means that e must have a finite $\mathcal{L}^2[0, \infty)$ norm. Suppose q is σ's final value. Then (17) governs the evolutions of x and e after switching stops. Because (17) is a detectable system and e has a finite $\mathcal{L}^2[0, \infty)$ norm, x must tend to zero as claimed.□

There are of course plenty of practical reasons why one would not want to seriously consider implementing the system just described. On the other hand, there are components of the preceding {e.g., the notion of detectability and how to use it} which will no doubt prove useful in the analysis of more meaningful algorithms.

One drawback of many "switched" control systems including the ones we've discussed so far, is that they make use of signal which grows monotonically with time. For the supervisor we've described this would be η. Since bounded monotone signals converge, switched systems which employ them tend to be fairly easy to analyze. The problem is that when $\mathcal{L}^\infty$ bounded noise and or exogenous disturbances signals are present, monotone signals tend to blow up. To get around this, it is generally necessary to eliminate monotone signals altogether, usually by introducing "forgetting factors" or "exponential weighting" of some form [8, 38, 24]. What this means is that with such modifications in place, switching can no longer be expected to terminate in finite time. As a result one is usually confronted with an analysis problem which is very much more challenging than that encountered in the noise-free case when monotone convergence could be counted on. Because of this *there is a specific need for technical results appropriate to the analysis of systems within which switching never terminates.*

Perhaps the most serious criticism of the nonestimator approach is its reliance on prerouted tuning. Clearly if $\mathcal{Q}$ is a large set, one should not expect a prerouted supervisory control system to perform very well.

4.3 Estimator-Based Supervisors

The overall responsibility of any multi-controller supervisor can be divided into a scheduling task - deciding *when* to switch controllers - and a routing task - deciding *which* controller to switch to next. Nonestimator-based supervisors have the routing question decided for them and are thus designed to deal only with scheduling. It is natural to expect that improved overall performance can be achieved by employing a supervisor endowed with the capability of making *both* scheduling and routing decisions in real time. An important class of supervisors possessing this capability are those which are estimator-based. Estimator-based supervisors utilize a form of certainty equivalence and as such are in some ways quite similar to conventional estimator-based tuners encountered in parameter adaptive control.

Since an estimator-based supervisor is responsible for both scheduling and routing, it is not surprising that defining one should require a more detailed description of $\mathcal{C}_P$ then we've assumed so far. For illustrative purposes suppose $\mathcal{C}_P$ to be of the form

$$\mathcal{C}_P = \bigcup_{p \in \mathcal{P}} \mathcal{C}(p)$$

where $\mathcal{P}$ is a closed, bounded {possibly finite} subset of a real, finite-dimensional linear space. Here $\mathcal{C}(p)$ denotes the subclass

$$\mathcal{C}(p) = \{\nu_p + \delta : ||\delta||_\infty \leq \epsilon_p\}$$

where ν_p is a preselected, reduced, strictly proper *nominal transfer function*, ϵ_p is a real non-negative number and δ is a stable, strictly proper norm-bounded perturbation representing unmodelled dynamics of the additive type; $||\cdot||_\infty$ denotes the shifted infinity norm

$$||\delta||_\infty \stackrel{\Delta}{=} \sup_{s \in C(\lambda_u)} |\delta(s)|,$$

where λ_u is a prespecified positive number called the *unmodelled dynamics stability margin*, and $C(\lambda_u)$ is the subset of the complex plane consisting of all points on and to the right of the vertical line $s = -\lambda_u$. Assume for each $p \in \mathcal{P}$, that the allowable values of δ exclude transfer functions for which $\nu_p + \delta$ has unstable poles and zeros in common. All transfer functions in $\mathcal{C}_P$ are thus strictly proper, but not necessarily stable rational functions.

As before, we take as given a parameterized a family of admissible controller transfer functions $\mathcal{K}$ which has the property that for each transfer function τ in $\mathcal{C}_P$ there is at least one controller transfer function $\kappa \in \mathcal{K}$ which internally stabilizes the interconnection shown in Figure 3. Because estimator-based supervisors base decision-making on the idea of certainty equivalence, to configure such a supervisor it is necessary to first specify a well-defined function F from the nominal process model transfer function class $\mathcal{N} \triangleq \{\nu_p : p \in \mathcal{P}\}$ to $\mathcal{K}$ in such a way that the assignment $\nu_p \longmapsto F(\nu_p)$ meets prescribed specifications. Given F, a natural way to make this assignment explicit is to stipulate that $\mathcal{P}$ be a subset of $\mathcal{K}$'s parameter space $\mathcal{Q}$ and then to define $\kappa_p \triangleq F(\nu_p)$ for each $p \in \mathcal{P}$. For the present we shall actually take $\mathcal{Q} = \mathcal{P}$. The reader should realize however that there are situations in which it is advantageous to choose $\mathcal{Q}$ larger than $\mathcal{P}$. For example, picking $\mathcal{Q}$ larger than $\mathcal{P}$ makes it possible to define generalized supervisors whose controller selection strategies are not based just on certainty equivalence alone {c.f. §4.4}.

Assume that the transfer functions in $\mathcal{K}$ satisfy the

Stability Margin Requirement: For each $p \in \mathcal{P}$ the real parts of the closed-loop poles of the feedback interconnection shown in Figure 5 are less than $-\lambda_S$ where λ_S some prespecified positive number called a *stability margin*.

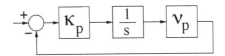

Fig. 5. Feedback Interconnection

In concept, an estimator-based supervisor can be explained in terms of the "multi-estimator" architecture shown in Figure 6

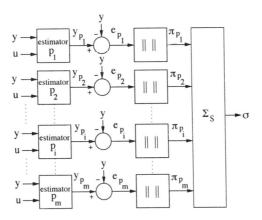

Fig. 6. Multi-Estimator Configured Supervisor

where each y_p is a suitably defined estimate of y which would be asymptotically

4 Self-Adjusting Control

correct if ν_p were the process model's transfer function. For each $p \in \mathcal{P}$,

$$e_p \triangleq y_p - y \qquad (18)$$

denotes the pth *output estimation error*; π_p is a "normed" value of e_p or a "performance signal" which is used by the supervisor assess the potential performance of controller p. Σ_S is a switching logic whose function is to determine σ on the basis of the current values of the π_p.

The underlying decision making strategy used by an estimator-based supervisor of the 'non-prerouted' type is basically this: From time to time select for v, that candidate control signal v_p whose corresponding performance signal π_p is the smallest among the π_p, $p \in \mathcal{P}$. What makes a non-prerouted supervisor such as this distinctly different from a prerouted one is thus the philosophy underlying the method it uses to carry out its task. In particular, a non-prerouted supervisor decides which controller to put in the feedback loop, not by search along a predetermined route in $\mathcal{K}$, but rather by continuously comparing in real time suitably defined normed output estimation errors or performance signals associated with the admissible nominal process models. Motivation for this idea is obvious: the process model whose associated performance signal is the smallest, "best" approximates what the process is and thus the candidate controller designed on the basis of that model ought to be able to do the best job of controlling the process. The origin of this idea is of course the concept of certainty equivalence from parameter adaptive control.

By an estimator of y, based on transfer function ν_p, is meant a linear system of the form

$$\dot{x}_p = A_p x_p + d_p y + b_p u \qquad (19)$$
$$y_p = c_p x_p \qquad (20)$$

where $\{A_p + d_p c_p, b_p, c_p\}$ is a realization of ν_p and A_p is a stability matrix. It is easy to verify that any such realization necessarily fulfills the requirement that y_p be an asymptotically correct estimate of y if the process model transfer function were ν_p. Notice that such realizations are invariably detectable because of A_p's stability. For the present we are only going to consider realizations which are stabilizable as well, even though by doing so we are sidestepping some subtle but important issues{cf., §4.4}. There are many ways to construct estimators which meet these requirements. For example, if n is an upper bound on ν_p's McMillan Degree, y_p can always be be generated by an observer-based estimator of the form

$$\dot{x}_p = A_O x_p + d_p y + b_p u$$
$$y_p = c_O x_p \qquad (21)$$

where (c_O, A_O) is an n-dimensional, parameter-independent observable, stable pair and $\{A_O + d_p c_O, b_p, c_O\}$ is a stabilizable realization of ν_p. It is also possible to generate y_p using an identifier-based estimator of the form

$$\dot{x}_I = \begin{bmatrix} A_I & 0 \\ 0 & A_I \end{bmatrix} x_I + \begin{bmatrix} b_I \\ 0 \end{bmatrix} y + \begin{bmatrix} 0 \\ b_I \end{bmatrix} u$$

$$y_p = c_p x_I$$

where (A_I, b_I) is a parameter-independent, n-dimensional siso, controllable pair with A_I stable and

$$\left\{ \begin{bmatrix} A_I & 0 \\ 0 & A_I \end{bmatrix} + \begin{bmatrix} b_I \\ 0 \end{bmatrix} c_p, \begin{bmatrix} 0 \\ b_I \end{bmatrix}, c_p \right\}$$

is a stabilizable realization of ν_p. Note that the state of this estimator is independent of p, whereas the state of the observer-based estimator in (21) is not. What this means is that if n is an upper bound on the McMillan Degrees of all of the nominal transfer functions in $\mathcal{N}$, then all of the y_p can be generated using a single estimator with shared state x_I and parameter-dependent readout map c_p.

There is a third way to generate y_p which is very similar to the second but which is especially well-suited to the set-point control problem under consideration. In this case one uses an identifier-based estimator Σ_E of the form

$$\dot{x}_E = \begin{bmatrix} A_E & 0 \\ 0 & A_E \end{bmatrix} x_E + \begin{bmatrix} b_E \\ 0 \end{bmatrix} y + \begin{bmatrix} 0 \\ b_E \end{bmatrix} v \qquad (22)$$

$$y_p = c_p x_E \qquad (23)$$

where (A_E, b_E) is a parameter-independent, $(n+1)$-dimensional siso, controllable pair with A_E stable and

$$\left\{ \begin{bmatrix} A_E & 0 \\ 0 & A_E \end{bmatrix} + \begin{bmatrix} b_E \\ 0 \end{bmatrix} c_p, \begin{bmatrix} 0 \\ b_E \end{bmatrix}, c_p \right\}$$

is a stabilizable realization of $\frac{1}{s} \nu_p$. A state-shared implementation based on this estimator would then appear as in Figure 7. Naturally this architecture can only be implemented as it stands if the number of output estimation errors is finite; i.e., if $\mathcal{P}$ is a finite set. It turns out however that such a supervisor can often be implemented using a simpler architecture - one which permits $\mathcal{P}$ to contain a continuum of points. To explain why this is so, it is useful to formalize the idea of a supervisor.

By an *estimator-based supervisor* {cf, Figure 8} is meant a specially structured hybrid dynamical system whose output σ is a switching signal taking values in $\mathcal{Q}$ and whose inputs are v and y. Internally such a supervisor consists of three subsystems: a state-shared estimator Σ_E, a *performance weight generator* Σ_W and a *switching logic* Σ_S. Σ_W is a causal dynamical system whose inputs are x_E and y and whose state and output W is a "weighting matrix" which takes values in a linear space $\mathcal{W}$. W together with a suitably defined *performance function* $\Pi : \mathcal{W} \times \mathcal{P} \to \mathbb{R}$ determine, for each $p \in \mathcal{P}$, scalar-valued *performance signals* of the form

$$\pi_p = \Pi(W, p), \ p \in \mathcal{P} \qquad (24)$$

4 Self-Adjusting Control

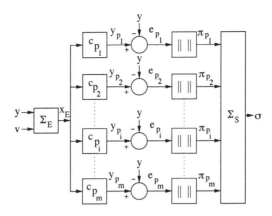

Fig. 7. State-Shared Estimator-Based Supervisor

These performance signals play the same role as before; i.e., π_p is considered to be a measure of the expected performance of control signal p. One possible pair of definitions for Σ_W and Π is

$$\dot{W} = -2\lambda W + \begin{bmatrix} x_E \\ y \end{bmatrix} \begin{bmatrix} x_E \\ y \end{bmatrix}' \qquad (25)$$

and

$$\Pi(W, p) = [\, c_p \quad -1\,]\, W \,[\, c_p \quad -1\,]' \qquad (26)$$

respectively where λ is a prespecified nonnegative number. In the light of (18) and (23) it is easy to see that these definitions imply that

$$\dot{\pi}_p = -2\lambda \pi_p + e_p^2, \quad p \in \mathcal{P} \qquad (27)$$

Although we will deal here exclusively with such "exponentially weighted $\mathcal{L}^2$" performance signal, it should be noted that it is possible to realize other types of performance signals by defining W and Π in other ways. For example, if $\mathcal{P}$ is a finite set {say $\mathcal{P} = \{1, 2, \ldots, m\}$} and if Σ_W is the dynamical system $\dot{w}_p = |e_p|$, $p \in \mathcal{P}$ with state $w \triangleq [\, w_1 \quad w_2 \quad \cdots \quad w_m\,]'$, defining $\Pi(w, p) \triangleq w_p$ would realize the $\mathcal{L}^1$ performance signal $\dot{\pi}_p = |e_p|$. Note however that if $\mathcal{P}$ were not finite, this particular performance signal could not be realized with $\mathcal{W}$ finite dimensional.

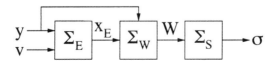

Fig. 8. Estimator-Based Supervisor

Hysteresis Switching: There are a number of different ways to define switching logic Σ_S. In the sequel we shall consider two. The first, called "Hysteresis Switching," was originally devised for switching between the members of a finite family of parameter adaptive controllers [11, 12, 13]. We shall explain this logic's basic attributes in the following manner.

Suppose $\{f_q : q \in \mathcal{Q}\}$ is a family of functions $f_q : \mathbb{R}^n \times [0, \infty) \to \mathbb{R}^n$. Our aim is to study the behavior of the dynamical system

$$\dot{x} = f_\sigma(x, t), \qquad x(0) = x_0 \qquad (28)$$

where σ is a switching signal taking values in $\mathcal{Q}$[5]. Suppose that $\mathcal{P} = \mathcal{Q}$, and that W is a function of x and t which takes values in $\mathcal{W}$; i.e.,

$$W = g(x, t) \qquad (29)$$

As before, suppose that $\Pi : \mathcal{W} \times \mathcal{P} \to \mathbb{R}$ is a performance function and that for $p \in \mathcal{P}$, $\pi_p \triangleq \Pi(w, p)$ is a performance signal. What we want to do is to explain how to generate a switching signal σ which under certain conditions, converges to a value $\bar{q} \in \mathcal{Q}$ at which $\pi_{\bar{q}}$ is a bounded signal. The algorithm which generates σ is called a "hysteresis switching logic."

By a *hysteresis switching logic* is meant a hybrid dynamical system Σ_H whose input is W and whose state and output are both σ.

$$W \longrightarrow \boxed{\Sigma_H} \longrightarrow \sigma$$

To specify Σ_H it is necessary to first pick a positive number $h > 0$ called a *hysteresis constant*. Σ_H's internal logic is then defined by the computer diagram shown in Figure 9 where for $X \in \mathcal{W}$, q_X denotes a value of $q \in \mathcal{Q}$ which minimizes $\Pi(X, q)$.

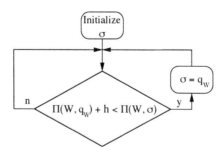

Fig. 9. Computer Diagram of Σ_H

---

[5] For the set-point control problem under consideration, x would represent the composite state $\{x_P, u, x_E, x_C, W\}$.

4 Self-Adjusting Control

In interpreting this diagram it is to be understood that σ's value at each of its switching times $\bar{t}$ is its limit from above as $t \downarrow \bar{t}$. Thus if $\bar{t}_i$ and $\bar{t}_{i+1}$ are any two successive switching times, then σ is constant on $[\bar{t}_i, \bar{t}_{i+1})$. Note that the definition of Σ_H implies that $\pi_{\sigma(t)}(t) \le \pi_q(t) + h$, $t \ge 0$, $q \in \mathcal{Q}$ and that $\pi_{\sigma(\bar{t})}(\bar{t}) \le \pi_q(\bar{t})$, $q \in \mathcal{Q}$ if $\bar{t}$ is a switching time.

The functioning of Σ_H is roughly as follows. Suppose that at some time t_0, Σ_H has just changed the value of σ to q. σ is then held fixed at this value unless and until there is a time $t_1 > t_0$ at which $\pi_p + h < \pi_q$ for some $p \in \mathcal{Q}$. If this occurs, σ is set equal to p and so on.

Note that since all the supervisor has to do is to compute values of $p \in \mathcal{P}$ which minimize $\Pi(W, p)$ at various times, there is in principle nothing to prevent $\mathcal{P}$ from containing a continuum of points. Of course the minimization problems to be solved must be tractable and the time it takes to compute these minima needs to be taken into account. We will discuss both of these points further in the sequel.

For the present our objective is to describe some of the properties of the closed-loop system determined by (28), (29) and Σ_H assuming that g and each f_q is at least locally Lipschitz in x and piecewise-continuous in t. Observe that because of the hysteresis constant h and the assumed smoothness of g and the f_q, there must exist an interval $(0, t_1)$ of maximal length on which σ is constant. Either this interval is the maximal interval of existence for x or it is not in which case x is bounded on $[0, t_1)$. If the latter is true, a switch must occur at t_1 and again because of the hysteresis constant h, the continuity of x and the smoothness of g and the f_q, there must be an interval $[t_1, t_2)$ of maximal length on which σ is constant. Continuing this reasoning we conclude that there must be an interval $[0, T)$ of maximal length on which there is a unique pair $\{x, \sigma\}$ with x continuous and σ piecewise constant, which satisfies (28) and (29). Moreover, on each proper subinterval $[0, \tau) \subset [0, T)$, σ can switch at most a finite number of times.

Our aim now is to characterize the limiting behavior of σ as $t \to T$. For this we need to make certain "open-loop" assumptions. Let $\mathcal{S}$ denote the class of all piecewise-constant functions $s : [0, \infty) \to \mathcal{Q}$. In what follows, for each $s \in \mathcal{S}$, T_s is the length of the maximal interval of existence for the equations

$$\dot{x} = f_{s(t)}(x, t), \qquad x(0) = x_0$$

and x_s is the corresponding solution. We make the following

Assumption 1 (Open-Loop)

1. For each $s \in \mathcal{S}$ and each $q \in \mathcal{Q}$, performance signal $\pi_q(t) = \Pi(g(x_s(t), t), q)$ has a limit (which may be infinite) as $t \to T_s$.
2. There exists at least one point $q^* \in \mathcal{Q}$ such that for each $s \in \mathcal{S}$, performance signal $\pi_{q^*}(t) = \Pi(g(x_s(t), t), q^*)$ is bounded on $[0, T_s)$.

These assumptions enable one to prove the following [12].

Lemma 1 Hysteresis Switching. *For fixed initial state $(x_0, \sigma_0) \in \mathbb{R}^n \times \mathcal{Q}$, let (x, σ) denote the unique solution to (28) and (29) with σ the output of Σ_H - and suppose $[0, T)$ is the largest interval on which this solution is defined. If the open-loop assumptions hold, there is a time $T^* < T$ beyond which σ is constant and no more switching occurs. Moreover, $\pi_{\sigma(T^*)}$ is bounded on $[0, T)$.*

Analysis: What we want to do next is to very briefly sketch how one might use the Hysteresis Switching Lemma to to analyze the closed-loop behavior of the supervisory control system shown in Figure 10.

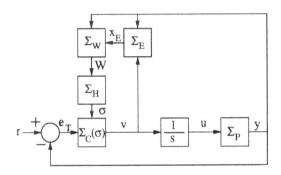

Fig. 10. Supervisory Control System Using Hysteresis Switching

Here $\Sigma_C(q)$ is a globally detectable/stabilizable realization of κ_q with state x_C, Σ_E is the globally detectable/stabilizable estimator defined by (22) and (23), and Σ_H is a hysteresis switching logic. Assume that $\lambda = 0$ and that Σ_W and Π are defined by (25) and (26) respectively. Therefore in this case, π_p is the $\mathcal{L}^2$ performance signal

$$\dot{\pi}_p = e_p^2, \quad p \in \mathcal{P} \tag{30}$$

Note that the Open-Loop Assumption 1 automatically holds because all of the π_p are monotone functions.

It can be shown [24] that there are constant vectors b and h, singly indexed matrices A_p, d_p, g_p, and $\bar{c}_p$ and doubly indexed matrices f_{qp} and c_{qp} such that for all constant r

$$\begin{bmatrix} x_E \\ x_C \end{bmatrix} = x + hr \tag{31}$$

where

$$\dot{x} = (A_l + bf_{\sigma l})x + d_l e_l \tag{32}$$

$$e_p = c_{pl}x + e_l \quad p \in \mathcal{P} \tag{33}$$

$$v = f_{\sigma l}x + g_\sigma e_l \tag{34}$$

$$e_T = e_l + \bar{c}_l x \tag{35}$$

4 Self-Adjusting Control

Because of the Stability Margin Requirement, it also turns out to be true that the matrix pairs $(c_{pl}, A_l + bf_{pl})$, $p, l \in \mathcal{P}$ are each detectable. These claims can be verified in a straight forward manner by direct analysis of the equations under consideration.

What we want to do next is to very briefly outline how one might use the Hysteresis Switching Lemma to analyze the closed-loop behavior of the supervisory control system shown in Figure 10 under the assumption that for some $p^* \in \mathcal{P}$, nominal transfer function ν_{p^*} matches or equals that of Σ_P. The exact matching assumption provides exactly one new piece of information, namely that e_{p^*} must go to zero as fast as $e^{-\lambda_E t}$ where $-\lambda_E$ is the largest of the real parts of the eigenvalues of A_E. Because of (30) this means that

$$\lim_{t \to \infty} \pi_{p^*}(t) \triangleq C^* < \infty$$

Thus Open-Loop Assumption 2 is satisfied.

In view of the Hysteresis Switching Lemma there there must be a time T^* beyond which σ is constant and no more switching occurs. Moreover, $\pi_{\sigma(T^*)}$ must be bounded on the maximal interval of existence $[0, T)$ for solution to the overall system of equations involved. Because switching has stopped, it can be shown that the solution in question in fact exists globally $\{$i.e., $T = \infty\}$.

Suppose that $\bar{q}$ is the final value of σ. Since $\pi_{\bar{q}}$ is bounded on $[0, \infty)$, $e_{\bar{q}}$ must have a finite $\mathcal{L}^2[0, \infty)$ norm because of (30). Next observe that for t sufficiently large and $l \triangleq p^*$, (32) can be written as

$$\dot{x} = (A_{p^*} + bf_{\bar{q}p^*})x + d_{p^*}e_{p^*} \tag{36}$$

In view of the detectability of $(c_{\bar{q}p^*}, A_{p^*} + bf_{\bar{q}p^*})$, there must exist a matrix k which stabilizes $A_{p^*} + bf_{\bar{q}p^*} + kc_{\bar{q}p^*}$. Thus because of (33), (36) can be rewritten as

$$\dot{x} = (A_{p^*} + bf_{\bar{q}p^*} + kc_{\bar{q}p^*})x - ke_{\bar{p}} + (k + d_{p^*})e_{p^*}$$

Since $A_{p^*} + bf_{\bar{q}p^*} + kc_{\bar{q}p^*}$ is a stability matrix and both $e_{\bar{q}}$ and e_{p^*} have finite $\mathcal{L}^2[0, \infty)$ norms, x must have a limit of zero as $t \to \infty$. Therefore x_E and x_C must have a finite limits because of (31). So also must v because of (34). Moreover, since x and e_{p^*} both tend to zero, so must e_T because of (35). Therefore $y \to r$. Since y and v have finite limits, and Σ_P's transfer function is nonzero at $s = 0$, u must have a finite limit as well. In other words, y, u, v, x_E, and x_C all tend to finite limits and $e_T \to 0$.

Note how detectability has once again played a central role in the analysis. Together with the Hysteresis Switching Lemma it has enabled us to establish the limiting behavior of y, u, v, x_E, and x_C in a very elementary way.

The preceding is less than satisfactory for at least four important reasons:

1. If $r \neq 0$, W will grow without bound.
2. If noise and disturbances are present W will almost certainly grow without bound.
3. The analysis fails to account for unmodelled process dynamics

4. The analysis fails to account for computation time; i.e., the time it takes the supervisor to carry out the calculations necessary to select a new control.

A possible remedy for the first two problems would be to introduce a forgetting factor or exponential weighting in the definition of Σ_W in (25). For example, one might pick $\lambda > 0$. Of course any such change would make the resulting system substantially more difficult to analyze than the one we've been considering since π_{p^*} would no longer be monotone and switching would not necessarily terminate in finite time. Add in a small amount of unmodelled dynamics, and the analysis problem would become even more difficult because it would no longer possible to presume at the outset that e_{p^*} tends to zero or even that it is bounded. Some progress in dealing with these difficulties has recently been announced in [42].

Taking into account computation time makes things even more difficult. On the other hand, the reality of a positive computation time - however small - to some extent mitigates the need for hysteresis, since the only reason for introducing hysteresis in the first place was to prevent unbounded chatter [11]. Rather than further pursue this topic, we turn instead to an alternative switching logic which takes computation time directly into account and which results in a supervisory control system which can be shown to perform its function in the face of unmodelled dynamics and exogenous disturbances [43].

Dwell-Time Switching: By a *dwell-time switching logic* [15] Σ_D, is meant a hybrid dynamical system whose input and output are W and σ respectively, and whose state is the ordered triple $\{X, \tau, \sigma\}$.

$$W \longrightarrow \boxed{\Sigma_D} \longrightarrow \sigma$$

Here X is a discrete-time matrix which takes on sampled values of W, and τ is a continuous-time variable called a *timing signal*. τ takes values in the closed interval $[0, \tau_D]$, where τ_D is a prespecified positive number called a *dwell time*. Also assumed prespecified is a *computation time* $\tau_C \leq \tau_D$ which bounds from above for any $X \in \mathcal{W}$, the time it would take a supervisor to compute a value $p = p_X \in \mathcal{P}$ which minimizes $\Pi(X, p)$. Between "event times" τ is generated by a reset integrator according to the rule $\dot{\tau} = 1$. Event times occur when the value of τ reaches either $\tau_D - \tau_C$ or τ_D; at such times τ is reset to either 0 or $\tau_D - \tau_C$ depending on the value of Σ_D's state. Σ_D's internal logic is defined by the computer diagram shown in Figure 11 where p_X denotes a value of $p \in \mathcal{P}$ which minimizes $\Pi(X, p)$.

The functioning of Σ_D can be explained as follows. Suppose that at some time t_0, Σ_D has just changed the value of σ to p. At this instant τ is reset to 0. After $\tau_D - \tau_C$ time units have elapsed, W is sampled and X is set equal to this value. During the next τ_C time units, a value $p = p_X$ is computed which minimizes $\Pi(X, p)$. At the end of this period, when $\tau = \tau_D$, if $\Pi(X, p_X)$ is smaller than $\Pi(X, \sigma)$, then σ is set equal to p_X, τ is reset to zero and the entire process is

4 Self-Adjusting Control

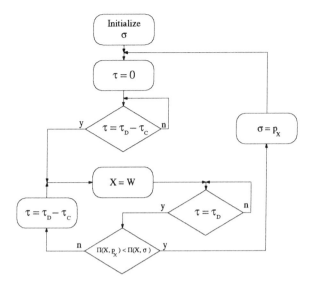

Fig. 11. Computer Diagram of Σ_D

repeated. If on the other hand, $\Pi(X,\sigma)$ is less than or equal to $\Pi(X,p_X)$, τ is reset to $\tau_D - \tau_C$, W is again sampled, X takes on this new sampled value, minimization is again carried out over the next τ_C time units..... and so on.

Note that Σ_D is *scale independent* in that its output σ remains unchanged if its performance function-weighting matrix pair (Π, W) is replaced by another performance function-weighting matrix pair $(\bar{\Pi}, \bar{W})$ satisfying $\bar{\Pi}(\bar{W},p) = \theta\Pi(W,p)$, $p \in \mathcal{P}$, where $\theta : [0,\infty) \to \mathbb{R}$ is a positive time function. This is because for any fixed t, the values of p which minimize $\Pi(W(t),p)$ are exactly the same as the values of p which minimize $\theta(t)\Pi(W(t),p)$.

Let us agree to call a piecewise-constant function $\sigma : [0, \infty) \to \mathcal{P}$ *admissible* if it either switches values at most once, or if it switches more than once and the set of time differences between each two successive switching times is bounded below by a positive number μ. The supremum of such values of μ is σ's *dwell time*. Because of the definition of Σ_D, it is clear its output σ will be admissible with dwell time no smaller than that of Σ_D. This means that switching cannot occur infinitely fast and thus that existence and uniqueness of solutions to the differential equations involved is not an issue.

Analysis: What we want to do next is to very briefly outline how one might analyze the closed-loop behavior of the supervisory control system shown in Figure 12 under the assumption that for some $p^* \in \mathcal{P}$, nominal transfer function ν_{p^*} matches or equals that of Σ_P. Unlike the supervisory control system considered in the last section, we will not {and probably cannot} prove that switching terminates in finite time.

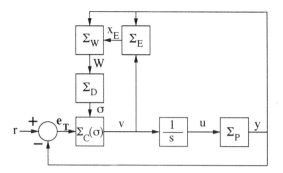

Fig. 12. Supervisory Control System Using Dwell-Time Switching

For the present we continue to assume $\lambda = 0$ and that Σ_W and Π are defined by (25) and (26) respectively. Thus

$$\dot{\pi}_p = e_p^2, \quad p \in \mathcal{P} \tag{37}$$

just as before. For simplicity we focus only on the case in which $\mathcal{P}$ is a finite set. Since (31)-(35) still hold, we can write

$$\begin{bmatrix} x_E \\ x_C \end{bmatrix} = x + hr \tag{38}$$

and

$$\dot{x} = (A_{p^*} + bf_{\sigma p^*})x + d_{p^*} e_{p^*} \tag{39}$$
$$e_p = c_{pp^*} x + e_{p^*} \quad p \in \mathcal{P} \tag{40}$$
$$v = f_{\sigma p^*} x + g_\sigma e_{p^*} \tag{41}$$
$$\mathbf{e_T} = e_{p^*} + \bar{c}_{p^*} x \tag{42}$$

Since the exact matching hypothesis implies that e_{p^*} goes to zero as fast as $e^{-\lambda_E t}$, it must be true that the set

$$\mathcal{P}^* \triangleq \left\{ p : \int_0^\infty ||e_p||^2 dt < \infty, \quad p \in \mathcal{P} \right\} \tag{43}$$

is nonempty. The assumption that $\mathcal{P}$ is a finite set can be used to prove that there must be a finite time t^* beyond which σ takes values only in $\mathcal{P}^*$ [24].

Let $\{c_{p_1 p^*}, c_{p_2 p^*}, \ldots, c_{p_m p^*}\}$ be a basis for the span of $\{c_{pp^*} : p \in \mathcal{P}^*\}$. Define $C = [c'_{p_1 p^*} \ c'_{p_2 p^*} \ \cdots \ c'_{p_m p^*}]'$ and

$$\bar{e} = Cx \tag{44}$$

These definitions together with (40) imply that $e_{p_i} - e_{p^*}$ is the ith entry of $\bar{e}$. Since each such entry has a finite $\mathcal{L}^2[0, \infty)$ norm, $\bar{e}$ must have a finite $\mathcal{L}^2[0, \infty)$ norm as well. Note also that the definition of C implies that there must be a bounded function $s : \mathcal{P}^* \to \mathbb{R}^{m \times 1}$ for which $s(p)C = c_{pp^*}$, $p \in \mathcal{P}^*$. In view of this and the previously noted detectability of the matrix pairs $(c_{pl}, A_l + bf_{pl})$, $p, l \in \mathcal{P}$, it must be that the matrix pair $(C, A_{p^*} + bf_{pp^*})$ is detectable for each $p \in \mathcal{P}^*$.

Note that for any appropriately sized, matrix $p \longmapsto K_p$ which is bounded on $\mathcal{P}^*$, (39) can be rewritten as

$$\dot{x} = (A_{p^*} + bf_{\sigma p^*} + K_\sigma C)x - K_\sigma \bar{e} + d_{p^*} e_{p^*}$$

for $t \geq t^*$. Suppose that such a function K_p can be shown to exist for which the time-varying matrix $A_{p^*} + bf_{\sigma p^*} + K_\sigma C$ is exponentially stable. Then because $\bar{e}$ and e_{p^*} have finite $\mathcal{L}^2[0, \infty)$ norms, x would tend to zero. Hence x_C and x_E would tend to finite limits because of (38). Moreover since e_{p^*} tends to zero, (41) and (42) would imply that v and e_T tend to zero as well. As a consequence, y would tend to r and u would tend to finite limit; the latter would be true because of the converging of y and v to constant values and because Σ_P's transfer function is nonzero at $s = 0$. In other words, to show that $y \to r$ and that x_C, x_E and u tend to finite limits its enough to show that $A_{p^*} + bf_{\sigma p^*} + K_\sigma C$ is exponentially stable for some suitably defined function K_p.

We claim that a function K_p exists provided τ_D is sufficiently large. To understand why this is so, first recall that $(C, A_{p^*} + bf_{pp^*})$ is detectable for each $p \in \mathcal{P}^*$. Thus for each such p there must be a constant matrix K_p which stabilizes $A_{p^*} + bf_{pp^*} + K_p C$. Therefore for each $p \in \mathcal{P}^*$ it is possible to find numbers $a_p \geq 0$ and $\lambda_p > 0$ for which

$$\left| e^{(A_{p^*} + bf_{pp^*} + K_p C)t} \right| \leq e^{(a_p - \lambda_p t)} \quad t \geq 0$$

Since $\frac{a_p}{\lambda_p}$ is an upper bound on the time it takes for $\left| e^{(A_{p^*} + bf_{pp^*} + K_p C)t} \right|$ to drop below one in value, it is not surprising that the state transition matrix of $A_{p^*} + bf_{\sigma p^*} + K_\sigma C$ will be exponentially stable provided

$$\tau_D > \sup_{p \in \mathcal{P}^*} \left\{ \frac{a_p}{\lambda_p} \right\}$$

This in fact can be shown to be true [24]. Thus we may conclude that if τ_D is chosen large enough, then u, x_C and x_E must converge to finite limits and and y must tend to r.

Performance Signals: One of the problems with the preceding is that W will not remain bounded if $r \neq 0$. One easy way to remedy this problem is as follows.

Under the exact matching hypothesis, $e_{p^*} \to 0$ as fast as $e^{-\lambda_E t}$. Thus there is a non-negative constant C_0 such that $e_{p^*}^2(t) \leq C_0 e^{-2\lambda_E t}$. Pick $\lambda \in (0, \lambda_E)$. Let Π and π_p be defined as in (24) and (26) respectively, but rather than using (25) to generate W, use the equation

$$\dot{W} = e^{2\lambda t} \begin{bmatrix} x_E \\ y \end{bmatrix} \begin{bmatrix} x_E \\ y \end{bmatrix}' \qquad (45)$$

instead. Clearly

$$\dot{\pi}_p = e^{2\lambda t} e_p^2$$

As defined, π_p has three crucial properties:

1. For each $p \in \mathcal{P}$, π_p is monotone nondecreasing.
2. $\lim_{t \to \infty} \pi_{p^*} \triangleq C^* \leq \pi_{p^*}(0) + \int_0^\infty C_0 e^{-2(\lambda_E - \lambda)t} dt < \infty$
3. If $\mathcal{P}^*$ is defined as before, then σ must take values only within $\mathcal{P}^*$ beyond some finite time.

These are precisely the properties needed to define C and $\bar{e}$ as in (44) so that $\bar{e}$ has a finite $\mathcal{L}^2[0, \infty)$ norm and that $(C, A_{p^*} + bf_{pp^*})$ is detectable for each $p \in \mathcal{P}^*$. In other words, if one were to use (45) to generate W, then the convergence properties of y, x_E, x_C and u would still hold.

Now consider replacing W with the "scaled" weighting matrix

$$\bar{W} \triangleq e^{-2\lambda t} W \tag{46}$$

Note that $\Pi(\bar{W}, p) = e^{-2\lambda t} \Pi(W, p)$, $p \in \mathcal{P}$. In the light of the scale independence property of Σ_D noted previously, it must be that replacing W with $\bar{W}$ has no effect on σ and consequently on y, x_E, x_C and u. The key point here is that the weighting matrix $\bar{W}$ defined by (46) can also be generated directly by the stable dynamical system

$$\dot{\bar{W}} = -2\lambda \bar{W} + \begin{bmatrix} x_E \\ y \end{bmatrix} \begin{bmatrix} x_E \\ y \end{bmatrix}' \tag{47}$$

Moreover, since y and x_E, tend to finite limits, it must be that $\bar{W}$ {and therefore its sampled state $\bar{X}$} tend to finite limits as well. Thus at this point we may conclude that if τ_D is chosen large enough, if λ is picked in $(0, \lambda_E)$, and if W is generated by (25), then u, x_C, x_E, and W must converge to to finite limits and and y must tend to r.

Fast Switching: A key step in the analysis just given was to show that for the family of detectable pairs $\{(C, A_{p^*} + bf_{pp^*}) : p \in \mathcal{P}^*\}$, there exists a a bounded, output injection function K_p and a dwell time τ_D for which $A_{p^*} + bf_{\sigma p^*} + K_\sigma C$ is exponentially stable for any admissible switching function σ with dwell time no smaller than τ_D. It turns out that for *any* given positive dwell time τ_D, it is possible to find a function K_p which exponentially stabilizes $A_{p^*} + bf_{\sigma p^*} + K_\sigma C$ for any admissible switching function σ with dwell time no smaller than τ_D [24].

To reader should realize that detectability of such matrix pairs is by itself *not* sufficient for the existence of a function K_p with the aforementioned property. To understand why, just consider the situation in which a family of detectable pairs of the form $\{(C, A_p) : p \in \mathcal{P}\}$ has a zero readout matrix C; in this case each A_p must be a stability matrix and $A_p + K_p C = A_p$ for all K_p. It is well known that if the A_p do not commute with each other, exponential stability of A_σ cannot in general be assured unless τ_D is large enough; for an example see [44]. In other words, there are families of detectable pairs of the form $\{(C, A_p) : p \in \mathcal{P}\}$ for which no stabilizing function K_p exists if τ_D is too small. What's especially interesting is that if $\{(C, A_p) : p \in \mathcal{P}\}$ is a family of *observable* matrix pairs, then no matter how small τ_D is, there does in fact exist a matrix function K_p

4 Self-Adjusting Control

with the required stabilizing property. This is an immediate consequence of the following result [14].

Squashing Lemma: *Let (C, A) be a fixed, constant, observable matrix pair, and let τ_0 be a positive number. For each positive number δ there exists a positive number λ and a constant output-injection matrix K for which*

$$|e^{(A+KC)t}| \leq \delta e^{-\lambda(t-\tau_0)}, \quad t \geq 0 \tag{48}$$

The way to construct K_p for a family of observable pairs such as $\{(C, A_p) : p \in \mathcal{P}\}$, is as follows. Pick $\delta \in (0,1)$, set $\tau_0 = \tau_D$ and for each $p \in \mathcal{P}$ use the Squashing Lemma to find a value of K_p for which

$$|e^{(A_p+K_pC)t}| \leq \delta e^{-\lambda(t-\tau)}, \quad t \geq 0$$

It can be shown that with K_p so chosen, $A_\sigma + K_\sigma C$ will be exponentially stable if σ is any admissible switching signal with dwell time no smaller than τ_0 [24].

Unfortunately, for the problems of interest in this paper, the matrix pairs in $\{(C, A_{p^*} + bf_{pp^*}) : p \in \mathcal{P}^*\}$ cannot be assumed to be observable without a definite loss of generality. On the other hand, observability is in general sufficient for stabilizability whereas detectability is not. The way out of this dilemma has been to make use of additional properties of the matrices under consideration. A typical result along these lines is the following.

Switching Theorem: *Let $\lambda_0 > 0$ and $\tau_0 > 0$ be fixed. Let $(C_{q_0 \times n}, A_{n \times n}, B_{n \times m})$ be a left invertible system. Suppose that $\{(C_p, F_p) : p \in \mathcal{P}\}$ is a closed, bounded subset of matrix pairs in $\mathbb{R}^{q \times n} \oplus \mathbb{R}^{m \times n}$ with the property that for each $p \in \mathcal{P}$, $(C_p, \lambda_0 I + A + BF_p)$ is detectable. There exist a constant $a \geq 0$ and bounded, matrix-valued output injection functions $p \longmapsto H_p$ and $p \longmapsto K_p$ on $\mathcal{P}$ which, for any admissible switching signal $\sigma : [0, \infty) \to \mathcal{P}$ with dwell time no smaller than τ_0, causes the state transition matrix of*

$$A + K_\sigma C_\sigma + H_\sigma C + BF_\sigma$$

to satisfies

$$|\Phi(t, \mu)| \leq e^{(a-\lambda_0(t-\mu))}, \quad t \geq \mu \geq 0$$

Using this theorem it has been possible prove that for any dwell time greater than zero and any value of $\lambda \in (0, \lambda_E)$, the supervisory control system we've been discussing achieves set-point regulation and global boundedness [24]. It has also been possible to show that these results continue to hold in the face of norm bounded unmodelled dynamics provided λ is further constrained to be smaller than both the stability margin λ_S and the unmodelled dynamics stability margin λ_u [43]. Moreover the introduction of $\mathcal{L}^\infty$ bounded noise and disturbance inputs cannot destabilize the system.

4.4 Cyclic Switching

As we have just explained, estimator-based supervisors generate control signals in accordance with the idea of certainty equivalence; i.e., at each instant of time the controller in feedback with the process is based on a current estimate of what the nominal process model transfer function is; such estimates are selected from a suitably defined admissible nominal process model transfer function set $\mathcal{N}$. Because $\mathcal{N}$ must be finitely parameterized, it can always be regarded as a subset of a finite dimensional linear space. In practice, $\mathcal{N}$ is typically chosen to best satisfy a number of conflicting requirements. For example, $\mathcal{N}$ should be "big" enough to ensure that $\mathcal{C}_P$ includes a transfer function model of the process. If $\mathcal{N}$ contains a continuum of transfer functions, then for on-line model estimation $\{$i.e., minimization of $\Pi(W,p)$ $\}$ to be tractable, $\mathcal{N}$ should be convex or at least the union of a finite number of convex sets. Since each transfer function in $\mathcal{N}$ is a candidate process model transfer function, for the formulated problem to make sense, each such transfer function should be at least stabilizable $\{$i.e., without any unstable poles and zeros in common$\}$.

It is not very difficult to see that these are conflicting requirements. In particular, stabilizability, convexity and largeness of $\mathcal{N}$ are at odds. If stabilizability and largeness are required, then convexity and consequently tractability must be sacrificed. If convexity and stabilizability are required, then $\mathcal{N}$ must be "small."

A way out of this dilemma, which enables one to achieve tractability while retaining stabilizability and largeness, is to embed $\mathcal{N}$ in a larger set of 'admissible' transfer functions $\bar{\mathcal{N}}$ which is convex, but which is not restricted to have only stabilizable transfer functions. Naturally those transfer functions in $\bar{\mathcal{N}}$ which are not stabilizable cannot be candidate process model transfer functions. Nevertheless, because of the tractability issue it is useful to consider such transfer functions to be *admissible for estimation purposes*. Therefore an alternative to certainty equivalence is needed for selecting controllers when such transfer functions are encountered during the on-line estimation process. Such an alternative, based on the concept of "cyclic switching," has recently been proposed for applications in parameter-adaptive control where the same problem also arises [45, 14]. The aim of this section is to explain what cyclic switching is within the context of the set-point problem we've been considering.

We will be concerned exclusively with the case when $\mathcal{N}$ contains a continuum of reduced transfer functions. For simplicity assume that each such transfer function has the same McMillan Degree n. This means that $\mathcal{N}$ can be viewed as a subset of the $2n$-dimensional linear space of strictly proper $\{$unreduced$\}$ rational functions whose denominators are monic and of degree n.

As before we assume that $\mathcal{P}$ is a closed, bounded subset of a finite dimensional linear space. Assume in addition that the coefficients of ν_p are defined on this space as affine linear functions. Assuming

$$\left\{ \begin{bmatrix} A_E & 0 \\ 0 & A_E \end{bmatrix} + \begin{bmatrix} b_E \\ 0 \end{bmatrix} c_p, \begin{bmatrix} 0 \\ b_E \end{bmatrix}, c_p \right\}$$

again realizes ν_p, this means that c_p will also be an affine linear function. As a consequence, the parameterized performance signal $\Pi(W,p)$ defined by

$$\Pi(W,p) = [\,c_p \quad -1\,]\, W\, [\,c_p \quad -1\,]' \qquad (49)$$

$$\dot{W} = -2\lambda W + \begin{bmatrix} x_E \\ y \end{bmatrix} \begin{bmatrix} x_E \\ y \end{bmatrix}' \qquad (50)$$

will be a quadratic function of p.

We are interested in the case when $\mathcal{P}$ is not necessarily convex since convexity of $\mathcal{P}$ would imply convexity of $\mathcal{N}$. To ensure a tractable minimization problem, we presume that $\mathcal{P}$ has been embedded in a conveniently chosen, closed, bounded convex subset $\bar{\mathcal{P}}$ {e.g., the convex hull of $\mathcal{P}$} and that the set of admissible nominal transfer functions has been enlarged to $\bar{\mathcal{N}} \triangleq \{\nu_p : p \in \bar{\mathcal{P}}\}$. This reduces the problem of minimizing $\Pi(W,p)$ over $\bar{\mathcal{P}}$ to a finite dimensional convex, quadratic programming problem. Such problems are highly tractable and many fast algorithms for solving them are known.

We shall assume that all of the points $p \in \bar{\mathcal{P}}$ {if any} at which $\frac{1}{s}\nu_p$ has a pole-zero cancellation are in the interior of a specified closed set $\mathcal{S} \subset \bar{\mathcal{P}}$, called a *singular region*. {Therefore $\frac{1}{s}\nu_p$ can't have any pole-zero cancellations on the closure of $\bar{\mathcal{P}} - \mathcal{S}$.} It is reasonable to require $\mathcal{C}_P$ and $\{\nu_p : p \in \mathcal{S}\}$ to be disjoint.

In the sequel we will define a generalized supervisor whose decision making strategy takes into account the possibility that there may be times at which the best possible admissible transfer function estimate, determined by minimizing $\Pi(W,p)$ over $\bar{\mathcal{P}}$, falls within the singular set $\{\nu_p : p \in \mathcal{S}\}$. To define such a supervisor two things are needed:

Controller Requirements:

1. A bounded set of controller transfer functions $\{\kappa_q : q \in (\bar{\mathcal{P}} - \mathcal{S})\}$ which satisfies the *Stability Margin Requirement* on $\bar{\mathcal{P}} - \mathcal{S}$; i.e., for each $p \in (\bar{\mathcal{P}} - \mathcal{S})$ the real parts of the closed-loop poles of the feedback interconnection shown in Figure 13 are less than $-\lambda_S$.

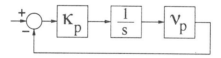

Fig. 13. Feedback Interconnection

Since $\frac{1}{s}\nu_p$ has no pole-zero cancellations on the closure of $\bar{\mathcal{P}} - \mathcal{S}$, such a family clearly exists.

2. A set of real gains $\{g_1, g_2, \ldots, g_{n_S}\}$ which fulfills the *Observation Requirement*; i.e., for each $l \in \mathcal{P}$ and each $p \in \mathcal{S}$, there is a value of $q \in \{1, 2, \ldots, n_S\}$ for which the feedforward interconnection of controllable, observable realizations of ν_p and ν_l shown in Figure 14 is observable through e_{ff}. It can be shown that such a family exists because of the assumed disjointness of $\mathcal{N}$ and $\mathcal{S}$ [14].

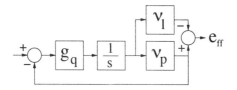

Fig. 14. Feedforward Interconnection

Assume that $\{\kappa_q : q \in (\bar{\mathcal{P}} - \mathcal{S})\}$ is a bounded set of controller transfer functions which satisfies the Stability Margin Requirement and that $\{\kappa_1, \kappa_2, \ldots, \kappa_{n_S}\}$ is a finite family of gains which satisfies the Observation Requirement. In addition, adopt the notation $\mathcal{I} \triangleq \{1, 2, \ldots, n_S\}$ and write $\mathcal{Q}$ for the disjoint union $\mathcal{Q} \triangleq (\bar{\mathcal{P}} - \mathcal{S}) \cup \mathcal{I}$. Suppose $\Sigma_C(q)$ is a globally detectable/stabilizable realization of κ_q on $\mathcal{Q}$.

The overall structure of the supervisory control system we want to consider is the same as before, except that now instead of Σ_D, the supervisor uses a yet-to-be-defined "dwell-time/cyclic logic" Σ_{DC}.

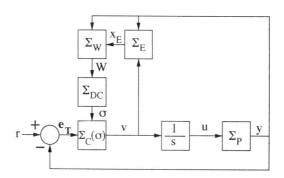

Fig. 15. Supervisory Control System Using Dwell-Time/Cyclic Switching

Σ_{DC} is essentially a combined version of Σ_D and the cyclic switching logic of [14]. The underlying strategy upon which Σ_{DC}'s logic is predicated can be explained roughly as follows. Consider again the by now familiar equations

$$\begin{bmatrix} x_E \\ x_C \end{bmatrix} = x + hr \tag{51}$$

$$\dot{x} = (A_l + bf_{\sigma l})x + d_l e_l \tag{52}$$

$$e_p = c_{pl}x + e_l \quad p \in \bar{\mathcal{P}} \tag{53}$$

$$v = f_{\sigma l}x + g_\sigma e_l \tag{54}$$

$$\mathbf{e_T} = e_l + \bar{c}_l x \tag{55}$$

4 Self-Adjusting Control

which hold for all constant r and all $l, p \in \mathcal{P}$. Σ_{DC}'s strategy stems from two facts, each a direct consequence of one of the two corresponding **Controller Requirement** stipulated above:

1. For each $l \in \mathcal{P}$ and each $p \in \bar{\mathcal{P}} - \mathcal{S}$ the matrix pair $(c_{pl}, A_l + bf_{ql})|_{q=p}$ is detectable.
2. For each $l \in \mathcal{P}$ and each $p \in \mathcal{S}$ there exists a $q \in \mathcal{I}$ such that $(c_{pl}, A_l + bf_{ql})$ is detectable.

Intuition: Here roughly is the idea upon which cyclic switching is based. Think of the supervisor as performing two separate tasks - one estimation and the other controller selection. The estimation task amounts to minimizing $\Pi(W, p)$ over $\bar{\mathcal{P}}$ and goes on over and over without interruption; this generates a sequence of values $\hat{p} \in \mathcal{P}$. Meanwhile the supervisor tries to select controller's in such as way as to maintain "detectability" through $e_{\hat{p}}$, at least on the average. Why? Because detectability through $e_{\hat{p}}$ implies smallness of x whenever $e_{\hat{p}}$ is small - and smallness of $e_{\hat{p}}$ ought to be a consequence of the estimation process. So here's how the supervisor achieves "detectability" through $e_{\hat{p}}$: If $\hat{p} \in \bar{\mathcal{P}} - \mathcal{S}$, the supervisor relies on property 1 above and certainty equivalence: detectability is achieved by setting $\sigma = \hat{p}$. On the other hand, if $\hat{p}$ enters $\mathcal{S}$, the supervisor relies on property 2: in this case "detectability" is achieved on the average by stepping σ through each of the values $\mathcal{I}$, holding fixed on each such value for a prespecified amount of time.

Formally a *Dwell-Time/Cyclic Switching Logic* Σ_{DC} is a hybrid dynamical system whose input and output are W and σ respectively, and whose state is the ordered quintuple $\{X, \hat{p}, \tau, \beta, \sigma\}$. X is a discrete-time matrix which takes on sampled values of W, $\hat{p}$ is a discrete-time variable taking values in $\bar{\mathcal{P}}$, τ is a continuous-time timing signal as before, and β is a logic variable taking values in $\{0, 1\}$. τ takes values in the closed interval $[0, \max\{n_S\tau_S, \tau_D\}]$, where τ_D and τ_S are a prespecified positive numbers called a *dwell time* and a *cycle dwell time* respectively. As before $\tau_C \leq \max\{n_S\tau_S, \tau_D\}$ is a prespecified computation time which bounds from above for any $X \in \mathcal{W}$, the time it would take the supervisor to compute a value $p = p_X \in \mathcal{P}$ which minimizes $\Pi(X, p)$. Between "event times" τ is generated by a reset integrator according to the rule $\dot{\tau} = 1$. Such event times occur for $\beta \in \{0, 1\}$, when the value of τ reaches either $T(\beta) - \tau_C$ or $T(\beta)$ where $T(0) \triangleq \tau_D$ and $T(1) \triangleq n_S\tau_S$; at such times τ is reset to either 0 or $T(\beta) - \tau_C$ depending on the value of Σ_{DC}'s state. Σ_{DC}'s internal logic is defined by the computer diagram shown in Figure 16 where p_X denotes a value of $p \in \mathcal{P}$ which minimizes $\Pi(X, p)$.

The functioning of Σ_{DC} can be explained as follows. Suppose that at some time t_0, Σ_S has just changed the value of $\hat{p}$. Depending on whether $\hat{p} \in \mathcal{S}$ or not, one of two different epochs can occur:

– Suppose $\hat{p} \notin \mathcal{S}$. In this case β is set equal to 0, σ is set equal to $\hat{p}$ and τ is reset to 0. After $\tau_D - \tau_C$ time units have elapsed, W is sampled and X

is set equal to this value. During the next τ_C time units, a value $p = p_X$ is computed which minimizes $\Pi(X, p)$. At the end of this period, when $\tau = \tau_D$, if $\Pi(X, p_X)$ is smaller than $\Pi(X, \widehat{p})$, then $\widehat{p}$ is set equal to p_X and the logic goes back to again test whether or not $\widehat{p} \in \mathcal{S}$. If, on the other hand, $\Pi(X, \widehat{p})$ is less than or equal to $\Pi(X, p_X)$, τ is reset to $\tau_D - \tau_C$, W is again sampled, X takes on this new sampled value, minimization is again carried out over the next τ_C time units..... and so on.
- Suppose $\widehat{p} \in \mathcal{S}$. In this case β is set equal to 1, τ is reset to 0, and two distinct sequences of events occur simultaneously, each lasting $n_S \tau_S$ time units:

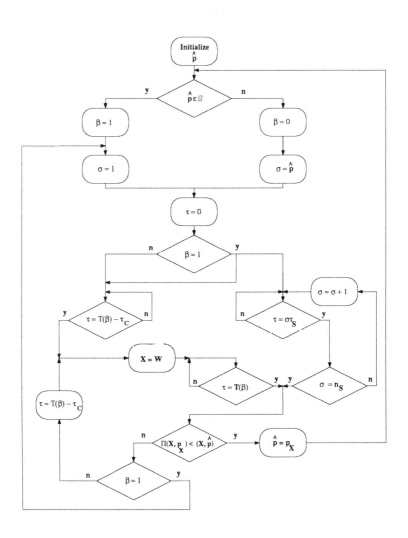

Fig. 16. Dwell-Time/Cyclic Switching Logic Σ_{DC}

1. At $\tau = 0$, a switching cycle is executed[6].
2. At $\tau = n_S\tau_S - \tau_C$, W is sampled and X is set equal to this value. During the next τ_C time units, a value $p = p_X$ is computed which minimizes $\Pi(X,p)$. At the end of this period, when $\tau = n_S\tau_S$, if $\Pi(X,p_X)$ is smaller than $\Pi(X,\widehat{p})$, then $\widehat{p}$ is set equal to p_X and the logic goes back to again test whether or not $\widehat{p} \in \mathcal{S}$. If, on the other hand, $\Pi(X,\widehat{p})$ is less than or equal to $\Pi(X,p_X)$, τ is reset to 0, another switching cycle is executed ....and so on.

An analysis of the supervisory control system just described can be found in [46]. For analyses in more traditional adaptive control contexts see [14, 22, 47]. The techniques used [14, 22] are similar to those outlined in the last section. The discrete time case is completely analyzed in [47].

5 Switched Linear Systems

Existing results concerned with the types of switched systems we've been discussing deal mainly with questions of stability, global boundedness and convergence. Interesting as they may be, these results are in many ways less than one might hope for. Especially lacking we think, are results of a more *quantitative* nature. What are needed are good norm bound estimates for allowable unmodelled dynamics. Also needed is a clearer understanding of the relationships between these estimates and design parameters.

Resolution of issues such as these calls for a better understanding of the basic properties of switched systems than we have at present. Needed is a catalog of basic results analogous to those for non-switched linear systems. In the sequel we briefly discuss some of the technical questions suggested by types of problems we've been discussing.

5.1 Stability of Switched Linear Systems

Let $\mathcal{P}$ be either a finite set or a closed, bounded subset of a finite dimensional linear space and let $\mathcal{A} = \{A_p : p \in \mathcal{P}\}$ be a closed, bounded set of real $n \times n$ matrices. Within this context one can formulate a number of different stability problems. For example, one can seek to find a switching logic Σ_S, with input x and piecewise-constant output σ which uniformly exponentially stabilizes

$$\dot{x} = A_\sigma x \tag{56}$$

in the sense that there are positive constants a and λ such that all solutions x to (56) {in closed-loop with Σ_S} exist and are norm bounded in time by $ae^{-\lambda t}$

---
[6] The supervisor *executes* a switching cycle at clock time $\tau = 0$ by setting $\sigma(t_0 + \tau) = s(\tau)$, $\tau \in [0, n_S\tau_S)$ where t_0 is the actual time τ was reset to 0 and $s : [0, n_S\tau_S) \to \mathcal{I}$ is the piecewise-constant function whose value is i on the subinterval $[(i-1)\tau_S, i\tau_S)$, $i \in \mathcal{I}$.

times the initial normed value of x. The synthesis of such a state driven switching logic seems to be very challenging. See [48] for a discussion of some recent results along these lines.

A somewhat less ambitious goal would be as follows: For a given set $\mathcal{A} = \{A_p : p \in \mathcal{P}\}$ and a given class of switching signals $\mathcal{S}$, find conditions under which there exist positive numbers λ and a such that for each $\sigma \in \mathcal{S}$ the state transition matrix of A_σ satisfies

$$|\Phi(t,\tau)| \leq ae^{-\lambda(t-\tau)}, \quad t \geq \tau \geq 0 \tag{57}$$

Interesting choices for $\mathcal{S}$ would include

1. $\mathcal{S}_1 \triangleq$ all piecewise constant switching signals
2. $\mathcal{S}_2 \triangleq$ all piecewise constant switching signals with dwell times no less than some positive number τ_D.
3. $\mathcal{S}_3 \triangleq$ the type of switching signals generated by Σ_{DC} {cf., [46]}.

Here are some easy to derive sufficient conditions under which there exist a and λ such that (57) holds for every piecewise constant $\sigma \in \mathcal{S}_1$

1. There exists a norm $||\cdot||$ on $\mathbb{R}^{n \times n}$ and a positive number λ such that

$$||e^{A_p t}|| \leq e^{-\lambda t}, \quad t \geq 0, \, p \in \mathcal{P}$$

2. The elements of $\mathcal{A}$ share a common quadratic Lyapunov function; i.e., there exists a positive definite matrix Q such that

$$QA_p + A_p'Q < 0, \quad \forall p \in \mathcal{P}$$

3. $\mathcal{P}$ {and therefore $\mathcal{A}$} are finite sets and $A_p A_q = A_q A_p$ for all $p, q \in \mathcal{P}$.
4. The matrices in $\mathcal{A}$ are row diagonally dominant with negative diagonal elements ; i.e., for $A = [a_{ij}] \in \mathcal{A}$,

$$2a_{ii} + \sum_{j=1}^{n} |a_{ij}| < 0, \, i \in \{1, 2, \ldots n\}$$

5. There exists an integer $\bar{n} \geq n$, a full rank $\bar{n} \times n$ matrix V and a family of $\bar{n} \times \bar{n}$ matrices $\bar{\mathcal{A}} = \{\bar{A}_p : p \in \mathcal{P}\}$ such that
   (a) $\bar{A}_p V = V A_p, \, p \in \mathcal{P}$
   (b) the matrix $\bar{A}_\sigma$ is exponentially stable for each $\sigma \in \mathcal{S}_1$

The sufficiency of conditions 1 and 2 are more or less obvious; the same is true of condition 3 since it implies that the associated matrix exponentials of the A_p commute. An explicit construction is given in [49] of a matrix Q satisfying condition 2 for a set of matrices $\mathcal{A}$ satisfying condition 3. Recently condition 1 was shown to be necessary for (57) to hold for all $\sigma \in \mathcal{S}_1$ [50]; of course one is still faced with the problem of deciding when such a norm exists. Condition 4 can be easily established using an simple estimate based on the idea of a matrix measure {c.f., [51], p. 47}. Condition 5 is a consequence of the fact that for any solution x to $\dot{x} = A_\sigma x$, $\bar{x} \stackrel{\Delta}{=} Vx$ is a solution to $\dot{\bar{x}} = \bar{A}_\sigma \bar{x}$

Note that conditions 4 and 5 imply that the following is sufficient for (57) to hold for all $\sigma \in \mathcal{S}_1$

Property: *There exists an integer $\bar{n} \geq n$, a full rank $\bar{n} \times n$ matrix V and a family of $\bar{n} \times \bar{n}$ matrices $\bar{\mathcal{A}} = \{\bar{A}_p : p \in \mathcal{P}\}$ such that*

1. $\bar{A}_p V = V A_p, \quad p \in \mathcal{P}$
2. *each matrix $\bar{A}_\sigma$ is row diagonally dominant with negative diagonal elements.*

Under certain conditions this property proves to be necessary as well [52]; of course one still needs to figure out when V and the $\bar{A}_p$ exist. For some results concerning the stability of switched *nonlinear* systems, see [44].

The kind of stability questions we've been discussing are very closely related to the problem of deciding when for a given class of matrices $\{M_p : p \in \mathcal{P}\}$ and a given class $\bar{\mathcal{S}}$ of infinite sequences mapping the nonnegative integers into $\mathcal{P}$,

$$\lim_{i \to \infty} \prod_{j=1}^{i} M_{\bar{\sigma}(j)} = 0, \quad \forall \bar{\sigma} \in \bar{\mathcal{S}}$$

This and related questions have been addressed in [53] and [54].

5.2 Other Questions

Here are several other questions involving switched linear systems:

- Given a family of switching signals $\mathcal{S}$ and a family of detectable pairs $\{(C_p, A_p) : p \in \mathcal{P}\}$ when does there exist a matrix function $p \longmapsto K_p$ on $\mathcal{P}$ for which $A_\sigma + K_\sigma C_\sigma$ is 'exponentially stable' for each $\sigma \in \mathcal{S}$?
- With reference to the preceding suppose $\mathcal{K}$ is a nonempty class of matrix functions $p \longmapsto K_p$ on $\mathcal{P}$ with the property that for each function $K_p \in \mathcal{K}$, $A_\sigma + K_\sigma C_\sigma$ is exponentially stable for each $\sigma \in \mathcal{S}$. For a given class of appropriately sized matrices $\{F_p : p \in \mathcal{P}\}$ compute {or at least tightly estimate} the sup over $\mathcal{S}$ of the inf over $\mathcal{K}$ of the induced $\mathcal{L}^2[0, \infty)$ norm of the linear operator

$$y \longmapsto \int_0^t F_{\sigma(t)} \phi(t,s) K_{\sigma(s)} y(s) ds$$

  where ϕ is the state transition matrix of $A_\sigma + K_\sigma C_\sigma$.
- Given a class of linear systems $\{(A_p, B_p, C_p, D_p) : p \in \mathcal{P}\}$, each with property **P**, and a class of switching signals $\mathcal{S}$, when is it true that for each $\sigma \in \mathcal{S}$ the switched system $(A_\sigma, B_\sigma, C_\sigma, D_\sigma)$ also has property **P**? Interesting choices for **P** include *stability, stabilizability, passivity*. Some findings related to the last of these appear in [55].

6 Concluding Remarks

Switching logics such as those discussed in this paper are typically derived without appealing to any formal notion of state or state transition. Explaining such logics informally can be beneficial and we have sought to do this throughout the paper. On the other hand we've found that it is also worthwhile to go through the exercise of formally modeling such logics as hybrid systems with inputs, states, state transitions and outputs. Formal modeling can clarify an algorithm's functioning by reducing ambiguity and in so doing can help to preclude erroneous conclusions. For example, when cyclic switching was first devised without the benefit of a formal model, it was thought to be a time-varying system which it clearly is not. The formal modeling of the hysteresis switching logic in [12] made it easier to explain. There are no doubt many existing switching logics whose behaviors could more easily grasped if they were both explained informally and modelled formally as hybrid dynamical systems.

In §2 it was emphasized that multi-controller architectures can usually be efficiently implemented as state-shared parameter dependent controllers. There are of course situations when it is advantageous *not* to explicitly compute off-line the parameter-dependent coefficient matrices of such a controllers. This is especially true if the controller in question is of the certainty equivalence type and if the associated family of nominal process models is a continuum. For example, suppose that the set $\mathcal{N}$ of nominal transfer functions considered in §4.3 is a continuum; suppose in addition that for each $p \in \mathcal{P}$, controller transfer function κ_p is to be designed via LQ-theory applied to ν_p. Because solutions to matrix Riccati equations depend on the equation's coefficient matrices in a complicated {nonrational} manner, the dependence of κ_p on p will be at least as complex, even if ν_p depends linearly on p. What this means is that the problem of explicitly parameterizing $\mathcal{K}$ assuming an LQ-based controller design is hopelessly intractable. For MIMO process models the problem is far worse even for simple pole-placement designs. There are at least two ways to avoid this problem. One is to compute controller coefficient matrices in real time; this is feasible with a dwell-time switched supervisory control system provided the computations can be carried out quickly enough.

Another way to avoid the parameterization problem is to settle for a smaller nominal process model transfer function class $\bar{\mathcal{N}}$ containing only finitely many elements. In fact a strong case can be made for doing this not just to avoid the parameterization problem, but for other reasons as well. For example, finiteness of $\bar{\mathcal{N}}$'s parameter space $\bar{\mathcal{P}}$ can greatly simplify the required minimization $\Pi(W, p)$ over $\bar{\mathcal{P}}$ even if $\bar{\mathcal{P}}$ is very large. We refer the reader to [56] for an interesting discussion of how one would go about covering a process model transfer function class of the form

$$\mathcal{C}_P = \bigcup_{p \in \mathcal{P}} \{\nu_p + \delta : ||\delta||_\infty \leq \epsilon_p\}$$

assuming $\mathcal{N} \triangleq \{\nu_p : \mathcal{P}\}$ is a compact continuum with a transfer function class

6 Concluding Remarks

$\bar{\mathcal{C}}_P \supset \mathcal{C}_P$ of the form

$$\bar{\mathcal{C}}_P = \bigcup_{p \in \bar{\mathcal{P}}} \{\nu_p + \delta : ||\delta||_\infty \leq \bar{\epsilon}_p\}$$

where $\bar{\mathcal{P}}$ is a finite subset of $\mathcal{P}$ and $\epsilon_p \leq \bar{\epsilon}_p$, $p \in \bar{\mathcal{P}}$.

One of the underlying ideas exploited in section 4 is that modeling uncertainty can be dealt with by switching between the members of a family of *fixed gain* controllers. The idea has been around for a long time. For example there is an extensive literature on the "multiple-model" {i.e., multi-estimator} approach to uncertainty which goes back almost thirty years; see for example [57] and the many references therein. The key feature of the estimator-based approach discussed in §4.3 which distinguishes it from the classical multiple-model approach is that switching is orchestrated by a supervisor using a logic which selects controllers on the basis of normed output estimation errors. Surprisingly, a provably correct version of this simple idea does not seem to have found its way into the literature until quite recently [15] - this despite the fact that the idea is a natural extension of the original concept of hysteresis switching [11] which appeared in 1988.

For the case when $\mathcal{P}$ contains a continuum of points, the dwell-time switched, estimator-based supervisory control discussed in §4.3 can be thought of as a form of estimator-based parameter adaptive control in which the supervisor plays the role of parameter tuner [58]. In this context, the concept of a supervisor represents a significant departure from more traditional estimator-based tuning algorithms which typically employ recursive or dynamical parameter tuning. Most closely related to what we've been discussing seems to be the type of adaptive algorithm studied by Naik, Kumar and Ydstie in [59]. Both the NKY algorithm and the dwell-time switched supervisor discussed in §4.3 search on compact parameter spaces; both are inherently robust to unmodelled dynamics in that dynamic normalization [60] is not employed. Perhaps the most significant differences between the two are 1. that the NKY algorithm employs recursive parameter tuning {i.e., pseudo-gradient/projection search} whereas the dwell-time switched supervisor does not and 2. the dwell-time switched supervisor allows time for computation whereas the NKY algorithm does not.

It is reasonable to suspect that many of the ideas covered in §4.3 can be successfully applied to specific classes of nonlinear systems. The well-known obstacles to generalization imposed by a limited nonlinear observer theory can almost certainly be side-stepped by focusing attention on systems whose states can be measured. It is quite likely that a supervised family consisting of a finite number of fixed nonlinear controllers will prove easier to analyze than a continuously parameterized family of controllers under the control of a parameter tuner. Understanding such switched systems calls for new methods of analysis which go beyond the partial Lyapunov function approach commonly used in the study of conventional parameter adaptive systems.

Simulations

The general concept of a dwell-time switched, estimator-based supervisory control system described in this paper has been tested in simulation under a variety of conditions. The reader wishing to experiment with these simulations can obtain Matlab Simulink files via the internet at the address

http://www.cis.yale.edu/~wchang/workshop.html

The simulations were designed and implemented by Wen-Chung Chang and João Hespanha.

References

1. V. I. Utkin. *Sliding Modes in Control Optimization.* Springer-Verlag, 1992.
2. M. I. Zelikin and V. F. Borisov. *Theory of Chattering Control.* Birkhauser, 1994.
3. G. F. Franklin, J. D. Powell, and M. L. Workman. *Digital control of Dynamic Systems.* Addison-Wesley, 1990.
4. C. R. Walli and I. S. Reed. Asynchronous finite-state machines - a novel control system class. *AIAA Journal,* 7(3):385–393, mar 1969.
5. C. L. Smith. *Digital Computer Process Control.* Intext Educational Publishers, 1972.
6. R. A. Hilhorst. *Supervisory Control of Mode-Switch Processes.* PhD thesis, University of Twente, 1992.
7. B. Mårtensson. The order of any stabilizing regulator is sufficient a priori information for adaptive stabilization. *Systems and Control Letters,* 6(2):87–91, 1985.
8. M. Fu and B. R. Barmish. Adaptive stabilization of linear systems via switching controls. *IEEE Transactions on Automatic Control,* pages 1079–1103, December 1986.
9. S. J. Cusumano and Poola. Adaptive control of uncertain systems: A new approach. In *Proceedings of the American Automatic Control Conference,* pages 355–359, June 1988.
10. D. E. Miller. *Adaptive Control of Uncertain Systems.* PhD thesis, University of Toronto, 1989.
11. R. H. Middleton, G. C. Goodwin, D. J. Hill, and D. Q. Mayne. Design issues in adaptive control. *IEEE Transactions on Automatic Control,* 33(1):50–58, jan 1988.
12. A. S. Morse, D. Q. Mayne and G. C. Goodwin. Applications of hysteresis switching in parameter adaptive control. *IEEE Transactions on Automatic Control,* 37(9):1343–1354, sep 1992.
13. S. R. Weller and G. C. Goodwin. Hysteresis switching adaptive control of linear multivariable systems. *IEEE Transactions on Automatic Control,* 39(7):1369–1375, july 1994.
14. F. M. Pait and A. S. Morse. A cyclic switching strategy for parameter-adaptive control. *IEEE Transactions on Automatic Control,* 39(6):1172–1183, jun 1994.
15. A. S. Morse. Dwell-time switching. In *Proceedings of the 2nd European Control Conference,* pages 176–181, 1993.
16. K. S. Narendra and J. Balakrishnan. Improving transient response of adaptive control systems using multiple models and switching. Technical report, Yale University, oct 1992.

17. R. W. Brockett. Hybrid models for motion control systems. In H. L. Ttentleman and J. C. Willems, editors, *Essays on Control: Perspectives in the Theory and its Applications*, pages 29–54. Birkhäuser, jul 1993.
18. M. S. Branicky, V. S. Borkar, and S. K. Mitter. A unified framework for hybrid control: Background, model, and theory. In *Proceedings of the 33rd Conference on Decision and Control*, pages 4228–4234. Control Systems Society, IEEE, dec 1994.
19. R. L. Grossman, A. Nerode, A. P. Ravn, and H. Rischel, editors. *Hybrid Systems*, volume 736 of *Lecture Notes in Computer Science*. Springer-Verlag, 1993.
20. K. T. Tan. *Maximal Output Admissible Sets and the Nonlinear Control of Linear Discrete-Time Systems with State and Control Constraints*. PhD thesis, University of Michigan, 1991.
21. J. P. Hespanha. Stabilization of the nonholonomic integrator via logic-based switching. Technical report, Yale, 1995.
22. A. S. Morse and F. M. Pait. Mimo design models and internal regulators for cyclicly-switched parameter-adaptive control systems. *IEEE Transactions on Automatic Control*, 39(9):1809–1818, sept 1994.
23. K. S. Narendra and J. Balakrishnan. Adaptive control using multiple models and switching. Technical report, Yale University, jul 1994.
24. A. S. Morse. Supervisory control of families of linear set-point controllers - part 1: Exact matching. submitted for publication.
25. A. R. Teel. Global stabilization and tracking for multiple integrators with bounded controls. *System and Control Letters*, pages 165–171, jan 1992.
26. H. J. Sussmann, E. D. Sontag, and Y. Yang. A general result on the stabilization of linear systems using bounded controls. *IEEE Transactions on Automatic Control*, pages 2411–2425, dec 1994.
27. E. G. Gilbert, I. Kolmanovsky, and K. T. Tan. Discrete-time reference governors and the nonlinear control of systems with state and control constraints. *Journal of Robust control and Nonlinear Systems*, aug 1995. to appear.
28. P. Kapasouris, M. Athans, and G. Stein. Design of feedback control systems for stable plants with saturating actuators. In *Proc. 27th IEEE Conference on Decision and Control*, pages 469–479, dec 1988.
29. G. F. Wredenhagen and P. R. Bélanger. Piecewise-linear lq control for systems with input constraints. *Automatica*, pages 403–416, mar 1994.
30. E. G. Gilbert and K. T. Tan. Linear systems with state and control constraints: The theory and application of maximal output admissible sets. *IEEE Transactions on Automatic control*, pages 1008–1020, sep 1991.
31. R. W. Brockett. Asymptotic stability and feedback stabilization. In R. Brockett, R. Millman, and H. Sussmann, editors, *Differential Geometric Control theory*, pages 181–191, 1983.
32. R. W. Brockett. Pattern generation and feedback control of nonholonomic systems. In *Proceedings of the IEEE Workshop on Mechanics, Holonomy and Control*, 1993.
33. O.J. Sødalen and O. Egeland. Exponential stabilization of nonholonomic chained systems. *IEEE Transactions on Automatic Control*, 40(1):35–49, 1995.
34. A. F. Filippov. Differential equations with discontinuous right-hand side. *Amer. Math. Soc. Translations*, pages 199–231, 1964.
35. J. Guckenheimer. A robust hybrid stabilization strategy for equilibria. *IEEE Transactions on Automatic Control*, 40(2):321–325, feb 1995.

36. A. S. Morse. Towards a unified theory of parameter adaptive control - tunability. *IEEE Transactions on Automatic Control*, 35(9):1002–1012, sep 1990.
37. D. E. Miller and E. J. Davison. An adaptive controller which provides an arbitrary good transient and steady-state response. *IEEE Transactions on Automatic Control*, 36(1):68–81, jan 1991.
38. D. E. Miller. Adaptive stabilization using a nonlinear time-varying controller. *IEEE Transactions on Automatic Control*, 39(7):1347–1359, jul 1994.
39. B. Mårtensson and J. W. Polderman. Correction and simplification to "the order of any stabilizing regulator is sufficient a priori information for adaptive stabilization". *Systems and Control Letters*, 20:465–470, 1993.
40. S. R. Kulkarni and P. J. Ramadge. Model and controller selection policies based on output prediction errors. Technical report, Princeton University, 1995.
41. S. Boyd, L. E. Ghaoui, E. Feron, and V. Balakrishnan. *Linear matrix Inequalities in Systems and Control Theory*. SIAM, 1994.
42. K. S. Narendra and J. Balakrishnan. Intelligent control using fixed and adaptive models. Technical report, Yale University, oct 1994.
43. A. S. Morse. Supervisory control of families of linear set-point controllers - part 2: Robustness. submitted for publication.
44. M. S. Branicky. Stability of switched and hybrid systems. In *Proceedings of the 33rd Conference on Decision and Control*, pages 3498–3503. Control Systems Society, IEEE, dec 1994.
45. F. M. Pait. *Achieving Tunability in Parameter Adaptive Control*. PhD thesis, Yale University, 1993.
46. A. S. Morse. Logic-based switching and control. In B. A. Francis and A. R. Tannenbaun, editors, *Feedback Control, Nonlinear Systems, and Complexity*, pages 173–195. Springer-Verlag, 1995.
47. L. Praly and B. E. Ydstie. *Adaptive Linear Control: Bounded Solutions and Their Properties*. to be published, ≈ 1995.
48. M. A. Wicks, P. Peleties, and R. A. Decarlo. Construction of piecewise lyapunov functions for stabilizing switched systems. In *Proceedings of the 1994 IEEE Conference on Decision and Control*, pages 3492–3497, December 1994.
49. K. S. Narendra and J. Balakrishnan. A common lyapunov function for stable lti systems with commuting a-matrices. *IEEE Trans. Auto. Control*, pages 2469–2470, December 1994.
50. L. Gurvits. Untitled manuscript. in preparation, jan 1995.
51. M. Vidyasagar. *Nonlinear Systems Analysis*. Prentice-Hall, 1993.
52. A. P. Molchanov and Ye.S. Pyatnitskiy. Criteria of asymptotic stability of differential and difference inclusions encountered in control theory. *Systems & Control Letters*, pages 59–64, 1989.
53. Ingrid Daubechies and J. C. Lagarias. Sets of matrices all infinite products of which converge. *Linear Algebra and Its Applications*, pages 227–263, 1992.
54. L. Gurvits. Stability of discrete linear inclusion. *Linear Algebra and Its Applications*, 1995. to appear.
55. W. A. Sethares, B. D. O. Anderson, and Jr. C. R. Johnson. Adaptive algorithms with filtered regressor and filtered error. *Mathematics of control, Signals, and Systems*, pages 381–403, feb 1989.
56. F. M. Pait. On the topologies of spaces on linear dynamical systems commonly employed as models for adaptive and robust control design. In *Proceedings of the Third SIAM Conference on Control and Its Applications*, 1995.

Bibliography

57. P. S. Maybeck and D. L. Pogoda. Multiple model adaptive controller for the stol f-15 with sensor/actuator failures. In *Proceedings of the 28th IEEE Conference on Decision and Control*, pages 1566–1572. IEEE, 1989.
58. A. S. Morse. Towards a unified theory of parameter adaptive control - part 2: Certainty equivalence and implicit tuning. *IEEE Transactions on Automatic Control*, 37(1):15–29, jan 1992.
59. S. M. Naik, P. R. Kumar, and B. E. Ydstie. Robust continuous-time adaptive control by parameter projection. *IEEE Transactions on Automatic Control*, 37(2):182–197, feb 1992.
60. L. Praly. Global stability of direct adaptive control schemes with respect to a group topology. In *Adaptive And Learning Systems*, pages 5–16, February 1986.

Daisy: A Large Flexible Space Structure Testbed for Advanced Control Experiments

Benoit Boulet[1*]*, Bruce A. Francis*[1**]*, Peter C. Hughes*[2]*, and Tony Hong*[2]

[1] Electrical and Computer Engineering, University of Toronto, Toronto, Ont., Canada M5S 1A4.
[2] Institute for Aerospace Studies, University of Toronto, 4925 Dufferin Street, North York, Ont., Canada M3H 5T6.

1 Introduction

Daisy is an experimental testbed facility at the University of Toronto's Institute for Aerospace Studies (UTIAS) whose dynamics are meant to emulate those of a real large flexible space structure (LFSS); see Figure 1 [8]. The purpose of the facility is to test advanced identification and multivariable control design methods. Modeled roughly to resemble the flower of the same name, Daisy consists of a rigid hub (the "stem") mounted on a spherical joint and on top of which are ten ribs (the "petals") attached through passive two-degree-of-freedom rotary joints and low-stiffness springs. Each rib is coupled to its two neighbors via low-stiffness springs. The hub would represent the rigid part of a LFSS, while the ribs would model its flexibilities.

Concerning Daisy's actuators, each rib is equipped with four unidirectional air jet thrusters that are essentially on-off devices, each capable of delivering a torque of 0.8 Nm at the rib joint. Pulse-width modulation (PWM) of the thrust is used to apply desired torques on the ribs. The four thrusters are aligned by pairs to implement two orthogonal bidirectional actuators. The hub actuators consist of three torque wheels driven by DC motors whose axes are orthogonal. Each can deliver up to 38.8 Nm.

Concerning the sensors, mounted at the tip of each rib is an infra-red emitting diode. Two hub-mounted infra-red CCD cameras measure the positions of these diodes via ten lenses. The cameras are linked to a computer that from the kinematics of Daisy computes the 20 rib angles relative to the hub in real-time (at a 30 Hz sampling rate) from the sampled infra-red video frames. This vision system, called DEOPS (Digital Electro-Optic Position Sensor), was developed at UTIAS [17]. Its resolution is approximately 0.1% of the cameras' field of view, which roughly translates to an angle measurement accuracy of 3.5×10^{-4} radians (0.02 degrees) in the ideal case. The hub orientation and angular velocity can be

---

[*] Supported through scholarships from the Natural Sciences and Engineering Research Council (NSERC) of Canada, the Fonds pour la formation de chercheurs et l'aide à la recherche, Québec, and a Walter C. Sumner Memorial Fellowship.
[**] Supported by NSERC.

measured with position and velocity encoders. There are also accelerometers on the ribs, but for this research only DEOPS and the hub position encoders were used as sensors.

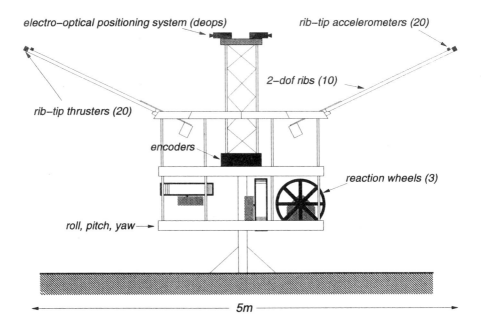

Fig. 1. Daisy LFSS experimental testbed.

In general, the dynamics of LFSSs are characterized by their high order and their significant number of closely-spaced, lightly-damped, clustered low-frequency modes. They pose a challenging problem to the control system designer, who must deal with those characteristics while ensuring a certain level of robustness in the face of significant model uncertainty. Mathematical linear dynamic models of LFSSs are usually obtained using finite-element (FE) methods, but these models are known to be accurate only for the first few modes of the structure. Moreover, these models do not provide the modal damping ratios, hence they are originally undamped. Model identification of LFSSs is often impractical because such structures are assembled in space and cannot be easily tested on earth due to problems caused by the gravity field and the atmosphere. Thus it would be desirable to have a design procedure that would directly use an FE model and a natural description of the uncertainty to produce a controller that could be implemented on real LFSSs with good confidence.

Daisy is a challenging testbed for the following reasons:

- The dynamic model available for Daisy is linear, of order 46, including 20 flexible modes and three rigid-body modes. The natural frequencies of the flexible modes are clustered around 0.6 rad/s, and model reduction is not possible, at least by conventional methods. An order of 46 presents a challenge, both for controller design and for subsequent digital implementation.
- There is significant uncertainty in modal parameters, for example, up to 50% uncertainty in damping constants.
- Daisy has some pronounced nonlinear characteristics, the most prominent being the PWM mode of the air-jet thrusters, together with their reaction delay.

Our program of research on Daisy includes the following issues:

- Getting a non-conservative linear uncertainty model.
- Testing the applicability of $\mathcal{H}_\infty$ optimization to control design for Daisy, and LFSSs in general.
- Developing a method for model validation.
- Testing the applicability of μ optimization.
- Testing the applicability of recent optimal sampled-data methods.
- Studying the control of Daisy with non-colocation, that is, without a full complement of actuators and sensors.
- Developing an effective tool for PWM control in the H-inf framework.

This paper summarises our results on the first two items. For a more complete report, the reader is referred to [4].

Notation

The norm of a complex matrix is taken to be its maximum singular value: $\|H\| = \bar{\sigma}(H)$. The ∞-norm of x in $\mathbb{R}^n$ is $\|x\|_\infty = \max_{i=1,\ldots,n} |x_i|$. We denote the open and closed right-half complex planes by $\mathbb{C}_+$ and $\overline{\mathbb{C}}_+$ respectively. The extension of $\overline{\mathbb{C}}_+$ to infinity is written as $\overline{\mathbb{C}}_+ \cup \{\infty\}$. For a normed space $\mathcal{X}$, $\mathcal{BX}$ denotes its open unit ball.

2 Modeling LFSSs

Uncertainty modeling in LFSSs is critical if one is to achieve an acceptable level of robustness with a practical controller. Some works [19], [1], [25] use norm-bounded additive or multiplicative perturbations of a nominal model in the frequency domain to account for uncertainty in the modal frequencies, damping ratios, and mode shape matrix of the model. Unmodeled modes of the structure and uncertain actuator dynamics can also be represented in this way [20]. Such approaches to uncertainty modeling in LFSSs do not handle modal parameter uncertainty very well: Slight variations in either the mode frequencies or damping ratios usually cause the associated dynamic perturbations to be large in the ∞-norm sense. Indeed, additive or multiplicative perturbations may contain large peaks in their frequency responses because of the inherently low damping ratios in LFSS dynamics. At the limit, undamped modes cannot have a representation as norm-bounded perturbations of these types. Covering unmodeled modes with such perturbations suffers from the same problem. As a result, one has to choose large weighting functions to bound perturbations that arise from small variations in the modal parameters. In an actual $\mathcal{H}_\infty$ controller design, this may lead to difficulties in making the closed-loop system robustly stable to all weighted perturbations of admissible ∞-norm while achieving some desired performance objective. Thus, in this case, the basic tradeoff between robust stability and nominal performance may be detrimental to the achievable performance level.

In this paper, it is suggested to transform real parameter uncertainty in the modes into unstructured uncertainty without getting too conservative in the sense that the uncertainty set in $\mathcal{H}_\infty$ has to be kept relatively small. This is motivated by the fact that many results and practical controller design techniques are available for this kind of uncertainty, whereas a useful frequency-based design method dealing explicitly with a large number of scalar real structured perturbations in a high-order dynamic model has yet to be developed. So far, μ-synthesis has proven to be one of the most effective ways to deal with complex structured uncertainty, and some authors (e.g. [19]) have used it to model real parametric uncertainty. Recently the mixed real/complex μ problem has been studied [11] and design methods based on minimizing an upper bound on the mixed μ function, such as the so-called Popov controller synthesis [14], have been developed. Application to a flexible structure has been reported in [15]. These methods are attractive but they quickly become numerically difficult (actually, they are NP-hard [5]) as the plant's order and the number of independent perturbations increase. Some of them also suffer from controller inflation. Hence they are of limited use for high-order LFSS models when many real scalar perturbations are modeled as individual scalar blocks. Furthermore, an unstructured complex uncertainty block must still be added to account for unmodeled dynamics, and this block usually represents additive or multiplicative uncertainty. This means that the attainable performance may be severely limited as previously discussed.

Yet another approach to the robust control of LFSSs is the passivity approach [16]. It is well known that an LFSS with colocated rate sensors and force/torque actuators and with the same number of inputs and outputs has a positive-real transfer matrix model. Then if a strictly positive real controller is designed, it follows that the closed-loop will be stable for all modal perturbations, regardless of the number of unmodeled modes. Although this result is of great importance, it only applies to structures of rather restricted configurations. Moreover, one cannot use this result for controller design achieving robust performance. Recently, dynamic embeddings have been proposed to turn a nonsquare/noncolocated LFSS model into a positive real system [18]. Even though this approach seems promising, it is not yet clearly known how robust the embedding technique is, i.e., small perturbations in the original plant may destroy the positive realness property of the embedded plant.

It appears that a different description of the uncertainty is needed. Some authors have argued (e.g. [26]) that coprime factor descriptions in $\mathcal{H}_\infty$ of nearly unstable plants, such as LFSSs, is a sound way to model these systems. This is the approach taken here, as introduced in [3]. The loopshaping technique of McFarlane and Glover [21] involves modeling the plant as a normalized coprime factorization and it has been successfully applied to design controllers for LFSSs [21]. Similar to the loopshaping method is the weighted-gap optimization technique that was tested on LFSS experimental facilities by Buddie et al. [6]. These techniques show the potential of modeling LFSS dynamics using coprime factorizations, but they don't address the problem of converting known bounds on perturbations of the modal parameters into norm bounds on factor perturbations. The difficulty comes from the fact that these methods rely on *normalized* coprime factorizations, which destroy the decoupled structure of the nominal modal state-space models.

Section 3 presents a very simple method to obtain a left coprime factorization (LCF) of LFSS dynamics in modal coordinates that preserves the decoupled structure. The plant uncertainty is described as stable perturbations of the coprime factors. The structure of the LCF allows one to go easily from modal parameter uncertainty to an unstructured description of the uncertainty as stable norm-bounded perturbations in the factors, as discussed in Section 4. This allows a better, less conservative description of the uncertainty set and hence should lead to better closed-loop performance and guaranteed robustness.

3 A Left Coprime Factorization of LFSS Dynamics

An FE method gives a high-order model of the flexible part of the structure consisting of perhaps thousands of ordinary differential equations. Rigid-body modes may be included to account for the attitude and position of rigid parts of the structure. In order to have a fixed model for our discussion, we consider three rigid-body modes accounting for the attitude of the main rigid part. (Daisy has these dynamics, although two of the rigid-body modes are pendulous, so they can be viewed as flexible modes.) The model is undamped, and it consists essentially of a positive definite mass matrix M and a positive semidefinite stiffness matrix K; the equations are

$$M\ddot{q} + Kq = B_0 u \tag{1}$$
$$y = C_0 q, \tag{2}$$

where $q(t) \in \mathbb{R}^{n_{FE}}$ is a vector of attitude coordinates for the rigid part and physical coordinates (displacements and rotations) of the flexible parts of the LFSS, the input $u(t) \in \mathbb{R}^m$ is a vector of actuator forces and torques applied to the structure, and $y(t) \in \mathbb{R}^p$ is the vector of measured outputs.

A real matrix E whose columns are eigenvectors of the matrix $M^{-1}K$ and such that it diagonalizes both M and K, i.e., $E^T M E = I$ and $E^T K E = \Lambda$, where Λ is diagonal with the squared mode frequencies on its main diagonal, always exists ([13], Theorem 4.5.15). It defines a coordinate transformation from the modal coordinate vector η to the physical coordinate vector q, i.e., $q = E\eta$. Such a matrix is called a *mode shape matrix* of the system and its columns are the *mode shapes* of the structure. Thus the mode shapes and mode frequencies are the eigenvectors and eigenvalues of $M^{-1}K$.

We start with LFSS dynamics in modal coordinates, reduced to a reasonable order by discarding the less significant flexible modes according to some measure of their input-output influence [24], [12]. The first three are the rigid-body modes. The modal frequencies of the $n-3$ remaining retained flexible modes, $\{\omega_i\}_{i=4}^n$, are given by the FE model; uncertainties will be introduced later. Damping is added to the nominal model, as it is known that damping ratios of flexible modes are nonzero, since flexibilities in any LFSS are dissipative in nature. So if $\{\overline{\zeta}_i\}_{i=4}^n$ are positive upper bounds and $\{\underline{\zeta}_i\}_{i=4}^n$ nonnegative lower bounds on the otherwise unknown damping ratios, we may take $\{\zeta_i := (\underline{\zeta}_i + \overline{\zeta}_i)/2\}_{i=4}^n$ as the nominal ones. Transforming (1) using the mode shape matrix E, truncating, and adding a diagonal damping matrix D, we get the nominal dynamic equations in modal coordinates:

$$\ddot{\eta} + D\dot{\eta} + \Lambda \eta = B_1 u \qquad (3)$$
$$y = C_1 \eta, \qquad (4)$$

where

$$D = \text{diag}\{0,0,0, 2\zeta_4\omega_4, \ldots, 2\zeta_n\omega_n\}$$
$$\Lambda = \text{diag}\{0,0,0, \omega_4^2, \ldots, \omega_n^2\}$$
$$B_1 = E_r^T B_0$$
$$C_1 = C_0 E_r$$

and E_r is composed of the columns of E corresponding to the modes kept in the model. Thus

$$\hat{\eta}(s) = [s^2 I + sD + \Lambda]^{-1} B_1 \hat{u}(s) \qquad (5)$$
$$\hat{y}(s) = C_1 [s^2 I + sD + \Lambda]^{-1} B_1 \hat{u}(s). \qquad (6)$$

The assumptions here are as follows:
(A1) The sensors have no dynamics.
(A2) No pole-zero cancellation at $s = 0$ occurs when the product $C_1 [s^2 I + sD + \Lambda]^{-1} B_1$ is formed.
(A3) The uncertainty in the output matrix C_1 can be lumped in with the input uncertainty.

The motivation behind assumptions (A1) and (A3) is that space sensors are usually accurate and fast while space actuators, which include torque wheels and gas jet thrusters, may add quite a bit of uncertainty in the torque and force inputs. Assumption (A2) is standard and just says that the unstable rigid-body modes must be controllable and observable with the set of actuators and sensors used.

Consider the matrix $[s^2 I + sD + \Lambda]$ in (5). It is diagonal, so its inverse is simply

$$[s^2 I + sD + \Lambda]^{-1} = \mathrm{diag}\left\{\frac{1}{s^2}, \frac{1}{s^2}, \frac{1}{s^2}, \frac{1}{s^2 + 2\zeta_4\omega_4 s + \omega_4^2}, \ldots, \frac{1}{s^2 + 2\zeta_n\omega_n s + \omega_n^2}\right\}. \tag{7}$$

The matrix B_1 in (5) is an $n \times m$ real matrix. Introduce a polynomial $s^2 + as + b$, Hurwitz with real zeros, and form the matrices $\tilde{M}(s)$, $\tilde{N}(s)$ as follows:

$$\tilde{M}(s) := \frac{1}{s^2 + as + b}\mathrm{diag}\left\{s^2, s^2, s^2, s^2 + 2\zeta_4\omega_4 s + \omega_4^2, \ldots, s^2 + 2\zeta_n\omega_n s + \omega_n^2\right\} \tag{8}$$

$$\tilde{N}(s) := \frac{1}{s^2 + as + b} B_1. \tag{9}$$

The complex argument s is dropped hereafter to ease the notation. Note that $\tilde{M}$ and $\tilde{N}$ belong to $\mathcal{RH}_\infty$ and the transfer function matrix from $\hat{u}$ to $\hat{\eta}$ is $G := \tilde{M}^{-1}\tilde{N}$, i.e., $\tilde{M}$ and $\tilde{N}$ form a left factorization of G in $\mathcal{RH}_\infty$. It can be proved that $\tilde{M}$ and $\tilde{N}$ are left coprime.

4 Uncertainty Modeling for LFSSs

Uncertainty in FE models is usually characterized by uncertainty in the modal parameters $\{\zeta_i\}_{i=4}^n$ and $\{\omega_i\}_{i=4}^n$, in the mode gains, and in the mode shape matrix E. Unmodeled modes can also be considered as perturbations changing the order of the model. Uncertainty in the modal parameters appears easier to characterize based on heuristics and experience, at least for the first few modes, than uncertainty in E. The uncertainty modeling process proposed here uses the a priori knowledge of the bounds for $\{\zeta_i\}_{i=4}^n$, $\{\omega_i\}_{i=4}^n$. For example, the structure designer might say with good certainty that the second mode has natural frequency between, say, 0.01 and 0.013 rd/s, and that its damping ratio ζ is almost surely less than 0.05. This information is used to derive a bound on the norm of the coprime factor perturbations at each frequency, which will be needed in the design process for robustness issues. Of course, some uncertainty is also present in the mode shape matrix E of the structure and will be accounted for as uncertainty in the entries of B_1. We will see that it is easy to go from parametric uncertainty to unstructured uncertainty in the coprime factors.

This section can be outlined as follows. First we start with the parametric uncertainty model (11); this induces stable perturbations in $\tilde{M}$ and $\tilde{N}$. These induced perturbations and their corresponding perturbed factors are given the subscript "rp" for *real parameter*. On the other hand, some of the results stated apply to more general perturbations in $\mathcal{RH}_\infty$, but then the subscript is dropped. Two scalings are performed on $\tilde{M}_{rp}$ and $\tilde{N}_{rp}$ so that the perturbations are better balanced. Finally, a third scaling normalizes the combined factor perturbations.

It is desired to lump the uncertainty in the mode frequencies, damping ratios, and mode gains into unstructured uncertainty in the coprime factors such that the perturbed LCF of G can be written as

$$G_p = (\tilde{M} + \Delta M)^{-1}(\tilde{N} + \Delta N), \qquad (10)$$

with $\Delta M, \Delta N \in \mathcal{RH}_\infty$. Perturbations in the modal parameters and the entries of B_1 are assumed bounded by nonnegative numbers as follows:

$$|\delta\omega_i| \leq l_\omega^i, \quad |\delta\zeta_i| \leq l_\zeta^i, \quad |\delta b_{1ij}| \leq l_b^{ij}, \quad i,j = 1,\ldots,n. \qquad (11)$$

The following matrices will be useful later on:

$$\Delta B_1 := \begin{bmatrix} \delta b_{11} & \cdots & \delta b_{1m} \\ \vdots & \ddots & \vdots \\ \delta b_{n1} & \cdots & \delta b_{nm} \end{bmatrix} \qquad (12)$$

$$L_B := \begin{bmatrix} l_b^{11} & \cdots & l_b^{1m} \\ \vdots & \ddots & \vdots \\ l_b^{n1} & \cdots & l_b^{nm} \end{bmatrix}. \qquad (13)$$

The uncertainty in the entries of B_1, which are also the numerators of the fractional entries of $\tilde{N}$, comes from different sources. First, uncertainty in a particular mode gain can be represented as an uncertain factor multiplying the corresponding row of B_1. Second, uncertainty in the mode shape matrix E affects B_1 because in the change from physical to modal coordinates in (1), the original input matrix B_0 gets premultiplied by E_r^T to form B_1. Third, the matrix B_0 itself is uncertain because the actuator gains are not known perfectly. Finally, by (A3), output uncertainty is transformed into input uncertainty.

Unmodeled modes, usually (but not necessarily) occurring at high frequencies, can be handled by adjusting the norm bound on the factor uncertainty (though not necessarily at high frequencies only). This can be done iteratively: Design a controller using the technique discussed in this paper and test it on a set of perturbed full-order evaluation models. If all closed loops are stable (while achieving some desired performance level in the robust performance case), stop—the controller is satisfactory. If not, increase the norm bound on factor uncertainty and redesign the controller.

Perturbations of the coprime factors resulting from perturbations of the real parameters only are easily computed: The perturbed factor $\tilde{M}_{rp}$ is defined as

$$\tilde{M}_{rp} := \tilde{M} + \Delta M_{rp}, \qquad (14)$$

where

$$\Delta M_{rp} := \mathrm{diag}\left\{0,0,0, \frac{[2\zeta_4\delta\omega_4 + 2\delta\zeta_4(\omega_4 + \delta\omega_4)]s + 2\omega_4\delta\omega_4 + \delta\omega_4^2}{s^2 + as + b}, \ldots, \right.$$
$$\left. \frac{[2\zeta_n\delta\omega_n + 2\delta\zeta_n(\omega_n + \delta\omega_n)]s + 2\omega_n\delta\omega_n + \delta\omega_n^2}{s^2 + as + b}\right\}, \qquad (15)$$

4 Uncertainty Modeling for LFSSs

and the perturbed factor $\tilde{N}_{rp}$ is defined as

$$\tilde{N}_{rp} := \tilde{N} + \Delta N_{rp}, \tag{16}$$

where

$$\Delta N_{rp} := \frac{\Delta B_1}{s^2 + as + b}, \tag{17}$$

Now let us consider closed-loop stability of the system in Figure 2, where a controller K is connected as a feedback around a perturbed LCF, with U and V arbitrary transfer matrices such that no pole-zero cancellation occurs in $\overline{\mathbb{C}}_+$ when the product $V(\tilde{M}+\Delta M)^{-1}(\tilde{N}+\Delta N)U$ is formed. Define the uncertainty matrix

$$\boldsymbol{\Delta} := [\Delta N \ -\Delta M]. \tag{18}$$

Clearly, if ΔM_{rp} and ΔN_{rp} are substituted in (18), the resulting $\boldsymbol{\Delta}_{rp}$ belongs to $\mathcal{RH}_\infty$. This matrix is defined because the result on stability of the feedback system in Figure 2 is expressed in terms of a norm bound on $\boldsymbol{\Delta}(j\omega)$ ([21], [26]).

Define the uncertainty set

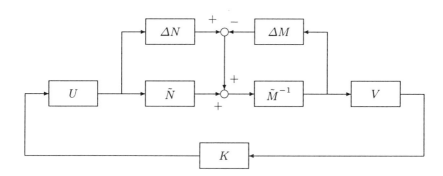

Fig. 2. Feedback control of a perturbed LCF model.

$$\mathcal{D}_r := \{\boldsymbol{\Delta} \in \mathcal{RH}_\infty \mid \|r^{-1}\boldsymbol{\Delta}\|_\infty < 1\}, \tag{19}$$

where r is a unit in $\mathcal{H}_\infty$. The small-gain theorem yields the following slightly modified result of [26] (see also [21]).

Theorem 1. *The closed-loop system of Figure 2 with controller K is internally stable for every $\boldsymbol{\Delta} \in \mathcal{D}_r$ iff*
(a) K internally stabilizes VGU, and
(b) $\left\| r \begin{bmatrix} UKV(I-GUKV)^{-1}\tilde{M}^{-1} \\ (I-GUKV)^{-1}\tilde{M}^{-1} \end{bmatrix} \right\|_\infty \leq 1.$

Given the parametric uncertainty in (11), a bound of the type $|r(j\omega)|$ that would tightly cover $\|\boldsymbol{\Delta}_{rp}(j\omega)\|$ must be found. But before this weighting function is constructed, different scalings must be performed on the factors and their perturbations to avoid any undue conservativeness and to "balance" the perturbations, i.e., to minimize the difference between the ∞-norms of $\Delta \boldsymbol{N}_{rp}$ and $\Delta \boldsymbol{M}_{rp}$. The first scaling aims at making the components of the rows and columns of B_1 have the same order of magnitude.

Let β_j denote the ∞-norm of the jth column of B_1 and form

$$J_2 := \text{diag}\{\beta_1, \ldots, \beta_m\}.$$

Now let α_i denote the ∞-norm of the ith row of $B_1 J_2^{-1}$ and form

$$J_1 := \text{diag}\{\alpha_1, \ldots, \alpha_n\}.$$

Let $\gamma := \|J_1^{-1} L_B J_2^{-1}\|$ and define the scaled matrices $B_{sc} := \gamma^{-1} J_1^{-1} B_1 J_2^{-1}$ and $\Delta B_{sc} := \gamma^{-1} J_1^{-1} \Delta B_1 J_2^{-1}$. For all ΔB_1 satisfying the inequalities in (11), $\|\Delta B_{sc}\| \leq 1$. Finally, we can define the scaled factor and its perturbation:

$$\tilde{\boldsymbol{N}}_0 := b\gamma^{-1} J_1^{-1} \tilde{\boldsymbol{N}} J_2^{-1} = \frac{b B_{sc}}{s^2 + as + b},$$

$$\Delta \boldsymbol{N}_{rp0} := b\gamma^{-1} J_1^{-1} \Delta \boldsymbol{N}_{rp} J_2^{-1} = \frac{b \Delta B_{sc}}{s^2 + as + b}.$$

The second scaling is performed on $\tilde{\boldsymbol{M}}$ to make sure that the norm of any perturbation of it induced by variations in the modal parameters is less than or equal to one, but close to one at low frequencies. Let $c_i := 2\zeta_i l_\omega^i + 2l_\zeta^i(\omega_i + l_\omega^i)$ and $d_i := 2\omega_i l_\omega^i + l_\omega^{i2}$ for $i = 4, \ldots, n$. These constants are the coefficients of the numerator of the (i, i) entry of $\Delta \boldsymbol{M}_{rp}$ when all modal perturbations are replaced by their upper bounds. Defining $c_{max} := \max_{i=4,\ldots,n} c_i$ and $d_{max} := \max_{i=4,\ldots,n} d_i$, we can now define the second scaled factor and its perturbation:

$$\tilde{\boldsymbol{M}}_0 := \frac{b}{d_{max}} \tilde{\boldsymbol{M}}, \quad \Delta \boldsymbol{M}_{rp0} := \frac{b}{d_{max}} \Delta \boldsymbol{M}_{rp}.$$

These two scalings are best illustrated by a sequence of block diagrams, Figure 3, showing the transformations performed on the coprime factors and their perturbations. We have included a block for the diagonal transfer matrix $\boldsymbol{T}_a$ that models actuator dynamics. Note that the properties of linearity and commutativity of diagonal matrices are used in order to move blocks around and get the desired final block diagram.

It is easy to check that $\tilde{\boldsymbol{N}}_0$ and $\tilde{\boldsymbol{M}}_0$ are still coprime and that they form an LCF of $\boldsymbol{G}_0 := d_{max} \gamma^{-1} J_1^{-1} \boldsymbol{G} J_2^{-1}$. It is our experience that these types of scalings help a lot in reducing the $\mathcal{H}_\infty$ norm of the generalized plant's weighted transfer matrix in an actual design. The last scaling performed on the perturbation $\boldsymbol{\Delta}_{rp0} := [\Delta \boldsymbol{N}_{rp0} \ -\Delta \boldsymbol{M}_{rp0}]$ normalizes it with the weighting function $r(s)$ to get $\|\hat{\boldsymbol{\Delta}}_{rp0}\|_\infty < 1$. This is illustrated in Figure 4.

4 Uncertainty Modeling for LFSSs

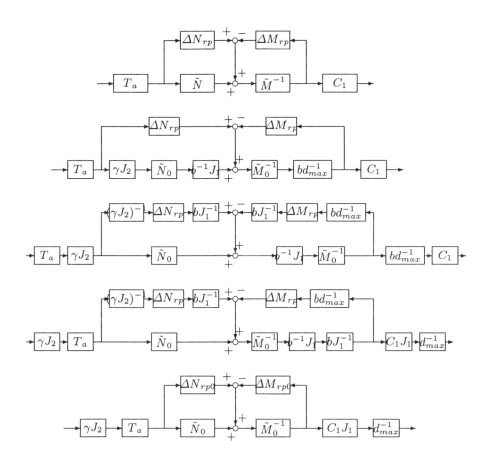

γJ_2 and d_{max}^{-1} may be absorbed by the controller.

Fig. 3. Sequence of transformations applied to the perturbed factors.

We are now ready to design a weighting function $R = rI$ for the scaled perturbation Δ_{rp0}. In so doing, the freedom provided by coefficients of the common denominator $s^2 + as + b$ will be used to advantage to keep the order of r as low as possible without paying the price of added conservativeness. Here is the result.

Theorem 2. *For $k > 0$, define a and b via $s^2 + as + b := \left(s + \frac{d_{max}}{c_{max}}\right)(s + k)$*

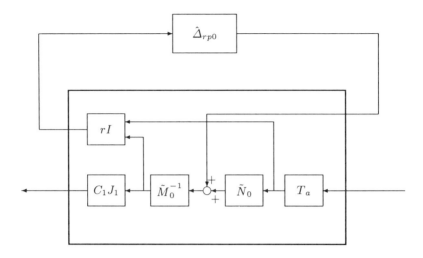

Fig. 4. Perturbed factorization after all three scalings.

and let the unit weighting function r be given by $r(s) = \dfrac{\epsilon_1 s + \sqrt{2} + \epsilon_0}{(s/k + 1)}$, where ϵ_0 and ϵ_1 are small positive numbers. Then $\|r^{-1}\mathbf{\Delta}_{rp0}\|_\infty < 1$.

This weighting function is of first order, which is a benefit considering that it will be duplicated $m + n$ times in the generalized plant of Figure 5. By construction, for small ϵ_0, ϵ_1, $|r(j\omega)|$ is a relatively tight bound on $\|\mathbf{\Delta}_{rp0}(j\omega)\|$, especially at low and high frequencies.

With the unit r given by Theorem 2, the factor perturbation $\mathbf{\Delta}_{rp0}$ belongs to the uncertainty set $\mathcal{D}_r$ and the normalized $\hat{\mathbf{\Delta}}_{rp0}$ belongs to $\mathcal{BRH}_\infty^{n \times (m+n)}$. Now introduce a normalized scaled perturbation $\hat{\mathbf{\Delta}}_0 \in \mathcal{BRH}_\infty^{n \times (m+n)}$. Then letting $\mathbf{\Delta}_0 := r\hat{\mathbf{\Delta}}_0$, one obtains that $\mathbf{\Delta}_0$ is a free perturbation in $\mathcal{RH}_\infty^{n \times (m+n)}$ with $\|\mathbf{\Delta}_0(j\omega)\| < |r(j\omega)|$, $\forall \omega \in \mathbb{R}$, i.e., $\mathbf{\Delta}_0$ is an arbitrary element in $\mathcal{D}_r$. In this way r can be included in the generalized plants of Figures 4 and 5. Figure 5 shows the scaled closed-loop system with all the weights for designing a controller K providing robust stability and nominal performance. Notice that the control input is u_{sc}, a scaled version of u. This block diagram and the associated control design problem will be discussed in the next section. According to Theorem 1, our robust stability objective will be to minimize the ∞-norm of the map $w \mapsto z_1$ to a value no more than 1 over all stabilizing controllers.

5 Robust $\mathcal{H}_\infty$ Design

Consider first the problems of attitude regulation and vibration attenuation. For these problems, it makes sense to ask for good torque/force disturbance rejection at low frequencies as a first requirement for nominal performance. For example, this may be required on a flexible space station on which there may be large robots or humans producing significant torque disturbances. As a second requirement for nominal performance, we will ask for good tracking of reference angle trajectories to allow accurate slewing maneuvers of the rigid part of the structure. These requirements can be translated into desired shapes for the norms of the sensitivity functions $\boldsymbol{S}_{rh} := r \mapsto e_h$ and $\boldsymbol{S}_{dh} := d \mapsto y_h$, where r is the vector of input references, e_h is the vector of attitude angle errors for the rigid part of the structure (the h subscript stands for hub, the rigid part of Daisy), d is the vector of external torque/force disturbances, and $y_h = [\theta_{hx}\ \theta_{hy}\ \theta_{hz}]^T$ is the vector of attitude angles of the rigid part. Note that if we define $\boldsymbol{S}_r := r \mapsto e$, where e is the vector of all position/angle errors, and $\boldsymbol{S}_d := d \mapsto y$, then $\boldsymbol{S}_{rh} = \begin{bmatrix} I_{3\times 3}\ 0_{3\times(p-3)} \end{bmatrix} \boldsymbol{S}_r$ and $\boldsymbol{S}_{dh} = \begin{bmatrix} I_{3\times 3}\ 0_{3\times(p-3)} \end{bmatrix} \boldsymbol{S}_d$. These frequency-domain specifications are well-suited for the $\mathcal{H}_\infty$ design method (see e.g. [10]) or a μ-synthesis [2]. Here we discuss the $\mathcal{H}_\infty$ approach.

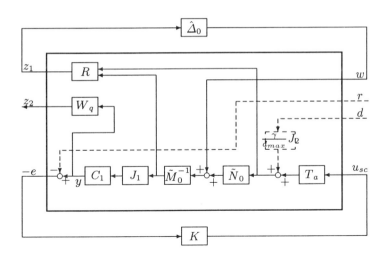

Fig. 5. Generalized plant with scaled perturbation and controller for $\mathcal{H}_\infty$ design.

The block diagram of Figure 5 shows the interconnections between the scaled factors and their perturbations, actuator dynamics, scaling and output matrices, controller, and weighting functions that form the controlled perturbed generalized plant used in the $\mathcal{H}_\infty$ design. The weighting function $\boldsymbol{W}_q$, taken to be of the form $\boldsymbol{W}_q := qw \begin{bmatrix} I_{3\times 3}\ 0_{3\times(p-3)} \end{bmatrix}$, where $q > 0$, will allow us to shape the sensitivity functions as desired. Note that the signals w, z_1, and z_2 do not have physical interpretations, but r, d and e do, as previously defined. The input

signal u_{sc} is just a scaled version of the physical signal u: $u_{sc} := \frac{\gamma}{d_{max}} J_2 u$.
In terms of Figure 5, we consider the design problem

> Design a finite-dimensional, proper, linear time-invariant controller K such that for $\hat{\Delta}_0 \equiv 0$, the nominal closed loop of Figure 5 achieves $\|w \mapsto z\|_\infty \leq 1$.

A solution to this achieves robust stability and nominal performance in the following sense.

Theorem 3. *Assume C_1 is right invertible and let $C_1^\dagger$ be a right inverse of C_1. Let $q = \max\left\{\|\tilde{M}_0 J_1^{-1} C_1^\dagger\|_\infty, d_{max}^{-1}\|J_1^{-1} B_1\|\right\}$. If the controller K is internally stabilizing and achieves*

$$\left\|w \mapsto \begin{bmatrix} z_1 \\ z_2 \end{bmatrix}\right\|_\infty \leq 1, \tag{20}$$

then the closed-loop system of Figure 5 is robustly stable to all perturbations $\hat{\Delta}_0 \in \mathcal{BRH}_\infty^{n \times (m+n)}$, and for every $\omega \in \mathbb{R}$ we have

$$\|S_{dh}(j\omega)\| \leq |w^{-1}(j\omega)|, \tag{21}$$
$$\|S_{rh}(j\omega)\| \leq |w^{-1}(j\omega)|. \tag{22}$$

The choice of q in Theorem 3 may be used for a first design to get insight into the tradeoff between robustness and performance, but smaller values of q may be tried to reach a satisfactory design achieving (21) and (22).

Other forms for W_q may be used to achieve other objectives such as weighting some or all of the outputs corresponding to the flexible part of the structure. Then it is easy to see that Theorem 3 remains basically the same, and particularly the expression for q is unchanged. In our experiments with Daisy, it was found that weighting all the outputs was asking too much given the uncertainty in the model and the actuator saturation levels. Hence only the hub angles were weighted with W_q as above. In any case, the modal coordinates are weighted by r (see Figure 5), which in our experiments on Daisy resulted in sufficient vibration attenuation.

To recap, if a stabilizing K satisfying (20) has been designed, it follows that (i) K provides robust stability to all perturbations of the modal parameters satisfying (11) and (ii) K provides nominal performance in the sense that inequalities (21) and (22) hold. Note that the controller K_p to be implemented on the real system is a scaled version of K, i.e., $K_p = \frac{d_{max}}{\gamma} J_2^{-1} K$.

6 $\mathcal{H}_\infty$ Design for Daisy

The dynamic model available for Daisy is of 46th order, including 20 flexible modes with frequencies ranging from approximately 0.56 rad/s to 0.71 rad/s and damping ratios from 0.015 to 0.06. The modal parameters are listed in Table 1

6 $\mathcal{H}_\infty$ Design for Daisy

with crude approximations of their uncertainties obtained simply from looking at time responses. Some of the modes are multiple. Two of the rigid-body modes are pendulous, so they can be considered as flexible modes; they both have nominal frequency 0.29 rad/s and their nominal damping ratios are 0.11 and 0.09. The model has the form of (3). The method described in Section 5 is illustrated by designing a robust controller for colocated and non-colocated configurations of Daisy using the $\mathcal{H}_\infty$ design method.

mode i	frequency ω_i (rad/s)	damping ratio ζ_i
1 (rigid)	0	0
2 (rigid)	$0.286 \pm 10\%$	$0.11 \pm 50\%$
3 (rigid)	$0.293 \pm 10\%$	$0.09 \pm 50\%$
4 (flex.)	$0.568 \pm 10\%$	$0.025 \pm 50\%$
5 (flex.)	$0.568 \pm 10\%$	$0.02 \pm 50\%$
6 (flex.)	$0.569 \pm 10\%$	$0.03 \pm 50\%$
7 (flex.)	$0.569 \pm 10\%$	$0.02 \pm 50\%$
8 (flex.)	$0.569 \pm 10\%$	$0.035 \pm 50\%$
9 (flex.)	$0.569 \pm 10\%$	$0.025 \pm 50\%$
10 (flex.)	$0.569 \pm 10\%$	$0.02 \pm 50\%$
11 (flex.)	$0.572 \pm 10\%$	$0.02 \pm 50\%$
12 (flex.)	$0.592 \pm 10\%$	$0.06 \pm 50\%$
13 (flex.)	$0.593 \pm 10\%$	$0.06 \pm 50\%$
14 (flex.)	$0.657 \pm 10\%$	$0.015 \pm 50\%$
15 (flex.)	$0.657 \pm 10\%$	$0.015 \pm 50\%$
16 (flex.)	$0.657 \pm 10\%$	$0.02 \pm 50\%$
17 (flex.)	$0.657 \pm 10\%$	$0.02 \pm 50\%$
18 (flex.)	$0.657 \pm 10\%$	$0.027 \pm 50\%$
19 (flex.)	$0.657 \pm 10\%$	$0.025 \pm 50\%$
20 (flex.)	$0.657 \pm 10\%$	$0.02 \pm 50\%$
21 (flex.)	$0.670 \pm 10\%$	$0.04 \pm 50\%$
22 (flex.)	$0.672 \pm 10\%$	$0.05 \pm 50\%$
23 (flex.)	$0.714 \pm 10\%$	$0.015 \pm 50\%$

Table 1. Modal parameters of Daisy's model.

Here we consider only colocation, by which we mean that all rotations and displacements produced by the actuators at their locations are measured. Thus 23 actuator/sensor pairs are used, namely the 20 bidirectional rib thrusters with the DEOPS system measuring the 20 rib angles, plus the three hub reaction wheels with the three corresponding angle encoders. In terms of system equations (3) and (4), the inputs are $u = [\tau_{hx}\ \tau_{hy}\ \tau_{hz}\ \tau_{r1}\ \tau_{r2} \cdots \tau_{r20}]^T$, where the first three are the hub torques around the x, y and z axes, and the last twenty inputs are the rib torques given by

$$\tau_{ri} = \begin{cases} \text{rib } (i+1)/2 \text{ out-of-cone torque,} & i \text{ odd,} \\ \text{rib } i/2 \text{ in-cone torque,} & i \text{ even.} \end{cases} \quad (23)$$

All torque inputs are expressed in Nm. The input matrix $B_1 \in \mathbb{R}^{23 \times 23}$ is assumed to have up to 8% uncertainty in its entries. Note that the torque wheels have some dynamics, i.e., for each wheel, the transfer function between the desired and produced torques is first-order and strictly proper. On the other hand, the PWM thrusters, which deliver average torques close to the desired ones, are modeled as pure gains. Overall, the transfer matrix T_a in Figure 5 is taken to be

$$T_a = \text{diag}\left\{\frac{0.01s+1}{0.36s+1}, \frac{0.01s+1}{0.36s+1}, \frac{0.01s+1}{0.36s+1}, 1, 1, \ldots, 1\right\}, \qquad (24)$$

where the the terms $0.01s$ are added in the numerators to regularize the generalized plant for the $\mathcal{H}_\infty$ problem, and the 0.36 time constants were measured experimentally. Note that T_a commutes with J_2. The outputs are the angles $y = [\theta_{hx} \theta_{hy} \theta_{hz} \theta_{r1} \theta_{r2} \cdots \theta_{r20}]^T$, which correspond to the input torques described above. The output matrix is just the mode shape matrix $C_1 = E \in \mathbb{R}^{23 \times 23}$, which is invertible. All angles are expressed in radians. Finally, $\Lambda = \text{diag}\{\omega_1^2, \ldots, \omega_{23}^2\}$ and $D = \text{diag}\{2\zeta_1\omega_1, \ldots, 2\zeta_{23}\omega_{23}\}$, where the modal parameters are those given in Table 1. A plot of the 23 singular values of $C_1 G(j\omega)$ is shown in Figure 6. It turns out that all the modes are significant and as a result it is very difficult to reduce the number of modes in the model. This was concluded from an analysis of the Hankel singular values of a normalized coprime factorization of the plant model $C_1 G$ [22]: They all lie between 0.2 and 0.9, which indicates that the model should not be reduced. Consequently, our design model includes all the modes in Table 1. It should be noted that this method of characterizing the input/output influence of the modes in the model seems appropriate for our control design method based on a coprime factorization. It avoids the singularity of measures such as modal costs [24] and Hankel singular values of the plant [12] when the damping ratios go to zero.

It is desired to control Daisy's model so that it remains stable for all bounded perturbations of the modal parameters in Table 1 and all perturbations of the entries of B_1 within 8% of their nominal values. We also want good torque/force disturbance rejection and good tracking in the sense of (21) and (22). The diagonal scaling matrices J_1 and J_2 are computed as explained in Section 4. The constants and weighting functions are

$$d_{max} = 0.107, \quad c_{max} = 0.046, \quad k = \frac{c_{max}}{d_{max}} = 0.43, \quad \gamma = 0.79, \quad q = 1000,$$

$$w(s) = \frac{100}{s^2/(0.01)^2 + 2 \times 0.7s/0.01 + 1}, \qquad (25)$$

$$r(s) = \frac{0.001s + 1.415}{2.33s + 1}. \qquad (26)$$

Computational delay and zero-order hold models were not included in the generalized plant even though both were present in the digital implementation of the controller on Daisy. It was anticipated that the design would be robust to these unmodeled dynamics; this was borne out by experiments. No antialiasing analog filters were available to filter the measured hub angle signals, nor

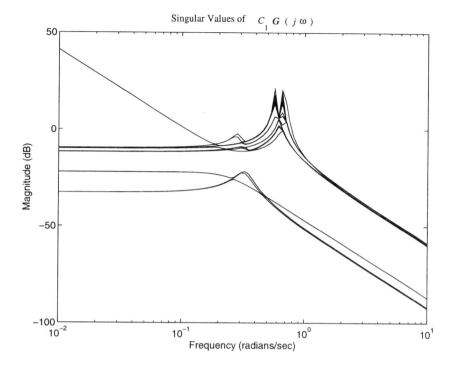

Fig. 6. Singular values of $C_1 G(j\omega)$.

for DEOPS signals, which are inherently digital. Even though this is rather undesirable, high-frequency noise levels seemed sufficiently small to avoid serious aliasing problems in the experiments. The $\mathcal{H}_\infty$ design was carried out in MATLAB^{TM} using the μ-Tools^{TM} [2] command *hinfsyn*. If a realization of the generalized plant is obtained using a computer, it will in general be nonminimal because pole-zero cancellations might not be carried out. Also note that the generalized plant is unstable, so one cannot use the balanced truncation method [23] to get rid of the unobservable and uncontrollable modes. Therefore we used the decentralized fixed-mode method [9] to obtain a minimal realization, reducing it from 147 to 78 state variables, which equals its McMillan degree. This method has the advantage of being computationally simple and hence more reliable for such large systems.

A stable suboptimal controller achieving $\|w \mapsto z\|_\infty = 0.94$ was obtained. Its order was the same as the order of the minimal generalized plant, i.e., 78, but a balanced truncation reduced it to 55 state variables without affecting the closed-loop ∞-norm. With this reduced controller K_1, Figure 7 shows that required performance has been attained, i.e., $\|S_{rh}(j\omega)\|$ and $\|S_{dh}(j\omega)\|$ are less than $|w^{-1}(j\omega)|$, as desired. The least-damped closed-loop mode has a damping ratio of 0.38.

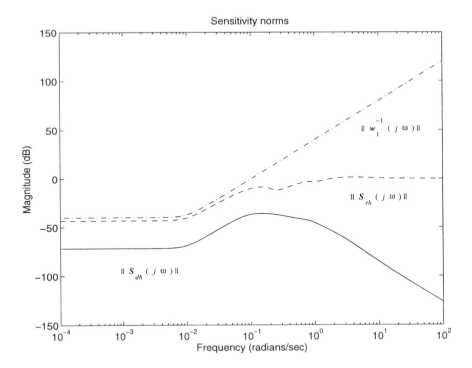

Fig. 7. Norms of $S_{dh}(j\omega)$ and $S_{rh}(j\omega)$ for K_1.

The 55th-order controller $\boldsymbol{K}_1$ was rescaled to $\boldsymbol{K}_{p1} = \frac{d_{max}}{\gamma} J_2^{-1} \boldsymbol{K}_1$, a controller using the actual rib and hub angles to compute actual rib and hub control torques. Then, since the implementation of the controller must be digital, $\boldsymbol{K}_{p1}$ was discretized at a sampling rate of 10 Hz using the bilinear transformation; call the resulting controller $\boldsymbol{K}_{p1d}$. Another discretized version of the controller was computed using the MATLAB^{TM} function *c2d* —it destabilized Daisy both in simulations and experiments. The *c2d* function performs the discretization by placing a sampler at the output of the controller and a zero-order hold at the input. The frequency responses of $\boldsymbol{K}_{p1}$ and $\boldsymbol{K}_{p1d}$ were close up to 10 rad/s, whereas for the controller discretized with *c2d*, the frequency responses started to differ significantly from 1 rad/s. The 10 Hz sampling rate was almost the highest achievable on the real-time control computer system with our control software. An earlier version of the program allowed a maximum sampling rate of 5 Hz only—when used to implement a discretized version of $\boldsymbol{K}_{p1}$, it destabilized the closed loop. This suggests that an $\mathcal{H}_\infty$-optimal sampled-data design might be warranted here, but this is left for future work.

We used the following hub torque disturbance profile for all tests:

$$\begin{cases} A_d, & \text{if } 0 \le t \le T \\ -1.5A_d, & \text{if } T < t \le 2T \\ 0, & \text{else.} \end{cases}$$

It can be applied by any of the three torque wheels, individually or in any combination. Notice that this disturbance is completely specified by three parameters: the amplitude of the first torque pulse, A_d; the duration of the first and second pulses, T; and the combination of hub axes around which the disturbance is applied, $axes$. This latter parameter can take on values in the set $\{x, y, z, xy, xz, yz, xyz\}$. With these definitions, let us denote the disturbance as $D(A_d, T, axes)$. The controller is switched on after the hub angle experiencing the largest deviation changes sign. Thus the disturbance has roughly the effect of a torque impulse applied to the hub because the controller starts when the hub angles are small while the angular velocities are large. However, the rib angles may not be small at switch-on time. Although experimental controller performance would be best assessed by performing frequency-response experiments and comparing with $S_{dh}(j\omega)$, these are certainly not practical for LFSSs. But the torque impulse response matrix is just the inverse Laplace transform of the sensitivity S_d. Hence this provides some motivation for judging and comparing controller performance using time responses of the rib and hub angles to the disturbance $D(A_d, T, axes)$. For all the plots, $t = 0$ corresponds to the instant at which the controller is turned on.

As a benchmark, an open-loop response of Daisy to $D(13.5\text{Nm}, 2s, x)$ is plotted in Figure 8 along with a simulated continuous-time response of the nominal model $C_1 G$. Discrepancies between some of the actual and nominal modal frequencies and damping ratios can be observed from these plots, illustrating the uncertainty in the model.

All simulations are linear and *discrete-time* with the plant model (including actuator dynamics) discretized at 10 Hz using *c2d*. As a typical test run, for the torque disturbance $D(13.5\text{Nm}, 2s, y)$, Figures 9 and 10 show respectively the hub and rib angle responses, while the hub control torques are plotted in Figure 11 and the rib control torques are in Figure 12. When compared with the response in Figure 8, it is clear that the $\mathcal{H}_\infty$ controller vastly improves the dynamics of Daisy. The experimental response of θ_{hy} has a slightly longer settling time than its simulated counterpart. The rib responses are quite consistent with the simulated ones, showing actual performance very close to the nominal. This is in spite of rib torque saturation, which occurred for the first two seconds.

In conclusion, the experimental data show that the $\mathcal{H}_\infty$ controller K_{p1d} designed using the coprime factorization method performed quite well. *No experimental tuning of the controller was necessary*, which shows evidence of robustness of the design.

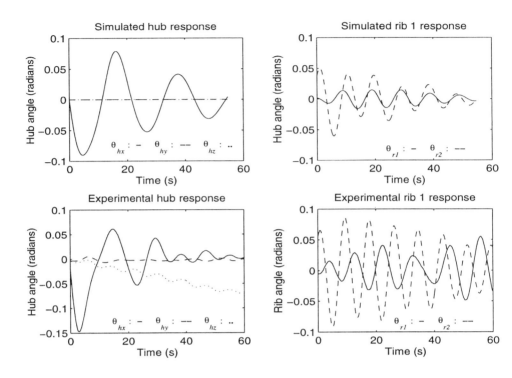

Fig. 8. Simulated and experimental open-loop responses of Daisy to $D(13.5\text{Nm}, 2s, x)$.

7 Conclusion

A new approach introduced in [3] to the robust control of LFSSs using a coprime factor description of the plant's dynamics was presented. This approach first involves the transformation of a natural description of the uncertainty as bounded perturbations of the modal parameters of an original FE model into norm-bounded stable perturbations of a nominal coprime factor pair. This new unstructured description of the uncertainty is not overly conservative in terms of real perturbations of the modal parameters, and it can also represent unmodeled dynamics. Moreover, the $\mathcal{H}_\infty$ and μ-synthesis controller design methods can be used with this type of uncertainty, and lead to computationally tractable problems despite the high order of LFSS dynamics. To illustrate the technique, an $\mathcal{H}_\infty$ controller was designed for Daisy. This model has significant parameter uncertainty, yet the controller designed using the coprime factorization technique was quite robust and achieved good performance levels in terms of rejection of hub torque disturbances. Furthermore, the controllers were stable, which is a desirable property. Extensive experimentations showed that digital implementations of the $\mathcal{H}_\infty$ controller performed very well without the need of any experimental tuning. Further work is underway to include nonlinearities and sampling issues

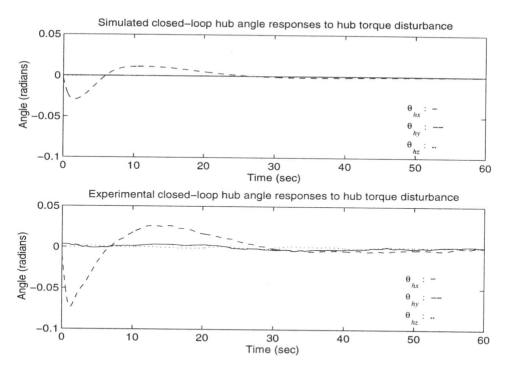

Fig. 9. Simulated and experimental closed-loop hub angle responses with K_{p1d}, $D(13.5\text{Nm}, 2s, y)$.

into the design technique.

8 Acknowledgements

Much thanks to Vince Pugliese, system manager, and Regina Sun Kyung Lee, graduate student, both from UTIAS, for help with the experiments. Regina also provided the drawing of Daisy in Figure 1.

References

1. G.J. Balas and J.C. Doyle, Robustness and Performance Tradeoffs in Control Design for Flexible Structures. *Proc. of the 1991 Amer. Control Conf.*
2. G.J. Balas, J.C. Doyle, K. Glover, A. Packard, and R. Smith, *µ-Analysis and Synthesis Toolbox: User's Guide*. The Mathworks Inc., 1991.
3. B. Boulet, B.A. Francis, P.C. Hughes, and T. Hong, Robust control of large flexible space structures using a coprime factor plant description. Proc. of the 1994 Amer. Cont. Conf., Baltimore, Maryland, pp. 265-266.

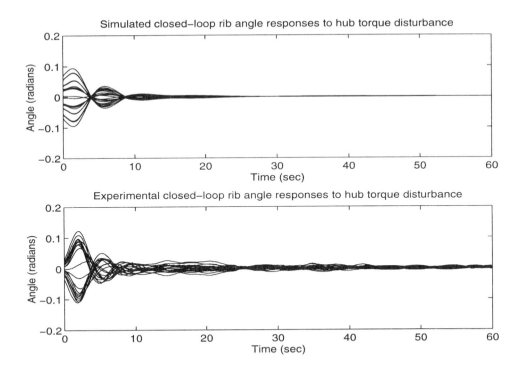

Fig. 10. Simulated and experimental closed-loop rib angle responses with K_{p1d}, $D(13.5\text{Nm}, 2s, y)$.

4. B. Boulet, B.A. Francis, P.C. Hughes, and T. Hong, Robust $\mathcal{H}_\infty$ Control of Large Flexible Space Structures Using a Coprime Factor Plant Description. *Systems Control Group Report No. 9401*, Dept. of Electrical and Computer Engineering, University of Toronto, 1994.
5. R.P. Braatz, P.M. Young, J.C. Doyle, and M. Morari, Computational Complexity of μ Calculation. *IEEE Trans. Aut. Control, Vol. 39, No. 5, pp. 1000-1002, May 1994.*
6. S.A. Buddie, T.T. Georgiou, U. Özgüner, and M.C. Smith, Flexible structure experiment at JPL and WPAFB: $\mathcal{H}_\infty$ controller designs. *Int. J. Control, Vol. 58, No. 1, pp. 1-19, 1993.*
7. E.G. Collins, Jr., D.J. Phillips, and D.C. Hyland, Robust Decentralized Control Laws for the ACES Structure. *IEEE Control Syst. Magazine, Vol. 11, No. 3, pp. 62-70, April 1991.*
8. G.W. Crocker, P.C. Hughes, and T. Hong, Real-Time Computer Control of a Flexible Spacecraft Emulator. *IEEE Control Syst. Magazine, Vol. 10, No. 1, pp. 3-8, January 1990.*
9. E.J. Davison, W. Gesing, and S.H. Wang, An Algorithm for Obtaining the Minimal Realization of a Linear Time-Invariant System and Determining if a System is Stabilizable-Detectable. *IEEE Trans. Aut. Control, Vol. 23, No. 6, pp. 1048-1054,*

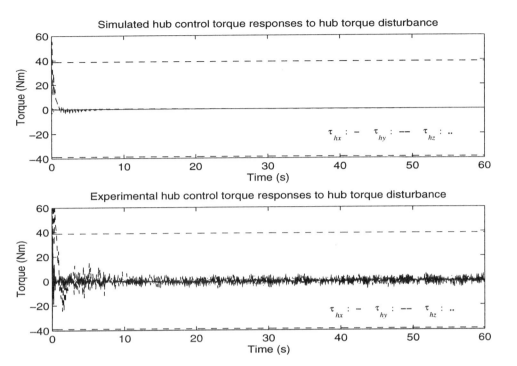

Fig. 11. Simulated and experimental computed hub control torques for K_{p1d}, $D(13.5\text{Nm}, 2s, y)$.

December 1978.
10. J.C. Doyle, K. Glover, P.P. Khargonekar, and B.A. Francis, State-Space Solutions to Standard $\mathcal{H}_2$ and $\mathcal{H}_\infty$ Control Problems. *IEEE Trans. Aut. Control*, Vol. 34, No. 8, pp. 831-847, August 1989.
11. M.K.H. Fan, A.L. Tits, and J.C. Doyle, Robustness in the Presence of Mixed Parametric Uncertainty and Unmodeled Dynamics. *IEEE Trans. Aut. Control*, Vol. 36, No. 1, pp. 25-38, January 1991.
12. C.Z. Gregory, Jr., Reduction of Large Flexible Spacecraft Models Using Internal Balancing Theory. *AIAA J. of Guid., Contr. and Dyn.*, Vol. 7, No. 6, Nov.-Dec. 1984.
13. R.A. Horn and C.R. Johnson, *Matrix Analysis*, Cambridge Univ. Press, 1990.
14. J.P. How, W.M. Haddad, and S.R. Hall, Robust Control Synthesis Examples with Real Parameter Uncertainty using the Popov Criterion. *Proc. of the Amer. Control Conf., San Francisco, June 1993.*
15. J.P. How, S.R. Hall, and W.M. Haddad, Robust Controllers for the Middeck Active Control Experiment using Popov Controller Synthesis. *IEEE Trans. Control Syst. Tech.*, Vol. 2, No. 2, pp. 73-87, June 1994.
16. S.M. Joshi, Control of Large Flexible Space Structures. *Lecture Notes in Control and Information Sciences*, Vol. 131, New-York: Springer-Verlag, 1989.

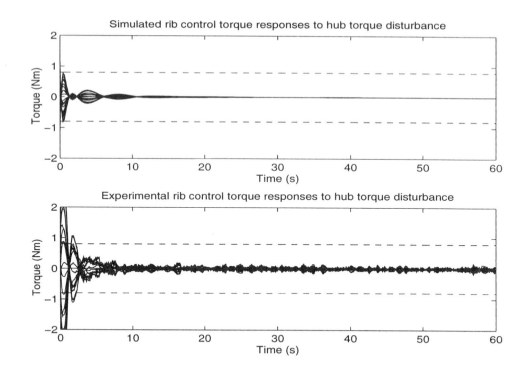

Fig. 12. Simulated and experimental computed rib control torques for K_{p1d}, $D(13.5\text{Nm}, 2s, y)$.

17. D.G. Laurin, Development of an Optical Imaging System for Shape Monitoring of Large Flexible Structures. Ph.D. Thesis, Dept. of Aerospace Eng., Univ. of Toronto, 1992.
18. F.C. Lee, H. Flashner, and M.G. Safonov, Positivity Embedding for Noncolocated and Nonsquare Flexible Structures. *Proc. of the Amer. Control Conf., June 1994, Baltimore, Maryland.*
19. K.B. Lim and G.J. Balas, Line-of-sight Control of the CSI Evolutionary Model: μ Control. *Proc. of the 1992 Amer. Control Conf.*
20. K.B. Lim, P.G. Maghami, and S.M. Joshi, Comparison of Controller Designs for an Experimental Flexible Structure. *IEEE Control Syst. Magazine, Vol. 12, No. 3, pp. 108-118, June 1992.*
21. D.C. McFarlane and K. Glover, *Robust Controller Design Using Normalized Coprime Factor Plant Descriptions,* Springer-Verlag, 1990.
22. D.G. Meyer, A Fractional Approach to Model Reduction. *Proc. of the Amer. Control Conf., pp. 1041-1047, 1988, Atlanta.*
23. B.C. Moore, Principal Component Analysis in Linear Systems: Controllability, Observability, and Model Reduction. *IEEE Trans. Aut. Control, Vol. 26, No. 1, pp. 17-32, February 1981.*

24. R.E. Skelton and P.C. Hughes, Modal Cost Analysis for Linear Matrix-Second-Order Systems. *ASME J. Dyn. Syst., Meas., and Cont., Vol. 102, pp. 151-158, September 1980.*
25. R.S. Smith, C.-C. Chu, and J.L. Fanson, The Design of $\mathcal{H}_\infty$ Controllers for an Experimental Non-collocated Flexible Structure Problem. *IEEE Trans. Control Syst. Tech., Vol. 2, No. 2, pp. 101-109, June 1994.*
26. M. Vidyasagar, *Control System Synthesis: A Coprime Factorization Approach. MIT press*, 1985.

Progress in Applied Robust Control

Keith Glover

Department of Engineering
University of Cambridge, UK.

1 Introduction

There have been many recent advances in the area of robust control and this paper discusses some of these results and indicates how they can be applied in practice with particular reference to a flight control example.

The central motivation for feedback systems is plant uncertainty and we begin with a discussion of types of models for uncertain systems. The general class considered are various forms of feedback configurations including fixed dynamics and uncertain blocks with bounded gain. Comparisons are made between microscopic and macroscopic descriptions.

This general model of an uncertain plant together with a performance measure constitute the basis of much of the current developments in robust control and is normally drawn as in Figure 1 where G is a given linear system; K is the controller to be designed; Δ represents the uncertainty; w is a disturbance vector; z is a vector of errors; y are the measurements available for the controller whose output is u.

Methods for the robust control of such systems depend on the class of signals from which w is taken and the class of uncertainty from which Δ is taken and this will next be discussed. Analysis and design are both considered and it is seen that the different assumptions can result in quite different control synthesis and analysis problems e.g. $\mathcal{H}_\infty$ loop shaping, μ-analysis and μ-synthesis.

The general methodology of minimizing the 'induced norm' in the face of uncertainty can be extended to sampled-data systems where some problems routinely translate (e.g. minimum induced norm without uncertainty), but others are more involved (e.g. robust stability to LTI perturbations). An additional concern in controller implementation is the typically high degree of the resulting controller and we next discuss methods for controller reduction.

The resulting system will have been designed on the basis of the earlier uncertain model classes and it is then necessary to validate these assumptions with (flight) test data. There have been a number of recent interesting results on uncertain model (in)validation which are finally outlined.

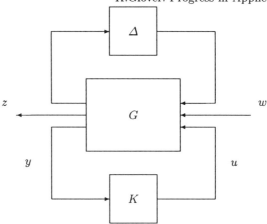

Fig. 1. General robust control block diagram

2 Harrier VSTOL example

A collaboration between the DRA, Bedford, UK, and the Control Group at Cambridge University, on flight control for a Harrier VSTOL aircraft has been in progress for the period 1989–1993, with R.A. Hyde as a research student and then post-doctoral researcher. This resulted in a successful series of flight tests in December 1993 and early 1994 (see Hyde, Glover and Shanks [HydGS] and the monograph by Hyde [Hyd] for more details). A simulation model had been supplied and was derived using basic aerodynamics and experimental results. It included both nonlinear dynamic equations and look-up tables for certain effects. The dynamics vary significantly with forward speed, operating as a conventional aircraft at high speed but being very unstable near the hover. The particular control problem considered was the longitudinal dynamics, controlling pitch, forward and vertical speed with the throttle, nozzle angle and tail plane.

3 Construction and Analysis of Uncertain Models

Classical control considered uncertainty in terms of stability margins and bandwidth. That is the desired bandwidth was chosen so that the model accuracy around the bandwidth was consistent with stability margins of, for example, 40° and 6 dB. For single loop systems this has been effective and straightforward since the desired stability margins are in some sense an aggregation of many forms of uncertainty and identification of all the contributing factors to this overall uncertainty may not be required.

We will discuss two approaches to representing uncertainty in multivariable systems: detailed microscopic modelling of each uncertain component and aggregated or macroscopic modelling of the overall system uncertainty. The standard

3 Construction and Analysis of Uncertain Models

approach in μ-analysis in Doyle et al ([Doy], [BDGPS], [PacD]) is to model all sources of uncertainty. For example an actuator might be modelled as

$$(1 + w_a(s)\delta_a(s)) g_a(s)$$

where $g_a(s)$ is the nominal transfer function, $\delta_a(s)$ is stable with $|\delta_a(j\omega)| < 1$ for all ω and $w_a(s)$ is a weighting function whose modulus reflects the uncertainty in $g_a(j\omega)$. Similarly an uncertain mass can be written as $(m + w_m\delta_m)$ where δ_m is real with $|\delta_m| < 1$ and w_m is a real number. This results in a block diagram of the form of Figure 2 that can be easily manipulated into the form of Figure 1 with a block diagonal Δ. Note that the construction of this uncertain model requires substantially greater effort than is usual since every parameter and component should be accompanied by an error estimate, including for example data derived from wind tunnel tests. It is however possible to use aggregated uncertainty descriptions for subsystems if necessary.

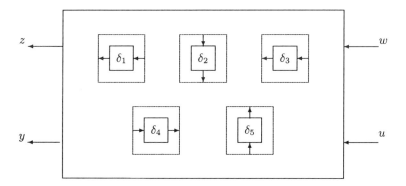

Fig. 2. Uncertain Model

The above is in contrast to aggregated methods where the multivariable transfer function could, for example, be written in a form such as,

$$(I + W_{01}(s)\Delta_0(s)W_{02}(s)) G(s) (I + W_{i1}(s)\Delta_i(s)W_{i2}(s)) + W_{a1}(s)\Delta_a(s)W_{a2}(s),$$

where $\|\Delta_0\|_\infty < 1$, $\|\Delta_i\|_\infty < 1$, $\|\Delta_a\|_\infty < 1$. Δ_0, Δ_i and Δ_a with their respective weighting matrices represent, multiplicative input uncertainty, multiplicative output uncertainty and additive uncertainty, respectively. It is generally necessary to scale inputs and outputs of G into units where similar variations in each of the inputs and each of the outputs are expected. Even with this 'broad brush' approach the number of weights to be chosen remains high and a compromise is required between making the uncertainty sufficiently large to include all likely plants but not so large as to make resulting stability/performance tests overly conservative.

The abstract question of the distance between two linear systems has attracted much interest in recent years and yielded a most elegant set of results.

The gap metric as proposed by Zames and El-Sakkary [ZamE] has been found to be particularly suitable for feedback systems. The results of Georgiou [Geo] on calculating the gap and of Georgiou and Smith [GeoS] on optimal controllers for robust stabilization with respect to plant perturbations in the gap metric form a most complete theory. If we denote the gap between P_0 and P_1 as $\delta(P_0, P_1)$ then a controller, K_0, will stabilize all P_1 with $\delta(P_0, P_1) < b(P_0, K_0)$ where

$$b(P_0, K_0) = \left\| \begin{bmatrix} I \\ K_0 \end{bmatrix} (I - P_0 K_0)^{-1} \begin{bmatrix} P_0 & I \end{bmatrix} \right\|_\infty^{-1}$$

This $\mathcal{H}_\infty$ norm is the norm of the closed loop transfer function, from w to z, in Figure 3.

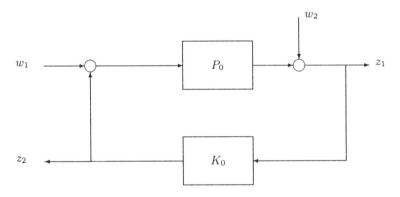

Fig. 3. Closed loop for gap metric robustness

Vinnicombe defines a new version of the gap, δ_ν, in [Vinn] that essentially gives that the set of plants, P, satisfying $\delta_\nu(P_0, P) < \alpha$ is given by those plants that are stabilized by all K_0 satisfying $b(P_0, K_0) \leq \alpha$.

We see here that the class of allowable perturbations in the open loop plant is intimately connected to the closed loop behaviour. With this aggregated description of uncertainty it is therefore necessary to specify the desired closed loop behaviour, in for example the bandwidth and loop gain, and uncertainty is described with this in mind. Generally frequency weights need to be introduced that give the desired loop shape, $W_0 P_0 W_i$ and perturbed plants, P, with $\delta_\nu(W_0 P_0 W_i, W_0 P W_i) < \alpha$ considered.

Finding a K_0 to maximize $b(W_0 P_0 W_i, K_0)$ is an $\mathcal{H}_\infty$ norm minimization problem with a special solution obtained by McFarlane and Glover [McFG1] when considering robust stabilization of plants subject to perturbations in the coprime factors $(N + \Delta_N)(M + \Delta_M)^{-1}$ with $\left\| \begin{bmatrix} \Delta_N \\ \Delta_M \end{bmatrix} \right\|_\infty < \alpha$.

The measure $b(W_0 P_0 W_i, K_0)$ can be viewed as either a robust stability condition or the signal gain (performance) criterion in Figure 3. A robust performance criterion, given by Vinnicombe [Vinn] and Qiu and Davison [QiuD], for the performance of a controller, K_1, applied to weighted plant, $W_0 P_1 W_i$, is

$$\arcsin b(W_0 P_1 W_i, K_1) \geq \arcsin b(W_0 P_0 W_i, K_0) - \arcsin \delta_\nu (W_0 P_0 W_i, W_0 P_1 W_i) \\ - \arcsin \delta_\nu (K_0, K_1). \quad (1)$$

In the VSTOL example *a priori* confidence limits were not available for the supplied model and hence microscopic modelling leading to μ-analysis was not feasible. The gap-based methods were however found to be appropriate especially with the robust performance result of (1).

In summary we have described three models for uncertain systems. The advantages of the structured uncertainty model used in μ-analysis is that it should provide an accurate model set at significant modelling effort. Care needs to be taken to ensure that all uncertainty sources are included, as well as all disturbance sources. The set from which the perturbations, Δ, are drawn is important, e.g. real scalars, transfer functions, time varying linear systems, nonlinear systems, as is the induced norm chosen to bound the size of Δ, e.g. $\|\Delta\|_\infty$ giving induced norm for $\mathcal{L}_2$ signals or the $\mathcal{L}_1$ norm of the impulse response giving the induced norm for $\mathcal{L}_\infty$ signals. Much recent progress has been made in deriving necessary and sufficient conditions for stability with varying assumptions on norms and perturbation classes (see Dahleh and Diaz-Bobillo [DahD] for a comprehensive discussion). When some of the diagonal elements of Δ are restricted to being real scalars robust stability conditions are more involved and bounds for the stability radius are given in Young [You]. The combination of multiplicative and/or additive perturbations needs care in scaling and typically the weights are chosen to be diagonal in the absence of a good methodology for choosing otherwise. The gap-based uncertainty descriptions tend to integrate the uncertainty modelling and control system design problems, and this turns out to be effective and straightforward as in classical control.

4 Controller Design

The synthesis of controllers to robustly stabilize the maximal set of plants with an unstructured uncertainty block Δ is a straightforward calculation with the $\mathcal{L}_2$ induced norms from $\mathcal{H}_\infty$ synthesis, [DGKF], and is also solvable in the case of $\mathcal{L}_\infty$ induced norms in the discrete time case using linear programming methods [DahD]. However in the case of structured uncertainty there are not guaranteed solutions, although in the case of μ-synthesis with LTI Δ-blocks an iteration between the D-scales of μ-analysis and $\mathcal{H}_\infty$ controller synthesis usually gives an effective procedure. If some of the parameters are restricted to being real then the synthesis is substantially harder [You].

In the Harrier design it was found to be straightforward to give a good design using $\mathcal{H}_\infty$ loop shaping. This procedure proposed in [McFG2] firstly specifies output scaling W_0 and a precompensator $W_i(s)$ so that the loop gains $\sigma_i(W_0 G W_i)$ are as desired. Then a feedback controller, K_∞, is used to maximise $b(W_0 G W_i, K_\infty) = b_{opt}(W_0 G W_i)$. An attractive feature is that the value of b_{opt} will be between 0 and 1 and if it is greater than approximately 0.25 the desired loop shape can be achieved with good robust stability. This then gives the designer an easy interaction with the design variables. Although the method appears to consider a somewhat restrictive problem formulation it nevertheless has been found to give a flexible design tool.

The above designs are all for the control of a family of plants in a neighbourhood of a fixed linear plant. In the VSTOL example the dynamics vary with speed to an extent that a single controller is not possible and the controller needs to be scheduled with speed. In this application the structure of the controllers were exploited to give their smooth variation with speed and to include desaturation of the actuators.

Recent alternative approaches to scheduling have been derived by Packard and Becker [Pack, Bec], where the system equations are considered to depend on the scheduling parameters, but that the scheduling parameters are then assumed to be external variables moving arbitrarily between their limits. This results in sufficient conditions for stability in this framework and the controller synthesis problem requires the solution of a Linear Matrix Inequality (LMI). This method is in its infancy but shows great promise.

5 Sampled-Data Systems

Since controllers for continuous time systems will be implemented on digital computers the subject of sampled data control has received renewed attention in recent years using the "norm-based" techniques. (See the monograph by Dullerud [Dul]). A solution of the induced $\mathcal{L}_2$-norm problem was derived in Bamieh et al [BPFT], using a lifting technique which essentially considers the sampled-data system as a periodically varying continuous system and then gives an equivalent discrete-time system for which the discrete-time controller is designed.

One of the motivations for this work was to avoid problems with intersample behaviour that can occur with for example assuming that the inputs are constant between samples and only considering the outputs at the sampling points.

In the continuous-time case with unstructured uncertainty the small gain theorem gives the same robust stability conditions independent of whether the perturbation is LTI or non-linear and time varying. This is not the case for sampled-data systems because the sampled-data maximum system gain will generally be achieved with signals that are not consistent with the input and output of a LTI perturbation but instead with a periodically time varying perturbation. This phenomena can result in the small gain theorem being arbitrarily conservative for LTI perturbations. Necessary and sufficient conditions robust stability with structured LTI perturbations have been derived in Dullerud and Glover [1993].

Although the theory of sampled-data systems is in a similar, albeit more involved, state of development as the continuous or purely discrete cases, design guidelines similar to those in $\mathcal{H}_\infty$ loop shaping are still under development. For example it is customary to always include $\mathcal{L}_2$ measurement noise in the continuous case which then prevents the controller from trying to differentiate the measurement. In the sampled-data case this is not so straightforward since a sampler has infinite gain from $\mathcal{L}_2 \to \ell_2$, and hence the measurement noise needs to be filtered and whether the filter should be a function of sampling period, or part of the controller design, or fixed needs to be determined. An alternative is to make the measurement noise discrete and act on the sampled signal.

Details of the VSTOL example discrete time aspects are given in [Hyd]. Originally the design had been based on a 20ms sampling period using the sampled-data equivalent system assuming that the inputs were constant between samples and only considering the sampled outputs. The performance could also be checked using the sampled-data induced norm as above. However when the controller came to be coded it was found that a 40 ms sampling period was required to complete the calculations. Some rather special rearrangements of the controller structure was then required to minimize the delay in the faster pitch loop, however the $\mathcal{H}_\infty$ loop-shaping cost was retained as the performance indicator.

6 Controller Reduction

Controllers derived using $\mathcal{H}_\infty$ or μ-synthesis tend to have high state dimension and it is generally advantageous to reduce this order if possible with only modest reduction in performance. In some specific problems intuitively appealing results are available such as those of Vinnicombe [Vinn] on model reduction in the nu-gap metric which can be combined with the robust performance result in (1).

In more general cases sufficient conditions for the closed-loop $\mathcal{H}_\infty$ norm or μ-value to be less than γ can be written as frequency weighted model reduction problems [AndL, ZSC, GoG]. The principal difficulty is that the frequency weighted model reduction problem is not as reliably solved as the unweighted case and various numerical approximations are required. However the final methods produce effective, albeit computationally demanding, procedures.

In the VSTOL example model reduction techniques of this type were not employed because the controller was to be scheduled against forward speed and the controller needed to vary smoothly with the scheduling parameter. With the above reduction techniques there would be no relation between the state space description of each reduced order controller.

7 Model Validation

The whole of the preceding development has been based on an uncertain model for the system and an important question is the validation of such a model class. There are several possible problem formulations given experimental data:

– Statistical model validation, with assumed stochastic assumptions on the noise signals and confidence limits on the parameter estimates. Then check for intersection or inclusion of this set with the given model class (nontrivial).

- Is the data consistent with $\|\Delta\| < 1$ and the noise assumptions?
- Do the data and noise assumptions imply that $\|\Delta\| < 1$?

These are fundamentally different problem formulations and we will be considering the second which only determines whether the data could have been produced by the model set. Results for the much stronger third formulation are typically discouraging; for example Dahleh et al [DahTT] show that in order to identify a finite impulse response system to within some hard bound and in the presence of worst case noise requires experiments of length that increases exponentially with the desired accuracy, giving impractical data lengths.

A number of results in the time domain model invalidation as in Polla et al [PKTKN] show that the problem can be reduced to solving a large LMI. This has been exploited in Davis and Glover [DavG] in the analysis of flight test data from the VSTOL example. Swept sinusoids were injected at the points w_1 and w_2 in Figure 3 with the weighted plant for this test with the pilot in a 'hands-off' mode. It was found that the class of uncertainty critically affected the required size of Δ to be consistent with the data. typically LTI perturbations needed to have significantly larger size than LTV and acausal LTV perturbations. Also the computational requirements were substantially greater for the LTI case. Nevertheless it is demonstrated that the flight test data was consistent with the prior model set.

8 Conclusions

We have discussed recent advances in uncertain system modelling, analysis, and controller synthesis together with results on sampled-data systems and model validation. The results and approaches have been illustrated with reference to a flight control case study that has been test flown. It is seen that all the methods have implications for application but that some are more fully developed into usable design and analysis tools.

9 Acknowledgements

The author is pleased to acknowledge that most of the work reported here was performed by former PhD students in particular Duncan McFarlane, Rick Hyde, Glenn Vinnicombe, Rob Davis, Phil Goddard and Sanjay Lall. In addition financial support from the EPSRC and NATO is acknowledged.

References

[AndL] Anderson, B.D.O., Liu, Y.: Controller reduction: concepts and approaches. IEEE Transactions on Automatic Control, **Vol. AC-34 no.8**, (1989), 802–812.

[BDGPS] Balas, G., Doyle, J.C., Glover, K., Packard, A., Smith, R.: μ-analysis and Synthesis Toolbox. The MathWorks, (1993).

Bibliography

[Bec] Becker, G.S.: Quadratic Stability and Performance of Linear Parameter Dependent Systems. Doctoral Dissertation, University of California at Berkeley, (1993).

[BPFT] Bamieh, B., Pearson, J.B., Francis, B.A., Tannenbaum, A.: A Lifting Technique for Linear Periodic Systems with Applications to Sampled-data Control. Systems and Control Letters, **8**, (1991).

[DahD] Dahleh, M.A., Diaz-Bobillo, I.J.: Control of Uncertain Systems: A linear programming approach. Prentice-Hall, (1995).

[DahTT] Dahleh, M.A., Theodosopoulos, T.V., Tsitsiklis, J.N.: The sample complexity of worst-case identification on FIR linear systems. Systems and Control Letters, **20**, (1993), 157–166.

[DavG] Davis, R.A., Glover, K.: An application of recent model validation techniques to flight test data. To be presented at the European Conference on Control, Rome, (1995).

[Doy] Doyle, J.C.: Analysis of feedback systems with structured uncertainties. IEE Proceedings, Part D, **133**, (1982), 45–56.

[DGKF] Doyle, J.C., Glover, K., Khargonekar, P. and Francis, B.: State-space solutions to standard $\mathcal{H}_2$ and $\mathcal{H}_\infty$ control problems. IEEE Transactions on Automatic Control, **34**, no.8, (1989).

[Dul] Dullerud, G.E.: Control of Uncertain Sampled-Data Systems. To appear Birkhäuser, (1995).

[Geo] Georgiou, T.T.: On the computation of the gap metric. Systems and Control Letters, **11**, (1988), 253–257.

[GeoS] Georgiou, T.T., Smith, M.C.: Optimal robustness in the gap metric. IEEE Transactions on Automatic Control, **35**, (1989), 673–686.

[GoG] Goddard, P.J., Glover, K.: Performance preserving frequency weighted controller approximation: a coprime factorization approach. Proceedings of the 33rd Conference on Decision and Control, Orlando, Florida, (December 1994).

[Hyd] Hyde, R.A.: $\mathcal{H}_\infty$ Aerospace control: A VSTOL Application. To appear Advances in Industrial Control Series, Springer-Verlag, (1995).

[HydGS] Hyde, R.A., Glover, K., Shanks, G.T.: VSTOL first flight of an $\mathcal{H}_\infty$ control law. IEE Computing and Control Engineering Journal, **6**, No. 1, (1995), 11–16.

[McFG1] McFarlane, D.C., Glover, K.: Robust Controller Design using Normalized Coprime Factor Plant Descriptions. Springer Verlag, Lecture Notes in Control and Information Sciences, **138**, (1989).

[McFG2] McFarlane, D.C., Glover, K.: A Loop Shaping Design Procedure using $\mathcal{H}_\infty$ Synthesis. IEEE Transactions on Automatic Control, **37**, no.6, (1989), 759–769.

[Pack] Packard, A.: Gain scheduling via linear fractional transformations. System and Control Letters, **22**, (1994), 79–92.

[PacD] Packard, A., Doyle, J.: The complex structured singular value. Automatica, **29**, (1993), 71–109.

[PKTKN] Poolla, K., Khargonekar, P., Tikku, A., Krause, J., Nagpal, K.:. A time-domain approach to model validation. IEEE Trans. Automat. Contr., **39(5)**, (1994),951–959.

[QiuD] Qiu, L., Davison, E.J.: Feedback stability under simultaneous gap metric uncertainties in plant and controller. Systems and Control Letters, **18(1)**, (1992), 9–22.

[Vinn] Vinnicombe, G.: Frequency domain uncertainty and the graph topology. IEEE Transactions on Automatic Control, **38**, no.9, (1993), 1371–1383.

[You] Young, P.M.: Robustness with Parametric and Dynamic Uncertainty. PhD Thesis, California Institute of Technology, (1993).

[ZamE] Zames, G., El-Sakkary: Unstable systems and feedback: The gap metric. Proceedings of the Allerton Conference, (October 1980), 380–385.

[ZSC] Zhou, K., D'Souza, C., Cloutier, J.R.: Structurally balanced controller order reduction with guaranteed closed loop performance. Systems and Control Letters, **23**, no.6, (1994).

Chain-Scattering Approach to Control System Design

Hidenori Kimura and Fumitake Okunishi

Department of Systems Engineering Osaka University,
1-3, Machikaneyama, Toyonaka 560, JAPAN

Abstract. This paper proposes a new design framework of control systems based on chain-scattering representation of the plant. The chain-scattering representation has several advantages over the conventional input-output description of the plant. It can represent the feedback connection in a simple way, which makes the role of factorization explicitly clear. It has a strong symmetry (duality) to its inverse. It clarifies the meaning of pole-zero cancellation. In this paper, we shall exploit some algebraic properties of the chain-scattering representation of the plant which are relevant to control system design, especially to H^∞ control.

1 Introduction

In addition to the familiar impedance and admittance matrices representing multiport electrical network, classical circuit theorists used a wide variety of circuit representations to deal with analysis and synthesis problems. Among them, a representation called *chain matrix* [Be1] proved to be useful to deal with the cascade connection of circuits arising in the design of filters. This representation is also called a *fundamental matrix* or simply *F-matrix*. It has a number of advantages compared with using the impedance or admittance. The most salient characteristic feature of the chain matrix is its cascade structure, namely, the cascade connection of the two multiport circuits results in the multiplication of each chain matrix. This is a remarkable property, because the bilateral power flows across the two circuits generated by the cascade connection are represented simply by the product of each chain matrix. In terms of control, the biliteral power flows can be interpreted as the existence of feedback loop. The chain matrix represents the feedback simply as a multiplication of a matrix. This property makes the analysis of closed-loop systems very simple and makes the role of factorization clearer. This is the fundamental motivation to bring forward the use of chain-scattering matrix in control system design [Ki1]. Another salient feature of the chain-scattering representation is the symmetry (duality) between the chain-scattering representation and its inverse. This property has not yet been exploited fully, but it is considered to be quite relevant to control system design. The chain-scattering representation, however, has a serious disadvantage that it may not exist for general plants and we need to augment the plant in order to compute its chain-scattering representation. But again, we can get a deep insight into the fundamental structure of control system by investigating the properties of augmentation. In this paper, we investigate algebraic properties of chain-scattering representations and clarify the fundamental structure of augmentation and factorization. The emphasis is placed on H^∞ control, but the use of the chain-scattering framework can be extended beyond H^∞ control. It should be noted that the chain-scattering representation is widely known in circuit theory and signal processing. A nice exposition of the chain-scattering theory is found in [DD1].

Notations:

- $BH^\infty = \{F(s) : \text{stable}, \|F\|_\infty < 1\}$.
- $F^\sim(s) := F^T(-s)$, $F^*(s) = \bar{F}^T(\bar{s})$.
- $\left[\begin{array}{c|c} A & B \\ \hline C & D \end{array}\right] := D + C(sI - A)^{-1}B$
- $Ric(H)$ for $H = \begin{bmatrix} A & W \\ Q & -A^T \end{bmatrix}$ denotes the solution of $XA + A^T X + XWX - Q = 0$ for which $A + WX$ is stable.

2 Chain-Scattering Representation of Plants

Consider a plant P of Fig.2.1 with two kinds of inputs (w, u) and the two kinds of outputs (z, y) represented as

$$\begin{bmatrix} z \\ y \end{bmatrix} = P \begin{bmatrix} w \\ u \end{bmatrix} = \begin{bmatrix} P_{11} & P_{12} \\ P_{21} & P_{22} \end{bmatrix} \begin{bmatrix} w \\ u \end{bmatrix} \quad (2.1)$$

Assuming that P_{21} is invertible, we have the *chain-scattering representation* (CSR) of P as

$$\begin{bmatrix} z \\ w \end{bmatrix} = CHAIN(P) \begin{bmatrix} u \\ y \end{bmatrix} \quad (2.2)$$

where

$$CHAIN(P) = \begin{bmatrix} P_{12} - P_{11}P_{21}^{-1}P_{22} & P_{11}P_{21}^{-1} \\ -P_{21}^{-1}P_{22} & P_{21}^{-1} \end{bmatrix} \quad (2.3)$$

If P represents a usual input/output relation of a system, $CHAIN(P)$ represents the characteristic of power ports reflecting more or less physical structure of the plant. The chain-scattering representaion describes the plant as a *wave scatterer* between (u, z)-wave and the (w, y)-wave that travel oppositely to each other (Fig.2.2).

The main reason of using the CSR lies in its ability of representing the feedback connection as a cascade one, as shown in Fig.2.3. The cascade connection of two CSRs G_1 and G_2 is actually a feedback connection because the loops across the two systems G_1 and G_2 exist. However, the resulting CSR is just the product $G_1 G_2$ of the two CSRs. This property greately simplifies the analysis and synthesis of feedback connection.

2 Chain-Scattering Representation of Plants

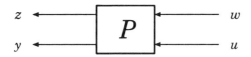

Fig. 2.1. Input/Output Representation

Fig. 2.2. Chain-Scattering Representation

This property is obvious by eliminating the intermediate valuables (z_1, w_1) from the relations
$$\begin{bmatrix} z \\ w \end{bmatrix} = G_1 \begin{bmatrix} z_1 \\ w_1 \end{bmatrix}, \quad \begin{bmatrix} z_1 \\ w_1 \end{bmatrix} = G_2 \begin{bmatrix} z_2 \\ w_2 \end{bmatrix}$$
If $G_i = CHAIN(P_i)$, $i = 1, 2$, the cascade connection represents the feedback connection represented in Fig.2.4. This connection is called a *star product* in Redheffer [Re1]. The use of CSR simply represent this connection by the product of the CSRs.

Another interesting property of CSR is that its inverse (if exists) is dually represented, i.e.,

$$H = CHAIN(P)^{-1}$$
$$= \begin{bmatrix} P_{12}^{-1} & -P_{12}^{-1}P_{11} \\ P_{22}P_{12}^{-1} & P_{21} - P_{22}P_{12}^{-1}P_{11} \end{bmatrix} \tag{2.4}$$

The representation (2.4) exists if P_{12} is invertible. It is called the *dual chain-scattering representation* (DCSR) of P which is denoted by

$$DCHAIN(P) = \begin{bmatrix} P_{12}^{-1} & -P_{12}^{-1}P_{11} \\ P_{22}P_{12}^{-1} & P_{21} - P_{22}P_{12}^{-1}P_{11} \end{bmatrix}. \tag{2.5}$$

It is the inverse of $CHAIN(P)$ if it exists. The duality between $CHAIN(P)$ and $DCHAIN(P)$ is explicitly represented in the following identity :

$$\begin{bmatrix} 0 & I \\ -I & 0 \end{bmatrix} CHAIN(P)^T \begin{bmatrix} 0 & -I \\ I & 0 \end{bmatrix} = DCHAIN(P^T)$$

If P is a symmetric, i.e.,
$$P(s) = P(s)^T \quad \forall s, \tag{2.6}$$

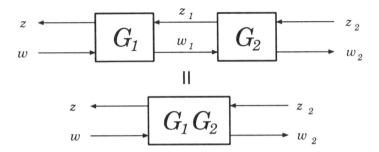

Fig. 2.3. Cascade Property of CSR

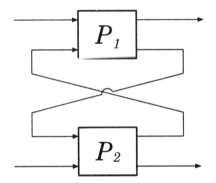

Fig. 2.4. Star Product

then $G = CHAIN(P)$ has a *symplectic structure* represented by

$$\begin{bmatrix} 0 & I \\ -I & 0 \end{bmatrix} G^T \begin{bmatrix} 0 & -I \\ I & 0 \end{bmatrix} = G^{-1} = H$$

Therefore, we see that

$$det\, G = \pm 1$$

This is a well-known property of the F-matrix for reciprocal circuits. A symmetric plant characterized by (2.6) has some interesting properties from the viewpoint of robust control.

Now, we derive the state space forms of CSR and DSCR. Let

$$P = \left[\begin{array}{c|cc} A & B_1 & B_2 \\ \hline C_1 & D_{11} & D_{12} \\ C_2 & D_{21} & D_{22} \end{array}\right]$$

be a state-space realization of the plant (2.1). The state-space form of $CHAIN(P)$ can be obtained based on the representation (2.3), but actually it is derived directly from the state equation

$$\dot{x} = A\,x + B_1\,w + B_2\,u \tag{2.7a}$$
$$z = C_1\,x + D_{11}\,w + D_{12}\,u \tag{2.7b}$$
$$y = C_2\,x + D_{21}\,w + D_{22}\,u \tag{2.7c}$$

In order that a state space representation of $CHAIN(P)$ exists, P_{21} is invertible and the inverse is proper. Therefore, we must assume that D_{21}^{-1} exists. In that case, the relation (2.7c) can be solved with respect to w yielding

$$w = D_{21}^{-1}(-C_2\,x - D_{22}\,u + y)$$

Substituting this relation in (2.7a)(2.7b), we have a realization of $CHAIN(P)$ as

$$CHAIN(P) = \left[\begin{array}{c|cc} A - B_1 D_{21}^{-1} C_2 & B_2 - B_1 D_{21}^{-1} D_{22} & B_1 D_{21}^{-1} \\ C_1 - D_{11} D_{21}^{-1} C_2 & D_{12} - D_{11} D_{21}^{-1} D_{22} & D_{11} D_{21}^{-1} \\ -D_{21}^{-1} C_2 & -D_{21}^{-1} D_{22} & D_{21}^{-1} \end{array}\right] \tag{2.8}$$

In the same way, $DCHAIN(P)$ is given by

$$DCHAIN(P) = \left[\begin{array}{c|cc} A - B_2 D_{12}^{-1} C_1 & B_2 D_{12}^{-1} & B_1 - B_2 D_{12}^{-1} D_{11} \\ -D_{12}^{-1} C_1 & D_{12}^{-1} & -D_{12}^{-1} D_{11} \\ C_2 - D_{22} D_{12}^{-1} C_1 & D_{22} D_{12}^{-1} & D_{21} - D_{22} D_{12}^{-1} D_{11} \end{array}\right] \tag{2.9}$$

The duality between $CHAIN(P)$ and $DCHAIN(P)$ is amplified in the state space forms (2.8) and (2.9).

3 Factorization of CSR and DCSR

In this section, we investigate the structure of CSR and DCSR through the factorization. First, we note the following factorizations which can be verified easily by direct computations:

$$CHAIN(P) = \begin{bmatrix} P_{12} & P_{11} \\ 0 & I \end{bmatrix} \begin{bmatrix} I & 0 \\ P_{22} & P_{21} \end{bmatrix}^{-1} \tag{3.1}$$

$$= \begin{bmatrix} I & -P_{11} \\ 0 & P_{21} \end{bmatrix}^{-1} \begin{bmatrix} P_{12} & 0 \\ -P_{22} & I \end{bmatrix} \tag{3.2}$$

In the same way, we can derive the corresponding factorization of $DCHAIN(P)$ as

$$DCHAIN(P) = \begin{bmatrix} I & 0 \\ P_{22} & P_{21} \end{bmatrix} \begin{bmatrix} P_{12} & P_{11} \\ 0 & I \end{bmatrix}^{-1} \tag{3.3}$$

$$= \begin{bmatrix} P_{12} & 0 \\ -P_{22} & I \end{bmatrix}^{-1} \begin{bmatrix} I & -P_{11} \\ 0 & P_{21} \end{bmatrix} \tag{3.4}$$

These factorizations are all coprime and hence it can be regarded as coprime factorizations over the stable matrices, provided that P is stable.

Now, we introduce the following four subplants associated with the original plant (2.1):

$$P_z := \begin{bmatrix} P_{11} & P_{12} \\ I & 0 \end{bmatrix} = \left[\begin{array}{c|cc} A & B_1 & B_2 \\ \hline C_1 & D_{11} & D_{12} \\ 0 & I & 0 \end{array}\right] \tag{3.5}$$

$$P_y := \begin{bmatrix} 0 & I \\ P_{21} & P_{22} \end{bmatrix} = \left[\begin{array}{c|cc} A & B_1 & B_2 \\ \hline 0 & 0 & I \\ C_2 & D_{21} & D_{22} \end{array}\right] \tag{3.6}$$

$$P_w := \begin{bmatrix} P_{11} & I \\ P_{21} & 0 \end{bmatrix} = \left[\begin{array}{c|cc} A & B_1 & 0 \\ \hline C_1 & D_{11} & I \\ C_2 & D_{21} & 0 \end{array}\right] \tag{3.7}$$

$$P_u := \begin{bmatrix} 0 & P_{12} \\ I & P_{22} \end{bmatrix} = \left[\begin{array}{c|cc} A & 0 & B_2 \\ \hline C_1 & 0 & D_{12} \\ C_2 & I & D_{22} \end{array}\right] \tag{3.8}$$

The subplant P_z represents the dynamics generating z and take $y = w$. The subplant P_y is just the replacement of z by y in P_z. The subplant $P_w(P_u)$ emphasizes the influence of $w(u)$ in the outputs z and y. According to the definition (2.3) of CSR, we have

$$CHAIN(P_z) = \begin{bmatrix} P_{12} & P_{11} \\ 0 & I \end{bmatrix} = \left[\begin{array}{c|cc} A & B_2 & B_1 \\ \hline C_1 & D_{12} & D_{11} \\ 0 & 0 & I \end{array}\right]$$

$$=: \left[\begin{array}{c|c} A & B \\ \hline C_z & D_z \end{array}\right] \tag{3.9}$$

$$CHAIN(P_y) = \begin{bmatrix} I & 0 \\ P_{22} & P_{21} \end{bmatrix}^{-1} = \left[\begin{array}{c|cc} A & B_2 & B_1 \\ \hline 0 & I & 0 \\ C_2 & D_{22} & D_{21} \end{array}\right]^{-1}$$

$$=: \left[\begin{array}{c|c} A & B \\ \hline C_y & D_y \end{array}\right]^{-1} \tag{3.10}$$

Note that the order of B_1 and B_2 is reversed in the definition of B, i.e.,

$$B = \begin{bmatrix} B_2 & B_1 \end{bmatrix}$$

Due to (3.1), we have the following factorization of $CHAIN(P)$:

$$CHAIN(P) = CHAIN(P_z) \cdot CHAIN(P_y) \tag{3.11}$$

3 Factorization of CSR and DCSR

The meaning of this factorization is illustrated in Fig.3.1 using star product of Fig.2.4. The factorization (3.11) brings an interesting state space form of $CHAIN(P)$ given in (2.8). From (3.1) (3.9) and (3.10), we have

$$CHAIN(P) = \begin{bmatrix} A & B \\ \hline C_z & D_z \end{bmatrix} \begin{bmatrix} A & B \\ \hline C_y & D_y \end{bmatrix}^{-1}$$

$$= \begin{bmatrix} A - BD_y^{-1}C_y & BD_y^{-1} \\ \hline C_z - D_z D_y^{-1} C_y & D_z D_y^{-1} \end{bmatrix} \qquad (3.12)$$

The above form implies that the CSR is obtained by applying the state feedback with gain $F = -D_y^{-1} C_y$ and the transformation of input space by $U = D_y^{-1}$ to $CHAIN(P_z)$.

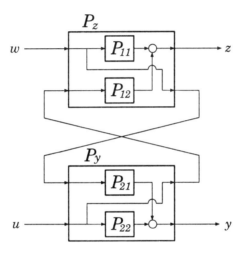

Fig. 3.1. Illustration of Factorization (3.11)

We can factorize $CHAIN(P)$ in an alternative way using P_w and P_u in (3.7) and (3.8), respectively. Due to (2.3), we have

$$CHAIN(P_w) = \begin{bmatrix} I & -P_{11} \\ 0 & P_{21} \end{bmatrix}^{-1} = \begin{bmatrix} A & 0 & B_1 \\ \hline -C_1 & I & -D_{11} \\ C_2 & 0 & D_{21} \end{bmatrix}^{-1}$$

$$=: \begin{bmatrix} A & B_w \\ \hline C & D_w \end{bmatrix}^{-1} \qquad (3.13)$$

$$CHAIN(P_u) = \begin{bmatrix} P_{12} & 0 \\ -P_{22} & I \end{bmatrix} = \begin{bmatrix} A & B_2 & 0 \\ \hline C_1 & D_{12} & 0 \\ -C_2 & -D_{22} & I \end{bmatrix}$$

$$=: \begin{bmatrix} A & B_u \\ \hline -C & D_u \end{bmatrix} \qquad (3.14)$$

Note that the sign of C_1 is reversed in the definition of C above, i.e.,

$$C = \begin{bmatrix} -C_1 \\ C_2 \end{bmatrix}$$

Due to (3.2), we have an alternative factorization of $CHAIN(P)$ as

$$CHAIN(P) = CHAIN(P_w) \cdot CHAIN(P_u) \qquad (3.15)$$

It follows, from (3.2) (3.13) and (3.14), that

$$CHAIN(P) = \left[\begin{array}{c|c} A & B_w \\ \hline C & D_w \end{array}\right]^{-1} \left[\begin{array}{c|c} A & B_u \\ \hline -C & D_u \end{array}\right]$$

$$= \left[\begin{array}{c|c} A - B_w D_w^{-1} C & B_u + B_w D_w^{-1} D_u \\ \hline -D_w^{-1} C & D_w^{-1} D_u \end{array}\right] \qquad (3.16)$$

This representation implies that $CHAIN(P)$ is obtained by applying the insertion of output with gain $L = -B_w D_w^{-1}$ and the transformation of output space by $V = D_w^{-1}$ to the subsystem $CHAIN(P_u)$. The structure of the factorization (3.15) is illustrated in Fig.3.2.

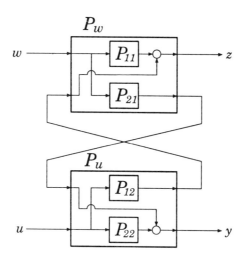

Fig. 3.2. Illustration of Factorization (3.16)

Dualizing the above arguments, we have analogous factorizations for $DCHAIN(P)$ based on (3.3) and (3.4) as follows :

$$DCHAIN(P) = DCHAIN(P_u) \cdot DCHAIN(P_w)$$

$$= \left[\begin{array}{c|c} A & B_u \\ \hline -C & D_u \end{array}\right]^{-1} \left[\begin{array}{c|c} A & B_w \\ \hline C & D_w \end{array}\right]$$

$$= \left[\begin{array}{c|c} A + B_u D_u^{-1} C & B_w + B_u D_u^{-1} D_w \\ \hline D_u^{-1} C & D_u^{-1} D_w \end{array}\right] \qquad (3.17)$$

$$DCHAIN(P) = DCHAIN(P_y) \cdot DCHAIN(P_z)$$
$$= \left[\begin{array}{c|c} A & B \\ \hline C_y & D_y \end{array}\right] \left[\begin{array}{c|c} A & B \\ \hline C_z & D_z \end{array}\right]^{-1}$$
$$= \left[\begin{array}{c|c} A - BD_z^{-1}C_z & BD_z^{-1} \\ \hline C_y - D_y D_z^{-1} C_z & D_y D_z^{-1} \end{array}\right] \quad (3.18)$$

4 Stability and Feedback

The transformation $CHAIN$ does not preserve the stability, namely, stability of P doesn't imply that of $CHAIN(P)$, and vise versa. Now consider a controller

$$u = K y \quad (4.1)$$

for the plant (2.1). The closed-loop transfer function Φ from w to z is represented by

$$\Phi = P_{11} + P_{12}K(I - P_{22}K)^{-1}P_{21} \quad (4.2)$$

We can represent Φ in terms of

$$G = CHAIN(P) = \begin{bmatrix} G_{11} & G_{12} \\ G_{21} & G_{22} \end{bmatrix}$$

From (4.1) and (4.2), it follows that

$$\Phi = (G_{11}K + G_{12})(G_{21}K + G_{22})^{-1} \quad (4.3)$$

We use a simplified notation

$$\Phi = HM(G; K) \quad (4.4)$$

to represent (4.3). The transformation (4.4) is a kind of linear fractional transformations which was extensively investigated by Siegel [Si1]. We list up some important properties of HM as follows:

Lemma 4.1. $HM(G; K)$ satisfies the following properties:

(i) $HM(I; K) = K$
(ii) $HM(G_1; HM(G_2; K)) = HM(G_1 G_2; K)$
(iii) If $HM(G; K) = L$ and G^{-1} exists, then $K = HM(G^{-1}; L)$.

The closed-loop system is illustrated in Fig.4.1, in which controller K appears as a "terminating load". In other words, feedback appears as a *termination* in chain-scattering framework.

The internal stability of the closed-loop system of Fig.4.1 is defined in [Ki1] in terms of state-space realizations of G and K. It is important to note the following result:

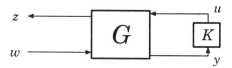

Fig. 4.1. Feedback as a Termination

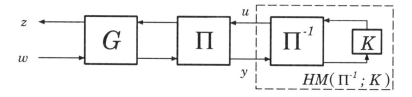

Fig. 4.2. Cancellation of Unimodular System

Lemma 4.2. If $HM(G; K)$ is internally stable, then $HM(G\Pi; HM(\Pi^{-1}; K))$ is also internally stable for unimodular Π.

This lemma implies that the stable pole-zero cancellations between CSRs depicted in Fig.4.2 does not destruct stability. This observation enables us to simplify the control problem by extracting unimodular portion from the plant description.

Lemma 4.3. Assume that $CHAIN(P)$ is represented as a product

$$CHAIN(P) = CHAIN(P_1) \cdot \Pi$$

where Π is a unimodular matrix, i.e., both Π and Π^{-1} are stable. Then, a control problem for P is solvable, if and only if the same problem for P_1 is solvable.

Proof. Assume that a solution K to the control problem for P exists. Then, $K_1 = HM(\Pi^{-1}; K)$ is a solution to the same control problem because, due to (ii) of Lemma 4.1, $HM(CHAIN(P); K) = HM(CHAIN(P_1); K_1)$ and the internal stability holds. The converse argument is also valid. □

The fact shown by Lemma 4.2 is a generalization of well-known fact in unity feedback system where the plant is represented as the product of a simpler plant P_1 and a unimodular system Π. Then, we can ignore Π and reduce the problem to that of P_1, because we can always cancel out this part by a controller.

When the state-space specification is imposed, the cancellation of unimodular portion may cause a serious problem, particularly when a pole/zero cancellation occurs near the imaginary axis. In that case, a state space performance of closed-loop system may be damaged because of the non-zero initial state of the cancelled modes.

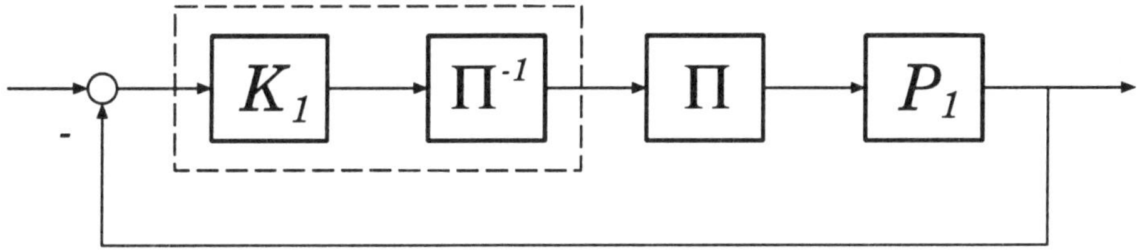

Fig. 4.3. Unity Feedback Systems

5 (J, J')-Lossless Systems and Related Factorizations

In this paper, we state several results concerning the (J, J')-losslessness which plays an important role in H^∞ control theory. Let J_{mr} be a signature matrix

$$J_{mr} := \begin{bmatrix} I_m & 0 \\ 0 & -I_r \end{bmatrix}. \tag{5.1}$$

A matrix $\Theta(s)$ with $(m+r)$ rows and $(p+r)$ columns satisfying

$$\Theta^\sim(s) J_{mr} \Theta(s) = J_{pr} \tag{5.2}$$

is said to be (J_{mr}, J_{pr})-unitary, or simply, (J, J')-unitary when the size of $\Theta(s)$ is clear from the context, or is irrelevant. If $m = p$, then it is simply called J_{mr}-unitary, or J-unitary. A (J_{mr}, J_{pr})-unitary system $\Theta(s)$ is said to be (J_{mr}, J_{pr})-lossless if it satisfies the inequality

$$\Theta^*(s) J_{mr} \Theta(s) \leq J_{pr}, \quad \forall Re[s] \geq 0. \tag{5.3}$$

It is well-known that if $\Theta = CHAIN(P)$, then, Θ is (J, J')-lossless iff P is lossless, i.e.,

$$P^\sim P = I, \quad P : \text{stable}.$$

Thus, the (J, J')-lossless system is regarded as the chain-scattering characterization of lossless systems. The following result exihibits an interesting property of (J, J')-unitary systems concerning its termination [Ki1].

Lemma 5.1. Assume that $\Theta(s)$ is (J, J')-unitary. There exists a termination S such that $HM(\Theta; S) \in BH^\infty$, iff $\Theta(s)$ is (J, J')-lossless. In that case, $HM(\Theta; S) \in BH^\infty$ iff $S \in BH^\infty$.

If a matrix $G(s)$ is represented as a product

$$G(s) = \Theta(s) \Pi(s) \tag{5.4}$$

where $\Theta(s)$ is (J, J')-lossless and $\Pi(s)$ is unimodular, then $G(s)$ is said to have a (J, J')-lossless factorization. This factorization includes the inner-outer factorization of stable matrices and Wiener factorization of positive matrices as special cases. It is of fundamental importance in linear system theory.

The computation of (J, J')-lossless factorization is carried out in the state space in a simple way based on the theory of conjugation [Ki2]. Let

$$G(s) = \left[\begin{array}{c|c} A & B \\ \hline C & D \end{array}\right] \tag{5.5}$$

be a state space realization of $G(s)$. State space realization of $G^\sim(s)JG(s)$ is given by

$$G^\sim(s)JG(s) = \left[\begin{array}{cc|c} A & 0 & B \\ -C^T JC & -A^T & -C^T JD \\ \hline D^T JC & B^T & D^T JD \end{array}\right] \tag{5.6}$$

$$= \left[\begin{array}{cc|c} -A^T & -C^T JC & -C^T JD \\ 0 & A & B \\ \hline B^T & D^T JC & D^T JD \end{array}\right] \tag{5.7}$$

Assuming that $D^T JD$ is invertible, we write

$$f(\left[\begin{array}{c|c} A & B \\ \hline C & D \end{array}\right]) := \left[\begin{array}{cc} A & 0 \\ -C^T JC & -A^T \end{array}\right] + \left[\begin{array}{c} -B \\ C^T JD \end{array}\right](D^T JD)^{-1}\left[\begin{array}{cc} D^T JC & B^T \end{array}\right], \tag{5.8}$$

which is equal to an A-matrix of $(G^\sim JG)^{-1}$ associated with the realization (5.6). Also, we write

$$g(\left[\begin{array}{c|c} A & B \\ \hline C & D \end{array}\right]) := \left[\begin{array}{cc} A^T & C^T JC \\ 0 & -A \end{array}\right] \tag{5.9}$$

which is equal to an A-matrix of $G^\sim JG$ associated with the realization (5.7) with signature being reversed. It should be noted that both $f(\left[\begin{array}{c|c} A & B \\ \hline C & D \end{array}\right])$ and $g(\left[\begin{array}{c|c} A & B \\ \hline C & D \end{array}\right])$ are Hamiltonian matrices, i.e.,

$$\left[\begin{array}{cc} 0 & -I \\ I & 0 \end{array}\right] f(\left[\begin{array}{c|c} A & B \\ \hline C & D \end{array}\right]) \left[\begin{array}{cc} 0 & I \\ -I & 0 \end{array}\right] = -f(\left[\begin{array}{c|c} A & B \\ \hline C & D \end{array}\right])^T$$

$$\left[\begin{array}{cc} 0 & -I \\ I & 0 \end{array}\right] g(\left[\begin{array}{c|c} A & B \\ \hline C & D \end{array}\right]) \left[\begin{array}{cc} 0 & I \\ -I & 0 \end{array}\right] = -g(\left[\begin{array}{c|c} A & B \\ \hline C & D \end{array}\right])^T$$

The stabilizing solutions to the Riccati equations associated with $f(\left[\begin{array}{c|c} A & B \\ \hline C & D \end{array}\right])$ and $g(\left[\begin{array}{c|c} A & B \\ \hline C & D \end{array}\right])$ are written as

$$Ric\, f(\left[\begin{array}{c|c} A & B \\ \hline C & D \end{array}\right]), \quad Ric\, g(\left[\begin{array}{c|c} A & B \\ \hline C & D \end{array}\right]),$$

respectively.

The following result gives a state space existence condition of (J, J')-lossless factorization [Ki1].

5 (J, J')-Lossless Systems and Related Factorizations

Theorem 5.2. Let (5.5) be a state space realization of $G(s)$. It has a (J, J')-lossless factorization (5.4), iff

(i) There exists a non-singular matrix E such that

$$D^T J D = E^T J' E \qquad (5.10)$$

(ii) $X = Ric\, f(\left[\begin{array}{c|c} A & B \\ \hline C & D \end{array}\right]) \geq 0$ and $\bar{X} = Ric\, g(\left[\begin{array}{c|c} A & B \\ \hline C & D \end{array}\right]) \geq 0$ exist satisfying

$$\sigma(X\bar{X}) < 1$$

The dual form of the factorization (5.4) is given by

$$G(s) = \Omega(s) \Psi^\sim(s) \qquad (5.11)$$

where $\Psi(s)$ is (J, J')-lossless matrix and $\Omega(s)$ is unimodular. The factorization of the form (5.11) is called a *dual (J, J')-lossless factorization*. In order to state the existence condition for the dual (J, J')-lossless factorization, we dualize (5.8) and (5.9) as

$$f^\sim(\left[\begin{array}{c|c} A & B \\ \hline C & D \end{array}\right]) := -f(\left[\begin{array}{c|c} A & B \\ \hline C & D \end{array}\right]^\sim) = -f(\left[\begin{array}{c|c} -A^T & -C^T \\ \hline B^T & D^T \end{array}\right])$$

$$= \left[\begin{array}{cc} A^T & 0 \\ BJB^T & -A \end{array}\right] - \left[\begin{array}{c} C^T \\ BJD^T \end{array}\right](DJD^T)^{-1} \left[\begin{array}{cc} DJB^T & -C \end{array}\right] \qquad (5.12)$$

$$g^\sim(\left[\begin{array}{c|c} A & B \\ \hline C & D \end{array}\right]) := -g(\left[\begin{array}{c|c} A & B \\ \hline C & D \end{array}\right]^\sim)$$

$$= \left[\begin{array}{cc} A & -BJB^T \\ 0 & -A^T \end{array}\right] \qquad (5.13)$$

It is clear that both $f^\sim(\left[\begin{array}{c|c} A & B \\ \hline C & D \end{array}\right])$ and $g^\sim(\left[\begin{array}{c|c} A & B \\ \hline C & D \end{array}\right])$ are also Hamiltonian matrices. The stabilizing solutions of the Riccati equations corresponding to these Hamiltonian matrices are denoted respectively by

$$Ric\, f^\sim(\left[\begin{array}{c|c} A & B \\ \hline C & D \end{array}\right]), \quad Ric\, g^\sim(\left[\begin{array}{c|c} A & B \\ \hline C & D \end{array}\right]).$$

Now, we are ready to state the dual version of Theorem 5.2 [Ki1].

Theorem 5.3. Let (5.5) be a state space realization of $G(s)$. It has a dual (J, J')-lossless factorization iff

(i) there exists a non-singular matrix E such that

$$DJD^T = EJ'E^T \qquad (5.14)$$

(ii) $Y = Ric\, f^\sim(\left[\begin{array}{c|c} A & B \\ \hline C & D \end{array}\right]) \geq 0$ and $\bar{Y} = Ric\, g^\sim(\left[\begin{array}{c|c} A & B \\ \hline C & D \end{array}\right]) \geq 0$ exist satisfying

$$\sigma(Y\bar{Y}) < 1.$$

Now, we shall exploit some properties of $f(\left[\begin{array}{c|c} A & B \\ \hline C & D \end{array}\right])$, $g(\left[\begin{array}{c|c} A & B \\ \hline C & D \end{array}\right])$ and their duals. Let λ be a zero of (5.5), i.e.,

$$\left[\begin{array}{cc} A - \lambda I & B \\ C & D \end{array}\right] \left[\begin{array}{c} x \\ u \end{array}\right] = 0$$

for some $(x, u) \neq 0$. Then, it is easy to see that

$$f(\left[\begin{array}{c|c} A & B \\ \hline C & D \end{array}\right]) \left[\begin{array}{c} x \\ 0 \end{array}\right] = \lambda \left[\begin{array}{c} x \\ 0 \end{array}\right].$$

In other words, each zero of $\left[\begin{array}{c|c} A & B \\ \hline C & D \end{array}\right]$ is an eigenvalue of $f(\left[\begin{array}{c|c} A & B \\ \hline C & D \end{array}\right])$. Therefore, if $\left[\begin{array}{c|c} A & B \\ \hline C & D \end{array}\right]$ has a zero on the $j\omega$-axis, then, $f(\left[\begin{array}{c|c} A & B \\ \hline C & D \end{array}\right])$ has an eigenvalue on the $j\omega$-axis. In that case, $Ric\, f(\left[\begin{array}{c|c} A & B \\ \hline C & D \end{array}\right])$ doesn't exist. Therefore, in order that $G(s)$ has a (J, J')-lossless factorization, $G(s)$ should be free from a $j\omega$-axis zero.

Lemma 5.4. The following identities hold for any K, L and invertible U, V:

$$f(\left[\begin{array}{c|c} A & B \\ \hline C & D \end{array}\right]) = f(\left[\begin{array}{c|c} A + BK & BU \\ \hline C + DK & DU \end{array}\right]) \tag{5.15}$$

$$f^\sim(\left[\begin{array}{c|c} A & B \\ \hline C & D \end{array}\right]) = f^\sim(\left[\begin{array}{c|c} A + LC & B + LD \\ \hline VC & VD \end{array}\right]) \tag{5.16}$$

If D is invertible,

$$g(\left[\begin{array}{c|c} A & B \\ \hline C & D \end{array}\right]^{-1}) = f^\sim(\left[\begin{array}{c|c} A & B \\ \hline C & D \end{array}\right]), \tag{5.17}$$

$$g^\sim(\left[\begin{array}{c|c} A & B \\ \hline C & D \end{array}\right]) = f(\left[\begin{array}{c|c} A & B \\ \hline C & D \end{array}\right]^{-1}). \tag{5.18}$$

The identities (5.17) and (5.18) can be extended to the case where D is not invertible but is of full rank. In that case, we can augment $G(s)$ in (5.5) by appending fictitious inputs or outputs, i.e.,

$$\hat{G}(s) = \left[\begin{array}{c|c} A & B \\ \hline C & D \\ C' & D' \end{array}\right] = \left[\begin{array}{c|c} A & B \\ \hline \hat{C} & \hat{D} \end{array}\right] \tag{5.19}$$

5 (J, J')-Lossless Systems and Related Factorizations 165

or
$$\hat{G}(s) = \left[\begin{array}{c|cc} A & B' & B \\ \hline C & D' & D \end{array}\right] = \left[\begin{array}{c|c} A & \hat{B} \\ \hline C & \hat{D} \end{array}\right] \tag{5.20}$$

where D' is chosen such that $\hat{D}$ is invertible in either of (5.19) and (5.20). Simple manipulations yield

$$f^{\sim}(\left[\begin{array}{c|c} A & B \\ \hline C & D \end{array}\right]) - g(\left[\begin{array}{c|c} A & B \\ \hline C & D \\ C' & D' \end{array}\right]^{-1}) = f^{\sim}(\left[\begin{array}{c|c} A & B \\ \hline C & D \end{array}\right]) - f^{\sim}(\left[\begin{array}{c|c} A & B \\ \hline \hat{C} & \hat{D} \end{array}\right])$$

$$= \left[\begin{array}{c} \hat{C}^T \\ BJ\hat{D}^T \end{array}\right] (\hat{D}J\hat{D}^T)^{-1} [\, \hat{D}JB^T \; -\hat{C}\,] - \left[\begin{array}{c} C^T \\ BJD^T \end{array}\right] (DJD^T)^{-1} [\, DJB^T \; -C\,]$$

$$= \left[\begin{array}{c} \hat{C}^T \hat{D}^{-T} J \\ B \end{array}\right] (J - JD^T(DJD^T)^{-1}DJ) [\, B^T \; -J\hat{D}^{-1}\hat{C}\,] \tag{5.21}$$

Analogously, we have

$$f(\left[\begin{array}{c|c} A & B \\ \hline C & D \end{array}\right]) - g^{\sim}(\left[\begin{array}{c|cc} A & B' & B \\ \hline C & D' & D \end{array}\right]^{-1})$$

$$= \left[\begin{array}{c} -\hat{B}\hat{D}^{-1}J \\ C^T \end{array}\right] (J - JD(D^T JD)^{-1}D^T J) [\, C \; J\hat{D}^{-T}\hat{B}^T\,] \tag{5.22}$$

Write
$$\left[\begin{array}{c} D \\ D' \end{array}\right]^{-1} = [\, D^{\dagger} \; D^{\perp}\,] \tag{5.23}$$

$$[\, D' \; D\,]^{-1} = \left[\begin{array}{c} D^{\perp} \\ D^{\dagger} \end{array}\right] \tag{5.24}$$

Obviously, $D^{\perp}$ in (5.23) is a right inverse of D, while $D^{\perp}$ in (5.24) is a left inverse of D. Taking the inverse of (5.19) and (5.20), we obtain a right annihilator $G^{\perp}$ of G given by (5.5) as

$$G^{\perp} = \left[\begin{array}{c|c} A - B\hat{D}^{-1}\hat{C} & B \\ \hline -\hat{D}^{-1}\hat{C} & I \end{array}\right] D^{\perp} \tag{5.25}$$

and a left annihilator $G^{\perp}$ of G as

$$G^{\perp} = D^{\perp} \left[\begin{array}{c|c} A - \hat{B}\hat{D}^{-1}C & -\hat{B}\hat{D}^{-1} \\ \hline C & I \end{array}\right] \tag{5.26}$$

Lemma 5.5. *Consider an augmentation (5.19) which satisfies*

$$(D^{\perp})^T JD^{\perp} < 0 \tag{5.27}$$

Then, for any augmentation satisfying (5.23),

$$Ric\, g(\left[\begin{array}{c|c} A & B \\ \hline C & D \\ C' & D' \end{array}\right]^{-1}) \leq Ric\, f^{\sim}(\left[\begin{array}{c|c} A & B \\ \hline C & D \end{array}\right]). \tag{5.28}$$

Moreover, the equality in (5.28) holds only for the augmentation which satisfies
$$(G^\perp)^\sim J G^\perp = (D^\perp)^T J D^\perp. \tag{5.29}$$

Dually, if
$$D^\perp J (D^\perp)^T > 0 \tag{5.30}$$
in an augmentation (5.20), then, for any augmentation satisfying (5.23),
$$Ric\, g^\sim\left(\left[\begin{array}{c|cc} A & B' & B \\ \hline C & D' & D \end{array}\right]^{-1}\right) \leq Ric\, f\left(\left[\begin{array}{c|c} A & B \\ \hline C & D \end{array}\right]\right). \tag{5.31}$$

The equality holds only for the augmentation which satisfies
$$G^\perp J (G^\perp)^\sim = D^\perp J (D^\perp)^T. \tag{5.32}$$

Proof. Let
$$Y := Ric\, f^\sim\left(\left[\begin{array}{c|c} A & B \\ \hline C & D \end{array}\right]\right), \quad \bar{X} := Ric\, g\left(\left[\begin{array}{c|c} A & B \\ \hline \hat{C} & \hat{D} \end{array}\right]^{-1}\right).$$

We can write
$$J - JD^T(DJD^T)^{-1}DJ = D^\perp R (D^\perp)^T$$
for some $\hat{R}$. The assumption (5.27) implies that $\hat{R} > 0$. Therefore, from (5.21) and
$$\begin{bmatrix} Y & -I \end{bmatrix} f^\sim\left(\left[\begin{array}{c|c} A & B \\ \hline C & D \end{array}\right]\right)\begin{bmatrix} I \\ Y \end{bmatrix} = 0,$$
we have
$$\begin{bmatrix} Y & -I \end{bmatrix} g\left(\left[\begin{array}{c|c} A & B \\ \hline C & D \\ C' & D' \end{array}\right]^{-1}\right)\begin{bmatrix} I \\ Y \end{bmatrix}$$
$$= (B - Y\hat{C}^T \hat{D}^{-T} J) D^\perp \hat{R}(D^\perp)^T (B^T - J\hat{D}^{-1}\hat{C} Y) \geq 0. \tag{5.33}$$

In a similar way used by Wimmer [Wi1], we can establish the inequality (5.28) based on (5.33). If $Y = \bar{X}$, the right hand side of (5.33) must vanish. Since $\hat{R} > 0$, we have
$$(B - Y\hat{C}^T \hat{D}^{-T} J) D^\perp = 0 \tag{5.34}$$

From (5.26), it follows that
$$G^\perp J (G^\perp)^\sim = D^\perp \left[\begin{array}{cc|c} A - \hat{B}\hat{D}^{-1}C & -B\hat{D}^{-1}J\hat{D}^{-T}B^T & -B\hat{D}^{-1} \\ 0 & -(A - \hat{B}\hat{D}^{-1}C)^T & C^T \\ \hline C & J\hat{D}^{-T}B^T & I \end{array}\right] J(D^\perp)^T$$

Taking the similarity transformation with
$$T = \begin{bmatrix} I & 0 \\ -\bar{X} & I \end{bmatrix}, \quad T^{-1} = \begin{bmatrix} I & 0 \\ \bar{X} & I \end{bmatrix}$$
and using (5.34) and $\bar{X} = Y$, we can establish (5.29).

The dual assertion can be established similarly. □

6 System Augmentations and H^∞ Control

Now, we come back to the CSR and DSCR of P introduced in Section 2. If neither of P_{12} and P_{21} is invertible, we cannot use CSR nor DSCR directly. In that case, we can use them by augmentating the plant introducing fictitious observation output or control input. Let us assume that

(A_1) D_{21} is row full rank and D_{12} is column full rank.

In that case, we can find D'_{21} and D'_{12} such that

$$\hat{D}_{21}^{-1} := \begin{bmatrix} D_{21} \\ D'_{21} \end{bmatrix}^{-1} = \begin{bmatrix} D_{21}^{\dagger} & D_{21}^{\perp} \end{bmatrix} \tag{6.1}$$

$$\hat{D}_{12}^{-1} = \begin{bmatrix} D'_{12} & D_{12} \end{bmatrix}^{-1} = \begin{bmatrix} D_{12}^{\perp} \\ D_{12}^{\dagger} \end{bmatrix} \tag{6.2}$$

exist. Using these matrices, we consider the two augmentations of the plants:

$$P_o := \left[\begin{array}{c|cc} A & B_1 & B_2 \\ \hline C_1 & D_{11} & D_{12} \\ C_2 & D_{21} & 0 \\ C'_2 & D'_{21} & 0 \end{array}\right] = \left[\begin{array}{c|cc} A & B_1 & B_2 \\ \hline C_1 & D_{11} & D_{12} \\ \hat{C}_2 & \hat{D}_{21} & 0 \end{array}\right] \tag{6.3}$$

$$P_i := \left[\begin{array}{c|ccc} A & B_1 & B'_2 & B_2 \\ \hline C_1 & D_{11} & D'_{12} & D_{12} \\ C_2 & D_{21} & 0 & 0 \end{array}\right] = \left[\begin{array}{c|cc} A & B_1 & \hat{B}_2 \\ \hline C_1 & D_{11} & \hat{D}_{12} \\ C_2 & D_{21} & 0 \end{array}\right]. \tag{6.4}$$

The augmentation (6.3) corresponds to (5.19), while the augmentation (6.4) corresponds to (5.20). Since both $\hat{D}_{21}$ and $\hat{D}_{12}$ are invertible, there exist

$$G_o := CHAIN(P_o), \tag{6.5}$$

$$G_i := DCHAIN(P_i). \tag{6.6}$$

Our purpose in this section is to derive a condition such that G_o has a (J_{mr}, J_{pr})-lossless factorization for any augmentation (C'_2, D'_{21}) satisfying (6.1) and G_i has a (J_{mr}, J_{mq})-lossless factorization for any augmentation (B'_2, D'_{12}). This condition turns out to be equivalent to the solvability of the normalized H^∞ control problem, i.e., the problem of finding a stabilizing controller K such that

$$\|P_{11} + P_{12}K(I - P_{22}K)^{-1}P_{21}\|_\infty < 1.$$

Due to (3.11) and (3.12), we can represent $CHAIN(P_o)$ as

$$CHAIN(P_o) = CHAIN(P_z) \cdot CHAIN(\hat{P}_y)$$

$$= \left[\begin{array}{c|c} A & B \\ \hline C_z & D_z \end{array}\right] \left[\begin{array}{c|c} A & B \\ \hline \hat{C}_y & \hat{D}_y \end{array}\right]^{-1}$$

$$= \left[\begin{array}{c|c} A - B\hat{D}_y^{-1}\hat{C}_y & B\hat{D}_y^{-1} \\ \hline C_z - D_z\hat{D}_y^{-1}\hat{C}_y & D_z\hat{D}_y^{-1} \end{array}\right] \tag{6.7}$$

where

$$\left[\begin{array}{c|c} A & B \\ \hline \hat{C}_y & \hat{D}_y \end{array}\right] = \left[\begin{array}{c|cc} A & B_2 & B_1 \\ 0 & I & 0 \\ \hline C_2 & 0 & D_{21} \\ C'_2 & 0 & D'_{21} \end{array}\right], \tag{6.8}$$

and C_z, D_z are given in (3.9). Any matrix with $\hat{}$ contains augmented part. Due to (5.15), we have

$$f(\left[\begin{array}{c|c} A & B \\ \hline C_z & D_z \end{array}\right]\left[\begin{array}{c|c} A & B \\ \hline \hat{C}_y & \hat{D}_y \end{array}\right]^{-1}) = f(\left[\begin{array}{c|c} A & B \\ \hline C_z & D_z \end{array}\right]),$$

which does not depend on augmentation. On the other hand, due to (3.16), we have

$$\begin{aligned} CHAIN(P_o) &= \left[\begin{array}{c|c} A & B_w \\ \hline \hat{C} & \hat{D}_w \end{array}\right]^{-1} \left[\begin{array}{c|c} A & B_u \\ \hline -C & D_u \end{array}\right] \\ &= \left[\begin{array}{c|c} A - B_w \hat{D}_w^{-1}\hat{C} & B_u + B_w \hat{D}_w^{-1} D_u \\ \hline -\hat{D}_w^{-1}\hat{C} & \hat{D}_w^{-1} D_u \end{array}\right] \end{aligned} \tag{0.9}$$

where

$$\left[\begin{array}{c|c} A & B_w \\ \hline \hat{C} & \hat{D}_w \end{array}\right] = \left[\begin{array}{c|cc} A & 0 & B_1 \\ -C_1 & I & D_{11} \\ \hline C_2 & 0 & D_{21} \\ C'_2 & 0 & D'_{21} \end{array}\right]. \tag{6.10}$$

Hence, we have

$$g(\left[\begin{array}{c|c} A & B_w \\ \hline \hat{C} & \hat{D}_w \end{array}\right]^{-1} \left[\begin{array}{c|c} A & B_u \\ \hline -C & D_u \end{array}\right]) = g(\left[\begin{array}{c|c} A & B_w \\ \hline \hat{C} & \hat{D}_w \end{array}\right]^{-1}).$$

Now, applying Theorem 5.2, we have the following result:

Lemma 6.1. $CHAIN(P_o)$ has a (J_{mr}, J_{pr})-lossless factorization, iff there exists a nonsingular matrix E such that

$$D_z^T J D_z = E^T J' E \tag{6.11}$$

and

$$X = Ric\, f(\left[\begin{array}{c|c} A & B \\ \hline C_z & D_z \end{array}\right]) \geq 0, \quad \bar{X} = Ric\, g(\left[\begin{array}{c|c} A & B_w \\ \hline \hat{C} & \hat{D}_w \end{array}\right]^{-1}) \geq 0 \tag{6.12}$$

exist such that

$$\sigma(X\bar{X}) < 1. \tag{6.13}$$

6 System Augmentations and H^∞ Control

Now, we dualize the above result. Due to (3.17), we have

$$DCHAIN(P_i) = DCHAIN(\hat{P}_u) \cdot DCHAIN(P_w)$$

$$= \left[\begin{array}{c|c} A & \hat{B}_u \\ \hline -C & \hat{D}_u \end{array}\right]^{-1} \left[\begin{array}{c|c} A & B_w \\ \hline C & D_w \end{array}\right]$$

$$= \left[\begin{array}{c|c} A + \hat{B}_u \hat{D}_u^{-1} C & B_w + \hat{B}_u \hat{D}_u^{-1} D_w \\ \hline \hat{D}_u^{-1} C & \hat{D}_u^{-1} D_w \end{array}\right] \quad (6.14)$$

where

$$\left[\begin{array}{c|c} A & \hat{B}_u \\ \hline -C & \hat{D}_u \end{array}\right] = \left[\begin{array}{c|ccc} A & B_2' & B_2 & 0 \\ \hline C_1 & D_{12}' & D_{12} & 0 \\ -C_2 & 0 & 0 & I \end{array}\right]. \quad (6.15)$$

Thus, due to (5.16), we have

$$f^\sim\left(\left[\begin{array}{c|c} A & \hat{B}_u \\ \hline -C & \hat{D}_u \end{array}\right]^{-1} \left[\begin{array}{c|c} A & B_w \\ \hline C & D_w \end{array}\right]\right) = f^\sim\left(\left[\begin{array}{c|c} A & B_w \\ \hline C & D_w \end{array}\right]\right).$$

Also, due to (3.18), we have

$$g^\sim\left(\left[\begin{array}{c|c} A & B \\ \hline C_y & D_y \end{array}\right]\left[\begin{array}{c|c} A & \hat{B} \\ \hline C_z & \hat{D}_z \end{array}\right]^{-1}\right) = g^\sim\left(\left[\begin{array}{c|c} A & \hat{B} \\ \hline C_z & \hat{D}_z \end{array}\right]^{-1}\right).$$

Hence, applying Theorem 5.3, we have the following result:

Lemma 6.2. $DCHAIN(P_i)$ has a dual (J_{mr}, J_{mq})-lossless factorization, iff there exist a nonsingular matrix E such that

$$D_w J_{mr} D_w^T = E J_{mq} E^T \quad (6.16)$$

and

$$Y = Ric \, f^\sim\left(\left[\begin{array}{c|c} A & B_w \\ \hline C & D_w \end{array}\right]\right) \geq 0, \quad \bar{Y} = Ric \, g^\sim\left(\left[\begin{array}{c|c} A & \hat{B} \\ \hline C_z & \hat{D}_z \end{array}\right]^{-1}\right) \geq 0 \quad (6.17)$$

satisfying

$$\sigma(Y\bar{Y}) < 1. \quad (6.18)$$

Due to (6.7), the zeros of $CHAIN(P_o)$ coincide with those of $CHAIN(P_z)$. As was remarked in Section 5, $CHAIN(P_o)$ must be free from $j\omega$-axis zero in order that it has a (J_{mr}, J_{pr})-lossless factorization. Therefore, we must impose the condition

$$rank \left[\begin{array}{cc} A - j\omega I & B \\ C_z & D_z \end{array}\right] = n + p + r, \quad \forall \omega.$$

From the definition of C_z and D_z in (3.9), the above condition is equivalent to

$$rank \left[\begin{array}{cc} A - j\omega I & B_2 \\ C_1 & D_{12} \end{array}\right] = n + p, \quad \forall \omega. \quad (6.19)$$

Dually, we impose the condition
$$rank \begin{bmatrix} A - j\omega I & B_1 \\ C_2 & D_{21} \end{bmatrix} = n + q, \quad \forall \omega. \tag{6.20}$$

Theorem 6.3. The following statements are equivalent :

(i) For any augmentation (C_2', D_{21}') in (6.3), $CHAIN(P_o)$ has a (J_{mr}, J_{pr})-lossless factorization, and for any augmentation (B_2', D_{12}') in (6.4), $CHAIN(P_i)$ has a (J_{mr}, J_{mq})-lossless factorization.

(ii) $X = Ric\, f(\left[\begin{array}{c|c} A & B \\ \hline C_z & D_z \end{array}\right]) \geq 0$ and $Y = Ric\, f^\sim(\left[\begin{array}{c|c} A & B_w \\ \hline C & D_w \end{array}\right]) \geq 0$ exist such that
$$\sigma(XY) < 1 \tag{6.21}$$

Remark. The statement (ii) is the well-known solvability condition of the normalized H^∞ control problem.

Proof. Assume that (i) holds. Let $D_z^\perp$ be a right annihilator of D_z which is explicitly given by
$$D_z^\perp = D_{12}^\perp \begin{bmatrix} I & -D_{11} \end{bmatrix}.$$
From the relation (6.11), it follows that
$$\begin{bmatrix} D_z^{l'} \\ D_z^\perp J_{mr} \end{bmatrix} J_{mr} \begin{bmatrix} D_z & J_{mr}(D_z^\perp)^T \end{bmatrix} = \begin{bmatrix} E^T J_{pr} E & 0 \\ 0 & D_z^\perp J_{mr}(D_z^\perp)^T \end{bmatrix}.$$
Due to the inertia theorem, we conclude that
$$D_z^\perp J_{mr}(D_z^\perp)^T > 0. \tag{6.22}$$
This corresponds to (5.30). Hence, Lemma 5.5 implies that
$$\bar{Y} \leq X \tag{6.23}$$
where $\bar{Y}$ and X are given by (6.17) and (6.12), respectively. Analogously, we have
$$D_w^\perp = \begin{bmatrix} D_{11} \\ I \end{bmatrix} D_{21}^\perp,$$
which must satisfy
$$(D_w^\perp)^T J D_w^\perp < 0,$$
due to (6.16). Therefore, according to Lemma 5.5, we have
$$\bar{X} \leq Y, \tag{6.24}$$
where $\bar{X}$ and Y are given by (6.12) and (6.17), respectively. Since (6.13) and (6.18) must hold for any augmentation, they must hold for the augmentations which give rise $\bar{X} = Y$ and $\bar{Y} = X$. Hence, (6.21) holds. Therefore, we have established the necessity of (6.21).

The proof of the converse can be done by using the fact that (ii) is a solvability condition for H^∞ control under the assumptions (6.19) and (6.20). Since the solvability of H^∞ control problem does not depend on the augmentations, we can establish (i). We can prove that (ii) implies (i) by more direct argument without any recourse to H^∞ control. □

7 Conclusion

Algebraic properties of the chain-scattering representation have been derived both in the frequency domain and in the state space. Several factorizations of the CSR have been derived which are elegant in their forms. The duality between the CSR and its inverse has been fully exploited. The notion of (J, J')-lossless factorization, which is closely related to H^∞ control problem, has been fully exploited. It has been shown that the (J, J')-lossless factorization is connected with the invariance of related Hamiltonian matrices. Finally, the relations between the system augmentations and the factorizations have been considered from the viewpoint of H^∞ control.

A number of problems are still to be solved in order that the CSR becomes a common tool for control system design. For instance, the order reduction of the CSR is an important problem which is directly related to the simplification of uncertainty description.

References

[Be1] Belevitch, V.: *Classical Network Theory.* Holden-Day. (1968) p.108

[DD1] Dewilde, P., Dym, H.: Lossless chain-scattering matrices and optimum linear prediction : The vector case. *Int.J. on Circuit Theory and Appl.* **9** (1981) 135-175

[Ki1] Kimura, H.: Chain-scattering representation, J-lossless factorization and H^∞ control, *J.of Mathematical Systems, Estimation and Control.* **5** (1995) 203-255

[Ki2] Kimura, H.: Conjugation, interpolation and model-matching in H^∞. *Int.J. Control.* **49** (1989) 243-275

[Re1] Redheffer, M., R.: On a certain linear fractional transformation. *J.of Mathematical Physics* **39** (1960) 269-286

[Si1] Siegel, L., C.: *Symplectic Geometry.* Academic Press. (1964)

[Wi1] Wimmer, K., H.: Monotonicity of maximal solutions of algebraic Riccati equations. *Systems & Control Letters.* **5** (1985) 317-319

Mixed H_2/H_∞ Control

Carsten Scherer

Mechanical Engineering Systems and Control Group, Delft University of Technology, Mekelweg 2, 2628 CD Delft, The Netherlands

Abstract. In this article we provide a solution to the mixed H_2/H_∞ problem with reduced order controllers for time-varying systems in terms of the solvability of differential linear matrix inequalities and rank conditions, including a detailed discussion of how to construct a controller. Immediate specializations lead to a solution of the full order problem and the mixed H_2/H_∞ problem for linear systems whose description depends on unknown but in real-time measurable time-varying parameters. As done in the literature for the H_∞ problem, we resolve the quadratic mixed H_2/H_∞ problem by reducing it to the solution of a finite number of algebraic linear matrix inequalities. Moreover, we point out directions how to overcome the conservatism caused by assuming a particular parameter dependence or by using constant solutions of the differential matrix inequalities. For linear time-invariant systems, we reveal how to incorporate robust asymptotic tracking or disturbance rejection as an objective in the mixed H_2/H_∞ problem. Finally, we address the specializations to the fully general pure H_∞ or generalized H_2 problem, and provide quadratically convergent algorithms to compute optimal values. Our techniques do not only lead to insights into the structure of the solution sets of the corresponding linear matrix inequalities, but they also allow to explicitly describe the influence of various system zeros on the optimal values.

Notation

$\mathcal{C} = \mathcal{C}^- \cup \mathcal{C}^0 \cup \mathcal{C}^+$ is the complex plain partitioned into open half planes and the imaginary axis. $\mathcal{R}^n$ is equipped with the Euclidean norm, and $\mathcal{R}^{n \times m}$ with the corresponding induced norm, both denoted as $\|.\|$. $\rho(A)$ denotes the spectral radius of the matrix A, and $A^{-1/2}$ the unique positive semidefinite square root of A if $A \geq 0$. L_p denotes the signal space $L_p^n[0,\infty)$ (for an appropriate n) and is equipped with the standard norm $\|.\|_p$. Functions are tacitly assumed to be continuous and bounded, and smooth functions are, in addition, continuously differentiable. Time functions are functions defined on $[0, \infty)$. For a symmetric valued function X defined on S, X is said to be strictly positive ($X \gg 0$) if there exists an $\epsilon > 0$ with $X(s) \geq \epsilon I$ for all $s \in S$. For the system or input output mapping $\dot{x} = Ax + Bu$, $x(0) = 0$, $y = Cx + Du$, we use the notation $\left[\begin{array}{c|c}A & B \\ \hline C & D\end{array}\right]$. If $\begin{pmatrix}A & B \\ C & D\end{pmatrix}$ is a constant matrix, the system is called LTI, if it is a time function, the system is called LTV. The time function A is exponentially stable if the system $\dot{x} = Ax$ has this property.

1 Introduction

Recently, linear systems which depend on time-varying a priorily unknown but on-line measurable parameters have gained a lot of interest [50, 38],[3]-[8]. These so-called linear parametrically-varying (LPV) systems appear in robustness problems, in gain-scheduling techniques for nonlinear systems, or in synthesis problems for nonlinear systems that can be described by a differential inclusion [11]. For a detailed discussion we refer to the literature [50, 4].

Given an LPV system, the goal is to construct a controller which not only uses a measured output but, in addition, the on-line measured actual parameters as information in order to exponentially stabilize the system and to provide good performance properties. Until now, the performance objective was specified as an L_2 disturbance attenuation problem with the standard interpretation such as guaranteeing robust stability or tracking [36],[3]-[8]. This is the so-called H_∞ problem for LPV systems.

In principle, an LPV system can be viewed to be time-varying, and any design technique which is available for a specific choice of the performance measure can be used for controller construction. However, the actual parameter curve is not known a priorily and many existing synthesis techniques (as e.g. H_2 and H_∞ control for LTV systems) instantaneously require the knowledge of the parameter values over the whole time interval of interest. For LPV systems and the H_∞ problem, it is not difficult to propose a (pretty conservative) way out of this dilemma: Assume that the parameters are contained in an a priorily given set. Then replace the time-varying solutions of differential Riccati equations along the actual parameter curve by constant solutions of algebraic Riccati inequalities over the whole set of possible parameters. Only under additional hypotheses on the structure of the parameter set (polytopic) and on the dependence of the system on the parameters (affine and partly constant), the verification of the existence of a suboptimal controller and its construction can be reduced to solving a set of linear matrix inequalities [3]-[8].

The purpose of this article is to show that these ideas can be extended to the so-called mixed H_2/H_∞ problem (only whose LTI version has been addressed previously [9, 17, 25, 28, 29, 33, 42, 43, 51, 60]) such that not only robustness specifications (in terms of an H_∞ constraint) but also performance specifications (measured in H_2 norm like criteria or upper bounds thereof) can be taken into account. In fact, we provide a full proof for our central result, a solution of the reduced order mixed H_2/H_∞ problem for LTV systems in terms of the solvability of differential linear matrix inequalities and rank coupling conditions, and we give explicit formulas for a full order controller.

Then we address various specializations of our main result. We point out how to recover results on the pure H_∞ problem for LTV systems [39, 55]. More importantly, we solve the mixed H_2/H_∞ problem for LPV systems not only in the spirit of previous work but also including possible refinements to avoid conservatism. This encompasses a solution of the H_2 problem for LPV systems. If the system is LTI, we obtain a new solution of the mixed H_2/H_∞ problem in terms of linear matrix inequalities where the underlying system is in no way restricted. Moreover, we show how to incorporate asymptotic tracking or disturbance rejection requirements by extending the system with a suitable internal model. This leads to a solution of the mixed H_2/H_∞ problem with robust regulation (for possibly large plant uncertainties) and extends [1, 2]. For the pure H_∞ problem, we generalize [47, 49] to systems having a nonzero direct feedthrough from the disturbances to the controlled outputs and provide a quadratically convergent algorithm to compute the optimal value. Finally, we reveal that such a computational scheme can be also obtained for the generalized H_2 problem [40], and we provide an explicit formula for the optimal value of the genuine H_2 problem, both for general LTI systems [52].

As auxiliary considerations, we investigate in detail the estimation inequality both for time-varying and for time-invariant data. In the latter case, we are not only able to gain insights into the structure of the solutions of this algebraic linear matrix inequality, but we can also explicitly display the influence of various system zeros on the solvability.

The article is organized as follows. In Section 2, we define the mixed H_2/H_∞ objective for time-varying systems and address the related analysis tests, including the role of scalings, for reasons of space mainly without proofs. Section 3 contains our main result, a solution to the reduced order controller mixed H_2/H_∞ control problem in terms of differential linear matrix inequalities, including explicit formulas for controllers. In Section 4, we show how the estimation differential inequality can be reduced to an initial value problem for a perturbed Riccati differential equation. For time-invariant data, we investigate the corresponding algebraic linear matrix inequality in full generality. Section 5 summarizes the consequences for our main result. In Section 6, we discuss linear parametrically-varying systems and demonstrate a controller construction by solving a finite number of algebraic linear matrix inequalities. In Section 7, we address the mixed H_2/H_∞ problem with robust regulation for LTI systems, and the Sections 8 and 9 are devoted to the pure H_∞ and (generalized) H_2 problems respectively.

Short proofs which provide insights into construction schemes are included in the text whereas more technical proofs are collected in the appendix.

2 Mixed H_2/H_∞ Performance Bounds

Consider the LTV system

$$z = Tw = \begin{pmatrix} z_1 \\ z_2 \end{pmatrix} = \begin{pmatrix} T_1 \\ T_2 \end{pmatrix} w = \left[\begin{array}{c|c} A & G \\ \hline H_1 & F_1 \\ H_2 & F_2 \end{array} \right] w. \tag{1}$$

We interpret $w \to z_1$ as the robustness channel and $w \to z_2$ as the performance channel. To be more specific, we assume that the uncertainty of the system is described by $w = \Delta z$ where Δ comprises the set of (possibly nonlinear) operators $L_2 \to L_2$ with incremental gain [13] not larger than $1/\gamma$. If A is exponentially stable, the small-gain theorem implies that stability is preserved if $\|T_1\|_\infty < \gamma$, where

$$\|T_1\|_\infty := \sup_{w \in L_2, \|w\|_2 = 1} \|T_1 w\|_2$$

defines the operator norm of T_1 induced by signals in L_2. In the LTI case, there is a well-known test on the state space matrices which characterizes stability of A and $\|T_1\|_\infty < \gamma$, the so-called Bounded Real Lemma (BRL) [57, 22, 26]. It is not difficult to prove the following generalization to LTV systems. Recall that time functions are bounded, and that a symmetric valued time function X is strictly positive ($X \gg 0$) if there exists an $\epsilon > 0$ with $X(t) \geq \epsilon I$ for all $t \geq 0$.

Theorem 1 *(LTV Strict Bounded Real Lemma) The time function A is exponentially stable and $\left\| \left[\begin{array}{c|c} A & G \\ \hline H_1 & F_1 \end{array} \right] \right\|_\infty < \gamma$ iff there exists a smooth time function X such that*

$$X \gg 0, \quad \begin{pmatrix} \dot{X} + A^T X + XA & XG & H_1^T \\ G^T X & -\gamma I & F_1^T \\ H_1 & F_1 & -\gamma I \end{pmatrix} \ll 0. \tag{2}$$

Hence norm bounds can be characterized by the existence of strictly positive solutions to a strict differential linear matrix inequality (DLMI). Note that asking for the existence of a constant solution is equivalent to the popular concept of quadratic H_∞ performance for an LTV system [3]-[8]. If T_1 is LTI, it is no loss of generality to confine X in Theorem 1 to be *constant* and the inequalities (2) become algebraic.

The channel $w \to z_2$ of (1) is used for describing performance specifications. Indeed, we have in mind to generalize certain H_2 norm like criteria for LTI systems [40] to the LTV system T_2. Among the several possibilities, we pay special attention to the deterministic criterion of assessing performance by the largest amplitude of z_2 for all w of finite and bounded energy. This is particularly useful if (components of) z_2 are interpreted as tracking errors. To quantify the gain of T_2 mapping L_2 into L_∞, we define the induced norm

$$\|T_2\| := \sup_{w \in L_2, \|w\|_2 = 1} \|T_2 w\|_\infty.$$

If A is exponentially stable, let Y denote the (bounded) solution of the initial value problem

$$\dot{Y} = AY + YA^T + GG^T, \quad Y(0) = 0. \tag{3}$$

If $F_2 = 0$, one has $\|T_2\|^2 = \sup_{t \geq 0} \|H_2 Y H_2^T(t)\|$. This allows to prove the following analogue of Theorem 1.

2 Mixed H_2/H_∞ Performance Bounds

Theorem 2 *A is exponentially stable and* $\left\| \begin{bmatrix} A & G \\ \hline H_2 & F_2 \end{bmatrix} \right\|^2 < \beta$ *iff* $F_2 = 0$ *and there exists a smooth Z with*

$$Z \gg 0, \quad \dot{Z} + A^T Z + ZA + ZGG^T Z \ll 0, \quad H_2 Z^{-1} H_2^T \ll \beta I. \qquad (4)$$

If defining the size of the amplitude of z_2 with the spatial norm $\max_j |x_j|$, the squared gain of T_2 equals $\sup_{t \geq 0} \max_j d_j [H_2 Y H_2^T(t)]$, where $d_j(M)$ denotes the j-th diagonal element of the square matrix M, and Y solves (3). Theorem 2 remains valid for this norm after replacing $H_2 Z^{-1} H_2^T \ll \beta I$ by $\max_j d_j(H_2 Z^{-1} H_2^T) \ll \beta$.

An alternative measure arises with a stochastic interpretation. If w is white noise, we recall (due to $x_0 = 0$ and hence $E(x_0 x_0^T) = 0$) that $E(z_2 z_2^T) = H_2 Y H_2^T$ [30]. Then

$$\|T_2\|_2^2 := \sup_{t \geq 0} E(z_2^T z_2) = \sup_{t \geq 0} \text{trace}[H_2 Y H_2^T(t)]$$

defines the maximal output variance and is a generalization of the genuine H_2 norm to LTV systems. Theorem 2 persists to hold for the H_2 norm with $H_2 Z^{-1} H_2^T \ll \beta I$ replaced by $\text{trace}(H_2 Z^{-1} H_2^T) \ll \beta$.

Even for LTI systems, the synthesis problem of optimizing $\|.\|$ or $\|.\|_2$ of over all stabilizing controllers which keep a bound on the norm $\|.\|_\infty$ for a different channel seems very hard [28, 42]. This motivates to replace the objective functional by an upper bound. Let us define

$$J(T) := \inf\{\alpha \mid \exists \text{ smooth time function } X \text{ with (2) and } H_2 X^{-1} H_2 \ll \alpha I\} \qquad (5)$$

(including, as usual, $J(T) = \infty$ if no solution of (2) exists.) Then $J(T) < \alpha$ implies $H_2 X^{-1} H_2 \ll \alpha I$ for some solution X of (2). Note that any such X satisfies $\dot{X} + A^T X + XA + \frac{1}{\gamma} XGG^T X \ll 0$. If $F_2 = 0$, Theorem 2 allows to infer $\|T_2\|^2 < \alpha\gamma$. We conclude

$$\|T_2\|^2 \leq \gamma J(T)$$

and $J(T)$ is indeed an upper bound of $\|T_2\|^2/\gamma$. If $\gamma = 1$, $H_1 = 0$, $F_1 = 0$, the solution sets of the DLMI in (2) and of the differential Riccati inequality in (4) are clearly identical, what implies $\|T_2\|^2 = J(T)$ and *recovers* $\|T_2\|^2$. Similar conclusions hold for the other norms of T_2.

One can clearly define $J(T)$ via (5) confining X to *constant* solutions of (2). This generalizes the so-called quadratic H_∞ performance specification to the quadratic mixed H_2/H_∞ specification and allows a specialization to the pure H_2 case. If T is LTI, it is well-known that this restriction to constant X causes no loss of generality, and we recover the definitions in [42, 40].

Let us finally comment on scalings. If one can take structural or other properties of the system uncertainty into account, the perturbation Δ is restricted to a certain subset $\boldsymbol{\Delta}_\gamma$ of all operators with incremental gain at most $1/\gamma$. It might then be possible to identify a class of scalings containing pairs of matrix valued time functions (S, S_1) with bounded inverses such that

$S \Delta S_1$ has incremental gain at most $1/\gamma$ for all $\Delta \in \Delta_\gamma$.

The existence of a pair of scalings (S, S_1) in this class with

$$\|S_1^{-1} T_1 S^{-1}\|_\infty < \gamma$$

is an obvious sufficient condition for stability robustness against uncertainties in Δ_γ, and it is weaker than $\|T_1\|_\infty < \gamma$. A most prominent specific example of this concept is the μ-upper bound with constant scalings. This gives a systematic tool to incorporate in this channel not only stability robustness requirements against structured uncertainties, but also robust performance specifications in the induced L_2-norm, as usually done for LTI systems [37]. One might as well specify a set of scalings S_2 for weighting the performance output and modeling alternative performance specifications [44]. The benefit of incorporating scalings S_2 for the channel z_2 remains to be explored.

Summarizing, if having fixed a class of scalings S consisting of triples (S_1, S_2, S) of time functions with bounded inverse, we can alternatively define

$$J_s(T) \text{ to be the infimal } \alpha$$

for which there exists an $(S_1, S_2, S) \in S$ and a solution to the BRL inequality corresponding to $\|S_1^{-1} T_1 S^{-1}\|_\infty < \gamma$ such that $S_2^{-1} H_2 X^{-1} H_2^T S_2^{-T} \ll \alpha I$.

3 The Mixed H_2/H_∞ Synthesis Problem

Suppose a specific control tasks (including the specification of weightings) leads to the generalized LTV plant

$$\begin{pmatrix} y \\ z_1 \\ z_2 \end{pmatrix} = \left[\begin{array}{c|cc} A & B & G \\ \hline C & 0 & D \\ H_1 & E_1 & F_1 \\ H_2 & E_2 & F_2 \end{array} \right] \begin{pmatrix} u \\ w \end{pmatrix} \tag{6}$$

where A is of size $n \times n$. With the LTV controller

$$u = Ry = \left[\begin{array}{c|c} K & L \\ \hline M & N \end{array} \right] y, \tag{7}$$

the closed loop system is described as

$$z = T(R)w = \begin{pmatrix} z_1 \\ z_2 \end{pmatrix} = \begin{pmatrix} T_1(R) \\ T_2(R) \end{pmatrix} w = \left[\begin{array}{c|c} \mathcal{A} & \mathcal{G} \\ \hline \mathcal{H}_1 & \mathcal{F}_1 \\ \mathcal{H}_2 & \mathcal{F}_2 \end{array} \right] w$$

ically implies that R is stabilizing. It is standard to approach this problem via
3 The Mixed H_2/H_∞ Synthesis Problem

where
$$\left[\begin{array}{c|c}\mathcal{A} & \mathcal{G} \\ \hline \mathcal{H}_1 & \mathcal{F}_1 \\ \mathcal{H}_2 & \mathcal{F}_2\end{array}\right] = \left[\begin{array}{cc|cc}A+BNC & BM & G+BND \\ LC & K & LD \\ \hline H_1+E_1NC & E_1M & F_1+E_1ND \\ H_2+E_2NC & E_2M & F_2+E_2ND\end{array}\right].$$

The controller R is called *stabilizing* if $\mathcal{A}$ is exponentially stable. We intend to minimize $J(T(R))$ over all LTV controllers R. Recall that $J(T(R)) < \infty$ automatically implies that R is stabilizing. It is standard to approach this problem via a suboptimality test: Characterize whether there exists an R with $J(T(R)) < \alpha$ or, equivalently, whether there exists an R and a smooth time function $\mathcal{X}$ such that

$$\mathcal{X} \gg 0, \quad \begin{pmatrix} \dot{\mathcal{X}} + \mathcal{A}^T\mathcal{X} + \mathcal{X}\mathcal{A} & \mathcal{X}\mathcal{G} & \mathcal{H}_1^T \\ \mathcal{G}^T\mathcal{X} & -\gamma I & \mathcal{F}_1^T \\ \mathcal{H}_1 & \mathcal{F}_1 & -\gamma I \end{pmatrix} \ll 0, \quad \mathcal{H}_2 \mathcal{X}^{-1} \mathcal{H}_2 \ll \alpha. \tag{8}$$

This not only guarantees robust stability (against perturbations with incremental gain at most $1/\gamma$) but also a performance level $\alpha\gamma$.

Recent approaches to the H_∞ problem for LTI systems are based on the following observations: The BRL inequality in (8) is, for a fixed $\mathcal{X}$, linear in the controller parameters. Hence one can eliminate these parameters using the so-called Projection Lemma (Lemma 7), what leads to a suboptimality test in terms of linear inequalities in (parts) of $\mathcal{X}$ and $\mathcal{X}^{-1}$ [22, 26, 49].

In the mixed problem (8), we have to fulfill *three* inequalities what makes it impossible to eliminate *all* controller parameters from the final characterization. Instead, we intend to keep as many (transformed) controller parameters as possible such that, still, matrix inequalities result which are linear in parts of $\mathcal{X}$ and $\mathcal{X}^{-1}$ *and* in the remaining (transformed) controller parameters. In fact, the key idea has its origins in [45, 49]: Eliminate K in (8) (which only affects the BRL inequality), transform L and M, and keep N to achieve the desired structure.

A central step in the proof is the following very simple explicit result for the solvability of a specially structured inequality. In fact, it will turn out in Section 4 that this is a version of the so-called 'completion of the squares argument' which is most suited for our purposes.

Lemma 3 *Let Q be a symmetric (partitioned) time function and consider the inequality*

$$\begin{pmatrix} Q_1 & Q_{21}^T & Q_{31}^T + X^T \\ Q_{21} & Q_2 & Q_{32}^T \\ Q_{31} + X & Q_{32} & Q_3 \end{pmatrix} \ll 0 \tag{9}$$

in the unstructured time function X. This inequality has a solution X iff

$$\begin{pmatrix} Q_1 & Q_{21}^T \\ Q_{21} & Q_2 \end{pmatrix} \ll 0 \quad \text{and} \quad \begin{pmatrix} Q_2 & Q_{32}^T \\ Q_{32} & Q_3 \end{pmatrix} \ll 0. \tag{10}$$

If (9) is solvable, one particular solution is given by

$$X = Q_{32}Q_2^{-1}Q_{21} - Q_{31}. \tag{11}$$

Proof. If (9) has a solution then (10) just follow from (9) by canceling the first or third block row/column.

Now suppose that (10) holds what implies $Q_2 \ll 0$. The central trick is to cancel in (9) the block Q_{21} by a congruence transformation. To be specific, (9) is equivalent to

$$\begin{pmatrix} I & -Q_{21}^T Q_2^{-1} & 0 \\ 0 & I & 0 \\ 0 & 0 & I \end{pmatrix} (9) \begin{pmatrix} I & 0 & 0 \\ -Q_2^{-1}Q_{21} & I & 0 \\ 0 & 0 & I \end{pmatrix} \ll 0$$

which rewrites to

$$\begin{pmatrix} Q_1 - Q_{21}^T Q_2^{-1} Q_{21} & 0 & Q_{31}^T - Q_{21}^T Q_2^{-1} Q_{32}^T + X^T \\ 0 & Q_2 & Q_{32}^T \\ Q_{31} - Q_{32} Q_2^{-1} Q_{21} + X & Q_{32} & Q_3 \end{pmatrix} \ll 0.$$

X defined in (11) is a solution since (10) implies $Q_1 - Q_{21}^T Q_2^{-1} Q_{21} \ll 0$. ∎

Now we are ready to formulate and prove our main result.

Theorem 4 *There exists a controller* $R := \left[\begin{array}{c|c} K & L \\ \hline M & N \end{array}\right]$ *with K of size $k \leq n$ which satisfies $J(T(R)) < \alpha$ iff there exist $\epsilon > 0$, time functions X, Y, Z with*

$$X \gg 0, \ Y \gg 0, \ X - Y^{-1} = ZZ^T, \ Z \text{ of size } n \times k, \ Z^T Z \gg 0, \tag{12}$$

and time functions F, J, N such that

$$\text{rank} \begin{pmatrix} F & NC \\ Y & I \\ I & X \end{pmatrix} = k + n, \ \text{rank} \begin{pmatrix} BN & Y & I \\ J & I & X \end{pmatrix} = k + n, \tag{13}$$

and such that the following DLMIs are satisfied:

$$\begin{pmatrix} \dot{X} + A^T X + XA + JC + (JC)^T & XG + JD & (H_1 + E_1 NC)^T \\ (XG + JD)^T & -\gamma I & (F_1 + E_1 ND)^T \\ H_1 + E_1 NC & F_1 + E_1 ND & -\gamma I \end{pmatrix} \ll 0, \tag{14}$$

$$\begin{pmatrix} -\dot{Y} + AY + YA^T + BF + (BF)^T & G + BND & (H_1 Y + E_1 F)^T \\ (G + BND)^T & -\gamma I & (F_1 + E_1 ND)^T \\ H_1 Y + E_1 F & F_1 + E_1 ND & -\gamma I \end{pmatrix} \ll 0, \tag{15}$$

$$\begin{pmatrix} \alpha - \epsilon I & H_2 Y + E_2 F & H_2 + E_2 NC \\ (H_2 Y + E_2 F)^T & Y & I \\ (H_2 + E_2 NC)^T & I & X \end{pmatrix} \geq 0. \tag{16}$$

3 The Mixed H_2/H_∞ Synthesis Problem

Proof of necessity. Suppose, for some controller of size $k \leq n$, there exists a smooth time function $\mathcal{X}$ with (8). We fix $\mathcal{X}$ and view the BRL inequality in (8) as an inequality in K. To be specific, define $\mathcal{I} := \begin{pmatrix} 0 & I \end{pmatrix}$ to get

$$\mathcal{A} = \begin{pmatrix} A + BNC & BM \\ LC & 0 \end{pmatrix} + \begin{pmatrix} 0 \\ I \end{pmatrix} K \begin{pmatrix} 0 & I \end{pmatrix} = \tilde{\mathcal{A}} + \mathcal{I}^T K \mathcal{I}.$$

Then the BRL inequality reads as

$$\begin{pmatrix} \dot{\mathcal{X}} + \tilde{\mathcal{A}}^T \mathcal{X} + \mathcal{X}\tilde{\mathcal{A}} + \mathcal{I}^T K^T \mathcal{I} \mathcal{X} + \mathcal{X}\mathcal{I}^T K \mathcal{I} & \mathcal{X}\mathcal{G} & \mathcal{H}_1^T \\ \mathcal{G}^T \mathcal{X} & -\gamma I & \mathcal{F}_1^T \\ \mathcal{H}_1 & \mathcal{F}_1 & -\gamma I \end{pmatrix} \ll 0. \quad (17)$$

The matrix $\mathcal{J} := \begin{pmatrix} I \\ 0 \end{pmatrix}$ clearly satisfies $\mathcal{I}\mathcal{J} = 0$. Hence (17) implies

$$\begin{pmatrix} \mathcal{J}^T[\dot{\mathcal{X}} + \tilde{\mathcal{A}}^T \mathcal{X} + \mathcal{X}\tilde{\mathcal{A}}]\mathcal{J} & \mathcal{J}^T \mathcal{X}\mathcal{G} & \mathcal{J}^T \mathcal{H}_1^T \\ \mathcal{G}^T \mathcal{X} \mathcal{J} & -\gamma I & \mathcal{F}_1^T \\ \mathcal{H}_1 \mathcal{J} & \mathcal{F}_1 & -\gamma I \end{pmatrix} \ll 0. \quad (18)$$

Similarly, with

$$\mathcal{Y} := \mathcal{X}^{-1},$$

we infer $(\mathcal{I}\mathcal{X})\mathcal{Y}\mathcal{J} = 0$ and thus

$$\begin{pmatrix} \mathcal{J}^T[-\dot{\mathcal{Y}} + \mathcal{Y}\tilde{\mathcal{A}}^T + \tilde{\mathcal{A}}\mathcal{Y}]\mathcal{J} & \mathcal{J}^T \mathcal{G} & \mathcal{J}^T \mathcal{Y}\mathcal{H}_1^T \\ \mathcal{G}^T \mathcal{J} & -\gamma I & \mathcal{F}_1^T \\ \mathcal{H}_1 \mathcal{Y} \mathcal{J} & \mathcal{F}_1 & -\gamma I \end{pmatrix} \ll 0. \quad (19)$$

Let us partition $\mathcal{X}$, $\mathcal{Y}$ according to $\mathcal{A}$ into n and k rows/columns as

$$\mathcal{X} = \begin{pmatrix} X & U \\ U^T & \hat{X} \end{pmatrix}, \quad \mathcal{Y} = \begin{pmatrix} Y & V \\ V^T & \hat{Y} \end{pmatrix}, \quad (20)$$

and let us recall

$$\mathcal{Y} = \begin{pmatrix} [X - U\hat{X}^{-1}U^T]^{-1} & -[X - U\hat{X}^{-1}U^T]^{-1}U\hat{X}^{-1} \\ -\hat{X}^{-1}U^T[X - U\hat{X}^{-1}U^T]^{-1} & [\hat{X} - U^T X^{-1} U]^{-1} \end{pmatrix}. \quad (21)$$

We can assume without loss of generality that

$$U \text{ (of dimension } n \times k\text{) satisfies } U^T U \gg 0;$$

if not true, just perturb $\mathcal{X}$ suitably without violating (8). The formula (21) reveals $Y^{-1} = X - U\hat{X}^{-1}U^T$. If Cholesky factorizing $\hat{X} = W^T W$ such that W is smooth, bounded, and has a bounded inverse, we arrive at (12) for $Z := UW^{-1}$. A simple computation shows, with the identities

$$F = NCY + MV^T \quad \text{and} \quad J = XBN + UL, \quad (22)$$

that the left-hand sides of (18), (14) and (19), (15) are *identical*. We remark that we define F and J via these equations for proving necessity, and we view (22) as equations in L and M for constructing a controller in the sufficiency proof. Under the hypothesis (12), it is now not difficult to see that (22) are solvable as equations in L and M iff the rank conditions (13) hold true. We clarify this for the first equation: It is solvable iff $\ker(V^T) \subset \ker(F - NCY)$. The (1,1) block of $\mathcal{X}\mathcal{Y} = I$ implies $XY + UV^T = I$ which reveals $\ker(I - XY) = \ker(V^T)$ since U has full column rank. Hence, solvability is equivalent to $\ker(I - XY) \subset \ker(F - NCY)$. By (12), $I - XY$ has rank k and, therefore, this inclusion is equivalent to the first condition in (13) since

$$\begin{pmatrix} F & NC \\ Y & I \\ I & X \end{pmatrix} \begin{pmatrix} I & 0 \\ -Y & I \end{pmatrix} = \begin{pmatrix} F - NCY & NC \\ 0 & I \\ I - XY & X \end{pmatrix}.$$

Finally, there is an $\epsilon > 0$ with $\mathcal{H}_2 \mathcal{X}^{-1} \mathcal{H}_2^T \leq (\alpha - \epsilon)I$ which is equivalent to

$$\begin{pmatrix} (\alpha - \epsilon)I & \mathcal{H}_2 \\ \mathcal{H}_2^T & \mathcal{X} \end{pmatrix} \geq 0. \tag{23}$$

With the $(n+k) \times 2n$ function

$$\mathcal{Z} := \begin{pmatrix} Y & I \\ V^T & 0 \end{pmatrix}, \tag{24}$$

the inequality (23) implies

$$\begin{pmatrix} (\alpha - \epsilon)I & \mathcal{H}_2 \mathcal{Z} \\ \mathcal{Z}^T \mathcal{H}_2^T & \mathcal{Z}^T \mathcal{X} \mathcal{Z} \end{pmatrix} \geq 0. \tag{25}$$

Recalling the definition of F and computing $\mathcal{Z}^T \mathcal{X} \mathcal{Z} = \mathcal{Z}^T \begin{pmatrix} I & X \\ 0 & U^T \end{pmatrix} = \begin{pmatrix} Y & I \\ I & X \end{pmatrix}$ leads to (16). This proves necessity. ∎

Constructive proof of sufficiency. Define $U := Z$, $\hat{X} := I$, and $\mathcal{X} := \begin{pmatrix} X & U \\ U^T & I \end{pmatrix}$. Due to $X - UU^T \gg 0$, $\mathcal{X}$ is smooth, bounded, and strictly positive, and the same holds for $\mathcal{Y} := \mathcal{X}^{-1}$. Again because of (12) and (21), $\mathcal{Y}$ has Y as its left-upper block. If using the partitions (20), we have $U^T U \gg 0$ by hypothesis and, since $V = -YU$ from (21), $V^T V \gg 0$ as a consequence. Let us now define the time functions

$$M := (F - NCY)V(V^T V)^{-1} \text{ and } L := (U^T U)^{-1} U^T (J - XBN).$$

Since U and V have full column rank, these are the unique solutions of the equations (22) if they are solvable at all. The latter, however, is assured by (13) as clarified in the necessity proof.

With L, M, N given so far, we can define $\tilde{\mathcal{A}}$, $\mathcal{G}$, $\mathcal{H}_j$, $\mathcal{F}_j$, $j = 1, 2$. If we introduce $\mathcal{Z}$ as in (24), we infer that (16) is the same as (25). Since $\mathcal{Z}$ has full row rank and

3 The Mixed H_2/H_∞ Synthesis Problem

thus a right-inverse, we can get back to (23) which leads to the third inequality in (8). Moreover, (14) and (15) are identical to (18) and (19) respectively. Hence it remains to find a time function K which satisfies (17) and, therefore, leads to the BRL inequality in (8).

We use $\mathcal{X}\mathcal{I}^T K\mathcal{I} = \begin{pmatrix} Z \\ I_k \end{pmatrix} K \begin{pmatrix} 0 & I_k \end{pmatrix}$ where we display the size of the identity blocks by using the index. Due to $Z^T Z \gg 0$, there exists a smooth time function Z_e such that $\begin{pmatrix} Z & Z_e \end{pmatrix}$ has a smooth bounded inverse. With the first k rows S_1 and the last $n-k$ rows S_2 of this inverse, we conclude that

$$\begin{pmatrix} S_1 \\ S_2 \end{pmatrix} \text{ has a smooth bounded inverse and } \begin{pmatrix} S_1 \\ S_2 \end{pmatrix} Z = \begin{pmatrix} I_k \\ 0 \end{pmatrix}.$$

With

$$S := \begin{pmatrix} S_1 & 0 \\ S_2 & 0 \\ -S_1 & I_k \end{pmatrix} \text{ we get } S \begin{pmatrix} Z \\ I_k \end{pmatrix} = \begin{pmatrix} I_k \\ 0 \\ 0 \end{pmatrix} \text{ and } S \begin{pmatrix} 0 \\ I_k \end{pmatrix} = \begin{pmatrix} 0 \\ 0 \\ I_k \end{pmatrix}.$$

Note that both S and S^{-1} are smooth time functions. We can left-multiply the first row of (17) with S and right-multiply the first column of (17) with S^T. After this congruence transformation, (17) is equivalent to

$$\begin{pmatrix} Q_1 & Q_{12} + K & Q_{13} \\ Q_{21} + K^T & Q_2 & Q_{23} \\ Q_{31} & Q_{32} & Q_3 \end{pmatrix} \ll 0 \tag{26}$$

for some computable Q. Now we just note that (18), (19) are equivalent to $\begin{pmatrix} Q_1 & Q_{13} \\ Q_{31} & Q_3 \end{pmatrix} \ll 0$ and $\begin{pmatrix} Q_2 & Q_{23} \\ Q_{32} & Q_3 \end{pmatrix} \ll 0$. Lemma 3 then reveals that $K := Q_{13} Q_3^{-1} Q_{32} - Q_{12}$ is indeed a time function which leads to (26) and hence to (17). The controller construction is complete. ∎

The following comments on Theorem 4 also apply, without explicitly mentioning, to all the other problems that will be considered in this article.

Remark on the dual problem. Consider the LTV system

$$\begin{pmatrix} y \\ z_1 \\ z_2 \end{pmatrix} = \left[\begin{array}{c|ccc} A & B & G & G_2 \\ \hline C & 0 & D & D_2 \\ H_1 & E_1 & F_1 & F_{12} \\ H_2 & E_2 & F_2 & F_{22} \end{array} \right] \begin{pmatrix} u \\ w \\ w_2 \end{pmatrix}.$$

The existence of a controller (7) such that the closed loop system described by

$$\begin{pmatrix} z_1 \\ z_2 \end{pmatrix} = \left[\begin{array}{c|cc} \mathcal{A} & \mathcal{G} & \mathcal{G}_2 \\ \hline \mathcal{H}_1 & \mathcal{F}_1 & \mathcal{F}_{12} \\ \mathcal{H}_2 & \mathcal{F}_2 & \mathcal{F}_{22} \end{array} \right] \begin{pmatrix} w \\ w_2 \end{pmatrix} \text{ admits a smooth } \mathcal{X} \text{ with (8) and } \mathcal{G}_2^T \mathcal{X} \mathcal{G}_2 \ll \alpha \text{ is}$$

characterized exactly as in Theorem 4 if just *adding*

$$\begin{pmatrix} (\alpha-\epsilon)I & (XG_2+JD_2)^T & (G_2+BND_2)^T \\ XG_2+JD_2 & X & I \\ G_2+BND_2 & I & Y \end{pmatrix} \geq 0.$$

Then the solution of the *dual* problem is obtained by canceling (16).

Remark on the controller construction. It is not difficult to see that any $\mathcal{X}$ satisfying (8) can be transformed by a coordinate change in the controller state (which amounts to a congruence transformation on $\mathcal{X}$) such that its right-lower block is identical to I. Hence this choice in the controller construction can be made without loss of generality.

Remark on scalings. Since scalings only change the data matrices, we can directly derive the corresponding characterization for $J_s(T(R)) < \alpha$ without new proof. In Theorem 4, we just need to require, in addition, the existence of scalings $(S_1, S_2, S) \in \mathbf{S}$ and replace the (2,2) identity blocks in (14) and (15) by $S^T S$, the (3,3) identity blocks by $S_1 S_1^T$, and the (1,1) identity block in (16) by $S_2 S_2^T$. Hence the DLMIs are *linear* in $(S^T S, S_1 S_1^T, S_2 S_2^T)$ for $(S_1, S_2, S) \subset \mathbf{S}$. It might be possible to reparametrize $\{(S^T S, S_1 S_1^T, S_2 S_2^T) \mid (S_1, S_2, S) \in \mathbf{S}\}$ and to transform the inequalities (14)-(16) such that the new parameters enter linearly [37, 38, 36, 3]. Generally, however, one has to turn to heuristic ways out [41].

Remark on other performance measures. Let e_j denote the standard unit vector. Clearly, $\max_j d_j(\mathcal{H}_2 \mathcal{X}^{-1} \mathcal{H}_2^T) \ll \alpha$ is equivalent to $\begin{pmatrix} \alpha - \epsilon\, e_j^T \mathcal{H}_2 \\ \mathcal{H}_2^T e_j & \mathcal{X} \end{pmatrix} \geq 0$
for all j and some $\epsilon > 0$. Hence, if we replace the performance inequality in (8) by this alternative one, we conclude without new proof that Theorem 4 remains valid if replacing (16) with

$$\forall j: \begin{pmatrix} \alpha-\epsilon & e_j^T(H_2Y+E_2F) & e_j^T(H_2+E_2NC) \\ (H_2Y+E_2F)^T e_j & Y & I \\ (H_2+E_2NC)^T e_j & I & X \end{pmatrix} \geq 0.$$

If H_2 has l rows, the same holds for $\operatorname{trace}(\mathcal{H}_2 \mathcal{X}^{-1} \mathcal{H}_2^T) \ll \alpha$ if replacing (16) with

$$\begin{pmatrix} \alpha-\epsilon & * & * & \cdots & * & * \\ (H_2Y+E_2F)^T e_1 & Y & I & & 0 & 0 \\ (H_2+E_2NC)^T e_1 & I & X & & 0 & 0 \\ & \vdots & & \ddots & & \\ (H_2Y+E_2F)^T e_l & 0 & 0 & & Y & I \\ (H_2+E_2NC)^T e_l & 0 & 0 & & I & X \end{pmatrix} \geq 0.$$

Note that one can easily extend all this to a system with more than one performance output and different performance specifications on each of these outputs; one just needs to add the corresponding inequalities in Theorem 4.

3 The Mixed H_2/H_∞ Synthesis Problem

Remarks concerning direct feedthroughs. Just by setting $N = 0$, we can extract a suboptimality test for controllers of the form $\begin{bmatrix} K & L \\ \hline M & 0 \end{bmatrix}$. Due to the H_2 nature of the performance specification, one might wish to include the requirement $F_2 + E_2 ND = 0$ on N. This puts another *linear* restriction on N without destroying the structure of the DLMIs.

Remark on special problems. The state-feedback problem is obtained by setting $C = I$ and $D = 0$. As immediate necessary conditions for suboptimality, there exist time functions Y (smooth) and F with

$$\begin{pmatrix} -\dot{Y} + AY + YA^T + BF + (BF)^T & G & * \\ G^T & -\gamma I & F_1^T \\ H_1 Y + E_1 F & F_1 & -\gamma I \end{pmatrix} \ll 0, \quad \begin{pmatrix} \alpha I & H_2 Y + E_2 F \\ * & Y \end{pmatrix} \gg 0.$$

If these two inequalities hold, it is easily seen directly that $N := FY^{-1}$ provides a *static* suboptimal feedback controller. An analoguous specialization tackles the full information problem $C = (I\ 0)^T$, $D = (0\ I)^T$. Similarly, we remark that the inequality (15) is related to an H_∞ estimation problem [34, 46].

Remark on quadratic performance. For solving the quadratic mixed H_2/H_∞ control problem, one needs to characterize the existence of a controller with (8) where $\mathcal{X}$ is restricted to be *constant*. Trivial simplifications of our proof lead to the conclusion that Theorem 4 fully applies if just asking for *constant X, Y, and Z*. Hence, one ends up having to solve *algebraic linear matrix inequalities (LMIs) in constant X, Y, Z and in time functions F, J, N along the parameter curve defined by the system data.*

Remark on LTI systems. If the system and the controllers are LTI, it causes no loss of generality to confine $\mathcal{X}$ in (8) to be constant. Hence Theorem 4 remains valid by specializing to *constant X, Y, Z, and F, J, N*. This is an LMI solution to the reduced order [9] or full order mixed H_2/H_∞ problem [17, 25, 28, 29, 33, 42, 43, 51, 60]. In contrast to all earlier papers, no technical assumptions on the system data are required. The design of full order controllers can be directly based on Theorem 5: Find the parameters X, Y and F, J, N by solving three coupled *linear matrix inequalities* [35, 24]. Then compute L, M, and K according to the formulas given after Theorem 5. Note that we do not require to include the a priori hypotheses that

$$(A, B) \text{ is stabilizable and } (A, C) \text{ is detectable} \tag{27}$$

since they are obvious *necessary conditions* for the existence of positive definite solutions of (14) and (15).

Remark on the pure H_∞ Problem. The whole discussion includes the pure H_∞ problem obtained with $H_2 = 0$, $E_2 = 0$, $F_2 = 0$. By (12), this just amounts to canceling (16) in Theorem 4. In Theorem 5, one has to replace (28) by the coupling condition $\begin{pmatrix} Y & I \\ I & X \end{pmatrix} \gg 0$.

Remark on the pure H_2 problem. As mentioned in Section 2, we can recover the pure generalized H_2 problem by setting $H_1 = 0$, $E_1 = 0$, $F_1 = 0$, $\gamma = 1$. In Theorem 4, this just amounts to canceling the third block rows/columns in (14) and (15). If allowing for nonproper controllers, one should include the linear constraint $F_2 + E_2ND = 0$ on N. To the best of our knowledge, this gives for the first time a solution of the H_2 problem in terms of DLMIs or, in the LTI case, in terms of LMIs.

Let us now specialize Theorem 4 to the case without any a priori restriction on the controller size. In fact, a slight modification of the above proof (if the controller has size larger than n) leads to the following result.

Theorem 5 *There exists a controller R with $J(T(R)) < \alpha$ iff there exist smooth time functions X, Y and time functions F, J, N such that the DLMIs (14) and (15) and*

$$\begin{pmatrix} \alpha I & H_2Y + E_2F & H_2 + E_2NC \\ (H_2Y + E_2F)^T & Y & I \\ (H_2 + E_2NC)^T & I & X \end{pmatrix} \gg 0 \qquad (28)$$

are satisfied. If existing, the parameter K in R can be chosen of size n.

Hence, suboptimality is characterized in terms of the solvability of two differential inequalities and one algebraic inequality where all the unknowns X, Y, F, J, N enter linearly. This structural property will be essential for the discussions to follow.

Remark on the construction of a full order controller. We would like to provide explicit formulas for how to construct a full order controller. Given X and Y, Cholesky factorize $X - Y^{-1} = UU^T$ such that U and U^{-1} are smooth and bounded. Motivated by $XY + UV^T = I$, define $V = (I - YX)U^{-T}$ which is smooth and has a bounded inverse. Motivated by (22), define the controller parameters

$$M := (F - NCY)V^{-T} \quad \text{and} \quad L := U^{-1}(J - XBN). \qquad (29)$$

Finally, with

$$Q_{21} = \dot{X}Y + \dot{U}V^T + (A + BNC)^T + X(A - BNC)Y + JCY + XBF,$$

$$(Q_{31} \ Q_{32} | Q_3) = \begin{pmatrix} (G + BND)^T & (XG + JD)^T \\ H_1Y + E_1F & H_1 + E_1NC \end{pmatrix} \begin{vmatrix} -\gamma I & (F_1 + E_1ND)^T \\ F_1 + E_1ND & -\gamma I \end{vmatrix},$$

a suitable K is given by

$$K = U^{-1}(Q_{32}^T Q_3^{-1} Q_{31} - Q_{21})V^{-T}.$$

(This is shown literally as in the sufficiency proof of Theorem 4 by choosing $\mathcal{S} := \mathcal{Z}^T$ with $\mathcal{Z}$ defined in (24).) Note that we could as well start with $U := X - Y^{-1}$ implying $V = -Y$ or, dually, $V := Y - X^{-1}$ implying $U = -X$.

4 Discussion of the Estimation DLMI

In this section we would like gain some insight into the DLMI (14) which is related to an H_∞ estimation problem. We separate the time-varying case in Section 4.1 from the time-invariant situation in Section 4.2 since the latter leads to an *algebraic* linear matrix inequality. Most of the proofs are found in the appendix.

4.1 Time-Varying Data

In the LTV case we proceed under the hypothesis $DD^T \gg 0$ and $E_1^T E_1 \gg 0$. If not true, these conditions can be enforced by (small) perturbations since the inequality (14) is strict. Then we can perform coordinate changes in the u- and y-space and orthogonal coordinate changes in the w- and z-space to obtain

$$\left(\begin{array}{cc|c} A & B & G \\ C & 0 & D \\ \hline H_1 & E_1 & F_1 \end{array}\right) = \left(\begin{array}{cc|cc} A & B & G_1 & G_2 \\ C & 0 & 0 & I \\ \hline H_1 & E_1 & F_{\bullet 1} & F_{\bullet 2} \end{array}\right) = \left(\begin{array}{cc|c} A & B & G \\ C & 0 & D \\ \hline H_{11} & 0 & F_{1\bullet} \\ H_{21} & I & F_{2\bullet} \end{array}\right) = \left(\begin{array}{cc|cc} A & B & G_1 & G_2 \\ C & 0 & 0 & I \\ \hline H_{11} & 0 & F_{11} & F_{12} \\ H_{21} & I & F_{21} & F_{22} \end{array}\right). \quad (30)$$

Under these hypotheses, the following result reduces the solvability of the DLMI to the test of whether a perturbed initial value problem has a bounded solution. We note that the proof provides, just by applying Lemma 3, a pretty quick derivation of the corresponding formulas.

Theorem 6 *If X satisfies the DLMI (14) then*

$$\|F_{\bullet 1}\| \ll \gamma, \quad \|F_{1\bullet}\| \ll \gamma, \quad (31)$$

and there exists an $\epsilon > 0$ such that the solution of the initial value problem

$$Z(0) = \epsilon I, \quad \dot{Z} = (A - G_2 C) Z + Z(A - G_2 C)^T - \gamma Z C^T C Z + \epsilon I - \\ - \left(G_1 \ Z(H_1 - F_{\bullet 2} C)^T\right) \begin{pmatrix} -\gamma I & F_{\bullet 1}^T \\ F_{\bullet 1} & -\gamma I \end{pmatrix}^{-1} \begin{pmatrix} G_1^T \\ (H_1 - F_{\bullet 2} C) Z \end{pmatrix} \quad (32)$$

exists on $[0, \infty)$ and remains bounded; for all small $\epsilon > 0$, $0 \ll Z \ll X^{-1}$. Conversely, (31) implies that

$$N := -F_{22} + \begin{pmatrix} F_{21} & 0 \end{pmatrix} \begin{pmatrix} -\gamma I & F_{11}^T \\ F_{11} & -\gamma I \end{pmatrix}^{-1} \begin{pmatrix} 0 \\ F_{12} \end{pmatrix} \quad (33)$$

yields $\|F_1 + E_1 N D\| \ll \gamma$. If N is any time function with the latter property, and if the solution of (32) exists and is bounded on $[0, \infty)$ for some $\epsilon > 0$, then $Z \gg 0$, and (14) holds for $X := Z^{-1}$ and

$$J = -(X G_2 + \gamma C^T) + \begin{pmatrix} G_1^T X \\ H_1 - F_{\bullet 2} C \end{pmatrix}^T \begin{pmatrix} -\gamma I & F_{\bullet 1}^T \\ F_{\bullet 1} & -\gamma I \end{pmatrix}^{-1} \begin{pmatrix} 0 \\ F_{\bullet 2} + E_1 N \end{pmatrix}. \quad (34)$$

Remark. The proof reveals that the set of all X satisfying the DLMI (14) (for some (J, N)) and the solution set of the differential Riccati inequality

$$\dot{X} + (A - G_2 C)^T X + X(A - G_2 C) - \gamma C^T C -$$
$$- \begin{pmatrix} X G_1 \ H_1^T - C^T F_{\bullet 2}^T \end{pmatrix} \begin{pmatrix} -\gamma I & F_{\bullet 1}^T \\ F_{\bullet 1} & -\gamma I \end{pmatrix}^{-1} \begin{pmatrix} G_1^T X \\ H_1 - F_{\bullet 2} C \end{pmatrix} \ll 0 \quad (35)$$

are identical.

4.2 Time-Invariant Data

If the data matrices are constant, we characterize in this section whether there exist *constant* $X > 0, J, N$ which satisfy (14); hence we discuss an algebraic linear matrix inequality. As a necessary condition for solvability, we infer that (A, C) is detectable; this is assumed from now on.

For fixed (X, N), we can view (14) as an LMI in J only. Defining

$$\mathcal{L}(X, N) := \begin{pmatrix} A^T X + X A & X G & (H_1 + E_1 N C)^T \\ G^T X & -\gamma I & (F_1 + E_1 N D)^T \\ H_1 + E_1 N C & F_1 + E_1 N D & -\gamma I \end{pmatrix},$$

this LMI reads as

$$\mathcal{L}(X, N) + \begin{pmatrix} I \\ 0 \\ 0 \end{pmatrix} J \begin{pmatrix} C & D & 0 \end{pmatrix} + \begin{pmatrix} C^T \\ D^T \\ 0 \end{pmatrix} J^T \begin{pmatrix} I & 0 & 0 \end{pmatrix} < 0.$$

If we introduce a matrix

K_y whose columns form a basis of the kernel of $\begin{pmatrix} C & D & 0 \end{pmatrix}$,

two obvious necessary conditions for the existence of a solution J are

$$\begin{pmatrix} -\gamma I & (F_1 + E_1 N D)^T \\ F_1 + E_1 N D & -\gamma I \end{pmatrix} < 0 \quad (36)$$

and $K_y^T \mathcal{L}(X, N) K_y < 0$. The latter inequality is in fact independent of N since we have $K_y^T \mathcal{L}(X, N) K_y = K_y^T \mathcal{L}(X, 0) K_y$.

The so-called Projection Lemma [22, 26] reveals that these conditions are as well sufficient for the existence of J.

Lemma 7 *(Projection Lemma) For arbitrary A, B, $Q = Q^T$, the LMI*

$$A^T X B + B^T X^T A + Q < 0 \quad (37)$$

in the unstructured X has a solution iff

$$A x = 0 \text{ or } B x = 0 \ (x \neq 0) \text{ imply } x^T Q x < 0.$$

4 Discussion of the Estimation DLMI

We reproduce a proof since it provides a construction scheme for J. Moreover, it also reveals that, in suitable coordinates, Lemma 7 reduces to Lemma 3 if the kernels of A and B together span the whole space.

Proof. Necessity is trivial. For proving sufficiency, let $S = (S_1\ S_2\ S_3\ S_4)$ be a nonsingular matrix such that the columns of S_2 span $\ker(A) \cap \ker(B)$, those of $(S_1\ S_2)$ span $\ker(A)$, and those of $(S_2\ S_3)$ span $\ker(B)$. With $(0\ 0\ A_3\ A_4) := AS$ and $(B_1\ 0\ 0\ B_4) := BS$, we observe that $(A_3\ A_4)$ and $(B_1\ B_4)$ have full column rank. Hence, the equation $\begin{pmatrix} A_3^T \\ A_4^T \end{pmatrix} X (B_1\ B_4) = \begin{pmatrix} Z_{31} & Z_{34} \\ Z_{41} & Z_{44} \end{pmatrix}$ has a solution X for each right-hand side. Therefore, $S^T(37)S$ is equivalent to

$$\begin{pmatrix} Q_1 & Q_{21}^T & Q_{31}^T + Z_{31}^T & Q_{41}^T + Z_{41}^T \\ Q_{21} & Q_2 & Q_{32}^T & Q_{42}^T \\ Q_{31} + Z_{31} & Q_{32} & Q_3 & Q_{43}^T + Z_{34} \\ Q_{41} + Z_{41} & Q_{42} & Q_{43} + Z_{34}^T & Q_4 + Z_{44} + Z_{44}^T \end{pmatrix} < 0$$

with *free* blocks Z_{ij}. The hypotheses now just amount to (10) and, hence, we can find a (constant) Z_{31} such that the marked 3×3 block is negative definite. Since Z_{44} is free, it can be chosen to render the whole matrix negative definite. ■

We have fully proved the following result.

Theorem 8 *For fixed X and N, the LMI (14) in J has a solution iff (36) and*

$$K_y^T \begin{pmatrix} A^T X + XA & XG & H_1^T \\ G^T X & -\gamma I & F_1^T \\ H_1 & F_1 & -\gamma I \end{pmatrix} K_y < 0 \quad (38)$$

hold.

We infer that N is only restricted by (36) and does not influence (38). Moreover, for a fixed N with (36), the set of all X satisfying (14) (for some J) is *identical* to the solution set of (38). This decouples the construction of X from the determination of (J, N) in (14). As for J, testing the existence of N with (36) and computing a solution just amounts to applying Lemma 7.

Hence it remains to discuss how to test the existence of a positive definite solution of (38). If D has full row rank, one can reduce this LMI in exactly the same manner as in Section 4.1 to an algebraic Riccati inequality. In the general case, we need to transform the system data. For this purpose we look at

$$\begin{pmatrix} A & G \\ C & D \\ H_1 & F_1 \end{pmatrix} \to \begin{pmatrix} S & S_{12} & 0 \\ 0 & S_2 & 0 \\ 0 & 0 & I \end{pmatrix} \begin{pmatrix} A & G \\ C & D \\ H_1 & F_1 \end{pmatrix} \begin{pmatrix} S^{-1} & 0 \\ 0 & S_3 \end{pmatrix}$$

where S, S_2, S_3 are coordinate changes in the state-, y- and w-space (S_3 orthogonal) and S_{12} is an output-injection applied to (6). It is easily seen that

(X, J, N) satisfies the LMI (14) iff $(S^{-T}XS^{-1}, (S^{-T}J - S_{12})S_2^{-1}, N)$ satisfies the same LMI for the transformed data. Hence solutions of (38) transform as

$$X \to S^{-T}XS^{-1}.$$

As clarified in [46, 49], one can transform in this manner

$$\left(\begin{array}{c|c} A & G \\ \hline C & D \\ \hline H_1 & F_1 \end{array}\right) \text{ to } \left(\begin{array}{cc|ccc} A_r & 0 & G_r & 0 & 0 \\ G_s C_r & A_s & 0 & G_s & 0 \\ \hline 0 & C_s & 0 & 0 & 0 \\ D_r & D_s & 0 & 0 & I \\ \hline H_r & H_s & F_r & F_{r1} & F_{r2} \end{array}\right) \quad (39)$$

with the crucial property that

$$\begin{pmatrix} A_s - \lambda I & G_s \\ C_s & 0 \end{pmatrix} \text{ has full column rank for all } \lambda \in \mathcal{C}. \quad (40)$$

This transformation separates a

$$\text{regular subsystem } \begin{bmatrix} A & G_r & 0 & 0 \\ \hline C_r & 0 & I & 0 \\ D_r & 0 & 0 & I \\ \hline H_r & F_r & F_{r1} & F_{r2} \end{bmatrix} \text{ from a singular part } \begin{bmatrix} A_s & G_s \\ \hline C_s & 0 \end{bmatrix} \quad (41)$$

(what clarifies the indices s and r). For some more detailed explanations of this structure in the language of geometric control theory we refer to [49, 46]. It is important to observe that, on the basis of the structure algorithm, the regular subsystem can be computed in a numerically reliable manner [23]. For our purposes, the transformation is motivated by (40). It is well-known that left-invertible systems without any zeros have very nice properties. As an example, we mention a well-known fact from the theory of almost disturbance decoupling [58, 56]: if H_r and F_1 vanish, (38) has a positive definite solution *for all* $\gamma > 0$. This makes it plausible to expect that the singular part of the system can be factored out, and that the solvability of (38) can be characterized in terms of the regular subsystem only. The precise result reads as follows.

Lemma 9 *The inequality (38) has a (positive definite) solution iff*

$$\begin{pmatrix} A_r^T X_r + X_r A_r - \gamma(C_r^T C_r + D_r^T D_r) & X_r G_r & * \\ G_r^T X_r & -\gamma I & F_r^T \\ H_r - F_{r1}C_r - F_{r2}D_r & F_r & -\gamma I \end{pmatrix} < 0 \quad (42)$$

has a (positive definite) solution X_r.

Let us introduce the abbreviations

$$\tilde{H}_r := H_r - F_{r1}C_r - F_{r2}D_r, \quad F_\gamma := \begin{pmatrix} \gamma I & -F_r^T \\ -F_r & \gamma I \end{pmatrix}.$$

4 Discussion of the Estimation DLMI

Then (42) is equivalent to $F_\gamma > 0$ and

$$R(X_r) := A_r^T X_r + X_r A_r - \gamma(C_r^T C_r + D_r^T D_r) + \left(X_r G_r \; \tilde{H}_r^T \right) F_\gamma^{-1} \begin{pmatrix} G_r^T X_r \\ \tilde{H}_r \end{pmatrix} < 0.$$

Note that this is just the ARI which corresponds to (32) for the regular subsystem (41). If $(-A_r, G_r)$ is stabilizable, a standard result implies that this strict ARI has a (positive definite) solution iff the largest or antistabilizing solution of the corresponding algebraic Riccati equation (ARE) $R(X_r) = 0$ exists (and is positive definite) [48, Theorem 2].

If (A_r, G_r) has uncontrollable modes in $\mathcal{C}^-$ or $\mathcal{C}^0$, the validation test is more involved. Let us display the critical uncontrollable modes by choosing S in (39) (w.l.o.g.) such that

$$\left(A_r | G_r \right) = \begin{pmatrix} A_1 & A_{12} & A_{13} & | & G_1 \\ 0 & A_2 & 0 & | & 0 \\ 0 & 0 & A_3 & | & 0 \end{pmatrix}$$

where

$(-A_1, G_1)$ is stabilizable, $\sigma(A_2) \subset \mathcal{C}^0$, $\sigma(A_3) \subset \mathcal{C}^-$.

This separates the uncontrollable modes of (A_r, G_r) in the open right-half plane (in (A_1, G_1)) from those on the imaginary axis (in A_2) and in the open left-half plane (in A_3). Let us partition the columns of

$$\begin{pmatrix} C_r \\ D_r \end{pmatrix} = \left(C_1 \; C_2 \; C_3 \right), \quad \tilde{H}_r = \left(\tilde{H}_1 \; \tilde{H}_2 \; \tilde{H}_3 \right)$$

according to those of A_r. Finally, if A_2 has the eigenvalues $i\omega_j$, let E_j be a basis of the (complex) kernel of $A_2 - i\omega_j I$.

The following result provides a complete *verifiable* characterization of whether (38) has a positive definite solution.

Theorem 10 *The LMI (38) has a positive definite solution iff*

$$F_\gamma = \begin{pmatrix} \gamma I & -F_r^T \\ -F_r & \gamma I \end{pmatrix} > 0, \tag{43}$$

the unique solution X of

$$A_1^T X + X A_1 - \gamma C_1^T C_1 + \left(X G_1 \; \tilde{H}_1^T \right) F_\gamma^{-1} \begin{pmatrix} G_1^T X \\ \tilde{H}_1 \end{pmatrix} = 0 \tag{44}$$

with

$$\sigma \left(A_1 + \left(G_1 \; 0 \right) F_\gamma^{-1} \begin{pmatrix} G_1^T X \\ \tilde{H}_1 \end{pmatrix} \right) \subset \mathcal{C}^+ \tag{45}$$

exists and satisfies $X > 0$, and the unique solution Y of the linear equation

$$A_2^T Y + Y A_1 + A_{12}^T X - \gamma C_2^T C_1 + \left(Y G_1 \; \tilde{H}_2^T \right) F_\gamma^{-1} \begin{pmatrix} G_1^T X \\ \tilde{H}_1 \end{pmatrix} = 0 \tag{46}$$

satisfies

$$E_j^* \left[A_{12}^T Y^T + Y A_{12} - \gamma C_2^T C_2 + (Y G_1 \; \tilde{H}_2^T) F_\gamma^{-1} \begin{pmatrix} G_1^T Y^T \\ \tilde{H}_2 \end{pmatrix} \right] E_j < 0 \qquad (47)$$

for all j.

Remark. Although these formulas look complicated, they have a very simple origin: After partitioning

$$X_r = \begin{pmatrix} X & Y^T & * \\ Y & Z & * \\ * & * & W \end{pmatrix}, \quad R(X_r) = \begin{pmatrix} R_1(X_r) & R_{21}^T(X_r) & * \\ R_{21}(X_r) & R_2(X_r) & * \\ * & * & R_3(X_r) \end{pmatrix} \qquad (48)$$

according to A_r, the block $R_1(X_r)$ is indentical to the left-hand side of (44), $R_{21}(X_r)$ is that of (46), and the left-hand side of (47) is nothing else than $E_j^* R_2(X_r) E_j$.

Let us now comment on how to test these properties. Clearly, (43) is just a matter of verification. Since $(-A_1, G_1)$ is stabilizable, to test the existence of a solution X of (44) with (45) is a standard problem. We can apply the results in [18, Section 7.2] to infer that X exists iff the Hamiltonian matrix

$$H := \begin{pmatrix} A_1 + (G_1 \; 0) F_\gamma^{-1} \begin{pmatrix} 0 \\ \tilde{H}_1 \end{pmatrix} & (G_1 \; 0) F_\gamma^{-1} \begin{pmatrix} G_1^T \\ 0 \end{pmatrix} \\ \gamma C_1^T C_1 - (0 \; \tilde{H}_1^T) F_\gamma^{-1} \begin{pmatrix} 0 \\ \tilde{H}_1 \end{pmatrix} & -\left(A_1 - (G_1 \; 0) F_\gamma^{-1} \begin{pmatrix} 0 \\ \tilde{H}_1 \end{pmatrix} \right)^T \end{pmatrix}$$

does not have eigenvalues on the imaginary axis. If true, and if the columns of $\begin{pmatrix} X_1 \\ X_2 \end{pmatrix}$ span the generalized eigenspace of H with respect to its eigenvalues in $\mathcal{C}^+$, then X_1 is square and nonsingular, and $X = X_2 X_1^{-1}$ satisfies (44)-(45). After all, one can easily verify $X > 0$. Due to (45) and $\sigma(A_2) \subset \mathcal{C}^0$, we can solve the linear equation (46) for a unique Y, and (47) is, again, only a matter of verification.

Remarks.

(a) This theorem provides insight in which parts of the system are relevant for the solvability of (38). First, the singular part (41) does not play any role. Due to (40), it is easily seen that the zeros of

$$\begin{pmatrix} A - sI & G \\ C & D \end{pmatrix} \qquad (49)$$

are the uncontrollable modes of (A_r, G_r), which are separated according to the partition $\mathcal{C}^+ \cup \mathcal{C}^0 \cup \mathcal{C}^-$ into the uncontrollable modes of (A_1, G_1) and the eigenvalues of A_2, A_3 respectively. Hence, zeros in the open left-half plane are irrelevant (A_3 does not appear) and the influence of the zeros on the

4   Discussion of the Estimation DLMI   193

imaginary axis is explicitly displayed by (47). Similarly, the affect of the $\mathcal{C}^+$ zeros can be made explicit [46]. If $G_1 = 0$, we note that (44) is a *linear equation* in X and (45) trivially holds; the resulting simplifications for the following results are easy to extract [46, 49].

(b) The proofs of Theorems 9 and 10 provide explicit insights into the structure of the solution set of (38). In particular, one can extract which blocks of the solutions can be freely chosen or which ones can be made arbitrarily large.

(c) Suppose that D has full row rank and that $GD^T = 0$. Then we can assume w.l.o.g. $D = (0\ I)$ and conclude

$$\left(\begin{array}{c|c} A & G \\ \hline C & D \\ \hline H_1 & F_1 \end{array}\right) = \left(\begin{array}{c|cc} A_r & G_r & 0 \\ \hline D_r & 0 & I \\ \hline H_r & F_r & F_{r2} \end{array}\right). \tag{50}$$

Hence (39) does not have a singular part and $R(X_r) < 0$ is an algebraic Riccati inequality for the original matrices. Theorem 10 then characterizes the solvability of this general ARI without restrictions on the uncontrollable modes of (A, G). If $GD^T \neq 0$, one just needs to replace A by $A - GD^T C$ (see Section 4.1).

(d) If (49) has no zeros in $\mathcal{C}^0$, A_2 is empty (such that (46), (47) have to be cancelled) and one only needs to test the existence and positivity of X with (44)-(45). If (49) has no zeros in $\mathcal{C}^0 \cup \mathcal{C}^-$, an ARE in terms of the regular subsystem (41) results.

(e) Suppose $D = (0\ I)$ and $GD^T = 0$. If (49) has no zeros in $\mathcal{C}^- \cup \mathcal{C}^0$, then $X > 0$, (44)-(45) are conditions in the original data matrices. With $P = X^{-1}$, they can be rewritten as $P \geq 0$ and

$$AP + PA^T - \gamma PC^T CP + \left(G_r\ P(H_1^T - DF_1^T)\right) F_\gamma^{-1} \begin{pmatrix} G_r^T \\ (H_1 - F_1 D^T)P \end{pmatrix} = 0,$$

$$\sigma\left(A - \gamma PC^T C + \left(G_r\ P(H_1^T - DF_1^T)\right) F_\gamma^{-1} \begin{pmatrix} 0 \\ H_1 - F_1 D^T \end{pmatrix}\right) \subset \mathcal{C}^-.$$

If (49) has no $\mathcal{C}^0$ zeros, the same holds with $P = S^T \begin{pmatrix} X^{-1} & 0 \\ 0 & 0 \end{pmatrix} S$ (in the partition of A_r). This transforms the conditions (44)-(45) in special coordinates back to the original data matrices, and reveals the relation to the indefinite Riccati equations as appearing in [14, 15]. Again, if $GD^T \neq 0$, replace A by $A - GD^T C$.

Our main interest lies in the quick computation of the critical parameter

$$\gamma_c := \inf\{\gamma > 0 \mid \text{The LMI (38) has a positive definite solution.}\} \tag{51}$$

For this purpose we first clarify that one can give an explicit formula for the interval of those values γ for which (43) holds and a *symmetric* X with (44)-(45) exists. One needs to solve a standard LQ Riccati equation which is, due to the stabilizability of $(-A_1, G_1)$, always possible.

Theorem 11 *Let X_0 satisfy*

$$A_1^T X_0 + X_0 A_1 - C_1^T C_1 + X_0 G_1 G_1^T X_0 = 0, \quad \sigma(A_1 + G_1 G_1^T X_0) \subset \mathcal{C}^+.$$

Then (43) holds and there exists a symmetric X with (44) and (45) iff

$$\gamma > \gamma_e := \|F_r - (\tilde{H}_1 + F_r G_1^T X_0)(sI + A_1 + G_1 G_1^T X_0)^{-1} G_1\|_\infty.$$

Remark. Due to this formula, γ_e can be computed by quadratically convergent algorithms [10].

Here is now the central trick to determine γ_c: View the unique solutions of (44)-(46) as *functions X_γ and Y_γ of γ on the interval* (γ_e, ∞). Defining

$$U_\gamma = -\mathrm{diag}_j E_j^* \left[A_{12}^T Y_\gamma^T + Y_\gamma A_{12} - \gamma C_2^T C_2 + \begin{pmatrix} G_1^T Y_\gamma^T \\ \tilde{H}_2 \end{pmatrix}^T F_\gamma^{-1} \begin{pmatrix} G_1^T Y_\gamma^T \\ \tilde{H}_2 \end{pmatrix} \right] E_j, \tag{52}$$

we can abbreviate Theorem 10 as follows.

Corollary 12 *The LMI (38) has a positive definite solution iff $\gamma > \gamma_c$ iff $\gamma > \gamma_e$, $X_\gamma > 0$, $U_\gamma > 0$.*

Due to (45), the implicit function theorem implies that X_γ and, hence, also Y_γ, U_γ are analytic functions on (γ_e, ∞).

By differentiating (44) with respect to γ, it is not difficult to see that X_γ is *nonincreasing* and *concave*. A perturbation trick allows to show that U_γ shares these properties with X_γ.

Lemma 13 *X_γ and U_γ are analytic functions on (γ_e, ∞) which are positive definite for some large γ, and which are nondecreasing (the first derivative is positive semidefinie) and concave (the second derivative is negative semidefinite).*

With

$$f_\gamma := \mathrm{diag}(X_\gamma, U_\gamma),$$

Corollary 12 clearly implies $\gamma_c = \inf\{\gamma > \gamma_e \mid f_\gamma > 0\}$. Since $f_\gamma > 0$ for some large γ, we infer $\gamma_c < \infty$. It might happen that f_γ, although nondecreasing, remains positive if γ decreases to γ_e, what implies $\gamma_c = \gamma_e$. In the other case, there exists some $\gamma_1 > \gamma_e$ with $f_{\gamma_1} \not> 0$. Due to the properties in Lemma 13, there is exactly one $\gamma \in (\gamma_1, \infty)$ for which f_γ is positive semidefinite and singular, and this parameter in fact coincides with γ_c [47]. (It cannot happen that f_γ is positive semidefinite and singular for more than one values of γ.) This implies $\gamma_1 \leq \gamma_c$ and γ_1 can be taken as the starting point for the following Newton algorithm.

5 Consequences for the Mixed H_2/H_∞ Problem

Theorem 14 *Suppose f_γ has the properties as in Lemma 13. If $\gamma_e < \gamma_j \le \gamma_c$, there exists a unique γ_{j+1} such that*

$$f_{\gamma_j} + f'_{\gamma_j}(\gamma_{j+1} - \gamma_j) \text{ is positive semidefinite and singular,}$$

and γ_{j+1} satisfies $\gamma_j \le \gamma_{j+1} \le \gamma_c$. The inductively defined sequence (γ_j) converges montonically and quadratically to γ_c.

Hence, given γ_j, one just needs to solve a symmetric generalized eigenvalue problem in order to find a better approximation γ_{j+1} of γ_c. For a proof we refer to [47].

As a final result, we provide an important property of

$$\boldsymbol{X}_\gamma := \{X^{-1} \mid X > 0 \text{ satisfies } (38)\}$$

which is identical to the set of X^{-1} if X varies over all positive definite solutions of (14) (for some (J, N)). We show that there exists a P_γ which is a *strict lower bound* and, at the same time, a *limit point* of $\boldsymbol{X}_\gamma$. For brevity, P_γ is called a *strict lower limit point*. As easily seen, any such strict lower limit point is necessarily unique. Hence, existence is the nontrivial and crucial point in the following result.

Theorem 15 *Let $\boldsymbol{X}_\gamma$ be nonempty. With $X_\gamma > 0$ satisfying (44)-(45), define*

$$P_r(\gamma) := \begin{pmatrix} X_\gamma^{-1} & 0 & 0 \\ 0 & 0 & 0 \\ 0 & 0 & 0 \end{pmatrix} \text{ in the partition of } A_r \text{ and } P_\gamma := S^T \begin{pmatrix} P_r(\gamma) & 0 \\ 0 & 0 \end{pmatrix} S \text{ in the}$$

partition of A. Then $P_\gamma \ge 0$ is the unique strict lower limit point of $\boldsymbol{X}_\gamma$. P_γ is analytic on (γ_c, ∞) and satisfies $P'_\gamma \le 0$, $P''_\gamma \ge 0$.

Remark. If D has full row rank and (49) has no zeros in $\mathcal{C}^0$, P_γ just coincides with the solution of the estimation Riccati equation in [14, 15] (see Remark (e) after Theorem 10).

In this section we have generalized the results from [48, 49, 20] to arbitrary data and we considerably simplified some of the earlier proofs.

5 Consequences for the Mixed H_2/H_∞ Problem

We just observe that, for fixed (Y, F, N), the left-hand side of (28) is monotone in X. If combining Theorem 5 and 6, we arrive at the following corollary for LTV systems.

Corollary 16 *There exists a controller R with $J(T(R)) < \alpha$ iff there exists an $\epsilon > 0$ such that the solution of (32) exists on $[0, \infty)$ and remains bounded, and there exist time functions Y (smooth), F, N satisfying (15) and*

$$\begin{pmatrix} \alpha I & H_2 Y + E_2 F & H_2 + E_2 NC \\ (H_2 Y + E_2 F)^T & Y & I \\ (H_2 + E_2 NC)^T & I & Z^{-1} \end{pmatrix} \gg 0.$$

This decouples the construction of (J, X) from the construction of (Y, F, N) in Theorem 5. In addition, it is interesting to observe that Z can be *computed online* which will become important for linear parametrically-varying systems as discussed in Section 6.

If the system is LTI, we let all the unknowns in Theorem 5 be constant as well. Moreover, we note that (28) is equivalent to $X > 0$ and

$$\begin{pmatrix} \alpha I & H_2 Y + E_2 F \\ * & Y \end{pmatrix} - \begin{pmatrix} H_2 + E_2 N C \\ I \end{pmatrix} X^{-1} \begin{pmatrix} H_2 + E_2 N C \\ I \end{pmatrix}^T > 0. \quad (53)$$

Now we exploit Theorem 15 to infer that the set of all X^{-1} in Theorem 5 has the lower limit point P_γ. It is easily seen that we can replace X^{-1} in (53) by P_γ.

Corollary 17 *If the system is LTI, there exists an LTI controller R with $J(T(R)) < \alpha$ iff $\gamma > \gamma_c$ (as defined in (51)), and there exist a solution (Y, F, N) of (15) with*

$$\begin{pmatrix} \alpha I - (H_2 + E_2 N C) P_\gamma (H_2 + E_2 N C)^T & * \\ (H_2[Y - P_\gamma] + E_2[F - N C P_\gamma])^T & Y - P_\gamma \end{pmatrix} > 0. \quad (54)$$

Indeed, (53) leads to (54) by $-P_\gamma > -X^{-1}$, and (54) implies (53) since X can be chosen close to P_γ.

Note that (54) is equivalent (Schur complement) to the *linear* matrix inequality

$$\begin{pmatrix} I & W^T (H_2 + E_2 N C)^T & 0 \\ (H_2 + E_2 N C) W & \alpha I & * \\ 0 & (H_2[Y - P_\gamma] + E_2[F - N C P_\gamma])^T & Y - P_\gamma \end{pmatrix} > 0.$$

in (Y, F, N) if decomposing $P_\gamma = W W^T$. Hence, with the reformulation in Corollary 17, we have eliminated the variables (X, J) in the suboptimality test.

Note that [42] contains a separation result which converts, under technical hypotheses, the output feedback problem to a state-feedback problem. Corollary 17 is the full generalization to arbitrary LTI systems.

6 Linear Parametrically Varying Systems

A practically relevant specific time dependence in (6) arises if specifying (continuous bounded) functions

$$\begin{pmatrix} A(p) & B(p) & G(p) \\ C(p) & 0 & D(p) \\ H_1(p) & E_1(p) & F_1(p) \\ H_2(p) & E_2(p) & * \end{pmatrix}$$

6 Linear Parametrically Varying Systems

defined on some set $\boldsymbol{P} \subset \mathcal{R}^m$, and letting the system be described by some unknown parameter curve $p(t) \in \boldsymbol{P}$. The available a priori information consists of the set $\boldsymbol{P}$ and the on-line information is the actual parameter value $p(t)$ at the time instant t. This structure comprises linear systems with time-varying measurable parameters and gain scheduling structures for possibly nonlinear systems.

Suppose we can find *constant* X, Y and continuous *functions* F, J, N defined on $\boldsymbol{P}$ such that (14), (15), and (28) hold for all $p \in \boldsymbol{P}$. As described earlier, one can construct a (full order) controller as a function of p and a *constant* $\mathcal{X}$a such that, for the corresponding closed-loop matrices, (8) holds on $\boldsymbol{P}$. As a consequence, if the controller is scheduled along a specific parameter curve (which is measured on-line), it solves the mixed H_2/H_∞ control problem with bounds γ and α. Hence the original dynamic problem is reduced to a static problem of solving linear matrix inequalities over the parameter set $\boldsymbol{P}$.

Due to the (possibly) nonlinear dependence of the data on p and due to the unspecified structure of $\boldsymbol{P}$, solving these inequalities results in a nonlinear problem. If the parameter space has a moderate dimension, simple discretization methods reduce this nonlinear problem to the convex problem of testing the feasibility of a finite number of LMIs.

Let us briefly describe one possible paradigm. Collect the parameters as $a = (A, G, H_1, H_2, F_1)$, $b = (B, E_1, E_2, C, D)$ and the unknowns as $c = (F, J, N)$, $x = (X, Y)$. Define the block diagonal function $g(a, b, c, x)$ by putting the left-hand sides of (14) and (15) (without derivatives) and the negative of the left-hand side of (28) on the diagonal. For *constant* x, the three inequalities (14), (15), (28) can then be written as $g(a(p(.)), b(p(.)), c(p(.)), x) \ll 0$. The generally *quadratic* function g has the following properties: For fixed system data (a, b) it is linear in (c, x), and for fixed (b, x) it is linear in (a, c).

In our case, $a(p), b(p)$ are continuous functions on $\boldsymbol{P}$. The problem is to find a continuous function $c(p)$ and a constant x such that

$$g(a(p), b(p), c(p), x) \ll 0 \quad \text{on} \quad \boldsymbol{P}. \tag{55}$$

Let us now choose a finite number of points $p_j \in \boldsymbol{P}$. If $c(p)$ and x with (55) exist, one can find $\delta_j > 0$ such that, with $c_j := c(p_j)$, the inequalities

$$g(a(p_j), b(p_j), c_j, x) \leq -\delta_j I \tag{56}$$

are satisfied.

For a specific choice of p_j and δ_j, let us now assume that (56) hold for some c_j, x, and δ_j. In addition, suppose that $\boldsymbol{P}$ is open, that the data functions $(a(.), b(.))$ are smooth, and that their derivatives are bounded on $\boldsymbol{P}$. Let

$$g'(p, c, x) := \left(\|\partial_{p^k} g(a(p), b(p), c, x)\| \right)_k$$

denote the vector of the norms of the partial derivatives of $g(a(p), b(p), c, x)$ with respect to the components of p. Finally, let $\boldsymbol{P}_j$ denote (large) open convex subsets $\boldsymbol{P}_j$ of $\boldsymbol{P}$ with $p_j \in \boldsymbol{P}_j$ and

$$\sup\{\|g'(p, c_j, x)\| \|p - p_j\| \mid p \in \boldsymbol{P}_j\} < \delta_j. \tag{57}$$

Then the mean value theorem implies $g(a(p), b(p), c_j, x) \ll 0$ on $\boldsymbol{P}_j$. If the sets $\boldsymbol{P}_j$ cover $\boldsymbol{P}$, one can easily find a function $c(p)$ which 'interpolates' the controller parameters c_j such that (55) is verified. Just choose a smooth partition of unity ϕ_j which is subordinated to $\cup_j \boldsymbol{P}_j$, and define $c(p) := \sum_j c_j \phi_j(p)$ [32, Section 6.1]. (Indeed, there exists an $\epsilon > 0$ with $g(a(p), b(p), c_j, x) \leq -\epsilon I$ for all $p \in \boldsymbol{P}_j$ and all j. If p meets $\boldsymbol{P}_{j_k}$ but no other sets, we infer $\sum_k \phi_{j_k}(p) = 1$ and thus $g(a(p), b(p), c(p), x) = \sum_k \phi_{j_k}(p) g(a(p), b(p), c_{j_k}, x) \leq -\epsilon \sum_k \phi_{j_k}(p) I = -\epsilon I$.)

If the sets $\boldsymbol{P}_j$ do not cover $\boldsymbol{P}$, one has to include more points from $\boldsymbol{P}$ in the list p_j. If (55) is solvable, it is not difficult to see that there *always* exists a choice of (sufficiently dense) p_j and (sufficiently small) δ_j such that the corresponding $\boldsymbol{P}_j$ cover $\boldsymbol{P}$: Suppose x and the (bounded) function $c(p)$ satisfy $g(a(p), b(p), c(p), x) \leq -\delta I$ for $p \in \boldsymbol{P}$. Let us introduce the bounds $\hat{x}$ and $\hat{c}$ by $\|x\| < \hat{x}$ and $\sup\{\|c(p)\| \mid p \subset \boldsymbol{P}\} < \hat{c}$, and define $s := \sup\{\|g'(p, c, x)\| \mid p \in \boldsymbol{P}, \|c\| < \hat{c}, \|x\| < \hat{x}\}$. For an *arbitrary* choice of $p_j \in \boldsymbol{P}$ and with $\delta_j := \delta$, the LMIs (56) with the constraints $\|c\| < \hat{c}$, $\|x\| < \hat{x}$ have solutions c_j and x. Then $\|p - p_j\| < \delta/(s+1)$ implies $\|g'(p, c_j, x)\| \|p - p_j\| < \delta s/(s+1) < \delta$ and thus $\boldsymbol{P}_j$ satisfying (57) can be chosen to contain at least the ball $\{p \in \boldsymbol{P} : \|p - p_j\| < \delta/(s+1)\}$. Since the radius of this ball does not depend on the choice of p_j, it is clear that a sufficiently dense set of points p_j indeed leads to $\boldsymbol{P}_j$ covering $\boldsymbol{P}$. ∎

We have proved that we can indeed test whether (55) is solvable by some bounded continuous function $c(p)$ and by some x:

- For $p_j \in \boldsymbol{P}$, $\delta_j > 0$, and some bounds $\hat{c}$, $\hat{x}$, test the feasibility of (56) with the constraints $\|c\| < \hat{c}$, $\|x\| < \hat{x}$. If not feasible, decrease $\delta_j > 0$, increase the bounds $\hat{c}$, $\hat{x}$, and repeat. If (56) is never feasible then stop.
- Otherwise, construct $\boldsymbol{P}_j$ (as large as possible). If the sets $\boldsymbol{P}_j$ cover $\boldsymbol{P}$ then a $c(p)$ can be computed. Otherwise, refine the choice of p_j and return to the first step.

It is important to note that one does not need to start with a *uniform partition* of the parameter set but one can systematically vary the the density of the points p_j according to the sizes of $\boldsymbol{P}_j$, which takes the rate of variation (derivatives) of the system data with respect to the parameter into account. In practice, this might allow to keep the number of points (and hence of LMIs in (56)) reasonably small.

Under certain additional hypothesis on the system data and on the parameter set, one can exploit convexity. In the spirit of [11], let us just describe how to

6 Linear Parametrically Varying Systems

proceed if the (possibly nonlinear) system can be written as

$$\begin{pmatrix} \dot{x} \\ y \\ z_1 \\ z_2 \end{pmatrix} = \sum_j \lambda_j(t) \begin{pmatrix} A_j & B_j & G_j \\ C_j & 0 & D_j \\ (H_1)_j & (E_1)_j & (F_1)_j \\ (H_2)_j & (E_2)_j & (F_2)_j \end{pmatrix} \begin{pmatrix} x \\ u \\ w \end{pmatrix}$$

where the finitely many (constant) vertices (a_j, b_j) are a priorily available and the continuous convex combination coefficients $\lambda_j(t)$ with $\lambda_j(t) \geq 0$ and $\sum_j \lambda_j(t) = 1$ can be measured on-line. With this parameter dependence, a necessary condition for the solvability of (55) is the existence of c_j and x satisfying

$$g(a_j, b_j, c_j, x) < 0$$

for all vertices.

To reverse the arguments, we assume that (c_j, x) satisfy these LMIs. Now the reason for the distinction of the system parameters a and b becomes evident. Namely, if b_j does not vary and is identical to b, we infer by linearity that $g(\sum_j \lambda_j(t)a_j, b, \sum_j \lambda_j(t)c_j, x) = \sum_j \lambda_j(t)g(a_j, b, c_j, x) \ll 0$ and hence the parameters

$$c(t) := \sum_j \lambda_j(t) c_j$$

can be used to construct a controller which leads to (8). This is the *full generalization* of the results in [3]-[8] to the mixed H_2/H_∞ problem.

If the b_j also depend on j, our approach allows some way out: Try to solve the LMIs $g(a_j, b_j, c, x) < 0$ with a common (constant) parameter c. Then one can exploit $g(\sum_j \lambda_j(t)a_j, \sum_j \lambda_j(t)b_j, c, x) = \sum_j \lambda_j(t)g(a_j, b_j, c, x) \ll 0$ to infer that a controller leading to (8) can be constructed with c. In this case, N is constant, but the other parameters K, L, M are still scheduled along the parameter curve. If we fix $N = 0$, the formulas (29) reveal that L and M are constant as well. All this also works for mixtures: If only (C, D) is constant, schedule J and keep (F, N) constant. Dually, if (B, E_1, E_2) is constant, schedule F and keep (J, N) constant. Other specific structures might allow refinements to avoid conservatism, just by investigating the function g. These remedies have been made possible by our trick of keeping the (transformed) controller parameters c in our problem solution. Even for the pure H_∞ problem, this would not have been possible solely on the basis of the results in [3]-[8].

Let us remark that the nonlinear (discretization) scheduling technique can be combined with those based on convexity, possibly over different regions of the parameter space. There is no need to discuss all the details.

The implementation of hybrid structures can be pursued further along the following lines. As described in Corollary 16, the estimation differential inequality is related to testing whether an initial value problem has a bounded solution. For a specific LPV system, standard comparison results [19] could allow to guarantee the existence and boundedness of the solution for all possible parameter curves $p(.)$, perhaps taking a bound on the derivative $\dot p(.)$ into account. This is generally less conservative than using a *constant* and, along each parameter curve, globally valid solution of the algebraic inequality. Then the estimation DLMI is solved by a time function X whereas the control LMI is still solved with a constant Y. The controller is *nonlinear* since it includes the on-line solution of the corresponding Riccati initial value problem (along the actual parameter curve). The robustness and performance bounds γ and α are still guaranteed by Theorem 5, and our general controller construction scheme is valid without change.

Let us include one specific example. Suppose $R(p, X)$ is a Riccati map and we try to solve the differential Riccati inequality $\frac{d}{dt}X(p(t)) + R(p(t), X(p(t)) \ll 0$ with $X(p(t)) \gg 0$. Let us assume that the frozen parameter ARIs have solutions: There exist strictly positive functions $X(p)$ and $\delta(p)$ on $\boldsymbol{P}$ satisfying $R(p, X(p)) + \delta(p)I \ll 0$. Note that, in the case of interest (35), Theorem 10 allows to test the existences of a positive definite solution of the frozen parameter ARI. The chain rule implies that $X(p(t))$ satisfies the desired differential Riccati inequality for all parameter curves $p(.)$ with

$$\|\dot p(t)\| \le \delta(p(t))/\|X'(p(t))\|,$$

where $X'(p) := \left(\|\partial_{p^k} X(p)\|\right)_k$.

Summarizing, we have proposed some initial interpolation and approximation techniques in order to solve, on the basis of Theorem 5, the mixed H_2/H_∞ problem for LPV systems without particular structural dependence of the data on the scheduling parameter. Some observations indicated how to overcome the conservatism introduced by considering only constant solutions of the DLMIs in Theorem 5, as usually done in the literature [3]-[8].

7 Mixed $\boldsymbol{H_2/H_\infty}$ Control and Robust Regulation

In this section we would like to show how to incorporate asymptotic tracking or disturbance rejection objectives in the mixed H_2/H_∞ problem if the system is LTI. Consider

$$\begin{pmatrix} y \\ z_1 \\ z_2 \end{pmatrix} = \left[\begin{array}{c|ccc} A & B & G & G_2 \\ \hline C & 0 & D & D_2 \\ H_1 & E_1 & F_1 & F_{12} \\ H_2 & E_2 & F_2 & F_{22} \end{array}\right] \begin{pmatrix} u \\ w \\ w_2 \end{pmatrix} \tag{58}$$

and suppose that the measured output is partitioned as

$$y = \begin{pmatrix} y_1 \\ y_2 \end{pmatrix}, \; (C \; D \; D_2) = \begin{pmatrix} C_1 & D_1 & D_{12} \\ C_2 & D_2 & D_{22} \end{pmatrix}.$$

We say that the controller (7) achieves *regulation* if it is stabilizing and satisfies $\lim_{t\to\infty} y_2(t)$ for $w = 0$ and all signals w_2 in the class

$$\mathcal{W} := \{w \mid \dot w = Sw\}.$$

7 Mixed H_2/H_∞ Control and Robust Regulation

Here, S is the so-called signal generator which satisfies

$$\sigma(S) \subset \mathcal{C}^0 \cup \mathcal{C}^+. \tag{59}$$

The controller achieves *robust regulation* if it achieves regulation for all systems which result from (58) by small perturbations in the describing parameters.

This is the classical problem of asymptotically rejecting the disturbance $w_2 \in \mathcal{W}$ from y_2 or letting $C_2 x$ asymptotically track the reference input $-D_{22} w_2$ for all signals w_2 in $\mathcal{W}$. Since we can neglect decaying signals, the hypothesis (59) causes no loss of generality.

Any robust regulator is known to contain a reduplicate model of the signal generator. To be precise, let us construct $\tilde{S}$ as follows: Let

$$q \text{ be the number of components of } y_2.$$

Transform S to Jordan canonical form, choose for each eigenvalue of S the largest Jordan block, and collect q of these blocks on the diagonal of $\tilde{S}$.

If we assume (27) to assure that stabilizing controllers exist, the following result is well-known [12].

Theorem 18 *There exists a controller (7) which achieves robust regulation for (58) iff*

$$\begin{pmatrix} A - \lambda I & B \\ C_2 & 0 \end{pmatrix} \text{ has full row rank for all } \lambda \in \sigma(S).$$

A controller is a robust regulator iff it has a realization

$$\begin{bmatrix} K & L_1 & L & 0 \\ 0 & \tilde{S} & 0 & R \\ M & N_1 & N & 0 \end{bmatrix} = \begin{bmatrix} K & L_1 & L \\ M & N_1 & N \end{bmatrix} \begin{bmatrix} \tilde{S} & 0 & R \\ I & 0 & 0 \\ 0 & I & 0 \end{bmatrix} \tag{60}$$

for some R (with q columns) such that

$$(\tilde{S}, R) \text{ is controllable} \tag{61}$$

and some $u = \begin{bmatrix} K & L_1 & L \\ M & N_1 & N \end{bmatrix} \begin{pmatrix} y \\ v \end{pmatrix}$ *stabilizing*

$$\begin{pmatrix} y \\ v \\ z_1 \\ z_2 \end{pmatrix} = \begin{bmatrix} A & 0 & B & G \\ RC_2 & \tilde{S} & 0 & RD_2 \\ C & 0 & 0 & D \\ 0 & I & 0 & 0 \\ H_1 & 0 & E_1 & F_1 \\ H_2 & 0 & E_2 & F_2 \end{bmatrix} \begin{pmatrix} u \\ w \end{pmatrix}, \tag{62}$$

the system (58) postcompensated with $v = \begin{bmatrix} \tilde{S} & R \\ I & 0 \end{bmatrix} y_2$.

As a consequence, there exists a controller which achieves robust regulation *and* solves the mixed H_2/H_∞ problem for (58) iff there exists a controller solving the mixed H_2/H_∞ problem for the extended system (62).

Remark. One can in fact prove that the controller (60) achieves regulation for *any* LTI system which is stabilized by (60). Due to the H_∞ constraint implying robust stability, any controller (60) which achieves robust regulation and solves the mixed H_2/H_∞ problem for (58), indeed *achieves robust regulation in the large*: it is a regulator for all systems that result from (58) by perturbations $w = \Delta z_1$ with a (stable) LTI Δ satisfying $\|\Delta\|_\infty \leq \gamma$.

In order to design a robustly regulating mixed H_2/H_∞ controller, one has to solve the LMIs in Theorem 5 for the system (62) *and for some R satisfying (61)*. We get an additional variable R which does *not enter linearly*. However, there is a practically relevant special case in which we can always choose a *fixed* $\tilde{R}$ to circumvent this difficulty: S is diagonizable. In this case we can assume w.l.o.g.

$$S = \mathrm{diag}(\lambda_1 I, \ldots, \lambda_s I) \text{ such that } \tilde{S} = \mathrm{diag}(\lambda_1 I_q, \ldots, \lambda_s I_q)$$

where $\lambda_1, \ldots, \lambda_s$ are the pairwise different eigenvalues of S and I_q is the unit matrix of size q. If R satisfies (61) and is partitioned into $\begin{pmatrix} R_1^T & \cdots & R_s^T \end{pmatrix}^T$, the blocks R_j of size $q \times q$ are nonsingular. With $T = \mathrm{diag}(R_1, \ldots, R_s)$, we then infer $T^{-1}\tilde{S}T = \tilde{S}$ and $T^{-1}R = \begin{pmatrix} I_q & \cdots & I_q \end{pmatrix}^T =: \tilde{R}$. Hence *any* R satisfying (61) can be transformed by a coordinate change in the controller state to $\tilde{R}$, and we can indeed *fix R to $\tilde{R}$* in Theorem 18.

Hence it remains to apply Theorem 5 for the fixed internal model $(\tilde{S}, \tilde{R})$. The inequalities (14) and (15), (28) are defined with

$$\left(\begin{array}{cc|c} A - sI & 0 & G \\ \tilde{R}C_2 & \tilde{S} - sI & \tilde{R}D_2 \\ \hline C & 0 & D \\ 0 & I & 0 \end{array}\right) \text{ and } \left(\begin{array}{cc|c} A - sI & 0 & B \\ \tilde{R}C_2 & \tilde{S} - sI & 0 \\ \hline H_j & 0 & E_j \end{array}\right), \ j = 1, 2$$

respectively. Hence the extension from (58) to (62) adds the eigenvalues of S as zeros in $\mathcal{C}^0 \cup \mathcal{C}^+$ to the latter pencils. For (14), however, we conclude that, due to the identity block in the last row of the first pencil, the estimation inequalities (38) for the original data $\begin{pmatrix} A - sI & G \\ C & D \end{pmatrix}$ and for the extended data are *identical*; the extension does not influence the solvability of (14).

Summarizing, we have generalized the results in [1, 2] to the mixed H_2/H_∞ problem. Even the specialization to the pure problems are more general since we do not restrict the underlying plant by technical hypotheses and, still, get *explicit* suboptimality tests in the next two sections.

8 The Pure H_∞ Problem

For time-varying parameters we assume again (30) and specialize Corollary 16 by setting $H_2 = 0$, $E_2 = 0$, $F_2 = 0$. We can just dualize the remark after Theorem 6 and conclude that the set of all Y satisfying (15) (for some (F, N)) is identical to the solution set of the differential Riccati inequality

$$-\dot{Y} + (A - BH_{21})Y + Y(A - BH_{21})^T - \gamma BB^T -$$
$$- (G - BF_{2\bullet}\, YH_{11}^T) \begin{pmatrix} -\gamma I & F_{1\bullet}^T \\ F_{1\bullet} & -\gamma I \end{pmatrix}^{-1} \begin{pmatrix} G^T - F_{2\bullet}^T B^T \\ H_{11}Y \end{pmatrix} \ll 0. \quad (63)$$

For an arbitrary N with $\|F_1 + E_1 ND\| \ll \gamma$ (whose existence is characterized by (31) and which can be chosen as (33)) and for any solution Y of (63), (Y, F, N) satisfies (15) if

$$F = -(H_{21}Y + \gamma B^T) + (F_{2\bullet} + ND\ 0) \begin{pmatrix} -\gamma I & F_{1\bullet}^T \\ F_{1\bullet} & -\gamma I \end{pmatrix}^{-1} \begin{pmatrix} G^T - F_{2\bullet}^T B^T \\ H_{11}Y \end{pmatrix}.$$

Let us now consider the solution Q_T of the perturbed final value problem

$$Q_T(T) = 0, \quad \dot{Q} + (A - BH_{21})^T Q + Q(A - BH_{21}) - \gamma QBB^T Q + \epsilon I -$$
$$- (Q(G - BF_{2\bullet})\ H_{11}^T) \begin{pmatrix} -\gamma I & F_{1\bullet}^T \\ F_{1\bullet} & -\gamma I \end{pmatrix}^{-1} \begin{pmatrix} (G - BF_{2\bullet})^T Q \\ H_{11} \end{pmatrix} = 0.$$

Comparison arguments [30, Chapter 10] lead to $0 \le Q_{T_1}(t) \le Q_{T_2}(t)$ for $T_1 \le T_2$ and for t in the existence intervals of both solutions. We say that $Q_\infty(t)$ exists if t is in the existence interval of Q_T for all $T \ge t$ and if $Q_T(t)$ remains bounded for $T \to \infty$, and we define $Q_\infty(t) := \lim_{T \to \infty} Q_T(t)$. Note that the limit exists and is positive semidefinite by monotonicity; moreover, $Q_\infty(t)$ is *unique*.

If Y satisfies (63), comparison results [30, Chapter 10] allow to conclude the existence of $\epsilon > 0$ such that Q_∞ exists on $[0, \infty)$ and satisfies $Q_\infty \ll Y^{-1}$; in particular, Q_∞ is bounded. Conversely, if Q_∞ exists on $[0, \infty)$ and is bounded, one can prove (by perturbing the final value and using the techniques in [39]) that (63) has a solution $Y \gg 0$ for which $\sup_{t \ge 0} \|Q_\infty(t) - Y^{-1}(t)\|$ is arbitrarily small. Since the coupling condition in Corollary 16 reduces to $Z^{-1} \gg Y^{-1}$ or $\rho(ZY^{-1}) \ll 1$, we arrive at the following slightly generalized and more explicit version of a basically well-known result [55, 39].

Theorem 19 *There exists a stabilizing LTV controller R with $\|T_1(R)\|_\infty < \gamma$ iff there exists an $\epsilon > 0$ such that the solution Z of (32) and Q_∞ exist and are bounded on $[0, \infty)$, and such that they satisfy $\rho(ZQ_\infty) \ll 1$.*

Under additional hypotheses, a technical result about differential Riccati equations allows to get rid of the perturbation parameter ϵ [39].

If considering LTI systems, we can dualize Theorem 8 for (15) with a basis K_u of the kernel of $\begin{pmatrix} B^T & 0 & E_1^T \end{pmatrix}$. We arrive at the results of [22, 26].

Theorem 20 *There exists a stabilizing LTI controller R with $\|T_1(R)\|_\infty < \gamma$ iff there exists X and Y with*

$$\begin{pmatrix} X & I \\ I & Y \end{pmatrix} > 0 \tag{64}$$

and

$$K_y^T \begin{pmatrix} A^T X + XA & XG & H_1^T \\ G^T X & -\gamma I & F_1^T \\ H_1 & F_1 & -\gamma I \end{pmatrix} K_y < 0, \quad K_u^T \begin{pmatrix} AY + YA^T & G & YH_1^T \\ G^T & -\gamma I & F_1^T \\ H_1 Y & F_1 & -\gamma I \end{pmatrix} K_u < 0. \tag{65}$$

The only point which needs clarification: (65) imply the existence of N with (31). This follows from Lemma 7. Note that, as a consequence, one can construct a suboptimal controller on the basis of an *arbitrary* N with (31).

Let us now exploit our results in Section 4.2 to derive algebraically verifiable criteria. From now on, assume (27) and recall the relevance of X_γ, U_γ, $P_\gamma = S_X X_\gamma^{-1} S_X^T$ defined on (γ_e, ∞) for the first inequality in (65). Dually, one can construct functions Y_γ, V_γ defined on (δ_e, ∞) with a computable δ_e (Theorem 11) and sharing the properties of X_γ, U_γ in Lemma 13 such that the second inequality in (65) has a solution $Y > 0$ iff $\gamma > \delta_e$, $Y_\gamma > 0$, $V_\gamma > 0$. The set of all possible Y^{-1} has the lower limit point $Q_\gamma = S_Y Y_\gamma^{-1} S_Y^T$ with a (computable) S_Y (Theorem 15).

Finally, define the optimal value of the H_∞ control problem as

$$\gamma_{opt} := \inf\{\gamma > 0 \mid \text{Exists a stabilizing LTI controller } R \text{ with } \|T_1(R)\|_\infty < \gamma\}.$$

Then we arrive at the following central verifiable characterization of whether a parameter γ in the most general LTI H_∞ control problem is suboptimal.

Corollary 21 *The parameter γ satisfies $\gamma > \gamma_{opt}$ iff $\gamma > \gamma_e$, $\gamma > \delta_e$, and*

$$f_\gamma := \begin{pmatrix} Y_\gamma & S_Y^T S_X & 0 & 0 \\ S_X^T S_Y & X_\gamma & 0 & 0 \\ 0 & 0 & U_\gamma & 0 \\ 0 & 0 & 0 & V_\gamma \end{pmatrix} > 0. \tag{66}$$

Remark. The proof reveals that

$$\begin{pmatrix} Y_\gamma & S_Y^T S_X \\ S_X^T S_Y & X_\gamma \end{pmatrix} > 0 \iff \rho(P_\gamma Q_\gamma) < 1. \tag{67}$$

This gives a *coordinate independent* formulation of the coupling between (14), (15) or the two inequalities (65). With the critical parameters γ_c and δ_c for the LMIs (65) (defined by (51) and its dual), we can infer that $\gamma > \gamma_{opt}$ iff $\gamma > \gamma_c$, $\gamma > \delta_c$, and $\rho(P_\gamma Q_\gamma) < 1$. This characterization is closer to the standard solvability conditions in terms of LMIs (or ARIs for regular problems or AREs for regular problems without imaginary axis zeros) and a spectral radius coupling condition. We also note that U_γ and V_γ in (66) nicely display the influence of the imaginary axis zeros of $\begin{pmatrix} A - sI & G \\ C & D \end{pmatrix}$ or, dually, of $\begin{pmatrix} A - sI & B \\ H & E \end{pmatrix}$ on the optimal value. The functions X_γ, Y_γ or P_γ, Q_γ are not affected by these zeros. We finally stress that the five remarks after Theorem 10 have obvious dual versions; moreover, they highlight cases in which one can choose $S_X = I$ and/or $S_Y = I$.

Corollary 21 implies $\gamma_{\text{opt}} = \inf\{\gamma > \max\{\gamma_e, \delta_e\} \mid f_\gamma > 0\}$. Due to (27), a stabilizing controller exists and, hence, $f_\gamma > 0$ for some large γ. Therefore, f_γ shares the properties of X_γ in Lemma 13. If $f_\gamma > 0$ for all $\gamma > \max\{\gamma_e, \delta_e\}$, we infer $\gamma_{\text{opt}} = \max\{\gamma_e, \delta_e\}$. Otherwise, there exists a γ_1 with $f_{\gamma_1} \not> 0$ which necessarily satisfies $\max\{\gamma_e, \delta_e\} < \gamma_1 \leq \gamma_{\text{opt}}$. The parameter γ_1 is the starting point for the following iteration: Let $\max\{\gamma_e, \delta_e\} < \gamma_j \leq \gamma_{\text{opt}}$ and determine the unique γ_{j+1} such that

$$f_{\gamma_j} + f'_{\gamma_j}(\gamma_{j+1} - \gamma_j) \text{ is positive semidefinite and singular}$$

by solving a symmetric eigenvalue problem. Then $\gamma_j \leq \gamma_{j+1}$, and (γ_j) converges quadratically to γ_{opt} (see Theorem 14).

We have generalized [20] to singular systems with imaginary axis zeros and our previous results in [49] to systems with $F_1 \neq 0$. Hence, for the most general LTI system, we have not only derived a verifiable suboptimality test, but we also obtained a quadratically convergent algorithm to compute γ_{opt}.

9 The Pure H_2 Problem

As mentioned earlier, the pure H_2 problem is covered by Theorem 5 with $H_1 = 0$, $E_1 = 0$, $F_1 = 0$, and $\gamma = 1$. As in Section 8, one could reduce the DLMIs to perturbed Riccati differential equations if the problem is regular ($DD^T \gg 0$, $E_2^T E_2 \gg 0$) which is not pursued here. More importantly, this specialization of our main result leads to an LMI solution of the (generalized) output feedback H_2 control problem. Hence all the techniques in Section 6 discussed for LPV systems can be applied to robust H_2 performance problems, which seems impossible for alternative approaches.

Let us now address, in more detail, the generalized and standard H_2 problem for LTI systems. We concentrate on strictly proper controllers and assume $F_2 = 0$. Let us define the optimal value as

$$\alpha_{\text{opt}} := \inf\{\alpha \mid \text{Exists strictly proper stabilizing LTI } R \text{ with } \|T_2(R)\| < \alpha\}.$$

In fact, the techniques developed so far allow to determine α_{opt} for the generalized H_2 problem by a one parameter search, and they lead to an explicit formula for the genuine H_2 optimal value. We assume again w.l.o.g. (27) such that $\alpha_{\text{opt}} < \infty$.

Since (A, C) is detectable, the solution set

$$\{X^{-1} \mid X > 0 \text{ and } \exists J : \begin{pmatrix} A^T X + XA + JC + (JC)^T & XG + JD \\ (XG + JD)^T & -I \end{pmatrix} < 0\}$$

of the H_2 version of (14) is nonempty. Hence it has a lower limit point which we denote as P (Theorem 15). Exploiting the explicit formula for P, a direct computation shows $P \geq 0$ and $AP + PA^T + GG^T \geq 0$ such that we can write

$$P = UU^T \quad \text{and} \quad AP + PA^T + GG^T = VV^T.$$

Let us now reformulate the H_2 version of Corollary 5.

Corollary 22 *The parameter α satisfies $\alpha > \alpha_{\text{opt}}$ iff there exist $Z > 0$ and F with*

$$\begin{pmatrix} AZ + ZA^T + BF + (BF)^T & 0 & (H_2 Z + E_2 F)^T \\ 0 & -\alpha I & (H_2 U)^T \\ H_2 Z + E_2 F & H_2 U & -\alpha I \end{pmatrix} < 0 \tag{68}$$

and

$$\alpha I - V^T Z^{-1} V > 0. \tag{69}$$

If $\alpha^2 I > H_2 P H_2^T$, the dual version of Theorem 15 implies that there exists a lower limit point Q_α of Z^{-1} where Z varies in the set of all positive definite solutions of (68) (for some F). Again, we can replace Z^{-1} by Q_α.

Corollary 23 *The parameter α satisfies $\alpha > \alpha_{\text{opt}}$ iff $\alpha^2 > \|H_2 P H_2^T\|$ and $\alpha I - V^T Q_\alpha V > 0$.*

In contrast to [40], we have reduced the computation of α_{opt} to a *one parameter search* over α. Let us define $\alpha_e := \sqrt{\|H_2 P H_2^T\|}$. We extract from Theorem 15 the structure $Q_\alpha = S_Z Z_\alpha^{-1} S_Z^T$ and observe that $\alpha I - V^T Q_\alpha V > 0$ is equivalent to

$$f_\alpha := \begin{pmatrix} \alpha I & V^T S_Z \\ S_Z^T V & Z_\alpha \end{pmatrix} > 0.$$

By Corollary 23, $\alpha_{\text{opt}} = \inf\{\alpha \in (\alpha_e, \infty) \mid f_\alpha > 0\}$. Since f_α shares all the properties in Lemma 13 with Z_α on the interval (α_e, ∞), we can hence compute the optimal value by a quadratically convergent algorithm (Theorem 14). Apart from $F_2 = 0$ and $N = 0$, we stress again that we do not need to restrict the system by technicalities.

The genuine H_2 problem involves the trace operator. Since this functional is linear, we can even go further and provide an explicit formula for the optimal value. Indeed, a slight adaption of the proof of Corollary 22 leads to the following result.

Corollary 24 *There exists a strictly proper stabilizing LTI controller R with $\|T_2(R)\|_2^2 < \alpha$ iff there exist $Z > 0$ and F with*

$$\begin{pmatrix} AZ + ZA^T + BF + (BF)^T & (H_2Z + E_2F)^T \\ H_2Z + E_2F & -I \end{pmatrix} < 0 \qquad (70)$$

and $\alpha > \text{trace}(H_2 P H_2^T) + \text{trace}(V^T Z^{-1} V)$.

Let Q denote the lower limit point of the set of all Z^{-1} where $Z > 0$ solves (70) (for some F). Then we conclude that the optimal value of the H_2 control problem equals

$$\sqrt{\text{trace}[H_2 P H_2^T + (AP + PA^T + GG^T)Q]}.$$

This generalizes the (dual) results of [53] to system which are not restricted with respect to imaginary axis zeros. It is interesting to observe that, similarly as for the coupling condition in the H_∞ problem, these zeros do not influence the lower limit points P and Q and hence they don't have an effect on the optimal value.

10 Conclusion

We have provided a solution of the mixed H_2/H_∞ problem for time-varying systems and discussed various specializations to linear parametrically-varying and time-invariant systems, as well as to the pure (generalized) H_2 and H_∞ problems. For completely general linear time-invariant systems, we showed how to include a robust regulation objective in the mixed H_2/H_∞ problem. Moreover, in the pure problems, we have obtained algebraic solvability tests for the corresponding linear matrix inequalities, and arrived at pretty deep insights into the structure of their solutions. This enabled us to devise quadratically convergent algorithms for computing the optimal values, or to provide an explicit formula in the H_2 case.

Apart from our main results, Theorems 4 and 5, we believe that the technique of keeping transformed versions of the controller parameters in output feedback problems is the most relevant contribution of this article with further potentials.

References

1. Abedor, J., Nagpal, K., Khargonekar, P.P., Poolla, K.: Robust regulation in the presence of norm-bounded uncertainty. IEEE Trans. Automat. Control **40** (1995) 147–153
2. Abedor, J., Nagpal, K., Poolla, K.: Robust regulation with $\mathcal{H}_2$ performance. Systems Control Lett. **23** (1994) 431–443

3. Apkarian, P., Gahinet, P.: A convex characterization of gain-scheduled $\mathcal{H}_\infty$ controllers. IEEE Trans. Automat. Control (to appear)
4. Apkarian, P., Gahinet, P.: $\mathcal{H}_\infty$ control of linear parameter-varying systems: A design example. Preprint (1993)
5. Apkarian, P., Gahinet, P., Becker, G.: Self-scheduled $\mathcal{H}_\infty$ control of linear parameter-varying systems. Preprint (1993)
6. Basar, T., Bernhard, P.: H^∞-Optimal Control and Related Minimax Design Problems, A Dynamic Game Approach. Birkhäuser, Basel (1991)
7. Becker, G., Packard, A., Philbrick, D., Balas, G.: Control of parametrically-dependent linear systems: A single quadratic Lyapunov approach, Proc. Amer. Contr. Conf., San Francisco, CA (1993) 2795–2799
8. Becker, G., Packard A.: Robust performance of linear parametrically varying systems using parametrically dependent linear feedback. Systems Control Lett. **23** (1994) 205–215
9. Bernstein, D.S., Haddad, W.M.: LQG control with an H_∞ performance bound: A Riccati equation approach. IEEE Trans. Automat. Control **34** (1989) 293–305
10. Boyd, S., Balakrishnan, V.: A regularity result for the singular values of a transfer matrix and a quadratically convergent algorithm for computing its L_∞-norm. Proc. 28th IEEE Conf. Decision Contr. (1989) 954–955
11. Boyd, S.P., El Ghaoui, L., Feron, E., Balakrishnan, V.: Linear Matrix Inequalities in Systems and Control Theory. SIAM Studies in Applied Mathematics 15, SIAM, Philadelphia (1994)
12. Davison, E.J.: The robust control of a servomechanism problem for linear time-invariant multivariable systems. IEEE Trans. Automat. Control **21** (1976) 25–34
13. Desoer, C.A., Vidyasagar, M.: Feedback Synthesis: Input-Output Properties. Academic Press, New York (1975)
14. Doyle, J., Glover, K.: State-space formulae for all stabilizing controllers that satisfy an H_∞ norm bound and relations to risk sensitivity. Systems Control Lett. **11** (1988) 167–172
15. Doyle, J., Glover, K., Khargonekar, P., Francis, B.: State-space solutions to standard H_∞ and H_2 control problems. IEEE Trans. Automat. Control **34** (1989) 831–847
16. Doyle, J.C., Packard, A., Zhou, K.: Review of LFTs, LMIs, and μ. Proc. 30th IEEE Conf. Decision Contr. (1991) 1227–1232
17. Doyle J., Zhou, K., Glover, K., Bodenheimer, B.: Mixed $\mathcal{H}_2$ and $\mathcal{H}_\infty$ performance objectives II: Optimal control. IEEE Trans. Automat. Control **39** (1994) 1575–1586
18. Francis, B.A.: A Course in H_∞ Control Theory. Lect. N. Contr. Inform. Sci. No. 88, Springer-Verlag, Berlin (1987)
19. Freiling, G., Jank, G., Abou-Kandil, H.: On global existence of solutions to coupled matrix Riccati equations in closed loop Nash games. Preprint (1994)
20. Gahinet, P.: On the game Riccati equation arising in $\mathcal{H}_\infty$ control problmes. SIAM J. Control Optim. **32** (1994) 635–647
21. Gahinet, P.: Explicit controller formulas for LMI-based $\mathcal{H}_\infty$ control. Proc. Amer. Contr. Conf., Baltimore (1994) 2396–2400
22. Gahinet, P., Apkarian, P.: A linear matrix inequality approach to H_∞ control. Int. J. of Robust and Nonlinear Control **4** (1994) 421–448

23. Gahinet, P., Laub, A.J.: Numerically reliable computation of γ_{opt} in singular H_∞ control. Preprint (1994)
24. Gahinet, P., Nemirovskii, A., Laub, A.J., Chilali, M.: The LMI control toolbox. Proc. 33rd IEEE Conf. Decision Contr. (1994) 2038–2041
25. Haddad, W.M., Bernstein, D.S., Mustafa, D.: Mixed-norm $\mathcal{H}_2/\mathcal{H}_\infty$ regulation and estimation: The discrete time case. Systems Control Lett. **16** (1991) 235–247
26. Iwasaki, T., Skelton, R.E.: All controllers for the general $\mathcal{H}_\infty$ control problem: LMI existence conditions and state space formulas. Automatica **30** (1994) 1307–1317
27. Iwasaki, T., Skelton, R.E.: A unified approach to fixed order controller design via linear matrix inequalities. Proc. Amer. Contr. Conf., Baltimore (1994) 35–39
28. Khargonekar, P.P., Rotea, M.A.: Mixed $\mathcal{H}_2/\mathcal{H}_\infty$ control: a convex optimization approach. IEEE Trans. Automat. Control **36** (1991) 824–837
29. Khargonekar, P.P., Rotea, M.A., Sivashankar, N.: Exact and approximate solutions to a class of multiobjective controller synthesis problems. Proc. Amer. Contr. Conf., San Francisco, CA (1993) 1602–1606
30. Knobloch, H.W., Kwakernaak, H.: Lineare Kontrolltheorie. Springer-Verlag, Berlin (1985)
31. Limebeer, D.J.N., Anderson, B.D.O., Hendel, B.: A Nash game approach to mixed H_2/H_∞ control. IEEE Trans. Automat. Control **39** (1994) 69–82
32. Lu, W.M., Doyle, J.C., Robustness analysis and synthesis for uncertain nonlinear systems. Proc. 33rd IEEE Conf. Decision Contr. (1994) 787–792
33. Mustafa, D., Glover, K.: Minimum Entropy $\mathcal{H}_\infty$ Control. Lect. N. Contr. Inform. Sci. No. 146, Springer-Verlag, Berlin (1990)
34. Nagpal, K.M., Khargonekar, P.P.: Filtering and smoothing in an $\mathcal{H}_\infty$ setting. IEEE Trans. Automat. Control **36** (1991) 152–166
35. Nestereov, Y., Nemirovsky, A.: Interior point polynomial methods in convex programming: Theory and applications. SIAM Studies in Applied Mathematics 13, SIAM, Philadelphia (1994)
36. Packard, A.: Gain-scheduling via linear fractional transformations. Systems Control Lett. **22** (1994) 79–92
37. Packard, A., Doyle, J.: The complex structured singular value. Automatica **29** (1993) 71–109
38. Packard, A., Pandey, P., Leonhardson, J., Balas, G.: Optimal, constant I/O similarity scaling for full-information and state-feedback control problems. Systems Control Lett. **19** (1992) 271–280
39. Ravi, R., Nagpal, K.M., Khargonekar, P.P.: H^∞ control of linear time-varying systems: A state-space approach. SIAM J. Control Optim. **29** (1991) 1394–1413
40. Rotea, M.A.: The generalized $\mathcal{H}_2$ control. Automatica **29** (1993) 373–385
41. Rotea, M.A., Iwasaki, T.: An alternative to the $D - K$ iteration? Proc. Amer. Contr. Conf., Baltimore (1994) 53–57
42. Rotea, M.A., Khargonekar, P.P.: H^2-optimal control with an H^∞-constraint: The state feedback case. Automatica **27** (1991) 307–316
43. Rotea, M.A., Khargonekar, P.P.: Generalized $\mathcal{H}_2/\mathcal{H}_\infty$ control via convex optimization. Proc. 30th IEEE Conf. Decision Contr. (1991) 2719-2720
44. Rotea, M.A., Prasanth, R.K.: The ρ performance measure: A new tool for controller design with multiple frequency domain specifications. Proc. Amer. Contr. Conf., Baltimore (1994) 430–435

45. Sampei, M., Mita, T., Nakamichi, M.: An algebraic approach to H_∞ output feedback control problems. Systems Control Lett. **14** (1990) 13–24
46. Scherer, C.W.: The Riccati Inequality and State-Space H_∞-Optimal Control. Ph.D. thesis, University of Würzburg (1990)
47. Scherer, C.W.: H_∞-control by state-feedback and fast algorithms for the computation of optimal H_∞-norms. IEEE Trans. Automat. Control **35** (1990) 1090–1099
48. Scherer, C.W.: H_∞-control by state-feedback for plants with zeros on the imaginary axis. SIAM J. Control Optim. **30** (1992) 123–142
49. Scherer, C.W.: H_∞-optimization without assumptions on finite or infinite zeros. SIAM J. Control Optim. **30** (1992) 143–166
50. Shamma, J.F., Athans, M.: Guaranteed properties of gain-scheduled control for linear parameter-varying plants. Automatica **27** (1991) 559–564
51. Steinbuch, M., Bosgra, O.H.: Necessary conditions for static and fixed order dynamic mixed H_2/H_∞ optimal control. Proc. Amer. Contr. Conf. (1991) 1137–1143
52. Stoorvogel, A.A.: The H_∞ control problem: a state space approach. Prentice Hall, Hemel Hempstead, UK (1992)
53. Stoorvogel, A.A.: The singular H_2 control problem. Automatica **28** (1992) 627–631
54. Szanier, M.: An exact solution to general SISO mixed $\mathcal{H}_2/\mathcal{H}_\infty$ problems via convex optimization. IEEE Trans. Automat. Control **39** (1995) 2511–2517
55. Tadmor, G.: Worst-case design in the time domain: the maximum principle and the standard H_∞ problem. Math. Control Signals Systems **3** (1990) 301–324
56. Trentelman, H.L.: Almost invariant subspaces and high gain feedback. CWI Tract No. 29, Amsterdam (1986)
57. Willems, J.C.: Least-squares stationary optimal control and the algebraic Riccati equation. IEEE Trans. Automat. Control **21** (1971) 319–338
58. Willems, J.C.: Almost invariant subspaces: An approach to high gain feedback design-Part II: Almost conditioned invariant subspaces. IEEE Trans. Autom. Control **27** (1982) 1071–1085
59. Yeh, H., Banda, S., Chang, B.: Necessary and sufficient conditions for mixed $\mathcal{H}_2$ and $\mathcal{H}_\infty$ control. Proc. 29th IEEE Conf. Decision Contr. (1990) 1013-1017
60. Zhou, K., Glover, K., Bodenheimer, B., Doyle, J.: Mixed $\mathcal{H}_2$ and $\mathcal{H}_\infty$ performance objectives I: Robust performance analysis. IEEE Trans. Automat. Control **39** (1994) 1564–1574

Appendix: Proofs

Proof of Theorem 6. Due to $DD^T = I$, we can eliminate the matrix $JC + (JC)^T$ with the transformation

$$(14) \to \begin{pmatrix} I & -C^T D & 0 \\ 0 & I & 0 \\ 0 & 0 & I \end{pmatrix} (14) \begin{pmatrix} I & 0 & 0 \\ -D^T C & I & 0 \\ 0 & 0 & I \end{pmatrix}.$$

Appendix: Proofs

Since $(H_1 + E_1NC) - (F_1 + E_1ND)D^TC = H_1 - F_1D^TC$, we get

$$\begin{pmatrix} \dot{X} + A^TX + XA - XGD^TC - C^TDG^TX - \gamma C^TC & * & * \\ G^TX + D^TJ^T + \gamma D^TC & -\gamma I & * \\ H_1 - F_1D^TC & F_1 + E_1ND & -\gamma I \end{pmatrix} \ll 0.$$

Using the fine partitions (30) leads to

$$\begin{pmatrix} \dot{X} + (A - G_2C)^TX + X(A - G_2C) - \gamma C^TC & * & * & * \\ G_1^TX & -\gamma I & 0 & * \\ G_2^TX + \gamma C + J^T & 0 & -\gamma I & * \\ H_1 - F_{\bullet 2}C & F_{\bullet 1} & F_{\bullet 2} + E_1N & -\gamma I \end{pmatrix} \ll 0.$$

For fixed (X, N), we view this as an inequality in J and apply Lemma 3. We infer that a solution J exists iff

$$\begin{pmatrix} \dot{X} + (A - G_2C)^TX + X(A - G_2C) - \gamma C^TC & XG_1 & H_1^T - C^TF_{\bullet 2}^T \\ G_1^TX & -\gamma I & F_{\bullet 1}^T \\ H_1 - F_{\bullet 2}C & F_{\bullet 1} & -\gamma I \end{pmatrix} \ll 0 \quad (71)$$

and (after further refining the partition as in (30))

$$\begin{pmatrix} -\gamma I & 0 & F_{11}^T & F_{21}^T \\ 0 & -\gamma I & F_{12}^T & F_{22}^T + N^T \\ F_{11} & F_{12} & -\gamma I & 0 \\ F_{21} & F_{22} + N & 0 & -\gamma I \end{pmatrix} \ll 0. \quad (72)$$

A specific solution is given by J defined in (34). If viewing (72) as an inequality in N, we apply Lemma 3 to conclude that a solution exists iff

$$\begin{pmatrix} -\gamma I & F_{11}^T & F_{21}^T \\ F_{11} & -\gamma I & 0 \\ F_{21} & 0 & -\gamma I \end{pmatrix} \ll 0, \quad \begin{pmatrix} -\gamma I & 0 & F_{11}^T \\ 0 & -\gamma I & F_{12}^T \\ F_{11} & F_{12} & -\gamma I \end{pmatrix} \ll 0,$$

and a special solution is given by (33). These two inequalities are clearly equivalent to (31).

Let us assume (31) from now on. By $\begin{pmatrix} -\gamma I & F_{\bullet 1}^T \\ F_{\bullet 1} & -\gamma I \end{pmatrix} \ll 0$, (71) is equivalent to the differential Riccati inequality (35). Standard comparison arguments [30, Chapter 10] reveal that, for all sufficiently small $\epsilon > 0$, the solution of (32) satisfies $0 \ll Z \ll X^{-1}$ on the interval of existence. Consequently, the solution Z exists and the inequalities hold on the whole interval $[0, \infty)$. Conversely, with Z satisfying (32), we infer $Z \gg 0$ and $X = Z^{-1}$ satisfies (35). ∎

Proof of Lemma 9. Assume w.l.o.g. $S = I$. Given a basis K of the kernel of C_s, we choose

$$K_y = \begin{pmatrix} I & 0 & 0 & 0 & 0 \\ 0 & 0 & 0 & 0 & K \\ \hline 0 & I & 0 & 0 & 0 \\ -C_r & 0 & 0 & I & 0 \\ -D_r & 0 & 0 & 0 & -D_s K \\ \hline 0 & 0 & I & 0 & 0 \end{pmatrix}.$$

Partition $X = \begin{pmatrix} X_r & Y^T \\ Y & Z \end{pmatrix}$ to infer by a simple computation that (38) is given as

$$\begin{pmatrix} R_1(X_r) & R_2(Y)^T & R_3(Y)^T K \\ R_2(Y) & -\gamma I & G_s^T Z K \\ K^T R_3(Y) & K^T Z G_s & K^T [A_s^T Z + Z A_s - \gamma D_s^T D_s] K \end{pmatrix} < 0 \qquad (73)$$

with

$$\begin{pmatrix} R_1(X_r) \\ R_2(Y) \\ \hline R_3(Y) \end{pmatrix} = \begin{pmatrix} A_r^T X_r + X_r A_r - \gamma(C_r^T C_r + D_r^T D_r) & X_r G_r & * \\ G_r^T X_r & -\gamma I & F_r^T \\ H_r - F_{r1} C_r - F_{r2} D_r & F_r & -\gamma I \\ \hline G_s^T Y + \gamma C_r & 0 & F_{r1}^T \\ [A_s^T Y + Y A_r] - \gamma D_s^T D_r & Y G_r & (H_s - F_{r2} D_s)^T \end{pmatrix}.$$

If (73) has the (positive definite) solution X, we infer by canceling rows/columns that its left-upper block X_r satisfies $R_1(X_r) < 0$ what proves necessity.

Now suppose that there exists a (positive definite) X_r with $R_1(X_r) < 0$. Since G_s^T has full row rank, the first block of $R_2(Y)$ can be arbitrarily assigned by varying Y. Applying Lemma 3, some (explicitly given) Y leads to

$$\begin{pmatrix} R_1(X_r) & R_2(Y)^T \\ R_2(Y) & -\gamma I \end{pmatrix} < 0. \qquad (74)$$

Due to (74), a Schur complement argument reveals that (73) is equivalent to

$$K^T [\tilde{A}^T Z + Z \tilde{A} + \frac{1}{\gamma} Z G_s G_s^T Z + \tilde{Q}] K < 0 \qquad (75)$$

with $\tilde{A} := A_s - G_s R_2(Y) R_1(X_r)^{-1} R_3(Y)^T/\gamma$, $\tilde{Q} := -R_3(Y) R_1(X_r)^{-1} R_3(Y)^T + R_3(Y) R_1(X_r)^{-1} R_2(Y)^T R_2(Y) R_1(X_r)^{-1} R_3(Y)^T/\gamma - \gamma D_s^T D_s$. By (40), there exist Z and J such that $(\tilde{A} + J C_s)^T Z + Z(\tilde{A} + J C_s) + \frac{1}{\gamma} Z G_s G_s^T Z + \tilde{Q} < 0$. If varying J, the symmetric Z can be made *arbitrarily large*. (This follows from results in almost disturbance decoupling and is the dual version of a slight generalization of [46, Proposition 4.6 (b)]; see also [49].) If we recall that $C_s K = 0$ and that K has full column rank, Z obviously satisfies (75). If $X_r > 0$, Z can be made so large to get $X > 0$. ∎

Proof of Theorem 10. We first prove necessity and assume that $X_r > 0$ satisfies (42). Let use rewrite $Q := X_r$ for ease of notation, partition Q as A_r,

Appendix: Proofs

and define $\tilde{Q} := \begin{pmatrix} Q_1 & Q_{12} \\ Q_{21} & Q_2 \end{pmatrix}$, $\tilde{A} := \begin{pmatrix} A_1 & A_{12} \\ 0 & A_2 \end{pmatrix}$, $\tilde{I} := \begin{pmatrix} 0 & 0 \\ 0 & I \end{pmatrix}$, $\tilde{G} := \begin{pmatrix} G_1 \\ 0 \end{pmatrix}$, $\tilde{C} := (C_1 \ C_2)$, $\tilde{H} := (\tilde{H}_1 \ \tilde{H}_2)$, with $\tilde{I}$ partitioned as $\tilde{A}$. Obviously,

$$(-(\tilde{A} + \epsilon\tilde{I}), \tilde{G}) \text{ is stabilizable for all } \epsilon > 0. \tag{76}$$

Just by canceling the third block row/column of (42), we infer

$$\begin{pmatrix} (\tilde{A}+\epsilon\tilde{I})^T\tilde{Q}+\tilde{Q}(\tilde{A}+\epsilon\tilde{I})+\delta\tilde{I}-\gamma\tilde{C}^T\tilde{C} & \tilde{Q}\tilde{G} & \tilde{H}^T \\ \tilde{G}^T\tilde{Q} & -\gamma I & F_r^T \\ \tilde{H} & F_r & -\gamma I \end{pmatrix} < 0,$$

first for $\epsilon = 0$, $\delta = 0$ and, a posteriori, for some fixed $\delta > 0$ and all small $\epsilon > 0$. Consequently, $F_\gamma > 0$. Moreover, due to (76), there exists [48, Theorem 2] some $Q > \tilde{Q} > 0$ with

$$(\tilde{A}+\epsilon\tilde{I})^T Q + Q(\tilde{A}+\epsilon\tilde{I}) + \delta\tilde{I} - \gamma\tilde{C}^T\tilde{C} + (Q\tilde{G} \ \tilde{H}^T) F_\gamma^{-1} \begin{pmatrix} \tilde{G}^T Q \\ \tilde{H} \end{pmatrix} = 0 \tag{77}$$

and

$$\sigma\left((\tilde{A}+\epsilon\tilde{I}) + (\tilde{G} \ 0) F_\gamma^{-1} \begin{pmatrix} \tilde{G}^T Q \\ \tilde{H} \end{pmatrix}\right) \subset \mathcal{C}^+. \tag{78}$$

Partitioning $Q = \begin{pmatrix} X & Y^T \\ Y & Z \end{pmatrix}$ as $\tilde{A}$, the left-upper blocks of the left-hand sides of (77), (78) and (44), (45) coincide. Since the matrix in (78) is block triangular, we conclude (44)-(45). The left- and right-lower blocks of (77) are given by

$$(A_2+\epsilon I)^T Y + YA_1 + A_{12}^T X - \gamma C_2^T C_1 + (YG_1 \ \tilde{H}_2^T) F_\gamma^{-1} \begin{pmatrix} G_1^T X \\ \tilde{H}_1 \end{pmatrix} = 0 \tag{79}$$

and

$$(A_2+\epsilon I)^T Z + Z(A_2+\epsilon I) + A_{12}^T Y^T + YA_{12} - \gamma C_2^T C_2 + \begin{pmatrix} G_1^T Y^T \\ \tilde{H}_2 \end{pmatrix}^T F_\gamma^{-1} \begin{pmatrix} G_1^T Y^T \\ \tilde{H}_2 \end{pmatrix} = -\delta I. \tag{80}$$

Due to $E_j^*[(A_2+\epsilon I)^T Z + Z(A_2+\epsilon I)]E_j = 2\epsilon E_j^* Z E_j > 0$ (by $A_2 E_j = i\omega_j E_j$ and $Z > 0$), we obtain

$$E_j^* \left[A_{12}^T Y^T + YA_{12} - \gamma C_2^T C_2 + \begin{pmatrix} G_1^T Y^T \\ \tilde{H}_2 \end{pmatrix}^T F_\gamma^{-1} \begin{pmatrix} G_1^T Y^T \\ \tilde{H}_2 \end{pmatrix} \right] E_j = -\delta E_j^* E_j. \tag{81}$$

Now we let ϵ converge to zero. Clearly, for all small $\epsilon \geq 0$, (79) has a unique solution which depends on ϵ and converges for $\epsilon \to 0$ to that of (46). Due to $\delta E_j^* E_j > 0$, (81) leads to (47) in the limit.

For the sufficiency proof, we blockwise construct X_r with $R(X_r) < 0$ using (48) and exploiting the structural dependence of the blocks of $R(X_r)$ on those of X_r

as clarified in the remark after Theorem 10. Since $(-A_1, G_1)$ is stabilizable, we can find some X which is arbitrarily close to the stabilizing solution of (44) and such that $R_1(X_r) < 0$ [48, Theorem 2]. If X is sufficiently close to the ARE solution, $X > 0$ and (45) still hold. Hence (46) has a solution Y what amounts to $R_{21}(X_r) = 0$. Moreover, $R_2(X_r)$ equals the left-hand side of (80) for $\epsilon = 0$. Due to (47) and [48, Theorem 4], we can find arbitrarily large Z solving the Lyapunov inequality $R_2(X_r) < 0$. Assign the blocks $*$ in X_r arbitrarily. Since $R_3(X_r)$ has the same structure as $R_2(X_r)$, it equals $A_3^T W + W A_3 + R_3$ where neither R_3 nor any other blocks in $R(X_r)$ depend on W. Since $\sigma(A_3) \subset C^-$, we can render $R_3(X_r)$ so small that $R(X_r) < 0$. Since both Z and W can be chosen arbitrarily large, we can achieve $X_r > 0$ (by $X > 0$). ∎

Proof of Theorem 11. It is well-known that (44) has a solution with (45) iff the strict ARI which corresponds to (44) is solvable [48, Theorem 2]. Together with (43), we hence need to characterize those $\gamma > 0$ for which

$$\begin{pmatrix} A_1^T X + X A_1 - \gamma C_1^T C_1 & X G_1 & \tilde{H}_1^T \\ G_1^T X & -\gamma I & F_r^T \\ \tilde{H}_1 & F_r & -\gamma I \end{pmatrix} < 0 \quad (82)$$

has a symmetric solution. Let us replace $C_1^T C_1$ by $A_1^T X_0 + X_0 A_1 + X_0 G_1 G_1^T X_0$. Then

$$\begin{pmatrix} I & X_0 G_1 & 0 \\ 0 & I & 0 \\ 0 & 0 & I \end{pmatrix} (82) \begin{pmatrix} I & 0 & 0 \\ G_1^T X_0 & I & 0 \\ 0 & 0 & I \end{pmatrix}$$

is equivalent to (82), and a short computation results for $\Delta = \gamma X_0 - X$ in

$$\begin{pmatrix} -(A_1 + G_1 G_1^T X_0)^T \Delta - \Delta(A_1 + G_1 G_1^T X_0) & -\Delta G_1 & (\tilde{H}_1 + F_r G_1^T X_0)^T \\ -G_1^T \Delta & -\gamma I & F_r^T \\ \tilde{H}_1 + F_r G_1^T X_0 & F_r & -\gamma I \end{pmatrix} < 0.$$

Since $-(A_1 + G_1 G_1^T X_0)$ is exponentially stable, we can apply the LTI version of Theorem 2 and conclude that this inequality has a solution Δ iff the H_∞ norm of $F_r - (\tilde{H}_1 + F_r G_1^T X_0)(sI + A_1 + G_1 G_1^T X_0)^{-1} G_1$ is smaller than γ. ∎

Proof of Lemma 13. Since (A, C) is detectable, the LMI (14) has a positive definite solution for some large γ, what implies $X_\gamma > 0$, $U_\gamma > 0$ (Theorem 10).

To show the analytic properties, we use the same notations as in the proof of Theorem 10. We let $X = X_\gamma$ satisfy (44)-(45), $Y = Y_\gamma(\epsilon)$ solve (79), and $Z = Z_\gamma(\epsilon)$ be the unique solution of (80) for $\delta = 0$. Then $Q = Q_\gamma(\epsilon) := \begin{pmatrix} X_\gamma & Y_\gamma(\epsilon)^T \\ Y_\gamma(\epsilon) & Z_\gamma(\epsilon) \end{pmatrix}$ satisfies (77)-(78). Moreover, let us introduce $U_\gamma(\epsilon)$ by (52) with Y_γ replaced by $Y_\gamma(\epsilon)$.

Due to (45), the implicit function theorem implies that X_γ is analytic on (γ_c, ∞). There exists an $\epsilon_0 > 0$ such that $-(A_2 + \epsilon I)$ and the matrix in (45) do not have common eigenvalues for $\epsilon > -\epsilon_0$. Therefore, $Y_\gamma(\epsilon)$ and hence also $U_\gamma(\epsilon)$

Appendix: Proofs

are analytic on $(\gamma_c, \infty) \times (-\epsilon_0, \infty)$. However, $Z_\gamma(\epsilon)$ and hence $Q_\gamma(\epsilon)$ are only guaranteed to be analytic on $(\gamma_c, \infty) \times (0, \infty)$.

Let us abbreviate the partial derivatives $Q' = \partial_\gamma Q_\gamma(\epsilon)$, $Q'' = \partial_\gamma^2 Q_\gamma(\epsilon)$, and the matrix in (78) as $\hat{A} = \hat{A}_\gamma(\epsilon)$. Differentiating (44) two times with respect to γ leads to

$$\hat{A}^T Q' + Q' \hat{A} - \tilde{C}^T \tilde{C} - \begin{pmatrix} \tilde{G}^T Q \\ \tilde{H} \end{pmatrix}^T F_\gamma^{-2} \begin{pmatrix} \tilde{G}^T Q \\ \tilde{H} \end{pmatrix} = 0,$$

$$\hat{A}^T Q'' + Q'' \hat{A} + 2[F_\gamma^{-1} \begin{pmatrix} \tilde{G}^T Q \\ \tilde{H} \end{pmatrix} + \begin{pmatrix} G^T Q' \\ 0 \end{pmatrix}]^T F_\gamma^{-1} [F_\gamma^{-1} \begin{pmatrix} \tilde{G}^T Q \\ \tilde{H} \end{pmatrix} + \begin{pmatrix} G^T Q' \\ 0 \end{pmatrix}] = 0.$$

By $\sigma(\hat{A}) \subset \mathcal{C}^+$ and $F_\gamma > 0$, we get $Q' \geq 0$ and $Q'' \leq 0$.

This implies $X'_\gamma \geq 0$, $X''_\gamma \leq 0$ and $\partial_\gamma Z_\gamma(\epsilon) \geq 0$, $\partial_\gamma^2 Z_\gamma(\epsilon) \leq 0$. Now $E_j^*(80) E_j$ leads to $\mathrm{diag}_j(2\epsilon E_j^* Z_\gamma(\epsilon) E_j) = U_\gamma(\epsilon)$ (since $\delta = 0$) and hence $\partial_\gamma U_\gamma(\epsilon) \geq 0$, $\partial_\gamma^2 U_\gamma(\epsilon) \leq 0$. All this holds for $\epsilon > 0$. However, since $U_\gamma(\epsilon)$ is in fact analytic on $(\gamma_c, \infty) \times (-\epsilon_0, \infty)$, we can conclude $\partial_\gamma U_\gamma(0) \geq 0$, $\partial_\gamma^2 U_\gamma(0) \leq 0$ by taking the limit $\epsilon \to 0$. ∎

Proof of Theorem 15. Assume w.l.o.g. $S = I$ and recap the proofs of Lemma 9 and Theorem 10. In the necessity parts we showed: If X satisfies (38), then its left-upper block Q satisfies (42). Further, the left-upper block Q_1 of Q is a solution of the strict ARI which corresponds to (44). For X_γ satisfying (44)-(45) we infer $Q_1 < X_\gamma$ [48, Theorem 2] what implies $P_\gamma < X^{-1}$ [48, Lemma 14].

In the sufficiency proof of Theorem 10, the blocks Z and W can be chosen arbitrarily large such that X_r^{-1} is arbitrarily close to $P_r(\gamma)$ [48, Lemma 14]. Now X_r satisfies (42). In the construction of Lemma 9, Z can be taken arbitrarily large such that X^{-1} is arbitrarily close to X_r^{-1} [48, Lemma 14]. Hence X^{-1} can be chosen arbitrarily close to P_γ.

Analyticity of P_γ is a consequence of that of X_γ, and the inequalities for the derivates follow from $(X_\gamma^{-1})' = -X_\gamma^{-1} X'_\gamma X_\gamma^{-1}$, $(X_\gamma^{-1})'' = 2 X_\gamma^{-1} X'_\gamma X_\gamma^{-1} X'_\gamma X_\gamma^{-1} - X_\gamma^{-1} X''_\gamma X_\gamma^{-1}$ and $X_\gamma > 0$, $X'_\gamma \geq 0$, $X''_\gamma \leq 0$. ∎

Proof of Corollary 21. Let us first prove the relation (67). From $\rho(P_\gamma Q_\gamma) = \rho(S_X X_\gamma^{-1} S_X^T S_Y Y_\gamma^{-1} S_Y^T) = \rho(Y_\gamma^{-1/2} S_Y^T S_X X_\gamma^{-1} S_X^T S_Y Y_\gamma^{-1/2})$ we infer $\rho(P_\gamma Q_\gamma) < 1$ iff $Y_\gamma > S_Y^T S_X X_\gamma^{-1} S_X^T S_Y$. Taking Schur complement leads to (67).

Necessity is now proved as follows. If γ is suboptimal, there exist constant X, Y satisfying (64) and (65). Hence $\gamma > \gamma_e$, $\gamma > \delta_e$, and $X_\gamma > 0$, $U_\gamma > 0$, $Y_\gamma > 0$, $V_\gamma > 0$, and $\rho(X^{-1} Y^{-1}) < 1$. Since $P_\gamma < X^{-1}$ and $Q_\gamma < Y^{-1}$ (strict lower bound property), we obtain $\rho(P_\gamma Q_\gamma) < 1$ and hence (66) by (67).

For the sufficiency part, $\gamma > \gamma_e$, $\gamma > \delta_e$, and $X_\gamma > 0$, $U_\gamma > 0$, $Y_\gamma > 0$, $V_\gamma > 0$ imply that (65) have solutions $X > 0$, $Y > 0$. By (67) we infer $\rho(P_\gamma Q_\gamma) < 1$. Since X^{-1}, Y^{-1} can be chosen close to P_γ, Q_γ (by the limit point property), we have $\rho(X^{-1} Y^{-1}) < 1$ for suitable X, Y what implies (64). ∎

Proof of Corollary 22. If $\alpha > \alpha_{\text{opt}}$, Corollary 17 yields the existence of (Y, F) with $AY + YA^T + BF + (BF)^T + GG^T < 0$ and

$$\begin{pmatrix} \alpha^2 I - H_2 P H_2^T & H_2[Y - P] + E_2 F \\ (H_2[Y - P] + E_2 F)^T & Y - P \end{pmatrix} > 0.$$

With the transformation $\tilde{Y} := Y - P$, $\tilde{F} := F\tilde{Y}^{-1}$ and the abbreviation $S_\alpha := \alpha^2 I - H_2 P H_2^T$, these inequalities are rewritten to

$$(A + B\tilde{F})\tilde{Y} + \tilde{Y}(A + B\tilde{F})^T + VV^T < 0, \quad \begin{pmatrix} S_\alpha & (H_2 + E_2\tilde{F})\tilde{Y} \\ \tilde{Y}(H_2 + E_2\tilde{F})^T & \tilde{Y} \end{pmatrix} > 0$$

and the second one is equivalent to

$$S_\alpha > 0, \quad \tilde{Y} > 0, \quad I > S_\alpha^{-1/2}(H_2 + E_2\tilde{F})\tilde{Y}(H_2 + E_2\tilde{F})^T S_\alpha^{-1/2}.$$

We conclude that $A + B\tilde{F}$ is exponentially stable. Moreover, defining $M(t) := S_\alpha^{-1/2}(H_2 + E_2\tilde{F})e^{(A+B\tilde{F})t}V$, we infer $I > \int_0^\infty M(t)M(t)^T dt$. Going back to the integral definiton with Riemann sums, one shows $I > \int_0^\infty M(t)^T M(t) dt$. Hence there exists an $X > 0$ with

$$(A + B\tilde{F})^T X + X(A + B\tilde{F}) + (H_2 + E_2\tilde{F})^T S_\alpha^{-1}(H_2 + E_2\tilde{F}) < 0, \quad V^T X V < I.$$

With $Z := X^{-1}$, $F := \tilde{F}X^{-1}$ and recalling the definition of S_α, we obtain

$$\begin{pmatrix} AZ + ZA^T + BF + (BF)^T & (H_2 Z + E_2 F)^T \\ H_2 Z + E_2 F & H_2 UU^T H_2^T - \alpha^2 I \end{pmatrix} < 0, \quad V^T Z^{-1} V < I.$$

Dividing the first inequality by α leads to (68) and (69) for $(Z/\alpha, F/\alpha)$. The converse is obtained by reversing the arguments. ∎

Introduction to the Modelling, Control and Optimization of Discrete Event Systems

Christos G. Cassandras[1], *Stéphane Lafortune*[2], *Geert Jan Olsder*[3]

[1] Department of Electrical and Computer Engineering, University of Massachusetts at Amherst, USA. E-mail: cassandras@ecs.umass.edu
[2] Department of Electrical Engineering and Computer Science, University of Michigan, Ann Arbor, USA. E-mail: stephane@eecs.umich.edu
[3] Department of Technical Mathematics and Informatics, Delft University of Technology, P.O.Box 5031, 2600 GA Delft, the Netherlands. E-mail: g.j.olsder@math.tudelft.nl

Table of Contents

1 Introduction 218
2 Examples of Control and Optimization Problems for Discrete Event Systems 220
3 Modelling of Discrete Event Systems (DES) 228
   3.1 A General Language-Based Approach 228
   3.2 Languages and Automata 231
   3.3 Product and Parallel Composition of Automata 234
   3.4 Petri Nets and Event Graphs 236
4 Supervisory Control of DES 238
   4.1 The Feedback Loop of Supervisory Control 238
   4.2 A Word on Specifications: The "Legal Behavior" 240
   4.3 The Basic Controllability Theorem 241
   4.4 Realization of Supervisors 242
   4.5 Dealing with Uncontrollability 243
   4.6 Some Important Supervisory Control Problems and Their Solutions . . 245
   4.7 Comments on The $\uparrow$ and $\downarrow$ Operations 247
   4.8 Modular Supervisory Control 248
   4.9 Modeling and Design Issues 250
   4.10 Some Concluding Remarks 251
5 The Max-Plus Algebra Approach to DES 252
   5.1 The Stage for the Max-Plus Approach 252
   5.2 Formalization 255
   5.3 Periodic Behaviour 257
   5.4 The γ-Transform 259
   5.5 Some Extensions and Recent Literature 261
      Axiomatic Foundations. 261
      Minimal Realizations. 262
      Stochastic Discrete Event Systems. 263
      Min-Max-Plus Systems and Nonexpansive Mappings. 265
      Numerical Procedures. 265

'Continuous' Discrete Event Systems and the Fenchel Transform. 266
6 Sample Path Analysis and Performance Optimization of DES 268
   6.1 Stochastic Timed Automata 269
   6.2 Infinitesimal Perturbation Analysis (IPA) 271
       Event Triggering Sequences. 272
       Event Time Derivatives. 272
       Sample Function Derivatives. 275
       Unbiasedness and the Commuting Condition. 277
       Consistency of IPA Estimators. 279
   6.3 Extensions of IPA and Other Gradient Estimation Techniques 279
   6.4 The Constructability Problem and "Rapid Learning" 281
   6.5 The Standard Clock (SC) Approach 283
   6.6 Augmented System Analysis (ASA) 284
       ASA for optimal control of elevator systems. 287

1 Introduction

The theory of Discrete Event Systems (DES) is a research area of current vitality. The development of this theory is largely stimulated by discovering general principles which are (or are hoped to be) useful to a wide range of application domains. In particular, technological and/or 'man-made' manufacturing systems, communication networks, transportation systems, and logistic systems, all fall within the class of DES. There are two key features that characterize these systems. First, their dynamics are *event-driven* as opposed to *time-driven*, i.e., the behavior of a DES is governed only by occurrences of different types of events over time rather than by ticks of a clock. Unlike conventional time-driven systems, the fact that time evolves in between event occurrences has no visible effect on the system. Second, at least some of the natural variables required to describe a DES are discrete. Examples of *events* include the pushing of a button or an unpredictable computer failure. Examples of discrete variables involved in modelling a DES are descriptors of the state of a resource (e.g., UP, DOWN, BUSY, IDLE) or (integer-valued) counters for the number of users waiting to be served by a resource. Some authors use the acronym DEDS, for a discrete event dynamic system, rather than DES, to emphasize the fact that the behavior of such systems can, and usually will, change as time proceeds.

    Theoretical disciplines that support the study of DES include systems theory, operations research, computer science, and industrial engineering. With such a variety of fields of application and supporting disciplines it will come as no surprise that the theory of DES encompasses a variety of classes of problems and of modelling approaches. It is fair to say that no single approach dominates the others, nor is it necessarily desirable that it should be otherwise. In this paper, one will discover three main streams of activity. First, the 'logical' approach, based on automata and formal languages, in which the precise *ordering of the events* is of interest; this ordering must satisfy a given set of specifications imposed on it. Second, the max-plus algebra approach, in which, in addition to the ordering, the *timing of the events* plays an essential role. Lastly, the sample path

1 Introduction

analysis approach, in which once again the timing of the events is crucial for the purpose of evaluating and optimizing the performance of DES, especially in a stochastic environment. Historically, these three streams were developed more or less independently (and simultaneously), each of them coining the acronym DES or DEDS, but currently one discovers more and more interrelationships.

From a formal point of view, a DES can be thought of as a dynamical system, with a state space and a state-transition mechanism. The event-driven nature of a DES forces us to seek new mathematical frameworks for modelling and analysis, since differential/difference equations (developed for the analysis of time-driven systems) no longer provide an adequate setting. Moreover, one allows 'nondeterminism' in DES, that is, transitional choices have to be made, either by the system itself by means of some chance mechanism or by somebody from the outside world.

Pursuing a description of what DES are or are not would be somewhat abstract at this point, since a number of fundamental concepts have not yet been introduced; instead, we will continue with some brief historical remarks and an outline of the paper. Section 2 is devoted to three illustrative examples, one for each of the three main streams mentioned above. After having gone through these examples, it will be much easier to talk about various aspects of DES theory.

The 'first' stream, the logical approach, started within a system theory setting with the work of Ramadge and Wonham in 1982, [61]; see also [62]. Briefly, a DES is modelled as the generator of a formal language to which control aspects are added in the sense that certain events (i.e. transitions) can be disabled by an external controller. In the last 10 years, several researchers have built on and extended the seminal work of Ramadge and Wonham and their co-workers. The resulting systems and control theory for DES is known as 'Supervisory Control Theory'. This theory addresses the synthesis of controllers (or supervisors) for DES in order to satisfy a set of (qualitative) specifications on the admissible orderings of the events that can be executed by the system. One recent extension of the above work is the generalization to 'timed transition models', which incorporate real-time features.

The 'second' stream probably started with the publication of [22], though older traces surely can be found. This work remained largely unnoticed in the system theory society and a real boost was given by the publication of [20]. The starting point is a 'linear' model, linear not in the sense of the conventional algebra but in the sense of the so-called max-plus algebra. This has laid the foundation of a bona fide discrete event system theory, which has been shown to parallel the classical linear system theory in several ways. The notion of 'time' is a basic aspect in these models. Nowadays the relationship with Petri nets and stochastic extensions are well understood. Recently one has realized that similar algebraic techniques are used in the first and second stream (such as for instance techniques related to the 'closure' operator $(\cdot^*)$).

The 'third' stream is motivated by the fact that the state trajectory (sample path) of a DES observed under a given set of conditions contains a surprising amount of information about the behavior of the system under a spectrum of different conditions one might be interested in. This is particularly important when analytical models are unavailable or simply inadequate for DES of considerable complexity, especially in a stochastic environment. This stream has its roots in [38] (based on an idea first published in [39]), from which emerged the theory of *Perturbation Analysis* (PA) for DES. In its original form, this theory became known as Infinitesimal PA (IPA), a methodology for estimating sensitivities of performance measures of DES with respect to various real-valued controllable parameters from sample path data (see [35],[26]). In its broader and more recent form, its goal is the development of efficient learning schemes based on information extracted from sample paths. The ultimate aim in this third stream is to satisfy a set of quantitative specifications and to optimize the performance of a DES.

In this paper we treat the three streams in the order of their numbering. There is no significance to this numbering, other than going from untimed to timed to stochastic timed DES models (and even this characterization is not strict). It is interesting to point out that reference [29] identifies monotone structures as a unifying theme among these three streams; see also [59]. Readers interested in recent DES literature are referred to subsequent sections and to several books which have appeared recently, as well as to the proceedings of conferences and/or workshops held recently and more or less exclusively devoted to DES (at Sopron, Hungary, 1987, [75]; at Prague, 1992, [4]; at Sophia-Antipolis, 1994, [21]; and at Bristol, 1994, [33]).

The outline of this paper is as follows. In section 2 three basic examples will be given, each representing one of the three streams. This will help the reader to get a better feeling of what DES are and what the relative strengths and weaknesses of these approaches are. In section 3, the modeling of DES will be discussed in a manner that shows the complementarity of the three streams. Further details will be given about 'logical' or 'untimed' discrete event models and 'timed' discrete event models and some frequently used formalisms for modelling discrete event systems will be presented (automata, Petri nets, and event graphs). Section 4 will present an introduction to the theory of supervisory control, section 5 will give an introduction to the modelling based on the max-plus algebra and some variations of it, and finally, section 6 will consider the sample path analysis and perturbation analysis of discrete event systems.

2 Examples of Control and Optimization Problems for Discrete Event Systems

Example 1. Telephone System.
This example is concerned with purely logical properties of DES, i.e., properties that depend on the ordering of the events and not on the exact times at which these events occur. We model the behavior of a subscriber in a telephone network. Let us call this subscriber *user 0*. The purpose of building the model is to study the behavior of the telephone system under the constraints imposed on it for its

proper operation. We assume that the network has capabilities for forwarding of calls and also for the "call waiting" and "three-way calling" options. Call waiting is an option that enables a user to answer a call when the user is already talking on the phone; three-way calling allows a user to initiate a conference call involving three subscribers. This means that user 0 can be connected to up to two other subscribers at a given time. We denote these subscribers as *user 1* and *user 2*. Each user is connected to a corresponding local switch, where control is exerted to ensure proper behavior of the telephone system.

A logical (or untimed) discrete event model for user 0 is depicted in Figure 1. This figure represents an *automaton model* of user 0. A formal definition of automaton will be presented later; for now, it suffices to view an automaton as a directed graph where the nodes represent the states of the system (or system component) and the labeled arcs represent the transitions between these states due to the occurrence of events. Figure 1 models the "uncontrolled" behavior of user 0, i.e., before any control actions are applied. User 0 is initially in the state INIT, which is the meaning of the wedge pointing into this state in the figure. The events *offh0, onh0, fh0*, and *dfh0* model that user 0 picks up the phone, hangs up the phone, does a "flash-hook" (push and release the hook switch immediately), and does a double flash-hook, respectively. The latter two events arise when user 0 takes advantage of the call waiting and three-way calling options. The event *req0i* models the fact that user 0 requests the establishment of a connection with user i. A granting of this request by the local switch of user i is modeled by the event *con0i* and results in a ringing tone. A denied request is modeled by the event *nocon0i* and normally results in a busy tone. (Note that request *req00* is always denied in this model.)

The rest of the model implements call forwarding actions. The event *fwd0ij* signifies that the call to user i is forwarded to user j. In accordance with the "Signaling System 7" (SS7), which is the current set of protocol standards of the CCITT[4] for network signaling for the Integrated Services Digital Network (refer to [50]), the local switch at user 0 is notified of such a forwarding action and can in fact prevent it, if so desired. This is captured in the model by the events *req0j* that come out of the states FWD_TO_J; event *nocon0* represents the situation where the local switch of user 0 denies the forwarding of the call initiated by user 0.

It is important to observe that in this model, user 0 "cannot" hang up the telephone while his request is being processed. Only after a connection (*con0i*) or a no-connection (*nocon0i*) is established can the user hang up and return to the initial state. Thus the initial state also models the proper completion of a sequence of operations and for this reason this state is *marked*, which is indicated in the figure by the double-frame around it. This marking of states allows the study of *deadlock* and *livelock* in the behavior of the system. We will return to these issues later in this paper.

To complete the description of our model from the point of view of the local switch of user 0, we model the behavior of users 1 and 2, as seen by switch 0. In

---

[4] International Consultative Committee for Telephone and Telegraph

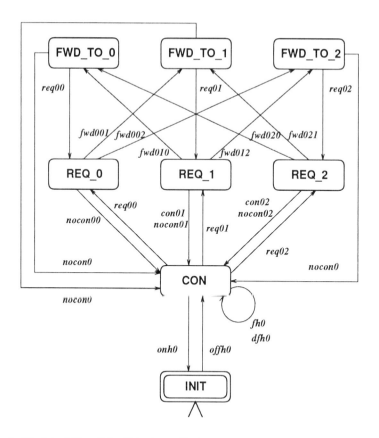

Fig. 1. Model of User 0 in Telephone System

general, switch 0 will only observe part of the behavior of these two users. The model of user 1 as seen by switch 0 is depicted by the automaton in Figure 2. The model of user 2 as seen by switch 0 is similar. These models assume, for the sake of simplicity, that switch 0 observes all the events involving the hook switches of users 1 and 2 in addition to the events that regard connections with user 0.

This model of telephone subscribers was originally developed in [72] and further refined in [73, 17] for the study of the logical interactions in the telephone system when users take advantage of the call forwarding, call waiting, and three-way calling options mentioned above as well as of several other available options such as, terminating and originating call screening, automatic call back, and completion of calls to busy subscribers, to mention but a few. These options are called *features* in the telephone industry. With the rapid introduction of new features in the telephone system, switch manufacturers have been faced with the problem of undesirable interactions between these features, resulting in

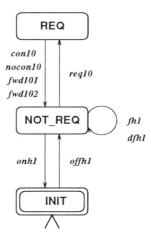

Fig. 2. Model of User 1 at Switch 0 in Telephone System

unanticipated side effects that interfere with the normal functionalities of existing features. This problem is known as the problem of *feature interactions*. The large size and increasing complexity of the software programs that are executed by the switches in the telephone network (millions of lines of code) have necessitated the use of formal model-based methods for the detection and resolution of feature interactions [49].

We will use this telephone system model as an illustrative example in our discussion on discrete event modeling and supervisory control of DES in sections 3 and 4, respectively. We refer the reader to [72, 73, 17] for a thorough discussion of the application of the concepts and techniques of the theory of supervisory control of DES to the detection and resolution of feature interactions.

We conclude our presentation of this example with a brief discussion of the control issues that arise for this system. The objective is to design controllers at the local switches that enforce the requirements on the normal behavior of the telephone system without any features – this is called Plain Old Telephone Service, or POTS, in the telephone industry – and on the behavior desired in the presence of one or a set of features. Furthermore, the joint operation of these controllers should not create feature interactions. Examples of POTS specifications are: (i) nobody can call user 0 successfully (i.e., receive a ringing signal) if user 0 has picked up the handset and (ii) if user 0 is answering a call, he cannot make another call. A specification for the feature "terminating call screening" would not allow user 0 to call user 1 if user 1 has user 0 listed on his screening list. An example of a feature interaction is the conflict between call forwarding and terminating call screening that may result in a deadlock (i.e., the system enters an unmarked state where no further event is possible) unless one feature is allowed to preempt the other (see [73] or [17] for further details). Once controllers have been designed and the behavior of the system has been deemed satisfactory (according to the formal model-based analysis), these controllers can be implemented in the software programs that are executed by the local switches. The theory of supervisory control that we introduce in section 4 provides system-theoretic concepts and techniques for the design of such controllers.

□

Example 2. Railway connections and time tables.
We will consider a closed network of railway connections and study questions related to the design of time tables according to which the trains run within this network. Specifically, we will study the intercity network of the Netherlands. This network, given in Figure 3, consists of 11 lines (routes along which the trains run in both directions). An example of such a line is the line with number 10 which starts (respectively terminates) at Amsterdam and terminates (respectively starts) at Vlissingen. There are no circular lines, but inclusion of circular lines into the model would not constitute any difficulties. We will not consider international connections: it is assumed in our model that trains arriving at the border return direction and continue in opposite direction (this is what actually happens with some trains at some border stations). Hence the name 'closed' network. On each line, both directions, there is a fixed number of trains which return at both endpoints in the opposite direction. The travelling times between stations is given and assumed to be deterministic.

Suppose there is no time table. A very simple policy for trains to run would be to drop and pick up passengers at stations and then directly continue. This is not very realistic since one would like trains to remain at a station for some time such as to wait for other arriving trains and allow change overs. Now suppose that realistic constraints are included which state which trains have to wait for each other, at each station, such as to allow change overs before the trains can continue on their respective lines.

Without an apriori time table, and with the constraints of trains waiting for each other, how fast can the system operate? The actual time table is based on a half hour schedule and hence from this fact of real life we know that without time table the system must be 'operable' in a time of at most half an hour (this time will be indicated by λ which turns out to be equal to about 27 minutes). The difference between the actual half hour and the theoretical λ minutes is the flexibility in the system which causes propagation of possible delays to disappear in finite time. Various questions can be posed now, some of which are:

- How do perturbations progagate through the system and how long does it take before they have completely disappeared?
- If one would add an extra five minutes to all changeover times, is it still possible to design a time table based on a half hour service? (answer: yes).
- What are the crucial parts of the system which determine the minimum operation time of λ minutes? Suppose one could add some extra trains to the system, on which lines should they be set in such as to reduce this minimum operation time as much as possible?

2 Examples of Control and Optimization Problems for Discrete Event Systems 225

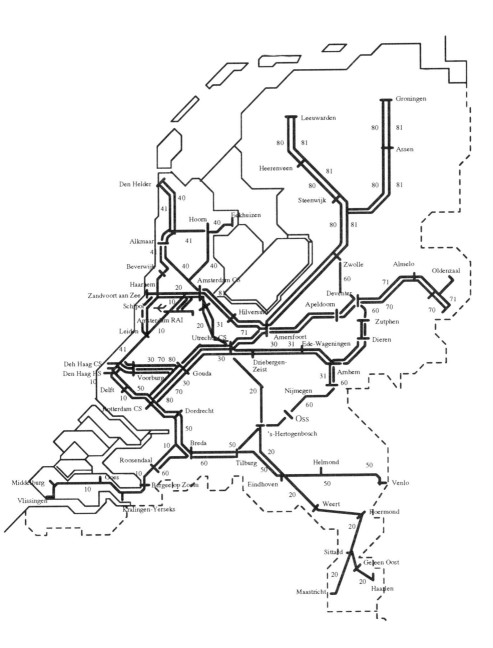

Fig. 3. The Dutch intercity network

- With the minimum operation time schedule, is it possible to have a 'regular' time table? With regular is meant here that if a train leaves station A in the direction of station B, the next train in the same direction will leave exactly λ minutes later.
- The lines along which the trains run were supposed to be given. Is it possible to design an 'optimal line structure'?

Answers to most of these questions will be given in section 5, in Example 5. □

Example 3. Dispatching control in an elevator system.
This example illustrates design and control issues that arise when the objective is to improve (ideally: optimize) the performance of a discrete event system, especially one that operates in a stochastic environment. Consider an elevator system serving a building with N floors. If N is large (typically 10-20 floors), the elevator system consists of several *cars* (typically, 4-8). Passengers arrive at the building lobby expecting to find a car to transport them to a destination floor. They also arrive at various floors and expect to be transported either to some other floor (further up or further down from their current location) or to the lobby so as to leave the building. The way in which passengers arriving at the lobby or any floor request service is by pressing a button ("up" or "down", depending on their location and destination); this is referred to as a *hall call*. In addition, when they enter a car, they indicate their desired destination by pressing a specific floor button; this is referred to as a *car call*. At any given point in time, the state of this system is a large vector including information regarding the position of cars (i.e., the floor last visited), the number of passengers waiting for a car at any one floor, and all active hall calls and car calls. State transitions take place as a result of a multitude of different events such as "car moving up arrives at floor i" or "hall call issued at floor i".

There are several measures of performance one can define for such a system. The most common is the *average waiting time* experienced by arriving passengers, which one obviously wishes to minimize. Modern elevator systems, however, also consider a level of "unsatisfactory service", such as a passenger having to wait for more than 1 minute for a car, and then aim at minimizing the *fraction of customers that experience poor service*. It is also clear that since passenger arrivals occur at random, it is necessary to make use of stochastic models for analyzing this system.

The basic control question one faces in this setting is the following. At any given point in time, if there is an active hall call at floor i, which car should be assigned to that floor? This seemingly simple question is extremely difficult to answer if the objective is to achieve truly optimal, or at least near-optimal, performance. Another viewpoint allowing us to better identify the set of feasible control actions at our disposal is the following. Suppose cars are constantly moving from floor to floor; when a car is at the lobby (floor 1) it always moves to floor 2 next, and when it is at the top floor (floor N), it moves to floor $N-1$ next. With every event "car arrives at floor i", there are now three actions to select from: (a) STOP at floor, (b) DO NOT STOP at floor, and (c) REVERSE

2 Examples of Control and Optimization Problems for Discrete Event Systems

DIRECTION at floor. Note that not all actions are fully controllable; for instance, STOP at floor i must always be executed if there is a passenger in the car whose destination is this floor. On the other hand, even if there are several passengers waiting at i to move up and a car arrives at i moving up, it may be preferable to elect DO NOT STOP; this may be because there are extremely long queues at floors $i+1$, $i+2$, etc. which need to be served first, especially if another car is right behind this one that could provide service to the hall calls at i.

This discussion serves to provide some insight into the dynamics of this discrete event system and the key issues related to its control from a performance optimization standpoint. To motivate the theoretical framework presented in section 6 and techniques that may be used to address these issues, let us consider next a special case of the general performance optimization problem for elevator systems. In particular, suppose a building is empty and passengers arrive only at the lobby so as to be transported to various floors indexed $2, \ldots, N$. This is referred to as the "uppeak case" and it is encountered every morning in a typical office building. Suppose there are M cars in the system and the objective is to minimize the average waiting time of passengers arriving at the lobby. In an uncontrolled environment, the first passenger enters car 1 and leaves, the second one enters car 2, and so on until all M cars are used up. This is obviously extremely inefficient, since a car with a capacity of 10-20 passengers ends up serving a single (or just a few) passengers. Moreover, it is easy to see that after the M cars become occupied, it may take a long time for a car to return to the lobby, causing significant waiting. It is possible to show (using a Markov decision process framework) that a very efficient (and in some cases optimal) scheme is the following. When more than a single car are present at the lobby, only one is made available. Moreover it is not allowed to leave until a certain threshold, K, of passengers in it is reached. When this threshold is reached, the car leaves and a second one is made available with a similar threshold-based control mechanism in place. The crucial question now is "what is the optimal value of K?" In other words, it is often the case that one can identify a parametric class of policies with some desirable (possibly optimal) properties, and the critical task is to identify the optimal parameter(s) within that class.

Even so, the problem is not easy to solve. At the heart of the problem is the lack of analytical models for specifying functional relationships between performance measures of interest and parameters that can be controlled. In the example above, let $W(K)$ denote the average waiting time of pasengers as a function of the threshold parameter K. Except for simple models of limited interest, $W(K)$ is simply unknown. Therefore, to determine the optimal value of K one must either resort to some form of *learning scheme* based on repeated trial-and-error or make use of *computer simulation* to estimate $W(K)$ over all admissible values of K. Both are prohibitively slow (if at all feasible) approaches. Here, however, is where one can take advantage of the event-driven dynamics of the systems we are considering in this paper. Suppose the set of admissible parameter values in our example is $\{K_1, \ldots, K_n\}$. If one selects K_1 as the ini-

tial value, the problem is to determine what would happen if K_1 were replaced by $K_2, \ldots, K_n$. This process normally requires n separate trials (e.g., simulation runs), one for each of the n values. However, as will be further discussed in section 6, it is possible to obtain the same information from a single trial under K_1 only. The basis for this approach lies on the theory of Perturbation Analysis [36], [26] and several recent extensions including the Standard Clock methodology [74] and Augmented System Analysis [15].
□

3 Modelling of Discrete Event Systems (DES)

3.1 A General Language-Based Approach

As was mentioned in the introduction and observed in the three preceding examples, a discrete event system (hereafter DES) is a dynamic system that possesses two distinguishing features: its state space is a discrete set and its dynamics are event-driven (as opposed to time-driven). In continuous-variable dynamic systems whose dynamics are time-driven and described by means of differential (continuous-time) or difference (discrete-time) equations, it is customary to describe the evolution of the system over time by plotting the state $x(t)$ as a function of t. If we do the same for a DES, we get state trajectories, or sample paths, of the form shown in Figure 4.[5] Instead of plotting such sample paths, it is more convenient to simply write the timed sequence of states

$$s_x = (x_2, 0)(x_5, t_1)(x_4, t_2)(x_1, t_4)(x_3, t_5)(x_4, t_6)(x_6, t_7)$$

or the timed sequence of events

$$s_e = (e_1, t_1)(e_2, t_2)(e_3, t_3)(e_4, t_4)(e_5, t_5)(e_6, t_6)(e_7, t_7)$$

to describe the sample path depicted in Figure 4. Note that the latter sequence of events is more general in the sense that the former sequence of states can be recovered from it (assuming a deterministic system and knowledge of the initial state) but not vice-versa, as events that do not cause state transitions, e.g., e_3 in state x_4, do not appear in the sequence of states.

Let us call a *timed language* a set of timed sequences of events of the above form and a *stochastic timed language* a timed language together with a probability distribution function defined over this set. Then in a very general sense a DES can be modeled by a stochastic timed language. The stochastic timed language represents all possible behaviors of the system together with the probability of each behavior. This type of modeling is the most detailed as it contains event information in the form of event occurrences and their orderings, timing information about the exact times at which the events occur (and not only their relative ordering), and statistical information about the probabilities of the various possible sample paths. If we omit the statistical information, then the corresponding

---
[5] This figure is adapted from Fig. 1.22 in [11], p. 42.

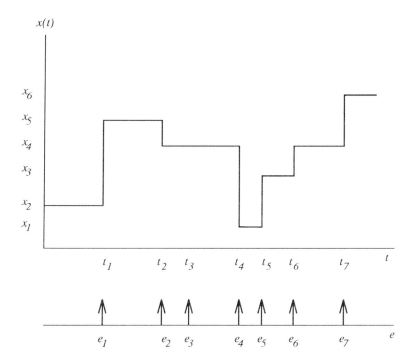

Fig. 4. Sample Paths of Discrete Event Systems

timed language enumerates all the possible sample paths of the DES, with timing information. Finally, if we project out the timing information from a timed language we obtain an *untimed language*, or simply *language*, which is the set of all possible orderings of events that could happen in the given system. The above projection means that the exact times of occurrence of events in a timed sequence are simply deleted; for example, the untimed sequence corresponding to the timed sequence of events s_e above is

$$e_1 e_2 e_3 e_4 e_5 e_6 e_7.$$

Languages, timed languages, and stochastic timed languages represent the three levels of abstraction at which DES are modeled and studied: untimed (or logical), timed, and stochastic. The choice of the appropriate level of abstraction clearly depends on the objectives of the analysis. In the telephone system example, since we are only concerned with logical properties of the behavior of the system, the untimed level of abstraction is adequate. In the train example, a timed model is necessary to answer the questions posed. However it is not required to introduce a probabilistic structure to answer these questions and therefore the DES need not be modeled at the stochastic level of abstraction. On the other hand, in the elevator system example, passengers at various floors arrive at random instants in time. It is, therefore, necessary to model the DES at the stochastic level of abstraction. These three levels of abstraction are complementary as they address different issues about the behavior of the DES. In fact, the literature in DES is quite broad and varied as extensive research has been done on modeling, analysis, control, optimization, and simulation at all three levels of abstraction.

This language-based approach to discrete event modeling is attractive to present modeling issues and discuss system-theoretic properties of DES. However, it is by itself not convenient to address verification, synthesis, or performance issues; what is also needed is a convenient way of *representing* languages, timed languages, and stochastic timed languages. If a language (or timed language or stochastic timed language) is finite, we could always list all its elements; but this is rarely practical. In certain cases, we can represent a (finite or infinite) language L by

$$L = \{s_e : s_e \text{ has property P}\}$$

where P could for instance specify that the sequence of events s_e should have the same number of e_1 events as e_2 events. This is often useful, but such a representation is not amenable to analysis when for instance calculations involving finding subsets or supersets of L have to be performed. More preferably, we would like to use *discrete event modeling formalisms* that would allow us to represent languages in a manner that highlights structural information about the system behavior and that is convenient to manipulate when addressing analysis and synthesis issues. Discrete event modeling formalisms can be untimed, timed, or stochastic, according to the level of abstraction of interest.

In this paper, we will present and use the following discrete event modeling formalisms: *automata* (or *state machines*) and *stochastic timed automata*, for the logical and stochastic levels of abstraction, respectively, and *timed event graphs*, a special class of *timed Petri nets*, for the timed and stochastic levels of abstraction. These formalisms have in common the fact that they represent languages by using a state transition structure. This makes them convenient for model building; recall how we specified the model of user 0 in Figure 1. They are also amenable to various composition operations, which allows the building of a discrete event model of a system from discrete event models of the system components. Analysis and synthesis issues are then typically addressed by making use of the structural properties of the transition structure in the model. This will become apparent in subsequent sections of this paper. We mention that other modeling formalisms exist, in particular, formalisms that do not rely explicitly on using a state transition structure. However these formalisms will not be discussed here.

In the next two subsections, we discuss in more detail the representation of languages by automata; this material will serve as introduction to the discussion of supervisory control in section 4. Stochastic timed automata represent stochastic timed languages and are used for the performance optimization of DES; they will be presented in section 6 where the topic of performance optimization is treated. Subsection 3.4 briefly discusses (timed) Petri nets and (timed) event graphs; this will lead to the max-plus algebra for timed modeling and analysis of DES treated in section 5.

3.2 Languages and Automata

In our discussion of DES at the logical level of abstraction, we will consider both the language domain and the automaton domain. For this reason, we start with a brief review of some concepts from *Formal Language Theory* [42].

Let Σ be a non-empty finite set of events (or "alphabet"). This set consists of all the events that can possibly be executed by the DES of interest. We will use both e and σ to denote elements of Σ. For instance, events *offh0*, *req01*, and *fwd012* are in the event set of user 0 in the telephone system example. A *trace* (or string, or word) is a finite sequence of events from Σ. The length of a trace s, denoted by $|s|$, is a non-negative integer corresponding to the number of events composing the trace (counting multiple occurrences of the same event). The empty trace, denoted ϵ (not to be confused with the generic event e), is the trace containing no events, i.e., $|\epsilon| = 0$. The concatenation of two traces s_1 and s_2 is the trace $s_1 s_2$ (i.e., s_1 followed by s_2). Thus the empty trace ϵ can be interpreted as the identity element for concatenation.

Denote by Σ^* the set of all finite traces of elements of Σ, including the empty trace ϵ; the * operation is called the Kleene closure. For example, if $\Sigma = \{a, b, c\}$, then

$$\Sigma^* = \{\epsilon, a, b, c, aa, ab, ac, ba, bb, bc, ca, cb, cc, aaa, \ldots\} \ .$$

A *language* is then formally defined as a *subset* of Σ^*. Thus $\emptyset, \Sigma, \Sigma^*$ are languages. (Note that $\epsilon \notin \emptyset$.) If $s't = s$ with $s, s', t \in \Sigma^*$, then s' is called a *prefix* of s; thus both ϵ and s are prefixes of s.

The usual set operations, such as union, intersection, difference, complement (with respect to Σ^*) are applicable to languages since languages are sets. In addition, we will also consider the following operations:[6]

- *Concatenation*: Let $L_1, L_2 \subseteq \Sigma^*$, then

$$L_1 L_2 := \{s \in \Sigma^* : (s = s_1 s_2) \wedge (s_1 \in L_1) \wedge (s_2 \in L_2)\} \ .$$

- *Prefix-closure*: Let $L \subseteq \Sigma^*$, then

$$\overline{L} := \{s \in \Sigma^* : (\exists t \in \Sigma^*) \ st \in L\} \ .$$

Thus the prefix-closure $\overline{L}$ of L is the language consisting of all the prefixes of all the traces in L. For example, if $L = \{abc, cde\}$ then $\overline{L} = \{\epsilon, a, ab, abc, c, cd, cde\}$. In general, $L \subseteq \overline{L}$. L is said to be *prefix-closed* (or simply *closed*) if $L = \overline{L}$. (Technical point: If $L = \emptyset$ then $\overline{L} = \emptyset$, and if $L \neq \emptyset$ then $\epsilon \in \overline{L}$.)

We now formally define automata and describe how they are used to represent (and manipulate – see section 3.3) languages.

---

[6] ":=" denotes "equal to by definition".

Definition 1. A *Deterministic Automaton*, denoted G, is a six-tuple

$$G = (X, \Sigma, f, \Sigma_G, x_0, X_m)$$

where

- X is the set of states of G.
- Σ is the set of events associated with the transitions in G.
- $f : X \times \Sigma \to X$ is the *partial* transition function of G: $f(x,e) = x'$ means that there is a transition labeled by event e from state x to state x'.
- $\Sigma_G : X \to 2^{\Sigma}$ is the active event function (or feasible event function): $\Sigma_G(x)$ is the set of all events e for which $f(x,e)$ is defined. $\Sigma_G(x)$ is called the *active event set* (or *feasible event set*) of G at x.
- x_0 is the *initial* state of G.
- $X_m \subseteq X$ is the set of *marked states* of X.

We make the following remarks about this definition.

- The words *generator* (which explains the notation G) and *state machine* are also extensively used in the literature to describe the above object.
- If X is finite, we call G a *deterministic finite automaton*, or DFA.
- The automaton is said to be *deterministic* because f is a function over $X \times \Sigma$. In contrast, the transition structure of a *nondeterministic* automaton is defined by means of a relation over $X \times \Sigma \times X$. Note that by default, the word automaton will refer to deterministic automaton in the sequel.
- The fact that we allow the transition function f to be partially defined over its domain $X \times \Sigma$ is a variation over the standard definition of a DFA that is quite important in DES theory.
- By designating certain states as marked, we record that the system, upon entering these states, has completed some operation or task.

For the sake of convenience, f is always extended from domain $X \times \Sigma$ to domain $X \times \Sigma^*$ in the following recursive manner (recall that ϵ denotes the empty trace):

$$f(x, \epsilon) = x$$
$$f(x, s\sigma) = f(f(x,s), \sigma) \quad \text{for } s \in \Sigma^* \text{ and } \sigma \in \Sigma .$$

Now think of the automaton G as a *directed graph* (recall Figure 1) and consider all the (directed) paths that can be followed from its initial state; consider among these all the paths that end in a marked state. This leads to the notions of the languages *generated* and *marked* by G.

Definition 2. The language *generated* by G is

$$\mathcal{L}(G) := \{ s \in \Sigma^* : f(x_0, s) \text{ is defined} \} .$$

The language *marked* by G is

$$\mathcal{L}_m(G) := \{ s \in \mathcal{L}(G) : f(x_0, s) \in X_m \} .$$

3 Modelling of Discrete Event Systems (DES)

The language $\mathcal{L}(G)$ is prefix-closed by definition, but $\mathcal{L}_m(G)$ need not be prefix-closed in general. If we denote the automaton in Figure 1 as G_{user0}, then the set of traces

$$\{\epsilon, offh0, offh0\ req01, offh0\ req01\ con01\ req02\ fwd021\}$$

is contained in $\mathcal{L}(G_{user0})$, since INIT is the initial state, while the set of traces

$$\{\epsilon, offh0\ onh0, offh0\ req01\ con01\ req02\ fwd021\ nocon0\ onh0\}$$

is contained in $\mathcal{L}_m(G_{user0})$, since INIT is a marked state.

Thus automaton G is a representation of *two* languages: $\mathcal{L}(G)$ and $\mathcal{L}_m(G)$. Two automata are said to be *equivalent* if they generate *and* mark the same languages. In general, $\mathcal{L}_m(G) \subseteq \overline{\mathcal{L}_m(G)} \subseteq \mathcal{L}(G)$. The automaton G is said to be *blocking* if $\mathcal{L}(G) \neq \overline{\mathcal{L}_m(G)}$ and *non-blocking* when $\mathcal{L}(G) = \overline{\mathcal{L}_m(G)}$.

The concept of blocking deserves further comments. The automaton G could reach a state x where $\Sigma_G(x) = \emptyset$ but $x \notin X_m$. This is commonly called a *deadlock* because no further event can be executed. Given our interpretation of marking, we say that the system blocks because it enters a deadlock state without having terminated the task at hand. Another cause of blocking is when there is a cycle of unmarked states in G with no transition going out of the cycle. If the system enters this cycle, then we get what is called a *livelock*: while the system is live in the sense that it can always execute an event, it can never complete the task started since no state in the cycle is marked and the system cannot leave the cycle. Again we refer to this as blocking. Blocking is an important issue when controlling a DES. It is usually desirable to minimize (in some sense) or completely avoid blocking in a controlled system. The notions of marked states and of the languages generated and marked by an automaton is an approach for considering deadlock and livelock that has proven useful in many applications.

From the definitions of $\mathcal{L}(G)$ and $\mathcal{L}_m(G)$, we see that we can delete from G all the states that are not *accessible* or *reachable* from x_0 by some trace in $\mathcal{L}(G)$, without affecting the languages generated and marked. Note that when we "delete" a state, this means also deleting all the transitions that are *attached* to that state. We will denote this operation by $Ac(G)$, where Ac stands for taking the accessible part. We will always assume that a given automaton is accessible. A state x of G is said to be *coaccessible* (to X_m) if there is a trace in $\mathcal{L}_m(G)$ that goes through x; this means that G can go from state x to a marked state. We denote the operation of deleting all the states of G that are not coaccessible by $CoAc(G)$. By definition, the $CoAc$ operation will not affect the marked language but it may reduce the generated language. The automaton G is said to be *trim* if $G = Ac[CoAc(G)] (= CoAc[Ac(G)])$.

We have now specified how automata can be used to represent languages. Clearly, given any language $K \subseteq \Sigma^*$, we can always construct an automaton –albeit not necessarily a finite automaton– that marks K: for instance, simply build the automaton as an infinite tree whose root is the initial state and where the nodes at layer n are entered by the traces of length n. However, it is well-known that not all subsets of Σ^* can be represented by *finite* automata. A language K is said to be *regular* if there exists a DFA G that marks it, i.e, $\mathcal{L}_m(G) = K$. The class of regular languages is denoted by $\mathcal{R}$.

We often need to represent languages with *finite* automata, e.g., when the representation has to be stored in memory for performing calculations on it or simply for storage of a control policy. For this reason, the class $\mathcal{R}$ is of special interest to us. This class has been well-studied in the computer science literature and it possesses many useful properties. In particular, it is closed under union, intersection, complementation, and concatenation of languages. The implementation of these operations is typically performed by manipulating the DFA representations of the given languages. For example, given G_1 and G_2, it is straightforward to build G_{12} such that $\mathcal{L}_m(G_{12}) = \mathcal{L}_m(G_1) \cap \mathcal{L}_m(G_2)$, as we will see in the next subsection.

It is important to note that $\mathcal{R} \neq 2^{\Sigma^*}$, i.e., not all languages are regular. For instance, the language $\{a^n b^n : n \geq 0\}$ is not regular. Intuitively, we need to count the number of a events so as to know when to mark the subsequent subtrace of b events; but since n can be arbitrarily large, this language cannot be marked by a DFA. For $K \in \mathcal{R}$, define $||K||$ to be the minimum of $|X|$ (the cardinality of X) among all DFA G that mark K. The DFA that achieves this minimum is called the *canonical recognizer* of K. It should be emphasized that $||\cdot||$ is not related to the size of the language. For instance, $||\Sigma^*|| = 1$: use one (marked) state and put a self-loop at that state for each event in Σ. Thus, in general, a larger language may be (but need not be) representable with fewer states than a smaller language.

3.3 Product and Parallel Composition of Automata

We define two operations on automata: the product, denoted $\times$, and the parallel composition, denoted $||$. (The parallel composition is often termed synchronous composition and the product, completely synchronous composition.) Consider $G_1 = (X_1, \Sigma_1, f_1, \Sigma_{G_1}, x_{01}, X_{m1})$ and $G_2 = (X_2, \Sigma_2, f_2, \Sigma_{G_2}, x_{02}, X_{m2})$. Then:

$G_1 \times G_2 := Ac(X_1 \times X_2, \Sigma_1 \cap \Sigma_2, f, \Sigma_{G_1 \times 2}, (x_{01}, x_{02}), X_{m1} \times X_{m2})$
where

$$f((x_1, x_2), e) := \begin{cases} (f_1(x_1, e), f_2(x_2, e)) & \text{if } e \in \Sigma_{G_1}(x_1) \cap \Sigma_{G_2}(x_2) \\ \text{undefined} & \text{otherwise} \end{cases}$$

and thus $\Sigma_{G_1 \times 2}(x_1, x_2) = \Sigma_{G_1}(x_1) \cap \Sigma_{G_2}(x_2)$;

$G_1 \parallel G_2 := Ac(X_1 \times X_2, \Sigma_1 \cup \Sigma_2, f, \Sigma_{G_{1||2}}, (x_{01}, x_{02}), X_{m1} \times X_{m2})$
where

$$f((x_1, x_2), e) := \begin{cases} (f_1(x_1, e), f_2(x_2, e)) & \text{if } e \in \Sigma_{G_1}(x_1) \cap \Sigma_{G_2}(x_2) \\ (f_1(x_1, e), x_2) & \text{if } e \in \Sigma_{G_1}(x_1) \setminus \Sigma_2 \\ (x_1, f_2(x_2, e)) & \text{if } e \in \Sigma_{G_2}(x_2) \setminus \Sigma_1 \\ \text{undefined} & \text{otherwise} \end{cases}$$

and thus $\Sigma_{G_{1||2}}(x_1, x_2) = [\Sigma_{G_1}(x_1) \cap \Sigma_{G_2}(x_2)] \cup [\Sigma_{G_1}(x_1) \setminus \Sigma_2] \cup [\Sigma_{G_2}(x_2) \setminus \Sigma_1]$.

3 Modelling of Discrete Event Systems (DES)

In the product, the transitions of the two automata must always be synchronized on a common event. It is easily verified that

$$\mathcal{L}(G_1 \times G_2) = \mathcal{L}(G_1) \cap \mathcal{L}(G_2)$$
$$\mathcal{L}_m(G_1 \times G_2) = \mathcal{L}_m(G_1) \cap \mathcal{L}_m(G_2) \, .$$

In fact, this last result proves the closure of $\mathcal{R}$ under the intersection of languages, as $G_1 \times G_2$ is a DFA if both G_1 and G_2 are DFA. In the parallel composition, a common event, i.e., an event in $\Sigma_1 \cap \Sigma_2$, can only be executed if the two automata both execute it simultaneously. Thus the two automata are "synchronized" on the common events. The other events, i.e., those in $(\Sigma_2 \backslash \Sigma_1) \cup (\Sigma_1 \backslash \Sigma_2)$, are not subject to such a constraint and can be executed whenever possible.

Let us define the *natural projections* $P_i : (\Sigma_1 \cup \Sigma_2)^* \to \Sigma_i^*$ for $i = 1, 2$, as follows:

$$P_i(\epsilon) := \epsilon$$
$$P_i(\sigma) := \begin{cases} \sigma & \text{if } \sigma \in \Sigma_i \\ \epsilon & \text{if } \sigma \notin \Sigma_i \end{cases}$$
$$P_i(s\sigma) := P_i(s)P_i(\sigma) \text{ for } s \in (\Sigma_1 \cup \Sigma_2)^*, \sigma \in (\Sigma_1 \cup \Sigma_2)$$

and the corresponding inverse maps $P_i^{-1} : \Sigma_i^* \to (\Sigma_1 \cup \Sigma_2)^*$ as follows:

$$P_i^{-1}(t) := \{s \in (\Sigma_1 \cup \Sigma_2)^* : P_i(s) = t\} \, .$$

The projections P_i and their inverses P_i^{-1} are extended to languages in the usual manner: for $L \subseteq (\Sigma_1 \cup \Sigma_2)^*$,

$$P_i(L) := \{t \in \Sigma_i^* : (\exists s \in L) P_i(s) = t\}$$

and for $L_i \subseteq \Sigma_i^*$,

$$P_i^{-1}(L_i) := \{s \in (\Sigma_1 \cup \Sigma_2)^* : (\exists t \in L_i) P_i(s) = t\} \, .$$

Then we have:
$$\mathcal{L}(G_1 || G_2) = P_1^{-1}[\mathcal{L}(G_1)] \cap P_2^{-1}[\mathcal{L}(G_2)]$$
$$\mathcal{L}_m(G_1 || G_2) = P_1^{-1}[\mathcal{L}_m(G_1)] \cap P_2^{-1}[\mathcal{L}_m(G_2)] \, .$$

Observe that $\times$ and $||$ are commutative (up to a renaming of the states) and associative. If $\Sigma_1 = \Sigma_2$, then the parallel composition reduces to the product, since all transitions are forced to be synchronized. If $\Sigma_1 \cap \Sigma_2 = \emptyset$, then there are no synchronized transitions and thus $G_1 || G_2$ is the *concurrent* behavior of G_1 and G_2. This is often termed the *shuffle* of G_1 and G_2. If we compose G_{user0} of Figure 1 with the automaton depicted in Figure 2 (let us denote it by $G_{user1at0}$), then $G_{user0} || G_{user1at0}$ is a shuffle and the resulting automaton has 24 states (8 × 3), all of which are accessible and coaccessible.

Finally, observe that we can define a corresponding parallel composition of languages. With $L_i \subseteq \Sigma_i^*$ and P_i defined as above,

$$L_1 || L_2 := P_1^{-1}(L_1) \cap P_2^{-1}(L_2) \, .$$

3.4 Petri Nets and Event Graphs

It is assumed that the reader is familiar with the basic properties of Petri nets, see [52] or [64]. It will be shown that the max-plus algebra is extremely suitable in describing the timed behaviour of tokens in so-called event graphs, which form a subclass of Petri nets. In order to set the notation and the stage, we do start with some formal definitions.

Definition 3 Petri net. A Petri net is a pair $(\mathcal{G}, b)$, where $\mathcal{G} = (\mathcal{E}, \mathcal{V})$ is a bipartite graph with a finite number of nodes (the set $\mathcal{V}$) which are partitioned into the disjoint sets $\mathcal{P}$ and $\mathcal{Q}$; $\mathcal{E}$ consists of pairs of the form (p_i, q_j) and (q_j, p_i) with $p_i \in \mathcal{P}$ and $q_j \in \mathcal{Q}$. The initial marking b is an m-vector, with m being the number of elements in $\mathcal{P}$, of nonnegative integers. The elements of $\mathcal{P}$ are called places, those of $\mathcal{Q}$ are called transitions. The number of elements in these sets are m and n respectively. The elements of the vector b denote the number of tokens in the respective places. One talks about a *timed Petri net* if time durations are associated with places and transitions.

Definition 4 Event graph. A (timed) Petri net is called a (timed) event graph if each place has exactly one upstream and one downstream transition.

Definition 5 State graph. A (timed) Petri net is called a (timed) state graph if each transition has exactly one upstream and one downstream place.

Note that the definitions of event graph and state graph are dual one to the other. In these definitions it was tacitly assumed that the networks are 'closed', i.e. all places (transitions) do have an upstream and a downstream transition (place). The definitions can be extended in the obvious way to include input transitions (places), so-called *sources*, which do not have upstream places (transitions) and output transitions (places), so-called *sinks*, which do not have downstream places (transitions).

A transition can fire (or: can start firing if there is a positive firing time, see definition 6) if all its (directly) upstream places contain at least one token (which must be 'enabled' – see definition 7). After the firing these tokens are removed and one token is added to each of the (directly) downstream places.

Definition 6 Firing time. The firing time of a transition is the time that elapses between the starting and the completion of the firing of the transition.

Definition 7 Holding time. The holding time of a place is the time a token must spend in the place before it can contribute to the enabling of the downstream transitions.

Theorem 8. *An event graph with both firing times and holding times is equivalent to an event graph with only holding times (i.e. the firing times are zero). This equivalence means that the time instants at which the transitions fire are the same in both event graphs.*

Proof See [2]. □

3 Modelling of Discrete Event Systems (DES)

Theorem 9. *The number of tokens in any circuit of an event graph is constant.*

From now on we will only consider event graphs with firing times which are zero. Each place connects precisely one transition with precisely one (possibly different) transition. One says in such a situation that the upstream transition, say q_j, is a predecessor of the downstream transition, say q_i. Equivalently one can say that q_i is a successor of q_j. One writes in such a case $j \in \pi^-(i)$ and $i \in \pi^+(j)$.

We make the explicit assumption that if a place connects two transitions such as just has been described, there is no other place with does exactly the same. In general event graphs there can be more 'parallel' places in between two transitions of which one is the successor of the other. The reason for this restriction is purely a notational issue. The theory to be given can handle the more general situation routinely. We also make the assumption that the underlying network is strongly connected.

If a place exists between the transitions q_j and q_i and q_j is upstream with regard to this place and q_i downstream, then the holding time of this place is indicated by a_{ij}. The holding times are nonnegative real numbers. The number of tokens in this place is indicated by b_{ij}. In the course of time, b_{ij} may change of course, but what is meant here, and also in the formulas to come, (1) and (2), is b_{ij} at the initial time. For some subtle issues with respect to initial conditions, see [2].

If $\tau_i(\chi)$ denotes the earliest time instant at which transition q_i has fired χ times, then

$$\tau_i(\chi) = \max_{j \in \pi^-(i)} a_{ij} + \tau_j(\chi - b_{ij}), i = 1, \ldots, n. \tag{1}$$

Rather than having used the conventional notation x for the state, we now used the symbol τ. This is to distinguish (1), sometimes referred to as the dater equations, from the so-called counter equations defined as

$$x_i(t) = \min_{j \in \pi^-(i)} b_{ij} + x_j(t - a_{ij}), \ i = 1, \ldots, n, \tag{2}$$

where $x_i(t)$ denotes the number of firings of transition q_i, $i = 1, \ldots, n$, which have taken place up to, and including, time t. Equations (1) and (2) describe the same underlying system and one equation is called the dual of the other. Note that χ and x_i are integer-valued. The functions $x_i(t)$ and $\tau_i(\chi)$ are each others inverse in a way.

If in the original event graph there would have been a positive firing time, then the equations above do not exclude the possibility that a transition 'works' simultaneously on two or more tokens. If one wants to exclude this, a loop, including one place with one token, around the transition concerned should be added. The holding time of this new place is defined to be equal to the original firing time of this transition. This loop now takes care of the fact that in the equivalent event graph with only zero firing times, the transition cannot work on two or more tokens simultaneously anymore.

Just as in conventional system theory, the product and parallel composition can also be defined for Petri nets by connecting the proper inputs with the proper outputs; we will not do that explicitly here. Recently evolution equations of timed Petri nets more general than event graphs have been studied; the reader is referred to [1] and [25].

4 Supervisory Control of DES

4.1 The Feedback Loop of Supervisory Control

Supervisory control is the situation where control is exerted on a given uncontrolled DES in order to satisfy the given specifications on the languages generated and marked by the system. The point of view adopted is that some of the uncontrolled behavior is *illegal* and thus must not be allowed under control. Illegal here refers to illegal states or illegal traces (or subtraces) of events. Thus control is exerted in order to reduce the set of traces of events that the system can generate to a legal subset and also to eliminate (or minimize) deadlock. This control paradigm was initiated by Ramadge & Wonham in 1982 [61] and studied extensively since then by themselves, their co-workers, and several other researchers. This body of work is known as *Supervisory Control Theory*. We will present some basic results of supervisory control theory, focusing on the issue of uncontrollability. (Our presentation is not meant to be a survey of this area of research; references to some papers in the literature are given in the last subsection of section 4.) It should be emphasized that the system-theoretic results of supervisory control theory are more naturally stated in the language domain, while the synthesis and computational results are more naturally stated in the (finite) automaton domain.

Consider a DES modeled at the logical level of abstraction by a pair of languages, L and L_m, where $L = \overline{L}$ is the set of all traces that the DES can generate and $L_m \subseteq L$ is the language of *marked* traces that is used to represent the completion of some operations or tasks. Languages L and L_m are defined over the event set Σ. Let $G = (X, \Sigma, f, \Sigma_G, x_0, X_m)$ be the automaton that represents these two languages (such a G always exists as mentioned earlier); that is, $\mathcal{L}(G) = L$ and $\mathcal{L}_m(G) = L_m$. Thus we will talk of the "DES G".

Let Σ be partitioned into two disjoint subsets

$$\Sigma = \Sigma_c \cup \Sigma_{uc}$$

where (i) Σ_c is the set of *controllable* events: these are the events that can be prevented from happening, or disabled, by control; and (ii) Σ_{uc} is the set of *uncontrollable* events: these events cannot be prevented from happening by control. There are many reasons why an event would be modeled as uncontrollable: it is inherently unpreventable (e.g., a failure event); it models a change of sensor readings; it cannot be prevented due to hardware limitations; or it is modeled as uncontrollable by choice, e.g., if the event has high priority and thus should not be disabled or if the event represents the tick of a clock.

4 Supervisory Control of DES

Let us now assume that the transition function of G can be controlled by an external agent in the sense that the controllable events can be enabled or disabled by an external controller. Automaton G, which represents the *uncontrolled behavior* of the DES, is connected in the feedback loop depicted in Figure 5 with a *controller* or *supervisor* S. Formally, supervisor S is a *function*

$$S : \mathcal{L}(G) \to \Gamma := \{\gamma \in 2^\Sigma : \Sigma_{uc} \subseteq \gamma\} .$$

For each $s \in \mathcal{L}(G)$ generated so far by G (under the control of S),

$$S(s) \cap \Sigma_G(f(x_0, s))$$

is the set of *enabled events* that G can execute at its current state $f(x_0, s)$. In other words, G cannot execute an event that is in its current active event set if that event is not also contained in $S(s)$. We call $S(s)$ the *control action* at s and S the *control policy*. By definition of Γ, the control action will always contain all of Σ_{uc}; this ensures that S never disables an uncontrollable event. Observe that this is a case of *dynamic feedback* in the sense that the domain of S is $\mathcal{L}(G)$ and not X; thus the control action may change on subsequent visits to a state.

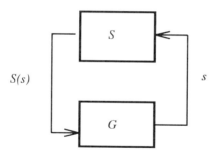

Fig. 5. The Feedback Loop

We denote the resulting closed-loop system by S/G. The *language generated* by S/G is defined recursively as follows:

1. $\epsilon \in \mathcal{L}(S/G)$
2. $[(s \in \mathcal{L}(S/G)) \land (s\sigma \in \mathcal{L}(G)) \land (\sigma \in S(s))] \Leftrightarrow [s\sigma \in \mathcal{L}(S/G)]$.

Clearly, $\mathcal{L}(S/G) \subseteq \mathcal{L}(G)$ and it is prefix-closed by definition. The *language marked* by S/G is defined as follows:

$$\mathcal{L}_m(S/G) := \mathcal{L}(S/G) \cap \mathcal{L}_m(G)$$

i.e., it consists exactly of the marked traces of G that survive under the control of S. Overall,

$$\emptyset \subseteq \mathcal{L}_m(S/G) \subseteq \overline{\mathcal{L}_m(S/G)} \subseteq \mathcal{L}(S/G) \subseteq \mathcal{L}(G) .$$

S is said to be *nonblocking* if S/G is nonblocking, i.e.,

$$\mathcal{L}(S/G) = \overline{\mathcal{L}_m(S/G)} \; ;$$

otherwise, S is said to be *blocking*. Since marked traces represent completed tasks or record the completion of some particular operation (by choice at modeling), blocking means that the controlled system cannot terminate the execution of the task at hand. As we mentioned earlier, the notions of marked traces and blocking allow to model deadlock and livelock and thus they are very useful.

4.2 A Word on Specifications: The "Legal Behavior"

The feedback supervisor S is introduced because the uncontrolled DES G is assumed to generate "illegal behavior". We will in the following describe the "legal behavior" as a subset of $\mathcal{L}(G)$, usually denoted by L_a where a stands for "admissible". In problems where blocking is of concern, then the legal behavior is given as a subset of $\mathcal{L}_m(G)$, usually denoted by L_{am}. One obtains L_a (or L_{am}) after accounting for all the *specifications* or *requirements* that are imposed on the system. These specifications are themselves described by one or more (possibly marked) languages $L_{spec,i}, i = 1, \ldots, m$. If the $L_{spec,i}$ languages are not given as subsets of $\mathcal{L}(G)$ (or $\mathcal{L}_m(G)$), then we take

$$L_a = \mathcal{L}(G) \cap (\cap_{i=1}^m L_{spec,i}) \quad \text{or} \quad L_{am} = \mathcal{L}_m(G) \cap (\cap_{i=1}^m L_{spec,i}) \; ,$$

accordingly.

For example, let us return to our telephone system example. Let $G_{user2at0}$ be an automaton like the one in Figure 2, but with 1 replaced by 2 and 2 by 1 in the subscripts of the events. The complete system model as observed at local switch 0 is then given by

$$G_{switch0} := G_{user0} \| G_{user1at0} \| G_{user2at0}$$

which has 72 states and 648 transitions. The event set of $G_{switch0}$ is denoted $\Sigma_{o,0}$ hereafter, for events *observable at 0*. Let us assume that user 0 wishes to use the terminating call screening (TCS) feature in order to screen out calls from user 2. An automaton that captures this specification is the automaton $Spec_{TCS\backslash 2}$ depicted in Figure 6. This automaton allows all the events in $\Sigma_{o,0}$ except the three events *con20*, *fwd201*, and *fwd202*. However, the language $L_{spec} = \mathcal{L}(Spec_{TCS\backslash 2})$ generated by this automaton is not a sublanguage of $\mathcal{L}(G_{switch0})$. Thus we take the product $G_{switch0} \times Spec_{TCS\backslash 2}$ to obtain an automaton that generates the legal language in the presence of this TCS specification, namely $L_a = \mathcal{L}(G_{switch0}) \cap L_{spec} \subseteq \mathcal{L}(G_{switch0})$.

Observe that depending on how a specification L_{spec} is given, we may have to take the parallel composition $L_{spec} \| \mathcal{L}(G)$ instead of the intersection $L_{spec} \cap \mathcal{L}(G)$. This will be the case if the events that appear in $\mathcal{L}(G)$ but not in L_{spec} are irrelevant to L_{spec} and thus should be allowed in L_a.

The question that we now address is: Under what conditions can a given L_a (or L_{am}) be *exactly* achieved by a supervisor S?

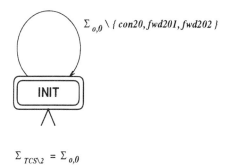

Fig. 6. Termination Call Screening of User 2 by User 0

4.3 The Basic Controllability Theorem

Theorem 10. *Consider a DES G where $\Sigma_{uc} \subseteq \Sigma$ is the set of uncontrollable events. Consider also the (marked) language $K \subseteq \mathcal{L}_m(G)$, where $K \neq \emptyset$. There exists a* nonblocking *supervisor S for G such that*

$$\mathcal{L}_m(S/G) = K$$

iff the two following conditions hold:
1. $\overline{K}\Sigma_{uc} \cap \mathcal{L}(G) \subseteq \overline{K}$
2. $K = \overline{K} \cap \mathcal{L}_m(G)$.

Condition 2 is technical in nature and has to do with the marking of the traces in K as compared with that in $\mathcal{L}_m(G)$. If it is satisfied, we say that K is $\mathcal{L}_m(G)$ – *closed*. Condition 1 is a fundamental concept of supervisory control. If it is satisfied, we say that *K is controllable with respect to $\mathcal{L}(G)$ and Σ_{uc}*. This condition is very intuitive. It means that a given language can be achieved by control iff there are no continuations by uncontrollable events of traces in the language to traces outside the language but in the uncontrolled behavior; in other words, "if you cannot prevent it, then it has to be in the controlled behavior". Observe that by definition, controllability is a property of the prefix-closure of a language. Thus K is controllable iff $\overline{K}$ is controllable.

We now present a useful corollary that explicitly states special cases of the Basic Controllability Theorem (theorem 10).

Corollary 11. *Consider a DES G where $\Sigma_{uc} \subseteq \Sigma$ is the set of uncontrollable events.*

1. *Let $K \subseteq \mathcal{L}(G)$, $K \neq \emptyset$. Then there exists S such that $\mathcal{L}(S/G) = K$ iff K is prefix-closed and controllable with respect to $\mathcal{L}(G)$ and Σ_{uc}.*
2. *Let $K \subseteq \mathcal{L}_m(G)$, $K \neq \emptyset$. Then there exists S such that $\mathcal{L}(S/G) = \overline{K}$ iff K is controllable with respect to $\mathcal{L}(G)$ and Σ_{uc}.*

We refer the reader to the literature for the proof of the Basic Controllability Theorem. It is important to note that this proof is *constructive* in the sense that if the controllability and $\mathcal{L}_m(G)$-closure conditions are satisfied, it gives us a supervisor that will achieve the required behavior. That supervisor is:

$$S(s) = \Sigma_{uc} \cup \{\sigma \in \Sigma_c : s\sigma \in \overline{K}\} \ .$$

We will discuss later what can be done when the legal language L_a (or L_{am}) is not controllable. But before that, we discuss the issue of building (finite) realizations of supervisors.

4.4 Realization of Supervisors

Let there exist a supervisor S such that $\mathcal{L}(S/G) = \overline{K}$ where $K \subseteq \mathcal{L}_m(G)$. Thus K is a controllable sublanguage of $\mathcal{L}(G)$. It suffices for now to consider $\mathcal{L}(S/G)$; if we are concerned with marked languages, then under the $\mathcal{L}_m(G)$ closure assumption, we get that $\mathcal{L}_m(S/G) = K$ and that S is nonblocking. We rule out the two trivial cases where $\overline{K} = \emptyset$ (controllable by definition but unachievable by control unless the system is "never turned on") and $\overline{K} = \mathcal{L}(G)$ (also controllable by definition but S plays no role so need not be there). Observe that the domain of S can be restricted to $\mathcal{L}(S/G)$ without loss of generality.

The issue here is that for implementation purposes, we need to build a convenient *representation* of the function S other than simply listing $S(s)$ for all $s \in \mathcal{L}(S/G)$. Given that we are using an automaton to represent the system, let us also use an automaton to represent the supervisor S. Then when we will be dealing with regular languages (here, the languages $\mathcal{L}_m(G)$, $\mathcal{L}(G)$, and K), the required representations will be *finite* and thus implementable. We will call an automaton representation of supervisor S a *realization* of S. It is important to emphasize that we are now concerned with building *off-line* a complete realization of S for all possible behaviors of the controlled system $\mathcal{L}(S/G)$. This realization will then be stored and at run time it will suffice to "read" the desired control action (i.e., the control action for the trace of events observed up to now). The issue of calculating $S(s)$ *on-line* is also of interest but it will not be discussed here.

It turns out that an easy way to build a realization of S is to build an automaton that marks the language $\overline{K}$. Let R be such an automaton, i.e., let

$$R = (X_R, \Sigma, f_R, \Sigma_R, x_{R,0}, X_R)$$

where R is trim and

$$\mathcal{L}_m(R) = \mathcal{L}(R) = \overline{K} \ .$$

Now if we "connect" R to G by the product operation, the result $R \times G$ is exactly the behavior that we desire for the closed-loop system S/G:

$$\mathcal{L}(R \times G) = \mathcal{L}(R) \cap \mathcal{L}(G) = \overline{K} \cap \mathcal{L}(G) = \overline{K} = \mathcal{L}(S/G)$$

and

$$\mathcal{L}_m(R \times G) = \mathcal{L}_m(R) \cap \mathcal{L}_m(G) = \overline{K} \cap \mathcal{L}_m(G) = \mathcal{L}(S/G) \cap \mathcal{L}_m(G) = \mathcal{L}_m(S/G) \ .$$

What this means is that the control action $S(s)$ is "encoded" in the transition structure of R, once we remove from $S(s)$ the uncontrollable events that are not in $\Sigma_G(f(x_0,s))$, the active event set of G after trace s (we have to do this because $S(s)$ was defined to enable all of Σ_{uc} and we only care about the enabled events that are actually possible in G after s). Namely,

$$S(s) \cap \Sigma_G(f(x_0,s)) = \Sigma_{R \times G}(f_{R \times G}((x_0, x_{R,0}), s)) = \Sigma_R(f_R(x_{R,0}, s))$$

where the last equality follows from the fact that $\overline{K} \subseteq \mathcal{L}(G)$. (Another way to see that $S(s) \cap \Sigma_G(f(x_0,s)) = \Sigma_R(f_R(x_{R,0}, s))$ is to consider the respective definitions of $S(s)$ and R, which are both based on $\overline{K}$.)

Of course, $R \times G$ is a composition of two automata that is defined without reference to a control mechanism à la S/G. The interpretation with our control paradigm is as follows: Let G be in state x and R in state x_R following the execution of $s \in \mathcal{L}(S/G)$. G generates an event σ that is currently "enabled". This means that this event is also present in the active event set of R at x_R. Thus R also executes the event, as a "passive observer" of G. Let x' and x'_R be the new states of G and R after the execution of σ. The set of enabled events of G at $s\sigma$ is now given by the active event set of R at x'_R.

Observe that the above interpretation does not exclude the possibility that the supervisor S *forces* G to execute a particular event, e.g., a command event. Proper modeling of G for this would include states where the active event set is composed exclusively of (controllable) command events, with the successor states having active event sets composed of the responses of the system to these commands (observed via the sensors). Then S would effectively force an event by disabling all the commands at the given state except the desired one.

Thus we have built a representation of S that, in the case of a *regular* K, will only require *finite* memory. We call the R derived by the above process the *standard realization* of S.

(The standard realization of S by automaton R raises the reverse question: If we are given automaton M and we form the product $M \times G$, can that be interpreted as controlling G by some supervisor? The answer need not be positive; it can be shown that a necessary and sufficient condition for a positive answer is that $\mathcal{L}(M)$ is controllable with respect to $\mathcal{L}(G)$ and Σ_{uc}.)

4.5 Dealing with Uncontrollability

We now address the question of how to deal with *uncontrollable* legal languages L_a and L_{am}. Suppose that a given $K \subseteq \mathcal{L}(G)$ is not controllable. We introduce the following two languages derived from K:

$K^\uparrow$: the *supremal controllable sublanguage of K*;
$K^\downarrow$: the *infimal closed and controllable superlanguage of K*.

The existence of these two languages is guaranteed by the following results.

Theorem 12. *The property of controllability.*
1. *If K_1 and K_2 are controllable, then $K_1 \cup K_2$ is controllable.*
2. *If K_1 and K_2 are controllable, then $K_1 \cap K_2$ need not be controllable.*
3. *Let us call K_1 and K_2 nonconflicting languages if $\overline{K_1 \cap K_2} = \overline{K_1} \cap \overline{K_2}$. If K_1 and K_2 are nonconflicting and both are controllable, then $K_1 \cap K_2$ is controllable.*
4. *If K_1 and K_2 are closed and controllable, then $K_1 \cap K_2$ is closed and controllable.*

In fact, parts 1 and 4 of the above theorem hold for *arbitrary* unions and intersections, which leads us to define the two classes of languages:

$$\mathcal{C}_{in}(K) := \{L \subseteq K : \overline{L}\Sigma_{uc} \cap \mathcal{L}(G) \subseteq \overline{L}\}$$
$$\mathcal{CC}_{out}(K) := \{L : (K \subseteq L \subseteq \mathcal{L}(G)) \wedge (\overline{L} = L) \wedge (\overline{L}\Sigma_{uc} \cap \mathcal{L}(G) \subseteq \overline{L})\} \ .$$

The class $\mathcal{C}_{in}(K)$ is a partially ordered set (or *poset*) that is closed under arbitrary unions. Thus it possesses a unique *supremal* element. Namely,

$$K^\uparrow := \bigcup_{L \in \mathcal{C}_{in}(K)} L$$

is a well-defined element of $\mathcal{C}_{in}(K)$. We call $K^\uparrow$ the *supremal controllable sublanguage* of K. In the worst case, $K^\uparrow = \emptyset$. If K is controllable, then $K^\uparrow = K$. Observe that $K^\uparrow$ need not be closed in general. This leads us to the following useful results.

Theorem 13. *The $\uparrow$ operation.*
1. *If K is closed, then so is $K^\uparrow$.*
2. *If $K \subseteq \mathcal{L}_m(G)$ is $\mathcal{L}_m(G)$-closed, then so is $K^\uparrow$.*
3. *In general, $\overline{K^\uparrow} \subseteq (\overline{K})^\uparrow$.*

The class $\mathcal{CC}_{out}(K)$ is a poset that is closed under arbitrary intersections (and unions). Thus it possesses a unique *infimal* element. Namely,

$$K^\downarrow := \bigcap_{L \in \mathcal{CC}_{out}(K)} L$$

is a well-defined element of $\mathcal{CC}_{out}(K)$. We call $K^\downarrow$ the *infimal closed and controllable superlanguage* of K. In the worst case, $K^\downarrow = \mathcal{L}(G)$. If K is controllable, then $K^\downarrow = \overline{K}$.

Overall, we have the following inequalities:

$$\emptyset \subseteq K^\uparrow \subseteq K \subseteq \overline{K} \subseteq K^\downarrow \subseteq \mathcal{L}(G) \ .$$

With these concepts defined and some of their properties stated, we now formulate a series of supervisory control problems and present their solutions.

4.6 Some Important Supervisory Control Problems and Their Solutions

Basic Supervisory Control Problem (BSCP): Given DES G, $\Sigma_{uc} \subseteq \Sigma$, and legal language $L_a = \overline{L_a} \subseteq \mathcal{L}(G)$, build supervisor S such that:

1. $\mathcal{L}(S/G) \subseteq L_a$
2. $\mathcal{L}(S/G)$ is *as large as possible*, i.e., for any other S' such that $\mathcal{L}(S'/G) \subseteq L_a$,
$$\mathcal{L}(S'/G) \subseteq \mathcal{L}(S/G) .$$

Solution of BSCP: Requirement 2 means that we wish the solution S to be optimal with respect to set inclusion. Such a solution is a *minimally restrictive solution* (MRS). Using the results of the preceding sections, the solution is to choose S such that
$$\mathcal{L}(S/G) = L_a^\uparrow$$
as long as $L_a^\uparrow \neq \emptyset$. If this language is regular, S can then be realized by building a DFA representation of $L_a^\uparrow$.

Nonblocking Version of BSCP (BSCP-NB): Given DES G, $\Sigma_{uc} \subseteq \Sigma$, and legal marked language $L_{am} \subseteq \mathcal{L}_m(G)$, with L_{am} assumed to be $\mathcal{L}_m(G)$-closed, build *nonblocking* supervisor S such that:

1. $\mathcal{L}_m(S/G) \subseteq L_{am}$
2. $\mathcal{L}_m(S/G)$ is *as large as possible*, i.e., for any other nonblocking S' such that $\mathcal{L}_m(S'/G) \subseteq L_{am}$,
$$\mathcal{L}(S'/G) \subseteq \mathcal{L}(S/G) .$$

Solution of BSCP-NB: Due to requirement 2, we call the desired solution S the *minimally restrictive nonblocking solution* (MRNBS). Using the results of the preceding sections, the solution is to choose S such that
$$\mathcal{L}(S/G) = \overline{L_{am}^\uparrow}$$
as long as $L_{am}^\uparrow \neq \emptyset$. Note here that since L_{am} is assumed to be $\mathcal{L}_m(G)$-closed, then $L_{am}^\uparrow$ is also $\mathcal{L}_m(G)$-closed, which guarantees that choosing S as above results in a nonblocking closed-loop system. If $L_{am}^\uparrow$ is regular, S can then be realized by building a DFA representation of $L_{am}^\uparrow$.

It is important to note that choosing
$$\mathcal{L}(S/G) = (\overline{L_{am}})^\uparrow$$
will satisfy requirement 1 but may yield a blocking S, unless $(\overline{L_{am}})^\uparrow$ and $\mathcal{L}_m(G)$ are nonconflicting (in which case $(\overline{L_{am}})^\uparrow = \overline{L_{am}^\uparrow}$).

"Dual" Version of BSCP (BSCP-D): Given DES G, $\Sigma_{uc} \subseteq \Sigma$, and *minimum required language* $L_{min} \subseteq \mathcal{L}(G)$, build supervisor S such that:

1. $\mathcal{L}(S/G) \supseteq L_{min}$

2. $\mathcal{L}(S/G)$ is *as small as possible*, i.e., for any other S' such that $\mathcal{L}(S'/G) \supseteq L_{min}$,

$$\mathcal{L}(S'/G) \supseteq \mathcal{L}(S/G) .$$

(Note that L_{min} need not be prefix-closed and could be given as a subset of $\mathcal{L}_m(G)$.)

Solution of BSCP-D: From results stated earlier, the desired optimal solution is to take S such that

$$\mathcal{L}(S/G) = L_{min}^{\downarrow}$$

which clearly meets requirements 1 and 2. Observe that if L_{min} was given as a subset of $\mathcal{L}_m(G)$, we would get

$$\mathcal{L}_m(S/G) = L_{min}^{\downarrow} \cap \mathcal{L}_m(G) \supseteq \overline{L_{min}} \cap \mathcal{L}_m(G) \supseteq L_{min} .$$

There would be no guarantee though that S be nonblocking. Indeed, the nonblocking version of this problem poses technical difficulties as the property of controllability is not preserved under intersection, unless other assumptions are made.

Finally, consider the following problem where instead of a legal language or minimum language, we are given a *desired* language (L_{des}) and a *tolerated* language (L_{tol}).

Supervisory Control Problem with Tolerance (SCPT): Given DES G, $\Sigma_{uc} \subseteq \Sigma$, desired marked language $L_{des} \subseteq \mathcal{L}_m(G)$ and tolerated legal language $L_{tol} = \overline{L_{tol}} \subseteq \mathcal{L}(G)$, where $\overline{L_{des}} \subseteq L_{tol}$, build supervisor S such that:

1. $\mathcal{L}(S/G) \subseteq L_{tol}$
   (this means that we can never exceed the tolerated language);
2. For all closed and controllable $K \subseteq L_{tol}$,

$$K \cap L_{des} \subseteq \mathcal{L}(S/G) \cap L_{des}$$

   (this means that we want S to achieve as much of the desired language as possible);
3. For all closed and controllable $K \subseteq L_{tol}$,

$$K \cap L_{des} = \mathcal{L}(S/G) \cap L_{des} \Rightarrow \mathcal{L}(S/G) \subseteq K$$

   (this means that 2 is achieved with the smallest possible solution).

Solution of SCPT: A little thought shows that the solution is obtained by taking S such that

$$\mathcal{L}(S/G) = (L_{tol}^{\uparrow} \cap L_{des})^{\downarrow} .$$

This S is not guaranteed to be nonblocking. The nonblocking version of this problem may not have an optimal solution and poses technical difficulties.

4.7 Comments on The ↑ and ↓ Operations

The solutions of the problems discussed in the preceding section motivate the detailed study of the ↑ and ↓ operations. We limit our discussion to the statement of some important results.

The most important result about these two operations is that they both preserve *regularity* of languages in the sense that if G is finite state and K is regular, then $K^\uparrow$ and $K^\downarrow$ are regular too. This is crucial from a synthesis viewpoint, since it means that if the DES is finite state and the (marked) legal language is regular, then the supervisors that solve the previously-stated supervisory control problems can be realized by *finite* automata. Let automaton H mark the given marked legal language L_{am}. The supervisor synthesis issues that need to be addressed are: Given G and H, calculate automata $H^\uparrow$ and $H^\downarrow$ such that

$$\mathcal{L}_m(H^\uparrow) = L_{am}^\uparrow \text{ and } \mathcal{L}(H^\downarrow) = L_{am}^\downarrow .$$

It turns out that the answer to this question for the ↓ operation is simply due to the fact that given $K \subseteq \mathcal{L}(G)$, $K^\downarrow$ is characterized by the following formula on languages:

$$K^\downarrow = \overline{K} \Sigma_{uc}^* \cap \mathcal{L}(G) .$$

This formula immediately proves that the ↓ operation preserves regularity of languages as the class $\mathcal{R}$ is closed under concatenation and intersection. Moreover, this formula suggests a simple algorithm for calculating $H^\downarrow$ from H and G: (i) add self-loops to each state of H for those uncontrollable events not in the active event set of the state (to implement the concatenation operation), and (ii) form the product of the modified H with G (to implement the language intersection). If the automaton that generates $\overline{K}$ has m states and G has n states, then this calculation is of $O(nm|\Sigma|)$ in the worst case.

The case of the ↑ is not so straightforward. First, the ↑ operation cannot be represented by a formula on languages unless the language it is applied to is prefix-closed, e.g., as in BSCP. Moreover, even in that case, the "direct" implementation of that formula by automata is not the most computationally efficient way of calculating $H^\uparrow$. The most efficient way of calculating $H^\uparrow$ is via an iterative algorithm that starts from a suitably modified version of automaton H that still marks the given language K, but that is a *subautomaton* of G in the sense that it is a *subgraph* of automaton G. (Observe that the original H need not be a subautomaton of G even if it marks a sublanguage of $\mathcal{L}_m(G)$.) Then the resulting automaton is pruned in an iterative manner until convergence is achieved, after a *finite* number of iterations. It is precisely the fact that convergence occurs in finite steps that proves that the ↑ operation preserves regularity. The work to perform at each step of the iterative procedure consists of (i) deleting the states of the automaton that violate the controllability condition, and (ii) deleting the states of the automaton that are not accessible or coaccessible. This iterative procedure is the consequence of the fact that $K^\uparrow$ can be characterized as the largest fixed point of a certain operator on languages. This suggests to compute the largest fixed point by iterative applications of the operator. Indeed, each iteration of the above-described algorithm corresponds to the implementation on automata of the application of the operator on languages.

It is important to emphasize that more than one iterations of the above algorithm may be necessary only when K is not prefix-closed. In this case, the worst-case computational complexity of obtaining $H^\uparrow$ from H and G is $O((mn)^2|\Sigma|)$ while it reduces to $O(mn|\Sigma|)$ when K is prefix-closed. Observe that it has been shown that $||K^\uparrow|| \leq ||K|| \cdot ||\mathcal{L}(G)|| + 1$ and that the bound is tight.

4.8 Modular Supervisory Control

By modular control, we refer to the situation where the control action of the supervisor S is given by some combination of the control actions of two or more supervisors. For simplicity, we consider the case of two supervisors and discuss their conjunction.

Given S_1 and S_2 each defined for DES G, define the *modular* supervisor denoted by $S_{1\wedge 2} : \mathcal{L}(G) \to \Gamma$ and corresponding to the *conjunction* of the two individual supervisors:

$$S_{1\wedge 2}(s) := S_1(s) \cap S_2(s) .$$

Then it is straightforward to verify that

$$\mathcal{L}(S_{1\wedge 2}/G) = \mathcal{L}(S_1/G) \cap \mathcal{L}(S_2/G)$$
$$\mathcal{L}_m(S_{1\wedge 2}/G) = \mathcal{L}_m(S_1/G) \cap \mathcal{L}_m(S_2/G) .$$

Given standard realizations R_1 and R_2 of S_1 and S_2, respectively, then the standard realization of $S_{1\wedge 2}$ could be obtained by building $R = R_1 \times R_2$. But the point here is precisely *not* to build this realization, but rather to use the existing R_1 and R_2 and realize the control action $S_{1\wedge 2}(s)$ by taking the *intersection of the active event sets of R_1 and R_2* at their respective states after the execution of s. We call this the *modular realization* of modular supervisor $S_{1\wedge 2}$. This modular approach saves on the size of the realization of $S_{1\wedge 2}$. If R_1 has n_1 states and R_2 has n_2 states, then we need only to store a total of $n_1 + n_2$ states for this modular realization instead of possibly as many as $n_1 n_2$ states if the above R were built. Note that we can *interpret* the supervision of G by $S_{1\wedge 2}$ as the product $R_1 \times R_2 \times G$.

It is a similar complexity argument that motivates the *synthesis* of a supervisor in modular form. If the admissible language L_a for BSCP is given as (or can be decomposed as) the intersection of two closed languages

$$L_a = L_{a1} \cap L_{a2}$$

then we would like to synthesize S_i for $L_{ai}^\uparrow$, $i = 1, 2$, and use these two supervisors in conjunction instead of doing the full calculation $L_a^\uparrow$. Using this modular approach, the total computational complexity for supervisor synthesis is reduced from $O(n_1 n_2 m)$ to $O(\max(n_1, n_2)m)$. This modular approach does work because in the case of closed languages,

$$(K_1 \cap K_2)^\uparrow = K_1^\uparrow \cap K_2^\uparrow .$$

4 Supervisory Control of DES   249

This discussion is formalized in the following modular version of BCSP.
BSCP-MOD: Basic Supervisory Control Problem: Modular Version
Given DES G, $\Sigma_{uc} \subseteq \Sigma$, and legal language $L_a = L_{a1} \cap L_{a2}$ where $L_{ai} = \overline{L_{ai}} \subseteq \mathcal{L}(G)$ for $i = 1, 2$, build a *modular* supervisor S_{mod} such that:

1. $\mathcal{L}(S_{mod}/G) \subseteq L_a$
2. $\mathcal{L}(S_{mod}/G)$ is optimal with respect to set inclusion.

Solution of BSCP-MOD: From the above discussion, we build the standard realizations R_i of S_i such that

$$\mathcal{L}(S_i/G) = L_{ai}^{\uparrow}$$

for $i = 1, 2$, and then take S_{mod} to be the modular supervisor $S_{1 \wedge 2}$. Then we get that

$$\mathcal{L}(S_{1 \wedge 2}/G) = L_{a1}^{\uparrow} \cap L_{a2}^{\uparrow} = (L_{a1} \cap L_{a2})^{\uparrow} = L_a^{\uparrow}$$

which is the desired optimal solution.

It is unfortunate that this modular approach cannot be extended to the nonblocking version BSCP, BSCP-NB. The problem is that the conjunction of two nonblocking supervisors need not be a nonblocking supervisor. Consider the following result.

Theorem 14. *Let S_i, $i = 1, 2$, be nonblocking supervisors for G. Then $S_{1 \wedge 2}$ is nonblocking iff $\mathcal{L}_m(S_1/G)$ and $\mathcal{L}_m(S_2/G)$ are nonconflicting languages, i.e,*

$$\overline{\mathcal{L}_m(S_1/G) \cap \mathcal{L}_m(S_2/G)} = \overline{\mathcal{L}_m(S_1/G)} \cap \overline{\mathcal{L}_m(S_2/G)} \ .$$

This result has the following implication. If we consider the modular version of BSCP-NB where

$$L_{am} = L_{am1} \cap L_{am2}$$

and where each $L_{ami} \subseteq \mathcal{L}_m(G)$ and both are $\mathcal{L}_m(G)$-closed (which implies that L_{am} itself is $\mathcal{L}_m(G)$-closed), then by synthesizing S_i such that

$$\mathcal{L}(S_i/G) = \overline{L_{ami}^{\uparrow}}$$

for $i = 1, 2$, and then by forming the modular supervisor $S_{1 \wedge 2}$, we get

$$\mathcal{L}(S_{1 \wedge 2}/G) = \overline{L_{am1}^{\uparrow}} \cap \overline{L_{am2}^{\uparrow}}$$
$$\mathcal{L}_m(S_{1 \wedge 2}/G) = \overline{L_{am1}^{\uparrow}} \cap \overline{L_{am2}^{\uparrow}} \cap \mathcal{L}_m(G)$$
$$= L_{am1}^{\uparrow} \cap L_{am2}^{\uparrow}$$
$$\supseteq (L_{am1} \cap L_{am2})^{\uparrow} = L_{am}^{\uparrow}$$

which means that the *modular supervisor could be blocking*, even though it is legal in the sense that $\mathcal{L}_m(S_{1 \wedge 2}/G) \subseteq L_{am}$.

By the above result, BSCP-NB has a nonblocking modular solution iff $L_{am1}^{\uparrow}$ and $L_{am2}^{\uparrow}$ are nonconflicting. The problem is that this condition cannot be verified *before* doing the $\uparrow$ calculations. Moreover, to verify this condition, we have to examine together both $L_{am1}^\uparrow$ and $L_{am2}^\uparrow$. In contrast, a "monolithic" or global (as opposed to modular) approach would require us to form the intersection $L_{am1} \cap L_{am2}$ and then do the $\uparrow$ operation on the result, a calculation that has essentially the same (worst case) computational complexity as the verification of the above nonconflicting condition. Moreover, the monolithic approach guarantees a nonblocking (global) supervisor, namely, $\mathcal{L}(S/G) = \overline{L_{am}^\uparrow}$, the MRNBS of BSCP-NB. Yet, if the modular solution is indeed nonblocking, then, as was mentioned earlier, it is still advantageous from an implementation viewpoint.

The conclusion of this discussion is that the issue of blocking is intrinsically a *global* one; it cannot in general be dealt with in a modular manner.

4.9 Modeling and Design Issues

The previous sections demonstrate that the supervisory control problems that we have presented can be solved *automatically* by algorithms of quadratic computational complexity in the worst case. The end result is the standard realization of the supervisor S that solves the supervisory control problem of interest. This is precisely one of the main goals of supervisory control theory: *the development of formal methods for synthesizing control laws that are guaranteed to satisfy a given set of specifications on event orderings*. In this sense, the theory has been very successful; it allows the treatment of large systems that cannot be solved simply by "good engineering intuition".

The two key ingredients for applying these synthesis results are the availability of automata that model the given system and the (marked) legal language. The system model G is usually obtained by the parallel composition of the individual models of the system components. Obtaining these individual models is in some sense the "usual" modeling task that has to be performed by a control engineer (for any class of model-based control of dynamic systems). The typical difficulties encountered in DES applications are: (i) choosing the right level of abstraction for the model, given the specifications that are imposed on the system (i.e., what events should Σ contain?); (ii) choosing the set of common events between system components that will appropriately capture the coupling between these components, when the parallel composition is performed (recall the role played by common events in $||$); and (iii) dealing with the resulting computational complexity of the complete model, which grows exponentially in the number of components (in the worst case; common events help considerably to contain that growth).

The second key ingredient in supervisory control theory is the automaton model H of the legal language. To a large extent, this issue is one of *design* rather than *synthesis* in the sense that the specifications on G are typically given in natural language with respect to $\mathcal{L}(G)$ or G and not immediately as automata. Thus one has to build the required automata and make sure that they capture all the specifications. There is no "magic recipe" for building these automata. Several *design techniques* that can handle certain types of specifications have been presented and used in the literature. The fact remains that, in

4 Supervisory Control of DES

our experience, building H has proven as difficult, if not more difficult, than building G.

As an illustration of this design process, we return to the telephone system example. Consider the automaton depicted in Figure 7 and denoted by $Spec_{CF\to 1}$. This automaton captures the specification for the call forwarding feature, when user 0 wishes to forward all incoming calls to user 1. This specification adopts the following strategy: once requests to call user 0 have been made, "remove" all possible outcomes except for the events that forward calls to user 1. Notice how this "removal" is implemented by the use of appropriate self-loops at each state. The corresponding legal language is the language generated by $G_{switch0}||Spec_{CF\to 1}$.

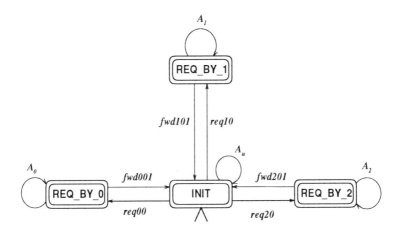

$A_u = \Sigma_{o,o} \setminus \{ req00, req10, req20 \} \setminus \{ fwd001, fwd002, fwd101, fwd102, fwd201, fwd202\}$
$A_0 = \Sigma_{o,o} \setminus \{ nocon00, fwd001, fwd002\}$
$A_1 = \Sigma_{o,o} \setminus \{ con10, nocon10, fwd101, fwd102\}$
$A_2 = \Sigma_{o,o} \setminus \{ con20, nocon20, fwd201, fwd202\}$

Fig. 7. Call Forwarding by User 0 to User 1

4.10 Some Concluding Remarks

Our discussion of Supervisory Control Theory in this section has only highlighted a *small subset* of the results that have been developed in this field of research. The majority of the results that we have stated are due to Ramadge and Wonham and can be found in [62], [78], [79], [60], or [77]. The results on the ↓ operation and SCPT can be found in [46] and [47], respectively. For further details, the reader is referred to the above references or to the two survey papers [60, 71] and the references therein, as well as to the recent book [45]. In particular, the two survey papers [60, 71] discuss the research on the topics of control under partial observation, decentralized control, hierarchical control, and the extension of supervisory control theory to timed discrete event models. We also mention that some recent papers have addressed on-line supervisory control with limited lookahead policies [18, 34] and the optimal control of discrete event systems (at the logical level of abstraction) [69].

We wish to make a few observations regarding control under partial observation. This is the situation where S does not "see" all the events of G but only a proper subset of *observable* events. Events in the system model G would not be observable by S due to the absence of sensors or due to limitations on communication. In this case, the control action after a trace s has to be based not on s alone, but on all the traces in $\mathcal{L}(G)$ that have the same observable projection as s. The generalization of the Basic Controllability Theorem to the case of partial event observation requires the introduction of a new requirement on the desired language, called the *observability* condition, which first appeared in [48, 19].

Various academic examples of the application of supervisory control theory in areas such as manufacturing, network protocols, database protocols, heating, ventilation and air conditioning units, telephony, and so forth, are available in the literature (again, refer to the survey papers [60, 71]). Actual implementations of supervisors synthesized by the techniques of supervisory control theory in the area of manufacturing are described in [3] (control implemented by PC/workstation) and [8] (control implemented by programmable logic controllers).

5 The Max-Plus Algebra Approach to DES

This section starts with the modelling of an introductory example in subsection 5.1, and subsequently this modelling is given a more abstract setting in subsections 5.2 and 5.3. In the latter subsection the intercity network example, as discussed in section 2, is revisited. Subsection 5.4 deals with the so-called γ-transform, which is the max-plus algebra equivalent of the z-transform in the conventional theory of linear difference equations. In subsection 5.5 various extensions are briefly considered.

5.1 The Stage for the Max-Plus Approach

In a metropolitan area there are two railway stations, S_1 and S_2, which are interconnected by a railway system as indicated in Figure 8. This railway system consists of an innercircle and of two outer circles. The trains on these outer circles deliver and pick up passengers in the suburbs. The stations in the suburbs have not been drawn since they do not play any role in the model to be formulated.

Suppose there are four trains (two at each station) and they leave the stations at time 0, one along each track. They reach the other (or the same) station after

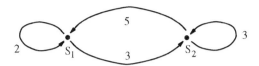

Fig. 8. The two stations example

a certain time which is indicated in the figure. The arriving trains at a station have to wait for each other such as to allow the passengers to change trains.

Figure 8 can easily be redrawn as an event graph. The two stations are transitions and in the four railway tracks one can put a place. If a train is running along a track, one puts a token in the place corresponding to this track.

Suppose that there is no time table and that the trains leave directly after the change over of the passengers at the stations and that the time needed for change overs has been incorporated in the travelling time. This 'travelling time' was called 'holding time' in the theory of Petri nets (see subsection 3.4). If this process of departing and arriving trains is continued, the departure time $x_i(k+1)$ for the $k+1$-st departure at station S_i satisfies

$$\begin{aligned} x_1(k+1) &= \max(x_1(k)+2,\ x_2(k)+5), \\ x_2(k+1) &= \max(x_1(k)+3,\ x_2(k)+3), \end{aligned} \quad (3)$$

for $k = 0, 1, 2, \ldots$.

Remark. Please note that in (1) we used $\tau_i(\chi)$ as a state variable when the i-th transition (here: station) became active for the χ-th time and that we now use $x_i(k)$ for essentially the same quantity (with χ replaced by k). The interpretation of the current $x_i(k)$ is therefore different from the one used in (2). The reason is that it is customary in the literature to write the evolution equations in dater form as (3) in spite of the fact that the notations in (1) and (2) seem more natural.

With $x_1 = 0$, $x_2 = 0$, the evolution of equations (3) becomes

$$\begin{pmatrix}0\\0\end{pmatrix} \to \begin{pmatrix}5\\3\end{pmatrix} \to \begin{pmatrix}8\\8\end{pmatrix} \to \begin{pmatrix}13\\11\end{pmatrix} \to \begin{pmatrix}16\\16\end{pmatrix} \to \cdots$$
$$x(0) \quad x(1) \quad x(2) \quad x(3) \quad x(4)$$

This pattern of departure times shows a periodic solution superimposed on a linear drift, the 'period' equals 2 and the average time between two subsequent departures is 4. From a time table point of view (time tables must be as 'regular' as possible), it is better to start with the initial departures $x_1 = 1$, $x_2 = 0$, since then the evolution becomes

$$\begin{pmatrix}1\\0\end{pmatrix} \to \begin{pmatrix}5\\4\end{pmatrix} \to \begin{pmatrix}9\\8\end{pmatrix} \to \begin{pmatrix}13\\12\end{pmatrix} \to \cdots$$
$$x(0) \quad x(1) \quad x(2) \quad x(3)$$

where the interdeparture time is now exactly 4 at each station and thus the departure times are very regular (they have 'period' 1). By trial and error it turns out that, whatever the initial condition, after possibly a short transient period of time, a periodic behavior of either period 1 or 2 is obtained with (average) interdeparture times 4. A solution with an (average) departure time smaller than 4 is not possible, since for a train to go around in the innercircle costs $3 + 5 = 8$ time units. There are two trains on the innercircle and therefore the (average) interdeparture time is limited from below by $8/2 = 4$.

With the above sketched railway system and trains it is not possible to design a time table with interdeparture times smaller than 4. If one wants a faster time table, one must change the problem. To this end, let us add a train on the innercircle, such that three trains will run along this circle all the time. Suppose that initially this extra train is situated at station S_1. The equations for the departure times now become

$$x_1(k+1) = \max(x_1(k) + 2,\ x_2(k) + 5), \qquad (4)$$
$$x_2(k+1) = \max(x_1(k-1) + 3,\ x_2(k) + 3), \qquad (5)$$

which can be rewritten as a set of first order equations as

$$\begin{aligned}x_1(k+1) &= \max(x_1(k) + 2,\ x_2(k) + 5),\\ x_2(k+1) &= \max(x_3(k) + 3,\ x_2(k) + 3),\\ x_3(k+1) &= x_1(k).\end{aligned} \qquad (6)$$

With the initial condition $x_1 = 0,\ x_2 = 0,\ x_3 = 0$ the evolution of the latter set of equations becomes

$$\begin{pmatrix}0\\0\\0\end{pmatrix} \to \begin{pmatrix}5\\3\\0\end{pmatrix} \to \begin{pmatrix}8\\6\\5\end{pmatrix} \to \begin{pmatrix}11\\9\\8\end{pmatrix} \to \begin{pmatrix}14\\12\\11\end{pmatrix} \to \cdots$$
$$x(0) \qquad x(1) \qquad x(2) \qquad x(3) \qquad x(4)$$

which shows, after a transient part, a regular behavior of 'period' 1 with interdeparture times 3. This interdeparture time is caused by the outerloop at station S_2; on this loop there is one train which needs 3 time units to travel around. The innerloop is not the bottleneck anymore: this innerloop itself would lead to a lower limit of $8/3$ (travelling time of the loop divided by the number of trains on this loop). In order to lower the interdeparture times even more, one should add the next extra train to the outerloop of S_2. If we do so, the equations for the departure times become

$$\begin{aligned}x_1(k+1) &= \max(x_1(k) + 2,\ x_2(k) + 5),\\ x_2(k+1) &= \max(x_3(k) + 3,\ x_2(k-1) + 3),\\ x_3(k+1) &= x_1(k),\end{aligned} \qquad (7)$$

which can be rewritten as a set of first order equations as

$$\begin{aligned}x_1(k+1) &= \max(x_1(k) + 2,\ x_2(k) + 5),\\ x_2(k+1) &= \max(x_3(k) + 3,\ x_4(k) + 3),\\ x_3(k+1) &= x_1(k),\\ x_4(k+1) &= x_2(k).\end{aligned} \qquad (8)$$

If we start again with zero initial conditions, the solution becomes

$$\cdots \to \begin{pmatrix}10\\8\\8\\3\end{pmatrix} \to \begin{pmatrix}13\\11\\10\\8\end{pmatrix} \to \begin{pmatrix}16\\13\\13\\11\end{pmatrix} \to \begin{pmatrix}18\\16\\16\\13\end{pmatrix} \to \begin{pmatrix}21\\19\\18\\16\end{pmatrix} \to \cdots \qquad (9)$$
$$x(3) \quad\quad x(4) \quad\quad x(5) \quad\quad x(6) \quad\quad x(7)$$

This solution has a 'period' 3 ($x_i(k+3) = x_i(k) + 8$, $k \geq 4$) and the average interdeparture time is 8/3, which is caused by the innercircle. Another solution results, with the same interdeparture time but with 'period' 1, if one starts with the initial condition $x_1(0) = 5$, $x_2(0) = 8/3$, $x_3(0) = 7/3$, $x_4 = 0$. One then has $x_i(k+1) = x_i(k) + 8/3$, $i = 1, 2, 3, 4$, and $k = 0, 1, \ldots$.

5.2 Formalization

The basic form of the systems we will study is

$$\begin{aligned}x_i(k+1) &= \max(a_{i1} + x_1(k), a_{i2} + x_2(k), \ldots, a_{in} + x_n(k))\\ &= \max_j(a_{ij} + x_j(k)), \quad i = 1, \ldots, n \ .\end{aligned} \qquad (10)$$

It is common practice to change the notation somewhat. Addition + will be written as $\otimes$ and max will be written as $\oplus$. This change of notation makes the resemblance with conventional linear difference systems visible:

$$x_i(k+1) = \bigoplus_j (a_{ij} \otimes x_j(k)), \quad i = 1, \ldots, n \ , \qquad (11)$$

which in vector notation will be written as

$$x(k+1) = A \otimes x(k). \qquad (12)$$

Of the latter equation one speaks as a linear (difference) equation in the max-plus algebra, this in clear analogy with linear difference equations in the conventional, 'plus-times', algebra. If it is clear from the context that the underlying algebra is the max-plus one, one even writes $x(k+1) = Ax(k)$ for (12). If the initial condition for (12) is $x(0) = x_0$, then

$$x(1) = A \otimes x_0 \ ,$$

$$x(2) = A \otimes x(1) = A \otimes (A \otimes x_0) = (A \otimes A) \otimes x_0 = A^2 \otimes x_0 \ .$$

It can be shown that indeed $A \otimes (A \otimes x_0) = (A \otimes A) \otimes x_0$. For the example given above it is easy to check this by hand. Instead of $A \otimes A$ we simply write A^2. We get, for the general case,

$$x(k) = \underbrace{(A \otimes A \otimes \cdots \otimes A)}_{k \text{ times}} \otimes x_0 = A^k \otimes x_0 \ .$$

The matrices A^2, A^3,..., can be calculated directly. Let us consider the A-matrix of (3),

$$A = \begin{pmatrix} 2 & 5 \\ 3 & 3 \end{pmatrix}$$

then

$$A^2 = \begin{pmatrix} \max(2+2, 5+3) & \max(2+5, 5+3) \\ \max(3+2, 3+3) & \max(3+5, 3+3) \end{pmatrix} = \begin{pmatrix} 8 & 8 \\ 6 & 8 \end{pmatrix}.$$

In general

$$(A^2)_{ij} = \bigoplus_l a_{il} \otimes a_{lj} = \max_l(a_{il} + a_{lj}) . \tag{13}$$

In terms of the railway example, the quantity $(A^2)_{ij}$ can be interpreted as the maximum (with respect to l) of all connections from station S_j via station S_l to station S_i. One speaks of paths of length two between the stations S_j and S_i. In graph-theory terminology, the stations are called nodes and the tracks between stations are called arcs. More generally, $(A^k)_{ij}$ denotes the maximum of all paths of length k, starting at node j and ending at node i.

In many networks such as a railway net there will not be an arc from each node to each other node. If there is no arc from node S_j to node S_i then the behaviour of node S_i is not directly influenced by that of node S_j. In such a situation it is useful to consider the element a_{ij} to be equal to $-\infty$. In (10) a term $-\infty + x_j(k)$ does not influence $x_i(k+1)$ as long as $x_j(k)$ is finite. The number $-\infty$ will occur frequently in the sequel and it will be indicated by ε.

Linear systems in the max-plus algebra with inputs and outputs are given by

$$\left.\begin{array}{l} x(k+1) = Ax(k) \oplus Bu(k) , \\ y(k) = Cx(k) , \end{array}\right\} \tag{14}$$

which is short-hand notation for

$$x_i(k+1) = \max(a_{i1} + x_1(k), \ldots, a_{in} + x_n(k),$$
$$b_{i1} + u_1(k), \ldots, b_{im} + u_m(k)), \; i = 1, \ldots, n ;$$
$$y_i(k) = \max(c_{i1} + x_1(k), \ldots, c_{in} + x_n(k)), \; i = 1, \ldots, p .$$

A seeming generalization of (12) is

$$x(k+1) = A_0 x(k+1) \oplus A_1 x(k) \oplus \cdots \oplus A_{l+1} x(k-l), \tag{15}$$

which is implicit in $x(k+1)$ and which has extra delays. By repeated substitution of the whole right-hand side of (15) for the term $x(k+1)$ in this right-hand side, one gets

$$x(k+1) = A_0^* A_1 x(k) \oplus \cdots \oplus A_0^* A_{l+1} x(k-l) \tag{16}$$

where

$$A_0^* := I \oplus A_0 \oplus A_0^2 \oplus A_0^3 \oplus \cdots$$

The notation I refers to the identity matrix in the max-plus algebra: it has zeros on the main diagonal and ε's elsewhere. Equation (16) only makes sense if A_0^* is well defined (its elements are finite or ε). This is for instance the case if the precedence graph of A, see subsection 5.3, does not contain circuits, because then $A^k = \varepsilon$ for $k \geq n$. Equation (16) can be rewritten as a first order difference equation by augmenting the state space. This is a standard trick in system theory and has already been used in subsection 5.1.

5.3 Periodic Behaviour

Given a square matrix A, we consider the problem of existence of eigenvalues and eigenvectors in the max-plus algebra, that is, the existence of λ and $v \neq \varepsilon$ such that:

$$Av = \lambda v \ . \tag{17}$$

This equation has to be interpreted in the max-plus algebra sense; the expression λv means that one adds λ to each component of v. We already have seen examples of eigenvalues and eigenvectors in subsection 5.1; v corresponds to an initial state resulting in a solution with 'period' 1 and λ is the interdeparture time.

Before formulating a theorem about eigenvalues, some graph theory must be recapitulated. In the following definition the starting point is a square matrix, the entries of which may again assume the 'value' ε.

Definition 15 Precedence graph. The precedence graph of an $n \times n$ matrix A is a weighted digraph with n nodes and an arc (j, i) if $a_{ij} \neq \varepsilon$, in which case the weight of this arc receives the numerical value of a_{ij}. The precedence graph is denoted $\mathcal{G}(A)$.

It is not difficult to see that any weighted digraph $\mathcal{G} := (\mathcal{V}, \mathcal{E})$, with $\mathcal{V}$ being the set of nodes and $\mathcal{E}$ being the set of arcs, is the precedence graph of an appropriately defined square matrix. The weight a_{ij} of the arc from node j to node i is defined as the ij-th entry of a matrix A. If an arc does not exist, the corresponding entry of A becomes ε. The matrix A thus defined has $\mathcal{G}$ as its precedence graph.

As we have seen before, the element (i, j) of $A^k = A \otimes \cdots \otimes A$, considered within the max-plus algebra denotes the maximum weight with respect to all paths of length k which go from node j to node i. If no such path exists, then $(A^k)_{ij} = \varepsilon$. The weight of a path ρ is denoted $|\rho|_{\text{w}}$ and its length is denoted $|\rho|_{\text{l}}$.

Definition 16. The *mean weight of a path* is defined as the sum of the weights of the individual arcs of this path, divided by the length of this path. If the path is denoted by ρ, then the mean weight equals $|\rho|_{\text{w}}/|\rho|_{\text{l}}$. If such a path is a circuit one talks about the mean weight of the circuit, or simply the *cycle mean*.

We are interested in the maximum of these cycle means, where the maximum is taken over all circuits in the graph. This number will be called the *maximum cycle mean*. If the cycle mean of a circuit equals the maximum cycle mean, then the circuit is called critical. The graph consisting of all critical circuits (if there happen to be more than one) is called the *critical graph* and denoted by $\mathcal{G}^c$. In the following theorem the notion 'strongly connected' (di-)graph is used. A graph is called strongly connected if there exists a path from any node to any other node. The matrix corresponding to a strongly connected graph is called *irreducible*.

Theorem 17. *We are given a square matrix A. If $\mathcal{G}(A)$ is strongly connected, then there exists one and only one eigenvalue and at least one eigenvector. The eigenvalue is equal to the maximum cycle mean of the graph:*

$$\lambda = \max_{\zeta} \frac{|\zeta|_w}{|\zeta|_l},$$

where ζ ranges over the set of circuits of $\mathcal{G}(A)$.

Definition 18 Cyclicity of a graph. Given a strongly connected graph, its cyclicity equals the greatest common divisor of the lengths of all its circuits. The cyclicity of an arbitrary graph (which may consist of several strongly connected subgraphs) equals the least common multiple of the cyclicities of all its maximal strongly connected subgraphs.

Definition 19 Cyclicity of a matrix. A matrix A is said to be cyclic if there exist scalars M, λ and d such that $\forall m \geq M$, $A^{m+d} = \lambda^d A^m$. The least such d is called the cyclicity of A. The quantity λ equals the maximum cycle mean of A.

The expression $A^{m+d} = \lambda^d A^m$ in the definition above must be interpreted in the max-plus algebra sense of course. Thus λ^d in the max-plus algebra means $d\lambda$ in the conventional algebra and $\lambda^d A^m$ refers to the addition of λ^d to each element of A^m.

Theorem 20. *Any irreducible matrix is cyclic. The cyclicity of the irreducible matrix A equals the cyclicity of $\mathcal{G}^c(A)$, being the critical graph corresponding to matrix A.*

Example 4. Consider the A-matrix of (8);

$$A = \begin{pmatrix} 2 & 5 & \varepsilon & \varepsilon \\ \varepsilon & \varepsilon & 3 & 3 \\ 0 & \varepsilon & \varepsilon & \varepsilon \\ \varepsilon & 0 & \varepsilon & \varepsilon \end{pmatrix}.$$

The corresponding precedence graph has three circuits, viz. from node 1 to node 1 with cycle mean $2/1 = 2$; from node 1 to 3 to 2 to 1 with cycle mean $(0+3+5)/3 = 8/3$; from node 2 to 4 to 2 with cycle mean $(0+3)/2 = 3/2$. (It is tacitly assumed here that node i corresponds to x_i.) The maximum cycle mean equals $8/3$. There is only one critical circuit. The cyclicity of the critical graph (which equals the critical circuit) equals 3. The quantities of definition 19 are $M = 5$, $\lambda = 8/3$ and $d = 3$;

$$A^8 = \begin{pmatrix} 20 & 23 & 24 & 24 \\ 19 & 20 & 21 & 21 \\ 18 & 21 & 20 & 20 \\ 15 & 18 & 19 & 19 \end{pmatrix} = \left(\frac{8}{3}\right)^3 A^5 = 8 \begin{pmatrix} 12 & 15 & 16 & 16 \\ 11 & 12 & 13 & 13 \\ 10 & 13 & 12 & 12 \\ 7 & 10 & 11 & 11 \end{pmatrix}.$$

□

Example 5 Intercity railway network (continued). The problem description has been given in Example 2. With the theory just given, answers to the various questions posed can be given now. The model which follows directly from the problem description is of the form (15), with the vector x being 53-dimensional. (It has been described explicitly in [6].) After rewriting it in the form of (12), a model with state vector of dimension 79 resulted. The matrix A will not be given here: it is made up of the different travelling times and standard change over times as set by the railway company. The answer to question 4 turns out to be $\lambda = 27\frac{1}{12}$. The critical circuit (this is not to be refused with a line of the net) turns out to be from Venlo to Eindhoven to Utrecht to Amsterdam to Zandvoort aan Zee and back. Hence if one wants a faster time table, one should add extra trains to a line (or lines) which forms part of this critical circuit (i.e. the lines numbered 20 and 50). If one would add $5\frac{2}{3}$ minutes to all change over times, to be incorporated into the model by adjusting the A matrix, then it turns out that eigenvalue $\lambda = 30$ minutes. Of course this eigenvalue is monotone with respect to the duration of the change over times. The critical circuit in this case with maximum change over time which allows a half hour time table is Venlo to Eindhoven to Utrecht to Amsterdam to Haarlem to Den Haag HS to Breda to Eindhoven to Venlo. If one would consider a very unfriendly way of handling passengers by not waiting for them, i.e. trains would simply stop at stations, deliver passengers and pick up passengers which happen to be there and depart again, then we can design a faster time table. It turns out that in this case $\lambda = 26\frac{2}{11}$. The critical circuit is now Venlo to Eindhoven to Breda to Den Haag CS and back.

The question of the design of an 'optimal line structure' cannot be answered directly by an application of the theory. One can of course compare different designs by calculating their critical circuits and minimum interdeparture times.

□

5.4 The γ-Transform

Conventional linear systems with inputs and outputs are of the form (14), though (14) itself has the max-plus algebra interpretation. This equation, now considered in the conventional way, is a representation of a linear system in the time domain. Its representation in the z-domain equals

$$Y(z) = C(zI - A)^{-1}BU(z) ,$$

where $Y(z), U(z)$ are defined by

$$Y(z) = \sum_{i=0}^{\infty} y(i)z^{-i}, \quad U(z) = \sum_{i=0}^{\infty} u(i)z^{-i} ,$$

where it is tacitly assumed that the system was at rest for $t \leq 0$ and where I refers to the unit matrix in the conventional algebra. The matrix $H(z) \stackrel{\text{def}}{=} C(zI - A)^{-1}B$ is called the transfer matrix of the system.

In the max-plus algebra context, the z-transform also exists, but here it is customary to refer to it as the γ-transform where γ operates as z^{-1} and is assumed to be real-valued. For instance, the γ-transform of u is defined as

$$U(\gamma) = \bigoplus_{i=0}^{\infty} u(i) \otimes \gamma^i , \tag{18}$$

and $Y(\gamma)$ and $X(\gamma)$ are defined likewise. Multiplication of (14) by γ^k yields

$$\left.\begin{array}{l}\gamma^{-1} x(k+1) \gamma^{k+1} = A \otimes x(k) \gamma^k \oplus B \otimes u(k) \gamma^k, \\ y(k) \gamma^k = C \otimes x(k) \gamma^k .\end{array}\right\} \tag{19}$$

If these equations are summed with respect to $k = 0, \ldots$, and if we add $\gamma^{-1} x_0$ to both sides of the first equation thus obtained, then we obtain

$$\left.\begin{array}{l}\gamma^{-1} X(\gamma) = A \otimes X(\gamma) \oplus B \otimes U(\gamma) \oplus \gamma^{-1} x_0 , \\ Y(\gamma) = C \otimes X(\gamma) .\end{array}\right\} \tag{20}$$

The first of these equations can be solved by first multiplying (max-plus algebra), equivalently adding (conventional), left- and right-hand side by γ and then repeatedly substituting the right-hand side for $X(\gamma)$ within this right-hand side. This results in

$$X(\gamma) = (\gamma A)^*(\gamma B U(\gamma) \oplus x_0) .$$

Thus we obtain $Y(\gamma) = H(\gamma) U(\gamma)$, provided that $x_0 = \varepsilon$, and where the transfer matrix $H(\gamma)$ is defined by

$$H(\gamma) = C \otimes (\gamma A)^* \otimes \gamma \otimes B = \gamma C B \oplus \gamma^2 C A B \oplus \gamma^3 C A^2 B \oplus \cdots \tag{21}$$

The expression $Y(\gamma) = H(\gamma) U(\gamma)$ is the max-plus algebra equivalent of $Y(z) = H(z) U(z)$ in the conventional system theory. If one writes

$$H(z) = C(zI - A)^{-1} B = C(\frac{1}{\gamma} I - A)^{-1} B = \gamma C(I - \gamma A)^{-1} B =$$

$$\gamma C(I + \gamma A + \gamma^2 A^2 + \cdots) B,$$

one has obtained the equivalence of (21) in the conventional sense!

The transfer matrix (21) is defined by means of an infinite series and the convergence depends on the value of γ. If the series is convergent for $\gamma = \gamma'$, then it is also convergent for all γ's which are smaller than γ'. If the series does not converge, it still has a meaning as a formal series.

Exactly as in conventional system theory, the transfer matrix is especially useful when subsystems are combined to build larger systems, by means of parallel, series and feedback connections (compare subsection 3.3). For instance, the product of two transfer matrices (of which it is tacitly assumed that the sizes of these matrices are such that the multiplication is possible), is a new transfer matrix which refers to a system which consists of the original systems put into a series connection.

Suppose that $H(\gamma)$ is a scalar function, i.e. the system has one input and one output. The term $\gamma^k C A^{k-1} B$ in (21) can be written in conventional algebra as $c_{k-1} + k\gamma$ (where c_{k-1} represents the coefficient $CA^{k-1}B$) which is a straight line with slope k. The transfer function can be viewed as the maximum (of an infinite number) of such lines and hence is a continuous, piecewise linear and convex function of the variable γ.

5.5 Some Extensions and Recent Literature

In this subsection we will briefly mention the following specialisations and/or extensions; Axiomatic foundations, Minimal realizations, Stochastic DES, Min-max-plus systems and nonexpansive mappings, Numerical procedures, 'Continuous' DES and the Fenchel transform.

Axiomatic Foundations. The operations $\oplus$ and $\otimes$ defined on the set R can also be defined with respect to a more general set of elements $\mathcal{D}$. One then speaks of a *dioid* (sometimes also referred to as a semiring).

Definition 21 Dioid. A *dioid* is a set $\mathcal{D}$ endowed with two operations denoted $\oplus$ and $\otimes$ (called 'sum' or 'addition', and 'product' or 'multiplication') obeying the following axioms:

Axiom 22 Associativity of addition.

$$\forall a, b, c \in \mathcal{D}, (a \oplus b) \oplus c = a \oplus (b \oplus c) \ .$$

Axiom 23 Commutativity of addition.

$$\forall a, b \in \mathcal{D}, a \oplus b = b \oplus a \ .$$

Axiom 24 Associativity of multiplication.

$$\forall a, b, c \in \mathcal{D}, (a \otimes b) \otimes c = a \otimes (b \otimes c) \ .$$

Axiom 25 Distributivity.

$$\forall a, b, c \in \mathcal{D}, (a \oplus b) \otimes c = (a \otimes c) \oplus (b \otimes c) \ ,$$
$$c \otimes (a \oplus b) = c \otimes a \oplus c \otimes b \ .$$

This is right, *respectively* left, *distributivity of multiplication with respect to addition. One statement does not follow from the other since multiplication is not assumed to be commutative.*

Axiom 26 Existence of a zero element.

$$\exists \varepsilon \in \mathcal{D} : \forall a \in \mathcal{D}, a \oplus \varepsilon = a \ .$$

Axiom 27 Absorbing zero element.

$$\forall a \in \mathcal{D}, a \otimes \varepsilon = \varepsilon \otimes a = \varepsilon \ .$$

Axiom 28 Existence of an identity element.

$$\exists e \in \mathcal{D} : \forall a \in \mathcal{D}, a \otimes e = e \otimes a = a .$$

Axiom 29 Idempotency of addition.

$$\forall a \in \mathcal{D}, a \oplus a = a .$$

Definition 30 Commutative dioid. A dioid is *commutative* if multiplication is commutative.

With the noticeable exception of Axiom 29, most the axioms of dioids are required for rings too. Indeed, Axiom 29 is the most distinguishing feature of dioids. Because of this axiom, addition cannot be cancellative, that is, $a \oplus b = a \oplus c$ does not imply $b = c$ in general. Multiplication is not necessarily cancellative either (of course, because of Axiom 27, cancellation would anyway only apply to elements different from ε). For an example in which multiplication is not cancellative take $\mathcal{D} = R \cup \{-\infty\} \cup \{+\infty\}$ and define $\oplus$ as max and $\otimes$ as min.

It is easily shown that in dioids the distributivity with respect to matrices also holds, i.e. $A \otimes (B \otimes C) = (A \otimes B) \otimes C$, where these multiplications only make sense if the matrices have appropriate dimensions. For more information on this axiomatic approach, see Chapter 4 of [2].

Minimal Realizations. In subsection 5.4 it was shown how to derive the transfer matrix of a system if the representation of the system in the 'event domain' is given. This event domain representation is characterized by the matrices A, B and C. Now one could pose the opposite question; how to obtain an event domain representation, or equivalently, how to find A, B and C if the transfer matrix is given. In the conventional linear system theory the corresponding theory is known as the realization theory and one speaks of a minimal relization if the sizes of A, B and C are as small as possible, see [43].

The simplest formulation of the (minimal) realization problem in the max-plus algebra is probably as follows. Let G be a sequence of real numbers $\{g_j\}_{j=0}^{\infty}$ and let $A \in R^{n \times n}$, $x_0 \in R^{n \times 1}$, $C \in R^{1 \times n}$ be such that $g_j = C \otimes A^j \otimes x_0$, $j = 0, 1, \ldots$, then G is said to be reproduced by the discrete event system $x(k+1) = A \otimes x(k)$, $x(0) = x_0$ and $y(k) = C \otimes x(k)$. Given a sequence produced in this way, find its realization of the smallest dimension. This realization problem has attracted a lot of attention recently, but for the moment it remains unclear whether an exact algorithmic procedure of polynomial complexity can be found for the general case. This problem was originally formulated in [58]. For recent results the reader is referred to [23] and [68]. In most approaches the theorem of Cayley-Hamilton, suitably adapted to the max-plus algebra, see [2], plays a crucial role.

5 The Max-Plus Algebra Approach to DES

Stochastic Discrete Event Systems. The evolution equation studied in this part is
$$x(k+1) = A(k) \otimes x(k), \ k = 0, 1, 2, \ldots, \tag{22}$$
with some initial condition $x(0)$. Some (or all) entries of $A(k)$ are stochastic. We assume that

- the underlying distribution functions do not depend on k.
- the stochastic entries can assume only a finite number of different values. It will also be assumed that these values are finite, though the method to be described can be generalized to the case that $-\infty$ is also allowed as a value.
- $A(k)$ and $A(l)$ are independent stochastic matrices for $k \neq l$. (Extensions exist for problems where $A(k)$ and $A(k+1)$ are correlated.)
- no correlation between stochastic entries of $A(k)$ exists, though such correlations can be treated rather routinely.
- $\mathcal{G}(A(k))$ is strongly connected. (If this assumption is true for one k, it automatically is true for all k due to the second assumption above.)

The quantity of central interest is
$$\lim_{k \to \infty} E(x_i(k)/k), \tag{23}$$
for an arbitrary i, being the average cycle time for component i. This quantity is a kind of 'average cycle time'; it can been proved [2] that this average cycle time is independent of i. The method of calculation of the average cycle time will be shown by means of a simple example.

Example 6. Consider the case that $x \in R^2$ and that for each k the matrix A is one of the following two matrices
$$\begin{pmatrix} 3 & 7 \\ 2 & 4 \end{pmatrix}, \begin{pmatrix} 3 & 5 \\ 2 & 4 \end{pmatrix}.$$

Both matrices occur with probability $1/2$ and there is no correlation in time. Starting from an arbitrary $x(0)$-vector, say $x(0) = (0,2)'$, we will set up the reachability tree of all possible states x. This is indicated in Table 1, being a table of transitions. In order to get a concise notation, the different state vectors are indicated by n_i, $i = 1, \ldots$. The table has been obtained in the following way. The starting point is $n_1 := (0,2)'$. From there, two states can be reached in one step: $(9,6)'$ or $(7,6)'$, depending on which A-matrix occurs. The states will be normalized such that the first component equals zero. This results in $(0,-3)'$ and $(0,-1)'$. (Other normalizations are possible, and they will lead to the same results.) Both states are new and are therefore added to the list, as n_2 and n_3 respectively. Now we take n_2 as the starting point. Two states can be reached from there: $(4,2)'$ and $(3,2)'$, or, after normalization, $(0,-2)'$ and $(0,-1)'$. Only the first of these states is new and will be added to the list as the next state n_4. In this way we continue: from all states obtained sofar we construct the states which can be reached from there in one step. If a state is found which did not exist sofar, it is added to the list. For the current example it turns out that

Table 1. Transitions of stochastic states

initial state	$a_{12}=7$	$a_{12}=5$
$n_1 = (0,2)'$	$n_2 + 91$	$n_3 + 71$
$n_2 = (0,-3)'$	$n_4 + 41$	$n_3 + 31$
$n_3 = (0,-1)'$	$n_2 + 61$	$n_3 + 41$
$n_4 = (0,-2)'$	$n_2 + 51$	$n_3 + 31$

there exist four different states. The notation $n_i + j\mathbf{1}$ in the table refers to the state n_i of which all components are increased by the number j. One directly notices, by viewing the table, that the system never returns to n_1. Hence this node is a transient one. In the stationary situation a Markov chain results with the three states n_2, n_3 and n_4. Let us be slightly more explicit. The elements of this Markov chain, to be denoted by $z(k)$, are, by construction,

$$z(k) = \begin{pmatrix} 0 \\ x_2(k) - x_1(k) \end{pmatrix}.$$

It is easily shown that

$$z(k+1) = \begin{pmatrix} 0 \\ (Ax(k))_2 - (Ax(k))_1 \end{pmatrix} = \begin{pmatrix} 0 \\ (Az(k))_2 - (Az(k))_1 \end{pmatrix},$$

and hence the process $\{z(k)\}$ is indeed Markovian. The transition matrix of the Markov chain is

$$\begin{pmatrix} 0 & 1/2 & 1/2 \\ 1/2 & 1/2 & 1/2 \\ 1/2 & 0 & 0 \end{pmatrix}.$$

The stationary distribution of this chain is easily calculated to be

$$\Pr(n_2) = 1/3, \; \Pr(n_3) = 1/2, \; \Pr(n_4) = 1/6.$$

The average cycle time becomes

$$\Pr(n_2)(4\Pr(A_1) + 3\Pr(A_2)) + \Pr(n_3)(6\Pr(A_1) + 4\Pr(A_2)) \\ + \Pr(n_4)(5\Pr(A_1) + 3\Pr(A_2)) = 13/3,$$

where the coefficients are the appropriate numbers out of the table above. The first term in this expression for instance, $\Pr(n_2)(4\Pr(A_1) + 3\Pr(A_2))$, is obtained as follows. If the state is in n_2, then this happens with (stationary) probability $\Pr(n_2)$. The next step either leads to n_4, with probability $\Pr(A_1)$ and obtained after 4 time units (see Table 1), or it leads to n_3, with probability $\Pr(A_2)$ and obtained after 3 time units. The other terms are obtained similarly. It is the quantity at the right-hand side, 13/3, which equals the expression in (23). □

This example described a method to calculate the average cycle time. The crucial feature in this method is that the number of different normalized state vectors is finite. See [65] and [2] for extensions.

Min-Max-Plus Systems and Nonexpansive Mappings. Referring to the right-hand side of (10), one can define a max-plus expression as a (finite) set of $x_i + a_{ij}$ terms, connected by the max operator. Similarly, one defines a min-max-plus expression as a (finite) set of $x_i + a_{ij}$ terms, connected by both the max and min operators. An example of such an expression is

$$\max(x_1 + 7, \min(x_2 - 4, x_3 + 1, \max(x_1, x_4 + 2))).$$

With respect to the max operator, we introduced the neutral element $-\infty$. Since we now also deal with the min operator, it is convenient to introduce its neutral element $+\infty$ also. Exactly as one can define a max-plus system by means of max-plus expressions, as (10) can be viewed, one can define a min-max-plus system. Such systems are nonlinear in the max-plus algebra (because of the presence of the min operator) and also nonlinear in the min-plus algebra (because of the presence of the max operator). Notwithstanding this higher complexity, various results about min-max-plus systems are known; the reader is referred to [54], [56] and [32]. Specifically necessary and sufficient conditions are known under which the evolution of (subclasses of) min-max-plus systems show a regular pattern as the evolution of max-plus systems does.

Min-max-plus systems are quite naturally imbedded in the class of so-called nonexpansive mappings, which also are known to have the possibility of periodic behaviour. For more information on such mappings, the reader is referred to [53], [57]. We end this subsection by the definition of nonexpansive mappings.

Definition 31. A mapping f, which maps R^n into R^n is called *nonexpansive* if

$$\| f(z) - f(\bar{z}) \| \leq \| z - \bar{z} \|, \qquad (24)$$

for arbitrary $z, \bar{z} \in R^n$, and where the norm is an arbitrary $\| \cdot \|_p$ norm, with $1 \leq p \leq \infty$.

Numerical Procedures. In this subsection we will confine ourselves to numerical procedures which yield the eigenvalue and eigenvector of a matrix A as expressed by (17).

Theorem 32. *Given is an $n \times n$ matrix A with corresponding precedence graph $\mathcal{G} = (\mathcal{V}, \mathcal{E})$. The maximum cycle mean is given by*

$$\lambda = \max_{i=1,\ldots,n} \min_{k=0,\ldots,n-1} \frac{(A^n)_{ij} - (A^k)_{ij}}{n - k}, \quad \forall j. \qquad (25)$$

In this equation, A^n and A^k are to be evaluated in the max-plus algebra; the other operations (subtraction and division) are conventional ones.

This theorem is known as Karp's theorem after [44]. This theorem yields the eigenvalue but does not give information about the eigenvectors. For that purpose construct the matrix B by subtracting λ, obtained by Karp's theorem, from all elements of A. The maximum circuit weight of $\mathcal{G}(B)$ equals 0. Hence

$B^* = I \oplus B \oplus B^2 \oplus \cdots$ and $B^+ := BB^*$ exist. Matrix B^+ has some columns with diagonal elements equal to zero. To prove this, pick a node k of a circuit χ such that $\chi \in \arg\max_\zeta |\zeta|_w/|\zeta|_l$. The maximum weight of paths from node k to k is 0. Therefore $B^+_{kk} = 0$. Let $B_{\cdot k}$ denote the k-th column of B. Then, since $B^+ = BB^*$ and $B^* = I \oplus B^+$ (I is the identity matrix), for that k,

$$B^+_{\cdot k} = B^*_{\cdot k} \Rightarrow BB^*_{\cdot k} = B^+_{\cdot k} = B^*_{\cdot k} \Rightarrow AB^*_{\cdot k} = \lambda B^*_{\cdot k}.$$

Hence $v = B^+_{\cdot k} = B^*_{\cdot k}$ is an eigenvector of A corresponding to the eigenvalue λ.

A few other numerical approaches exist which calculate the eigenvalue and/or eigenvector:

- Study of the zero(s) of the characteristic equation in the max-plus algebra yields the eigenvalue. For the definition of this equation see [2].
- By means of Linear Programming techniques, see [51].
- Consider (9). Calculate $x(k)$, $k = 0, 1, \ldots$, starting from an arbitrary initial condition, until a state becomes linearly dependent on a state already calulated ($x(7) = 8 \otimes x(4)$). Now $8/(7-4)$ equals the eigenvalue and $(x(4) + x(5) + x(6))/(7-4) = (15\frac{2}{3}\ 13\frac{1}{3}\ 13\ 10\frac{2}{3})'$ is the eigenvector. See [7] for a full account of this method.

'Continuous' Discrete Event Systems and the Fenchel Transform. The central equations of this part are (2) and (1). It is assumed now that in addition to t and τ_i, also χ and x_i are real-valued, and so are the quantities b_{ij}. The interpretation of these equations is still a (strongly connected) network with n transitions (also called nodes now). These nodes can now fire continuously. The intensity of this firing is indicated by $v_i(t)$. Quantity $x_i(t)$ denotes again the total amount produced by node i up to (and including) time t. As initial condition it is assumed that $x_i(0) = 0$. The production of a continuously firing transition is sent with unit speed along the outgoing arcs to the downstream transitions. Thus along an arc there is a continuous flow. The intensity of this flow is $\varphi_i(t, l)$, where l is the parameter indicating the exact location along the arc; $l = 0$ coincides with the beginning of the arc, $l = a_{ji}$ coincides with the end of the arc, where it is assumed that the downstream transition is q_j. As long as the parameters lie in appropriate intervals, we have $\varphi_i(t, l) = \varphi_i(t + s, l + s)$. Moreover, $\varphi_i(t, l) = \varphi_i(t - l, 0) = v_i(t - l)$.

At time t the total amount of material along the arc from q_i to q_j equals

$$\int_{l=0}^{l=a_{ji}} \varphi_i(t, s) \, ds. \tag{26}$$

The quantities b_{ij} satisfy $b_{ij} = \int_0^{a_{ij}} \varphi(0, l) \, dl$. The integrand and the integral in (26) must be considered with some care. It is quite well possible that the integrand will contain δ- functions. This will particularly happen at the end of an arc, when material must wait there to be processed by the downstream transition because the other incoming arcs to the same transition have brought in less material sofar. If q_k is a downstream transition to both q_i and q_j and if

$x_i(t) < x_j(t)$, then φ_{kj} will start to build a δ-function at $l = a_{kj}$, from t onwards. Of course this δ-function can disappear again later on if $x_i(s) > x_j(s)$ for an s-value with $s > t$. The total amount of material along an arc, as expressed by (26), will in general be time dependent. Many standard results of subsection 5.3 on periodic behaviour remain valid for this continous version of flows on networks, see [55].

Theorem 33. *Along a circuit the total amount of material is constant. In formula, if the circuit ζ is characterized by the transitions $\{q_{i_1}, q_{i_2}, \ldots, q_{i_{k+1}} = q_{i_1}\}$, then*

$$\sum_{l=1}^{l=k} \int_0^{a_{i_{l+1},i_l}} \varphi_{i_l}(t,s)\, ds$$

is constant (it does not depend on time).

This is the 'continuous event' analogue of Theorem 9. Please note that the total amount of material in the network is not necessarily constant.

Definition 34 Cycle mean. Given a circuit $\zeta = \{q_{i_1}, q_{i_2}, \ldots, q_{i_{k+1}} = q_{i_1}\}$, its cycle mean is defined as $|\zeta|_w / |\zeta|_l$, where the weight $|\zeta|_w$ and the length $|\zeta|_l$ (assumed to be positive) are defined as

$$|\zeta|_w = \sum_{l=1,\ldots,k} a_{i_{l+1},i_l}, \quad |\zeta|_l = \sum_{i=1,\ldots,k} b_{i_{l+1},i_l}.$$

Definition 35 Critical circuit. The circuits which have the maximum cycle mean are called critical. The corresponding cycle mean is indicated by λ.

Theorem 36. *Equations (2) have a solution $x_i(t) = \frac{1}{\lambda}t + d_i$, with appropriately chosen constants d_i.*

Just as there are linear difference equations versus z-transforms, and similarly, linear differential equations versus Laplace transforms, we have in the max-plus algebra setting their counterparts. For the discrete event systems we had the max-plus algebra systems versus the γ transform. For the continuous flow variation we have the description above versus a slight variation of the Fenchel transform, called Fenchel* transform. This Fenchel* transform turns out to be the max-plus algebra variant of Laplace transforms, see [2]. The Fenchel transform of a function is also referred to as the conjugate function. We conclude with its definition.

Definition 37. The Fenchel transform $\mathcal{F}(f)$ of a function f in a Hilbert space H is a function in the dual space H^*;

$$\forall c \in H^* : \quad [\mathcal{F}(f)](c) := \sup_{z \in H}(<c,z> - f(z)),$$

where $<\cdot,\cdot>$ denotes the innerproduct. The Fenchel* transform $\mathcal{F}^*(f)$ is defined by

$$\forall c \in H^* : \quad [\mathcal{F}^*(f)](c) := \sup_{z \in H}(<c,z> + f(z)). \tag{27}$$

Usually one confines the definition to convex functions f. Note the resemblance between (18) and (27).

6 Sample Path Analysis and Performance Optimization of DES

In this section, we will concentrate on problems related to *quantitative*, as opposed to *qualitative*, specifications of DES, with particular emphasis on stochastic DES. We begin with two basic problems that one encounters in performance analysis and optimization:

(**P1**)   evaluate $J(\theta) = E[L(\theta)]$ over all $\theta \in \Theta$

(**P2**)   find $\theta \in \Theta$ to minimize $J(\theta) = E[L(\theta)]$

where θ is some parameter to be chosen from some set Θ, and $J(\theta)$ is the performance measure of interest. Note that we are careful to distinguish between $L(\theta)$, the performance obtained over a specific sample path of the system (e.g., a single simulation run or an observation of the real system over some period of time), and $J(\theta)$, the expectation over all possible sample paths. In (**P1**), our goal is simply to evaluate the response surface $J(\theta)$ over the entire parameter range Θ. On the other hand, it is sometimes possible to formulate a well-defined optimization problem (**P2**), which we then seek to solve. There are two key difficulties associated with these problems. First, the function $J(\theta)$ is often simply unknown due to the complexity of the DES. Second, even if one succeeds in obtaining an analytical model for the DES, it is generally difficult to obtain accurate values for the parameters on which it depends. As an example, models for packetized voice and video traffic needed in the analysis of high-speed communication networks involve a number of parameters that are hard to accurately estimate. In short, because of the lack of analytical expressions for $J(\theta)$, the solution of problems (**P1**) and (**P2**) generally requires repetitive simulation (or trial-and-error real-time experimentation) in order to estimate $J(\theta)$ over the set Θ.

When the parameter θ takes real values and $J(\theta)$ is differentiable, estimating derivatives of the form $\partial J/\partial \theta$ is a process that allows us to invoke a number of standard gradient-based optimization techniques that can be used to solve (**P2**). Moreover, derivative estimates can be used to approximate $J(\theta)$ in trying to obtain the global response surface required in (**P1**). An obvious way to estimate $\partial J/\partial \theta$ is to estimate finite-difference approximations of the form $\Delta J/\Delta \theta$ through either simulation or direct observation as follows: observe a sample path under θ and obtain $L(\theta)$. Next, perturb θ by $\Delta \theta$ and observe $L(\theta + \Delta \theta)$. Finally, $[L(\theta + \Delta \theta) - L(\theta)]/\Delta \theta$ is an estimate of $\partial J/\partial \theta$. However, this approach involves $(n+1)$ sample paths if there are n parameters of interest, and it quickly becomes prohibitively time-consuming. In addition, accurate gradient estimation requires "small" $\Delta \theta$; however, division by a small number in $\Delta J/\Delta \theta$ leads to a host of numerical problems. This has motivated the effort towards derivative estimation based on information from a single observed sample path, and has led to *Perturbation Analysis* (PA) and *Likelihood Ratio* (LR) techniques.

Turning our attention to discrete parameters, suppose that the decision space is some finite set $\Theta = \{\theta_1, \ldots, \theta_m\}$. This was the case in Example 3 of section 2, where the policy for scheduling elevators at the building lobby was based on a "threshold" parameter: as long as the number of passengers in an elevator does not exceed this threshold, the elevator is held at the lobby. Similar examples include buffer capacities in queueing models, numbers of "kanban" in so-called Just-In-Time (JIT) production control policies, etc. More generally, we can view Θ as a set of (potentially very large) "control actions" that affect the performance of a DES. In the spirit of extracting information from a single sample path, we can pose the following question: by observing a sample path under θ_1, is it possible to infer the performance of the system under all (or at least many) other parameters $\theta_2, \ldots, \theta_m$? More precisely, can we construct sample paths under $\theta_2, \ldots, \theta_m$ from the information contained in the observed sample path under θ_1? This is referred to as the "constructability" problem. When this problem can be solved, it is tantamount to generating m simulation runs in parallel.

The area of gradient estimation for DES is well-studied and well-documented (see [26], [36], [11], [10], [66]). In what follows we will concentrate on the PA methodology and provide an overview in subsections 6.2-6.3. In subsections 6.4-6.6, we will focus on the sample path constructability problem, describe it in a formal setting, and then review informally, in tutorial fashion, the two main approaches that can be used to solve problems (**P1**) and (**P2**) over a discrete set Θ. We begin, however, in subsection 6.1 with the DES modeling framework we will adopt for our purposes, which is based on automata models as introduced in subsection 3.2.

6.1 Stochastic Timed Automata

Let us return to the definition of an automaton in subsection 3.2 and extend it to a *timed automaton*. Recall that an automaton G was defined by: a state space X; an event set Σ; a partial transition function f such that $f(x, e) = x'$ where $x \in X$ is the current state and $x' \in X$ is the next state entered when an event $e \in \Sigma$ takes place; a feasible event set $\Sigma_G(x) \subseteq \Sigma$ for each state $x \in X$, so that f is defined only for events $e \in \Sigma_G(x)$; an initial state x_0; and a set of marked states $X_m \subseteq X$. For our purposes, we drop X_m from this definition, since we are often interested in the behavior of a DES over time intervals rather than over an interval defined by a desired marked state.

To obtain a timed automaton, we associate to every event $e \in \Sigma_G(x)$ a *clock value* (or *residual lifetime*) y_e, which represents the amount of time required until event e occurs, with the clock running down at unit rate. The clock value of event e always starts with a *lifetime*, which is an element of an externally provided clock sequence $\mathbf{v}_e = \{v_{e,1}, v_{e,2}, \ldots\}$; we view this as an input to the automaton. In other words, our model is endowed with a *clock structure*, defined as the set of event lifetime sequences $V = \{\mathbf{v}_e, e \in \Sigma\}$. A timed automaton is the six-tuple $(X, \Sigma, f, \Sigma_G, x_0, V)$.

Let us informally describe how a timed automaton operates (for details, see [11]). If the current state is x, we look at all clock values y_e, $e \in \Sigma_G(x)$. The *triggering event* e' is the event which occurs next at that state, i.e., the event with the smallest clock value:

$$e' = \arg\min\ \{y_e,\ e \in \Sigma_G(x)\} \qquad (28)$$

Once this event is determined, the next state, x', is specified by $x' = f(x, e')$, where f is the given state transition function. More generally, f can be replaced by transition probabilities $p(x'; x, e)$ where $e \in \Sigma_G(x)$. The amount of time spent at state x defines the *interevent time* (between the event that caused a transition into x and the event e'):

$$y^* = \min\ \{y_e,\ e \in \Sigma(x)\} \qquad (29)$$

Thus, time is updated through $t' = t + y^*$. Clock values are updated through $y'_e = y_e - y^*$, except for e' and any other events which were not feasible in x, but become feasible in x'. For any such event, the clock value is set to a new lifetime obtained from the next available element in the event's clock sequence.

Finally, in a *stochastic* timed automaton $(X, \Sigma, f, \Sigma_G, x_0, F)$ the clock structure is replaced by a set of probability distribution functions $F = \{F_e,\ e \in \Sigma\}$. In this case, whenever a lifetime for event e is needed, we obtain a sample from F_e. The state sequence generated through this mechanism is a stochastic process known as a *Generalized Semi-Markov Process* (GSMP) (see also [26], [36]). This provides the framework for generating sample paths of stochastic DES. A sample path is defined by a sequence $\{e_k, t_k\}$, $k = 1, 2, \ldots$, where e_k is the kth event taking values from Σ, and t_k is its occurrence time; or, alternatively, by $\{x_k, t_k\}$, $k = 1, 2, \ldots$, where x_k is the kth state entered when an event occurs at time t_k.

Example 7. Stochastic timed automaton model of single-server queueing system.
A single-server queueing system is defined by the combination of a *server* preceded by a *queue*. Customers (e.g., jobs, messages, production parts) arrive from the outside world at random time instants and request processing at the server. When the server is busy, the customers wait in the queue. In a simple system, they are subsequently processed on a First-In-First-Out (FIFO) basis. After a customer is served, it departs from the system. This is a basic building block of many DES involving resource contention issues which need to be controlled and is used in modelling a wide variety of systems including communication networks, manufacturing processes, etc.

Let $\Sigma = \{a, d\}$ be the event set for this DES, where a denotes an arrival and d denotes a departure. The state space is $X = \{0, 1, \ldots\}$, where $x \in X$ represents the total number of customers in the system (either in queue or in process). Observe that a events are always feasible, whereas d events are feasible only if $x > 0$, since no departure is obviously possible from an empty system. Thus, $\Sigma_G(x) = \{a, d\}$ for all $x > 0$ and $\Sigma_G(0) = \{a\}$.

The dynamics of this system are quite simple and they are captured by the following state transition functions:

$$f(x, a) = x + 1 \quad \text{for all } x, \qquad f(x, d) = x - 1 \quad \text{for } x > 0$$

Finally, a clock structure is specified by associating with a, d lifetime distributions F_a and F_d respectively. A state trajectory can then be obtained, given some initial state $x_0 \in X$, by sampling from these distributions and using (28)-(29) in conjunction with the state transition equations above.

□

6.2 Infinitesimal Perturbation Analysis (IPA)

In what follows, we will denote a *sample path* of a DES by (θ, ω) where θ is a vector of parameters characterizing the state transition function and/or the clock structure V, and ω represents all random occurrences in the system of interest. In particular, we let the underlying sample space be $[0, 1]^\infty$, and ω a sequence of independent random variables uniformly distributed on $[0, 1]$. Given a sample path (θ, ω), we can evaluate a sample performance function $L(\theta, \omega)$ via a statistical experiment, i.e., a discrete-event simulation run or data collected from an actual system in operation. This, then, serves as an estimate of the actual performance measure of interest, which we consider to be the expectation $J(\theta) = E[L(\theta, \omega)]$. In Perturbation Analysis (PA), we are interested in questions of the form "what would the effect of a perturbation $\Delta \theta$ be on the system performance?" Thus, we use the adjectives *nominal* and *perturbed* to qualify the performance measures $J(\theta)$ and $J(\theta + \Delta \theta)$, as well as the corresponding sample paths (θ, ω) and $(\theta + \Delta, \omega)$.

In a narrow sense, IPA is a technique for the efficient computation of the n-dimensional gradient vector of a performance measure, $J(\theta) = E[L(\theta, \omega)]$, of a DES with respect to the n parameters constituting the vector θ, using information from a single sample path of the system. In particular, IPA provides the means for evaluating a *sample derivative* $\partial L / \partial \theta$. The crucial question then becomes: when can $\partial L / \partial \theta$ be used as a "good" estimate of $\partial J / \partial \theta$? By "good" we normally mean properties of an estimator, such as *unbiasedness* and *consistency*.

In the next few sections, we overview IPA for any DES modeled as a stochastic timed automaton. Let us assume that θ is a parameter which characterizes one or more lifetime distributions $G_e(t; \theta)$ of some event $e \in \Sigma$; in particular, θ does not affect the state transition mechanism. For simplicity, we will assume throughout the rest of this section that θ is scalar. Given a sample function $L(\theta, \omega)$, we are interested in deriving an expression for the derivative $dL/d\theta$. If we accomplish this goal, we will then study the conditions under which $dL/d\theta$ can be used as an unbiased estimator of $dJ/d\theta$.

We will make one simplifying assumption regarding the models to be considered: we will assume that once an event e is activated, i.e. $e \in \Sigma_G(x)$ for some state x, it cannot be deactivated. In other words, once a clock for some event starts running down, it cannot be interrupted. This is known as the non-interruption condition (see [26]). In a simulation setting, this means that a scheduled event cannot be deferred or cancelled. For most systems encountered in practice, this is not a serious limitation.

Event Triggering Sequences. To keep notation consistent, we will henceforth reserve Greek letters (usually α, β) to index events, and reserve Roman index letters (such as i, j, k, n, m) for counting event occurrences. We also define T_k to be the kth event occurrence time, $k = 1, 2, \ldots$ of an event in a sample path of our system (regardless of its type), and $T_{\alpha,n}$ to be the nth occurrence time of an event of type α. We can now immediately make the following simple observation: when event α takes place at time $T_{\alpha,n}$, it must have been activated at some point in time $T_{\beta,m} < T_{\alpha,n}$ by the occurrence of some event β (note that it is possible that $\beta = \alpha$). In turn, it is also true that event β at $T_{\beta,m}$ must have been activated by some event γ at time $T_{\gamma,k} < T_{\beta,m}$, and so on, all the way back to time $T_0 = 0$. Recalling that $v_{\alpha,k}$ denotes the kth lifetime of event α, it follows that since *(i)* α was activated at time $T_{\beta,m}$, *(ii)* α cannot be deactivated until it occurs (by the non-interruption assumption above), and *(iii)* α finally occurs at time $T_{\alpha,n}$, we have $v_{\alpha,n} = T_{\alpha,n} - T_{\beta,m}$. Similarly, we have $v_{\beta,m} = T_{\beta,m} - T_{\gamma,k}$, and so on. Therefore, we can always write

$$T_{\alpha,n} = v_{\beta_1,k_1} + \cdots + v_{\beta_s,k_s} \tag{30}$$

for some sequence of events $\beta_1, \ldots, \beta_s$. Clearly, a similar expression can be written for any event time T_k, since $T_k = T_{\alpha,n}$ for some $\alpha \in \Sigma$ and $n = 1, 2, \ldots$. The expression above can be rewritten in a more convenient form by introducing *triggering indicators*, i.e., functions $\eta(\alpha, n; \beta, m)$ taking values in $\{0, 1\}$ and defined by:

$\eta(\alpha, n; \beta, m) = 1$ if nth occurrence of α is triggered by mth occurrence of β

$\eta(\alpha, n; \alpha, n) = 1$ for all $\alpha \in \Sigma$, $n = 1, 2, \ldots$

$\eta(\alpha, n; \beta', m') = 1$ if $\eta(\alpha, n; \beta, m) = 1$ and $\eta(b, m; \beta', m') = 1$ for some $\beta \in \Sigma$

$\eta(\alpha, n; \beta, m) = 0$ otherwise

Then:

$$T_{\alpha,n} = \sum_{\beta,m} v_{\beta,m} \eta(\alpha, n; \beta, m) \tag{31}$$

The sequence of (β, m) pairs that leads to $T_{\alpha,n}$ with $\eta(\alpha, n; \beta, m) = 1$ defines the triggering sequence of (α, n).

Event Time Derivatives. Since the parameter θ affects one or more of the event lifetime distributions $F_\alpha(t; \theta)$, a change in θ generally affects all event lifetimes in the sequence $\{v_{\alpha,1}, v_{\alpha,2}, \ldots\}$. Therefore, we view lifetimes as functions $v_{\alpha,k}(\theta)$ of θ. Moreover, let us adopt the standard inverse transform technique (e.g., see [11]) for generating variates from a distribution $F_\alpha(t; \theta)$. In particular, if $\{u_1, u_2, \ldots\}$ is a sequence of random numbers uniformly distributed over $[0, 1]$, then a variate $v_{\alpha,k}$ generated by this method satisfies $u_k = F_\alpha(v_{\alpha,k}; \theta)$ or equivalently $v_{\alpha,k} = F_\alpha^{-1}(u_k; \theta)$ since F_α is non-decreasing. Omitting details (however, see Chapter 9 of [11]), one can then obtain, through elementary probabilistic calculations, the following expression for the derivative $dv_{\alpha,k}(\theta)/d\theta$:

$$\frac{dv_{\alpha,k}}{d\theta} = -\frac{[\partial F_\alpha(t;\theta)/\partial \theta]_{(v_{\alpha,k},\theta)}}{[\partial F_\alpha(t;\theta)/\partial t]_{(v_{\alpha,k},\theta)}} \quad (32)$$

where the notation $[\cdot]_{(v_{\alpha,k},\theta)}$ means that the term in brackets is evaluated at the point $(v_{\alpha,k},\theta)$. Thus, through (32), the derivative of an event lifetime is obtained from knowledge of the corresponding distribution and the observed lifetime along the sample path. It is worth pointing out that if θ is known to be either a scale or a location parameter of $F_\alpha(t;\theta)$, then knowledge of the distribution itself is not even required in (32).

Returning to (31), since $T_{\alpha,n}$ is expressed as a sum of event lifetimes, it is natural to expect that the derivative of $T_{\alpha,n}$ can be obtained in terms of the derivatives of these event lifetimes, under certain conditions. In particular, let us make the following assumptions:
(A1) For all $\alpha \in \Sigma$, $F_\alpha(t;\theta)$ is continuous in θ and $F_\alpha(0;\theta) = 0$.
(A2) For all $\alpha \in \Sigma$ and $k = 1, 2, \ldots$, $v_{\alpha,k}(\theta)$ is almost surely continuously differentiable in θ.

Under these conditions, we can guarantee that the event sequence leading to time $T_{\alpha,n}$ (assumed finite) remains unchanged for sufficiently small perturbations of θ. As a result, the sequence of triggering indicators also remains unchanged. Therefore, from (31), the event time derivatives $dT_{\alpha,n}(\theta)/d\theta$ are given by

$$\frac{dT_{\alpha,n}}{d\theta} = \sum_{\beta,m} \frac{dv_{\beta,m}}{d\theta} \eta(\alpha, n; \beta, m) \quad (33)$$

where the interchange of summation and differentiation is permitted since the sum above is finite. This expression captures the fact that event time perturbations are generated by the effect of θ on event lifetimes $v_{\beta,m}$, and subsequently propagate from (β, m) to (α, n) through the events contained in the triggering sequence of (α, n).

Example 8. IPA for a single-server queueing system.
To illustrate the precise way in which (33) can be used along an observed sample path of some system, let us consider a typical sample path of a single-server queueing system, as shown in Figure 9. In this case, there are two events, a (arrivals) and d (departures). Focusing on the departure event at $T_9 = T_{d,4}$, note that the 4th departure was activated at time $T_7 = T_{a,4}$ (when the second busy period starts), i.e., $\eta(d, 4; a, 4) = 1$. The 4th arrival, in turn, was activated at $T_4 = T_{a,3}$, the previous arrival, which in turn was activated at $T_2 = T_{a,2}$. Finally, the second arrival was activated at $T_1 = T_{a,1}$, and the first arrival was activated at time zero. We therefore get:

$$T_{d,4} = v_{d,4} + v_{a,4} + v_{a,3} + v_{a,2} + v_{a,1}$$

Thus, the triggering sequence of $(d, 4)$ is $\{(a, 1), (a, 2), (a, 3), (a, 4)\}$.

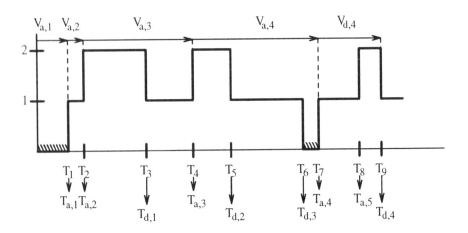

Fig. 9. Sample path of a single-server queueing system

Now, suppose the parameter θ is the mean of the service time distribution $F_d(t;\theta)$. Thus, all service times (departure event lifetimes) $v_{d,k}$, $k = 1, 2, \ldots$, are affected by perturbations in θ, whereas interarrival times are not (i.e., $dv_{a,k}/d\theta = 0$ for all $k = 1, 2, \ldots$). Therefore,

$$\frac{dT_{d,4}}{d\theta} = \frac{dv_{d,4}}{d\theta} + \frac{dv_{a,4}}{d\theta} + \frac{dv_{a,3}}{d\theta} + \frac{dv_{a,2}}{d\theta} + \frac{dv_{a,1}}{d\theta}$$

□

To complete the discussion of event time derivatives, we present below a general-purpose algorithm for evaluating such derivatives along an observed sample path. Let us define a *perturbation accumulator*, Δ_α, for every event $\alpha \in \Sigma$. Based on (33), an accumulator Δ_α is updated at event occurrences in two ways: *(i)* it is incremented by $dv_\alpha/d\theta$ whenever an event α occurs, and *(ii)* it becomes dependent on an accumulator Δ_β whenever an event β (possibly $\beta = \alpha$) occurs that activates an event α. Note that because of the non-interruption condition, the addition of $dv_{\alpha,k}/d\theta$ to Δ_α can be implemented either at the time of the actual occurrence of α or at the time of its activation. In what follows, it is assumed that the system starts out at some given state x_0. No particular stopping condition is specified, since this may vary depending on the problem of interest (e.g., stop after a desired total number of event occurrences, stop after a desired number of event type α occurrences, etc.) Thus, we have the following:

Algorithm for Evaluating Event Time Derivatives:

1. *Initialize*: If α is feasible at x_0, then $\Delta_\alpha := \frac{dv_{\alpha,1}}{d\theta}$; for all other $\alpha \in \Sigma$, $\Delta_\alpha := 0$

2. *For any event (say β) observed:* If α is activated by β with new lifetime v_α, then compute $\frac{dv_\alpha}{d\theta}$ through (32) and set $\Delta_\alpha := \Delta_\beta + \frac{dv_\alpha}{d\theta}$.

It is easy to verify that application of this algorithm to our single-server queueing system example above yields the same expression for $dT_{d,4}/d\theta$ as the one derived from first principles. Assuming the queue is initially empty, we set $\Delta_d = 0$ (because d is not feasible at this state) and $\Delta_a = 0$ (which is always true, since interarrival times are independent of θ). Then, the accumulator Δ_d, which keeps track of $dT_{d,k}/d\theta$ for all $k = 1, 2, \ldots$, is updated in one of two ways:

1. When d occurs, a new d event is activated as long as d remains feasible in the new state, i.e., as long as $x > 1$ (if $x = 1$, the queue becomes empty and d is no longer feasible). Thus, $\Delta_d := \Delta_d + \frac{dv_d}{d\theta}$.
2. When a occurs, a new d event is activated only if $x = 0$, i.e., a new busy period starts. In this case: $\Delta_d := \Delta_a + \frac{dv_d}{d\theta}$.

Sample Function Derivatives. We now proceed with the derivation of sample function derivatives. In many cases of interest, a sample performance function $L(\theta)$ can be expressed in terms of event times $T_{\alpha,n}$. Therefore, we may use (33) or, equivalently, the algorithm presented above, in order to obtain derivatives of the form $dL(\theta)/d\theta$. The sample performance functions we will consider in our analysis are all over a finite horizon. Letting $C(x(t, \theta))$ be a bounded cost associated with operating the system at state $x(t, \theta)$, we define (see also [26]):

$$L_T(\theta) = \int_0^T C(x(t,\theta))dt \qquad (34)$$

$$L_M(\theta) = \int_0^{T_M} C(x(t,\theta))dt \qquad (35)$$

$$L_{\alpha,M}(\theta) = \int_0^{T_{\alpha,M}} C(x(t,\theta))dt \qquad (36)$$

Here, $L_T(\theta)$ measures the total cost over an interval of time $[0, T]$ for some given finite T. On the other hand, $L_M(\theta)$ is the total cost measured over exactly M event occurrences, while $L_{\alpha,M}(\theta)$ is the corresponding cost over exactly M occurrences of some event type α. These functions cover many (but not all) useful performance measures encountered in practice. In queueing systems, for instance, $x(t, \theta)$ is usually the queue length, and by setting $C(x(t, \theta)) = x(t, \theta)$ in the first equation above we can estimate the mean queue length over $[0, T]$ as $L_T(\theta)/T$.

Under the non-interruption condition for all events in our models, and assumptions **(A1)** and **(A2)**, it is not difficult to see that the derivatives of $L_T(\theta)$, $L_M(\theta)$, and $L_{\alpha,M}(\theta)$ exist (with probability 1) and can be easily obtained. As an example, the derivative of $L_M(\theta)$ is given by:

$$\frac{dL_M}{d\theta} = \sum_{k=0}^{M-1} C(x_k) \left[\frac{dT_{k+1}}{d\theta} - \frac{dT_k}{d\theta} \right] \qquad (37)$$

A formal derivation can also be found in [26]. The crucial observation is that in DES the state remains unchanged at x_k in any interval $(T_k, T_{k+1}]$. This allows us to rewrite the integral defining the sample functions defined above in a simpler summation form. As an example,

$$L_M = \sum_{k=0}^{M-1} C(x_k)[T_{k+1} - T_k] \qquad (38)$$

In this form, it is not difficult to see that differentiation with respect to θ yields (37). We can now see how (33) may be used in conjunction with (37) to obtain sample derivatives: (33) allows us to evaluate the event time derivatives in the expression above, and hence transform perturbations in event times into perturbations in sample performance functions.

Example 9. IPA for a single-server queueing system (continued).
Let us return to the departure time perturbations of the form $dT_{d,k}/d\theta$ which were evaluated earlier for a single-server queueing system, where θ is the mean of the service time distribution $F_d(x; \theta)$. Let us see how to use this information to evaluate the derivative of the mean system time over M customer departure events (where M is fixed). Setting $C(x(t, \theta)) = 1$ in (36), we get the system time of the kth customer $S_k(\theta) = T_{d,k}(\theta) - T_{a,k}(\theta)$. Thus, the derivative $dS_k/d\theta$ can be obtained directly from the event time perturbations evaluated earlier using (33). In fact, since $dT_{a,k}/d\theta = 0$ for all arrivals, we have $dS_k/d\theta = dT_{d,k}/d\theta$, where $dT_{d,k}/d\theta$ is precisely the value of the accumulator Δ_d defined earlier. Thus, after i customer departures within a busy period, we get

$$\frac{dS_i}{d\theta} = \sum_{j=1}^{i} \frac{dv_{d,j}}{d\theta}$$

Suppose M happens to be the number of customers in a busy period. Then, since the mean system time over a busy period consisting of M customers is

$$L_{d,M}(\theta) = \frac{1}{M} \sum_{i=1}^{M} S_i,$$

its derivative is obtained by combining the last two equations:

$$\frac{dL_{d,M}}{d\theta} = \frac{1}{M} \sum_{i=1}^{M} \sum_{j=1}^{i} \frac{dv_{d,j}}{d\theta}$$

In addition, recall that $\Delta_d := \Delta_a + dv_d/d\theta = dv_d/d\theta$ when event a occurs initiating a new busy period. This allows us to evaluate the sample derivative, $dL_{d,M}/d\theta$, over a fixed number, M, of customers observed over several busy periods. Let n_b be the index of the last customer in the bth busy period, and $b = 1, 2\ldots$, with $n_0 = 0$. In general, the Mth customer falls within some busy period whose index we denote by B. To keep notation simple, we adopt the convention that in the last, generally incomplete, busy period n_B is the index that corresponds to the Mth customer. Then, the derivative of the mean system time over M customers, $dL_{d,M}/d\theta$, is given by

$$\frac{dL_{d,M}}{d\theta} = \frac{1}{M}\sum_{b=1}^{B}\sum_{i=n_{b-1}+1}^{n_b}\sum_{j=n_{b-1}+1}^{i}\frac{dv_{d,j}}{d\theta} \tag{39}$$

This expression is of special importance in the development of IPA, as it represents the first formal result derived from first principles (see also [36]).

Unbiasedness and the Commuting Condition. So far, we have discussed the first component of IPA, i.e., the derivation of a sample function derivative, $dL(\theta,\omega)/d\theta$, from information observed on a nominal sample path (θ,ω) alone, as discussed in the last section. We now address the second component of IPA: Assuming we have obtained an expression for $dL(\theta,\omega)/d\theta$, under what conditions can we guarantee that it is an unbiased estimate of the performance measure derivative $dJ/d\theta$? In other words, when is the interchange of expectation and differentiation below valid:

$$E\left[\frac{dL(\theta,\omega)}{d\theta}\right] = \frac{dE[L(\theta,\omega)]}{d\theta} \tag{40}$$

Before getting into details, one might immediately suspect that this interchange may be prohibited when $L(\theta,\omega)$ exhibits discontinuities in θ. Such discontinuities may arise when a change in θ causes various event order changes. However, as will be discussed next, some event order changes may in fact occur without violating the continuity of $L(\theta,\omega)$. Intuitively, we can characterize the effect of a perturbation $\Delta\theta$ on a nominal sample path under θ as discussed in the following.

No matter how small we make $\Delta\theta$, the nominal and the perturbed sample paths, (θ,ω) and $(\theta + \Delta\theta,\omega)$ respectively, must sooner or later differ. Equivalently, the ensemble of sample paths, $(\theta+\Delta\theta,\omega)$ will differ in at least one member from the ensemble (θ,ω) provided these ensembles are large enough. If we limit ourselves to finite horizon problems, the frequency of occurrence of such a difference is of order $\Delta\theta$. Furthermore, if the sample path difference is also small and of order $\Delta\theta$, then the average net effect of the difference in the nominal and the perturbed performance will be of order $(\Delta\theta)^2$, which is negligible for the first derivative calculation. In other words, if the differences caused by $\Delta\theta$ between the nominal and the perturbed sample paths are small and only temporary, then IPA will give an unbiased estimate of $dJ/d\theta$.

A precise characterization of this condition was provided by Glasserman [26] and is known as the *commuting condition*:

(C) Let $x,y,z_1 \in X$ and $\alpha,\beta \in \Sigma_G(x)$ such that $p(z_1;x,\alpha)p(y;z_1,\beta) > 0$. Then, there exists $z_2 \in X$, such that: $p(z_2;x,\beta) = p(y;z_1,\beta)$ and $p(y;z_2,\alpha) = p(z_1;x,\alpha)$. Moreover, for any $x,z_1,z_2 \in X$ such that $p(z_1;x,\alpha) = p(z_2;x,\alpha) > 0$, we have: $z_1 = z_2$.

In words, (C) requires that if a sequence of events $\{\alpha, \beta\}$ takes state x to state y, then the sequence $\{\beta, \alpha\}$ must also take x to y. Moreover, this must happen in such a way that every transition triggered by α or β in this process occurs with the same probability. The last part of (C) also requires that if an event α takes place at state x, the next state is unique, unless the transition probabilities to distinct states z_1, z_2 are not equal. The commuting condition is particularly simple to visualize through the diagram of Fig. 10.

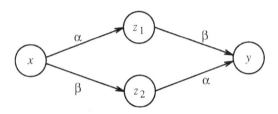

Fig. 10. The commuting condition (C)

Example 10. IPA for a single-server queueing system (continued).
Figure (11) depicts the state transition diagram of this system. Every state other than $x = 0$ has a feasible event set $\Sigma_G(x) = \{a, d\}$, and $p(x + 1; x, a) = p(x - 1; x, d) = 1$. It is immediately obvious from the diagram that the event sequence $\{a, d\}$ applied to any $x > 0$ results in the state sequence $\{x + 1, x\}$, whereas the event sequence $\{d, a\}$ results in the state sequence $\{x - 1, x\}$. Condition (C) is satisfied, since the final state, x, is always the same. □

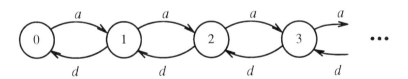

Fig. 11. State transition diagram for a single-server queueing system

The commuting condition plays a crucial role in IPA because it guarantees that sample functions $L(\theta)$ of the form (34)-(36) are continuous in θ. This is stated as a theorem below (for details see Glasserman [26]).

6 Sample Path Analysis and Performance Optimization of DES

Theorem 38. *Under assumptions* **(A1)**-**(A2)** *and condition* **(C)**, *the sample functions* $L_T(\theta)$, $L_M(\theta)$, *and* $L_{\alpha,M}(\theta)$ *(for finite* $T_{\alpha,M}$*) defined in (34)-(36) are (almost surely) continuous in* θ.

The fact that IPA derivative estimators are unbiased can also be formally stated (for details see [26]). For example, for the sample function of the form $L_T(\theta)$ we have:

Theorem 39. *Suppose assumptions* **(A1)**-**(A2)** *hold and condition* **(C)** *is satisfied. In addition, assume that* $\frac{dv_{\alpha,k}}{d\theta} \leq B(v_{\alpha,k}+1)$ *for some* $B > 0$. *Then, if* $E[\sup_\theta N(T)^2] < \infty$,

$$\frac{d}{d\theta} E[L_T(\theta)] = E\left[\frac{dL_T(\theta)}{d\theta}\right]$$

The commuting condition can be extended (see [30]) and stated as a *monotonicity condition*. This requires that if the nominal and perturbed event sequences differ only in the order and not in the total number of each event type present, then the system from which these event sequences are generated must be in the same state (or states that have the same feasible event set). This is also known as the *permutability condition*.

Consistency of IPA Estimators. Returning to the sample performance measures defined in (34)-(36), note that they all apply to a finite horizon, specified either through a given time instant or through a given number of events. Thus, strictly speaking, we should always explicitly write $L_N(\theta)$ when referring to a sample function based on N events. Adopting the notation $L'_N(\theta) = dL_N/d\theta$, then $L'_N(\theta)$ is a strongly consistent estimator of $dJ/d\theta$ if, with probability 1,

$$\lim_{N\to\infty} L'_N(\theta) = \frac{dJ}{d\theta} \qquad (41)$$

where $J(\theta)$ is some performance measure of the system at steady-state, such as the stationary mean system time of customers in a queueing system or its stationary mean queue length. It is reasonable to expect that if IPA provides an unbiased estimator for any finite N and if $L_N(\theta)$ is itself a strongly consistent estimator of $J(\theta)$, then (41) is also satisfied. The first formal proof of strong consistency for an IPA estimator is due to Suri and Zazanis [70] for the case of derivatives of the steady-state mean system time of an $M/G/1$ queueing system, in which case $dJ/d\theta$ can be analytically evaluated. In general, however, such analytical expressions are not available. Nevertheless, establishing strong consistency under certain technical conditions is possible. For details, the reader is referred to [28] or [76].

6.3 Extensions of IPA and Other Gradient Estimation Techniques

It should be clear from the previous section, and the commuting condition **(C)** in particular, that IPA can be applied to a limited class of DES and performance sensitivity estimation problems. If, for example, the parameter θ affects the state transition mechanism of the DES, rather than just the event lifetime distributions, then IPA will generally not yield unbiased estimates. Such limitations have motivated the effort to extend PA beyond the confines of the commuting condition. In what follows, we provide a very brief summary of extensions of IPA and other techniques that have emerged for gradient estimation.

1. IPA for sample functions with discontinuities. In general, when $L(\theta)$ is discontinuous in θ, IPA gives biased estimates. It is possible, however, to define constructions which provide sample paths equivalent to the original one while eliminating the discontinuities at the same time. When this is possible, IPA can be directly applied. A collection of such cases is provided in Chapter 9 of [11].

2. Finite and Extended Perturbation Analysis (FPA and EPA). In cases where IPA fails, it is possible to use a finite difference approximation to $\partial L/\partial \theta$, of the form $\Delta L/\Delta \theta$, where ΔL is still obtained from a single sample path observed under θ. This is referred to as *Finite* PA (FPA), and actually preceded IPA in the development of this field [37]. It should be clear, however, that in this case we can only obtain an approximation to the derivative required. In some cases, IPA can be used up to some event on the observed sample path, at which point one has to shift to FPA for some period of time. This is referred to as *Extended* PA [40].

3. Smoothed Perturbation Analysis (SPA). The most general way to solve the problem of sample path discontinuities that limits IPA is to replace $L(\theta)$ by a conditional expectation of the form $E[L(\theta)|z]$, where z is a "characterization" of the observed sample path [31], [27], [24]. Because of the smoothing property of the conditional expectation, $E[L(\theta)|z]$ is generally continuous in θ. The price to pay, however, is in determining the appropriate characterization for the problem of interest and extracting additional information from the sample path. It is also noteworthy that SPA may be used to derive second derivative estimates some classes of DES [5].

4. Rare Perturbation Analysis (RPA). In IPA, the effect of a perturbation $\Delta \theta$ is modelled through a sequence of infinitesimal perturbations in event times affected by this parameter. It can be shown, however, that a statistically equivalent way is to generate infrequent (rare) occurrences of finite perturbations. This idea lends its name to *Rare* Perturbation Analysis (RPA) [9], and a variety of related so called "marking" and "phantomizing" techniques [13].

5. The Likelihood Ratio (LR) approach. In this approach (also known as the "score function" methodology), a realization of a stochastic DES is viewed as remaining fixed when θ varies; the effect of θ is to change the probability measure that characterizes the realization. The LR approach is attractive because, compared to IPA, it yields unbiased estimates for a wider class of problems. Unfortunately, however, the variance of the LR derivative estimates increases with the length of a sample path. Thus, except for regenerative DES with reasonably short regenerative periods, this approach is limited in practice. In addition, to apply this technique, one needs to know precisely what the probability density functions of the event lifetimes with a dependence on θ are. This is not so in IPA, where, for instance, any $F_e(t;\theta)$ for which θ is a scale parameter gives $\partial v_{e,k}/\partial\theta = v_{e,k}/\theta$ in (32). Further details on the LR approach may be found in [63], [66] or [67].

6.4 The Constructability Problem and "Rapid Learning"

Let us now concentrate on our original problems (**P1**) and (**P2**) for a *discrete* parameter set $\Theta = \{\theta_1,\ldots,\theta_m\}$. Let some value from this set, θ_1, be fixed. A sample path of a DES depends on θ_1 and on ω, an element of the underlying sample space Ω, which (as in PA) is taken to be $[0,1]^\infty$. Let such a sample path be $\{e_k^1, t_k^1\}$, $k = 1, 2, \ldots$ Then, assuming all events and event times e_k^1, t_k^1, $k = 1, 2, \ldots$, are directly observable, the problem is to construct a sample path $\{e_k^j, t_k^j\}$, $k = 1, 2, \ldots$, for any θ_j, $j = 2, \ldots, m$, as shown in Fig. 1. We refer to this as the *constructability problem*. Ideally, we would like this construction to take place on line, i.e., while the observed sample path evolves. Moreover, we would like the construction of all $(m-1)$ sample paths for $j = 2, \ldots, m$ to be done concurrently.

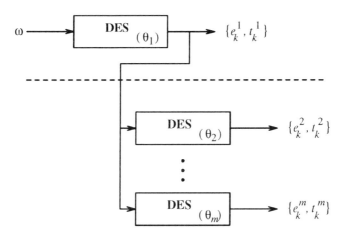

Fig. 12. The sample path constructability problem

The solution to this problem gives rise to what we call "rapid learning", since it enables us to learn about the behavior of a DES under all possible parameter values in Θ from a single "trial", i.e., a sample path under one parameter value. Returning to problem (**P2**), note that any sample performance metric $L(\theta)$ is obtained as a function of $\{e_k, t_k\}$, $k = 1, 2, \ldots$ Thus, from an optimization standpoint, if performance estimates $L(\theta_1), \ldots, L(\theta_m)$ are all available at the conclusion of one sample path, we can immediately select a candidate optimal parameter $\theta^* = \arg\min\{L(\theta),\ \theta \in \Theta\}$. This is potentially the true optimal choice, depending on the statistical accuracy of the estimates $L(\theta_1),\ldots,L(\theta_m)$ of $J(\theta_1),\ldots,J(\theta_m)$. If the accuracy at the end of one trial is not sufficient, the process may be repeated several more times, each time observing the sample path with the optimal parameter to date. It is, however, important to point out that what is of interest in this process is accuracy in the *order* of the estimates $L(\theta_1),\ldots,L(\theta_m)$, not their actual *cardinal* value. This idea can be further exploited to develop a new setting for solving optimization problems as recently described in [41]. It is also interesting to point out that this approach is inherently based on processing system data in parallel, as clearly illustrated in Fig. 12. Therefore, as a "parallel simulation" technique it is ideally suited for emerging massively parallel and distributed processing computer environments (see also a recent collection of papers in [12] for several perspectives on concurrent and parallel simulation of DES).

A natural question that now arises is "under what conditions can one solve the constructability problem?" This question was addressed in [16] and [15], and the constructability condition presented in [15] is briefly reviewed next. Suppose we are constructing a sample path $\{e_k^j, t_k^j\}$, $k = 1, 2, \ldots$, and the current state is x_k^j. In view of (28), to proceed with the construction, we need knowledge of (*i*) the feasible events defining $\Sigma_G(x_k^j)$, and (*ii*) the corresponding clock values to be compared, $y_{\sigma,k}$, $\sigma \in \Sigma_G(x_k^j)$. Observability is a property that addresses part (*i*) of this requirement. It applies to two state sequences, $\{x_k\}$, and $\{x_k^j\}$, $k = 0, 1, \ldots$, resulting from two sample paths:

Definition 40. A sample path $\{e_k^j, t_k^j\}$, $k = 1, 2, \ldots$, is *observable* with respect to $\{e_k, t_k\}$ if $\Sigma_G(x_k^j) \subseteq \Sigma_G(x_k)$ for all $k = 0, 1, \ldots$.

Intuitively, this condition guarantees that all feasible events required at state x_k^j to proceed with the sample path construction are "observable", since they are a subset of the feasible events at the observed state x_k. Unfortunately, this condition alone does not guarantee that part (*ii*) of the requirement above is satisfied, since the clock values in the constructed sample path are generally not identical to those of the observed sample path, not even in distribution. To address this issue, let v_σ be the lifetime of some event σ currently feasible, and let z_σ be its age, i.e., if y_σ is its clock value, we have $z_\sigma = v_\sigma - y_\sigma$. We can then define the conditional probability distribution of the kth clock value $y_{\sigma,k}$ in a sample path given the event age $z_{\sigma,k}$ (an observable quantity), and denote it by $H(\cdot, z_{\sigma,k})$. Using similar notation for the constructed sample path $\{e_k^j, t_k^j\}$, $k = 1, 2, \ldots$, the constructability condition presented in [15] is as follows.

Definition 41. A sample path $\{e_k^j, t_k^j\}$, $k = 1, 2, \ldots$, is *constructable* with respect to $\{e_k, t_k\}$ if:

$$\Sigma_G(x_k^j) \subseteq \Sigma_G(x_k) \quad \text{for all } k = 0, 1, \ldots \quad (42)$$

$$\text{and} \quad H^j(\cdot, z_{\sigma,k}^j) = H(\cdot, z_{\sigma,k}) \quad \text{for all } \sigma \in \Sigma_G(x_k^j) \quad (43)$$

One can then easily establish the following two results:

Theorem 42. *If all event processes are Markovian (generated by Poisson processes), then observability implies constructability.*

Theorem 43. *If $\Sigma_G(x_k^j) = \Sigma_G(x_k)$ for all $k = 0, 1, \ldots$, then constructability is satisfied.*

The first result follows from the memoryless property: condition (43) is satisfied regardless of the event ages of the feasible events at state x_k^j. The second result holds because (43) is now satisfied due to the fact that $z_{\sigma,k}^j = z_{\sigma,k}$ for all $k = 0, 1, \ldots$

The constructability condition (42)-(43) is naturally satisfied for some simple DES. For example, consider an $M/M/1/K$ queueing system and let the parameter of interest be the queueing capacity K. It can be easily verified that all sample paths of an $M/M/1/L$ system are constructable with respect to a sample path of an $M/M/1/K$ system for all L such that $1 \leq L < K$ (e.g., see Chapter 9 of [11]). However, constructability is not satisfied for $L > K$, because (42) is violated for at least one state. To solve the constructability problem when one or both of (42)-(43) are violated, two techniques have been proposed, the *Standard Clock* approach and *Augmented System Analysis*, which are overviewed next.

6.5 The Standard Clock (SC) Approach

In the Standard Clock approach, introduced in [74], we limit ourselves to Markovian DES, i.e., all event lifetime distributions F_σ in our model are assumed exponential with corresponding parameter λ_σ. By Theorem 42, the problem of constructability is then reduced to that of observability. This, in turn, is resolved by adopting the well-known uniformization approach which exploits the properties of Markov chains. In particular, the way that the SC approach gets around the observability problem is by generating a sample path of a "master" system with $\Sigma_G(x) = \Sigma$ for all states x, and then constructing sample paths of various systems of interest for which $\Sigma_G(x^j) \subseteq \Sigma$. Trivially, then, each such sample path satisfies the observability condition (42). A sample path of this "master" system is constructed as follows. Since at any state the total event rate is $\Lambda = \sum_{\sigma \in \Sigma} \lambda_\sigma$ (the maximal event rate possible), all interevent times are exponentially distributed with parameter Λ. Thus, the first step is to generate an i.i.d time sequence $\{t_k\}$, $k = 1, 2, \ldots$, with $(t_{k+1} - t_k)$ sampled from an exponential distribution with parameter Λ. This sequence is called a Standard Clock (SC). Note that it depends only on Λ. Next, since at any state event σ occurs with probability λ_σ/Λ, to determine the triggering event at any state we generate another i.i.d event sequence $\{e_k\}$, $k = 1, 2, \ldots$, with e_k determined through $P[e_k = \sigma] = \lambda_\sigma/\Lambda$. Note that this sequence is generated completely independently from the SC itself.

We now wish to construct a sample path of some system corresponding to a parameter θ_j for which the event set is Σ but where it is generally the case that

$\Sigma_G(x_k^j) \subseteq \Sigma$. To do so, we simply use the interevent times $(t_{k+1} - t_k)$ from the SC and the event sequence $\{e_k\}$ generated above. Whenever $e_{k+1} \in \Sigma_G(x_k^j)$, the state is updated using the state transition function f^j for this system. Whenever $e_{k+1} \notin \Sigma_G(x_k^j)$, this event is simply ignored; e_{k+1} is interpreted as a fictitious event that leaves the state unaffected, i.e., $f^j(x_k^j, e_{k+1}) = x_k^j$, and preserves the statistical validity of the sample path due to the memoryless property. This approach is well-suited for a simulation environment, where the SC sequence can be generated in advance, and then multiple sample paths can be constructed in parallel. If, on the other hand, an actual sample path $\{e_k, t_k\}$ of a DES is observed, then one observes only feasible events for the DES in question and must resort to a modified scheme described in [14].

Example 11. Concurrent sample path construction for an $M/M/1/K$ queueing system using the SC method.

To construct multiple sample paths of an $M/M/1/K$ queueing system for different values of $K = 1, 2, \ldots$ using the SC approach, we first generate a SC with rate $\lambda + \mu$, where λ is the arrival rate and μ is the service rate. Note that this is done independently of K. The event set for this DES is $E = \{a, d\}$, where a is an arrival and d is a departure. An event sequence $\{e_k\}$ is then generated as follows: use a random number u_k and set $e_k = a$ if $u_k \leq \lambda/(\lambda + \mu)$ and $e_k = d$ otherwise. Then, if a state sequence of the system corresponding to a value K is denoted by $\{x_k^K\}$, a sample path of this system is obtained by setting

$$x_{k+1}^K = x_k^K + 1 \quad \text{if } e_{k+1} = a \text{ and } x_k^K < K$$
$$x_{k+1}^K = x_k^K - 1 \quad \text{if } e_{k+1} = d \text{ and } x_k^K > 0$$

We can then see that arrivals when the queue is full and departures when the queue is empty are simply ignored and leave the state unaffected. However, the same SC and event sequence are used to construct as many sample paths in parallel as desired corresponding to $K = 1, 2, \ldots$ □

6.6 Augmented System Analysis (ASA)

Augmented System Analysis (ASA) was introduced in [16] for purely Markovian DES, and was subsequently extended in [15]. The scope of ASA is similar to that of the SC approach, with three basic differences: (*i*) While the SC technique applies to discrete or continuous parameters, ASA is specifically geared towards discrete parameters, (*ii*) ASA is always driven by *actually observed* event and time sequences, i.e., there is no "master" system and hence fictitious events as in the SC construction, and (*iii*) It is applicable to somewhat more general models, beyond Markovian ones.

Let us begin by limiting ourselves once again to purely Markovian DES. The main contribution of ASA is in providing a formal technique for overcoming the problem of "unobservable" events, i.e., instances along the construction of a sample path $\{e_k^j, t_k^j\}$, $k = 1, 2, \ldots$, where $\Sigma_G(x_k^j) \supset \Sigma_G(x_k)$, violating (42). The key idea of this technique is to "suspend" the sample path construction for the jth system until a state x_k where observability is satisfied is next encountered. The resulting procedure is known as *event matching*. In order to describe this procedure, let us define a variable associated with every constructed sample path indexed by j (corresponding to parameter value θ_j in problems (**P1**) and (**P2**)) called the "mode" of the jth system and denoted by $\mathcal{M}_k^j$, $k = 0, 1, \ldots$ This variable is always initialized to the value "active" and remains unaffected unless a state x_k is entered such that $\Sigma_G(x_k^j) \supset \Sigma_G(x_k)$. At this point, we set $\mathcal{M}_k^j = x_k^j$, i.e., store the constructed sample path's state. The construction is subsequently suspended until a new state x_k is entered such that $\Sigma_G(\mathcal{M}_k^j) \subseteq \Sigma_G(x_k)$. At this point, we set $\mathcal{M}_k^j =$ active, and continue the process. This procedure is summarized in the following:

Event Matching Algorithm:

- *Step 1*: Initialize: $x_0^j = x_0$ and $\mathcal{M}_0^j =$ active for all j

With every observed event e_{k+1}, $k = 0, 1, \ldots$, for every sample path j:

- *Step 2*:
  1. If $\mathcal{M}_k^j =$ active, update state: $x_{k+1}^j = f^j(x_k^j, e_{k+1})$
  2. If $\Sigma_G(x_{k+1}^j) \supset \Sigma_G(x_{k+1})$, set: $\mathcal{M}_{k+1}^j = x_{k+1}^j$
- *Step 3*: If $\mathcal{M}_k^j \neq$ active, and $\Sigma_G(\mathcal{M}_k^j) \subseteq \Sigma_G(x_k)$, set: $\mathcal{M}_{k+1}^j =$ active.
- *Step 4*: If $\mathcal{M}_k^j \neq$ active, and $\Sigma_G(\mathcal{M}_k^j) \supset \Sigma_G(x_k)$, wait for next observed event and go to *Step 2*.

It is worth pointing out the inherent parallelism of this procedure: whenever an event is observed, all sample path states are simultaneously updated, with the exception of those with $\mathcal{M}_k^j \neq$ active. Obviously, if observability is always satisfied for some j, then $\mathcal{M}_k^j =$ active and no suspension is needed. If, on the other hand, observability is violated at *Step 2.2* above, then at *Step 3*, when it is once again satisfied, we must in addition check that condition (43) is satisfied. This is always true for Markovian events, or it may be true in special cases where one can exploit the structure of the system, as in Example 13.

Example 12. Concurrent sample path construction for an $M/M/1/K$ queueing system using ASA.
Consider an $M/M/1/K$ queueing system for different values of $K = 1, 2, \ldots$ In contrast to the SC approach in Example 11, we must now base our analysis on an actual sample path observed under some value of K. Let us first assume that the nominal sample path corresponds to $K = 3$ and we are interested in constructing a sample path for the $M/M/1/2$ system. It is not difficult to verify that in this case the constructability condition (42)-(43) is satisfied. To see this, we construct what is referred to as an "augmented system", i.e., a hypothetical system whose state is of the form (x_k^2, x_k^3), where x_k^K denotes the queue length of the $M/M/1/K$ system, $K = 2, 3$. This hypothetical system is driven by the exact same event sequence as the observed sample path. In Figure 6.6 we show the state transition

diagrams of the $M/M/1/2$ and $M/M/1/3$ systems along with the state transition diagram of the augmented system. It is now easy to see that $\Sigma_G(x_k^2) \subseteq \Sigma_G(x_k^3)$ for all k. In fact, $\Sigma_G(x_k^2) = \Sigma_G(x_k^3)$ for all states with one exception: at the augmented system state $(0,1)$ we have $\Sigma_G(0) = \{a\} \subset \Sigma_G(1) = \{a,d\}$. In this case, however, observability is still satisfied. Thus, the mode of the $M/M/1/2$ system whose sample path we wish to construct is always "active", and steps 3 and 4 in the Event Matching Algorithm above are never invoked.

□

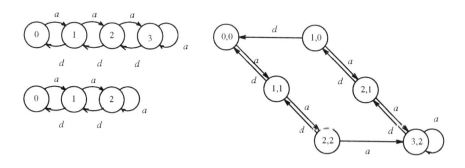

Fig. 13. State transition diagrams of $M/M/1/2$, $M/M/1/3$, and augmented system

Example 13. Concurrent sample path construction for an $M/G/1/K$ queueing system using ASA.
Consider an $M/G/1/K$ (i.e., allow the service time distribution to be arbitrary). Let us now reverse the role of the two systems in the last example: we observe a sample path of the $M/G/1/2$ system and wish to construct a sample path of the $M/G/1/3$ system. Interestingly, this role reversal causes the observability condition to be violated at the augmented system state $(0,1)$, since $\Sigma_G(1) = \{a,d\} \supset \Sigma_G(0) = \{a\}$. Intuitively, when the observed system is empty, there is no feasible d event available to be used in the construction of the sample path of the $M/G/1/3$ system beyond this point. Therefore, the construction is suspended and we set $\mathcal{M}_k^2 = 1$. However, since the only event feasible is a, the next transition leads the observed system to $x_{k+1}^2 = 1$ and observability is satisfied. Moreover, condition (43) is also satisfied regardless of the service time distribution, since the age of event d for both systems is always zero when a occurs at state $x_k^2 = 0$.

□

It should be emphasized that the augmented system is useful only in checking whether observability is satisfied or not. The Event Matching Algorithm, however, does not require constructing such a system.

It is possible to extend ASA to DES with at most one non-Markovian event. This requires the *Age Matching Algorithm* described in [15] and Chapter 9 of [11]. The key idea is to now store not only the state where a sample path construction is suspended, but also the age of the non-Markovian event when this happens (if such an event is feasible at that state). Then, to reset the mode to "active" requires waiting until a point in time where x_k is such that observability is satisfied and the age of the observed non-Markovian event matches that stored at the point of suspension. Extensions of ASA to DES with completely arbitrary event lifetime distributions are possible through techniques such as the "time warping" scheme described in Chapter 9 of [11]. However, additional storage and computational costs need to be imposed. The general sample path constructability problem remains open. In addition, it is necessary to develop systematic frameworks for comparing various schemes, such as the SC and ASA approaches, in terms of factors such as computational cost, variance properties, and suitability for various parallel processing architectures.

ASA for optimal control of elevator systems. Returning to Example 3 in section 2, recall that in the "uppeak" case, i.e., the case where passengers arrive at the buiding lobby only, our problem was reduced to determining an optimal threshold value. In particular, when more than a single car are present at the lobby, only one is made available and it is not allowed to leave until a certain threshold, K, of passengers in it is reached. To determine the optimal value of K, an approach based on ASA was adopted. First, any one value in the admissible set $\{K_1, \ldots, K_n\}$ was selected, say K_1. The system was then operated under this control policy with this threshold value. Based on the data observed while the system was in operation, sample paths were constructed under all other values $K_2, \ldots, K_n$, using an extension of the Event Matching Algorithm presented earlier. As a result, a number of performance measures were estimated and an optimal value ultimately selected (the details of the actual scheme for selecting the optimal value are omitted here). Note, however, that this approach never interfered with the system's normal operation and required a single "trial" period under K_1 only. The stochastic timed automaton model used in this case for a system with M cars was based on an event set $\Sigma = \{a, r_1, \ldots, r_M\}$, where a denotes a passenger arrival event, and r_i denotes the return of the ith car to the lobby after a trip to deliver passngers to various floors (note that events representing the car's stops at these floors are not necessary in the model). The state set X consisted of all $(x, y_1, \ldots, y_M)$ such that $x \in \{0, 1, \ldots\}$ represents the number of passengers at the lobby (possibly already waiting inside a car), and $y_i \in \{I, B\}$ denotes the state of car i, i.e., idle (I) or busy (B).

Acknowledgement The authors would like to thank their students who read earlier versions of this paper and provided many useful comments.

References

1. F. Baccelli, G. Cohen, and B. Gaujal. Recursive equations and basic properties of timed petri nets. *Journal of Discrete Event Dynamic Systems*, 2:415–439, 1992.
2. F. Baccelli, G. Cohen, G.J. Olsder, and J.P. Quadrat. *Synchronization and Linearity*. Wiley, 1992.
3. S. Balemi, G. J. Hoffmann, P. Gyugyi, H. Wong-Toi, and G. F. Franklin. Supervisory control of a rapid thermal multiprocessor. *IEEE Trans. Automatic Control*, 38(7):1040–1059, July 1990.
4. S. Balemi, P.Kozàk, and R. Smedinga, editors. *Discrete Event Systems: Modeling and Control*. Birkhäuser, Basel, 1993.
5. G. Bao and C.G. Cassandras. First and second derivative estimators for closed Jackson-like queueing networks using perturbation analysis techniques. In *Proceedings 31st IEEE Conf. on Decision and Control*, pages 698–703, 1993.
6. J.G. Braker. *Algorithms and applications in timed discrete event systems*. PhD thesis, Delft university of Technology, 1993.
7. J.G. Braker and G.J. Olsder. The power algorithm in max-algebra. *Linear Algebra and its Applications*, 182:67–89, 1993.
8. B. A. Brandin. The supervisory control of an experimental manufacturing cell. In *Proc. 32nd Allerton Conf. on Communication, Control, and Computing*, 1994.
9. P. Brémaud and F.J. Vazquez-Abad. On the pathwise computation of derivatives with respect to the rate of a point process: The phantom RPA method. *Queueing Systems*, 10:249–270, 1992.
10. X. Cao. *Realization probabilities - the dynamics of queueing systems*. Springer-Verlag, 1994.
11. C.G. Cassandras. *Discrete Event Systems: Modeling and Performance Analysis*. Irwin, 1993.
12. C.G. Cassandras. Special issue on parallel simulation and optimization of discrete event systems. *Journal of Discrete Event Dynamic Systems*, to appear, 1995.
13. C.G. Cassandras and V. Julka. Marked/phantom algorithms for a class of scheduling problems. *Queueing Systems*, to appear, 1995.
14. C.G. Cassandras, J.I. Lee, and Y.C. Ho. Efficient parametric analysis of performance measures for communications networks. *IEEE Journal on Selected Areas in Communications*, 8, 9:1709–1722, 1990.
15. C.G. Cassandras and S.G. Strickland. Observable augmented systems for sensitivity analysis of Markov and semi-Markov processes. *IEEE Trans. on Automatic Control*, AC-34, 10:1026–1037, 1989.
16. C.G. Cassandras and S.G. Strickland. On-line sensitivity analysis of Markov chains. *IEEE Trans. on Automatic Control*, AC-34, 1:76–86, 1989.
17. Y.-L. Chen, S. Lafortune, and F. Lin. Study of feature interactions in telecommunication systems using the supervisory control theory of discrete event systems. Technical Report CGR-94-14, College of Engineering Control Group Reports, University of Michigan, November 1994.
18. S. L. Chung, S. Lafortune, and F. Lin. Limited lookahead policies in supervisory control of discrete event systems. *IEEE Trans. Automatic Control*, 37(12):1921–1935, December 1992.
19. R. Cieslak, C. Desclaux, A. Fawaz, and P. Varaiya. Supervisory control of discrete-event processes with partial observations. *IEEE Trans. Automatic Control*, 33(3):249–260, March 1988.

20. G. Cohen, D. Dubois, J.P. Quadrat, and M. Viot. A linear system-theoretic view of discrete event processes and its use for performance evaluation in manufacturing. *IEEE Transactions on Automatic Control*, AC-30:210–220, 1985.
21. G. Cohen and J.-P. Quadrat, editors. *11th International Conference on Analysis and Optimization of Systems, Discrete Event Systems.* Springer Verlag, 1994.
22. R.A. Cuninghame-Green. *Minimax Algebra*. Number 166 in Lecture Notes in Economics and Mathematical Systems. Springer-Verlag, Berlin, 1979.
23. R.A. Cuninghame-Green and P.Butkovič. Discrete-event dynamic systems: the convex case. Technical Report 93/12, School of Mathematics and Statistics, The University of Birminghame, 1993.
24. M. Fu and J.Q. Hu. Extensions and generalizations of smoothed perturbation analysis in a generalized semi-Markov process framework. *IEEE Trans. on Automatic Control*, AC-37:1483–1500, 1992.
25. B. Gaujal. *Parallélisme et simulation des systèmes à évenements discrets*. PhD thesis, Université de Nice Sophia-Antipolis, 1994.
26. P. Glasserman. *Gradient estimation via perturbation analysis*. Kluwer, Boston, 1991.
27. P. Glasserman and W.B. Gong. Smoothed perturbation analysis for a class of discrete event systems. *IEEE Trans. on Automatic Control*, AC-35:1218–1230, 1990.
28. P. Glasserman, J.Q. Hu, and S.G. Strickland. Strong consistency of steady state derivative estimations. *Probability in the Engineering and Information Sciences*, 5:391–413, 1991.
29. P. Glasserman and D. D. Yao. *Monotone Structure in Discrete-Event Systems*. John Wiley and Sons, New York, 1994.
30. P. Glasserman and D.D. Yao. Algebraic structure of some stochastic discrete event systems with applications. *Journal of Discrete Event Dynamic Systems*, 1, 1:7–36, 1991.
31. W.B. Gong and Y.C. Ho. Smoothed perturbation analysis of discrete event systems. *IEEE Trans. on Automatic Control*, AC-32:858–866, 1987.
32. J. Gunawardena. Min-max functions. *Journal of Discrete Event Dynamic Systems*, 4:377–407, 1994.
33. J. Gunawardena, editor. *Idempotency*. Publications of the Newton Institute. Cambridge University Press, to appear.
34. N. Ben Hadj-Alouane, S. Lafortune, and F. Lin. Variable lookahead supervisory control with state information. *IEEE Trans. Automatic Control*, 39(12):2398–2410, December 1994.
35. Y.C. Ho and X. Cao. Perturbation analysis and optimization of queueing networks. *Journal of optimization theory and Applications*, 19:559–582, 1983.
36. Y.C. Ho and X. Cao. *Perturbation Analysis of Discrete Event Dynamic Systems*. Kluwer, Boston, 1991.
37. Y.C. Ho, X. Cao, and C.G. Cassandras. Infinitesimal and finite perturbation analysis for queueing networks. *Automatica*, 19:439–445, 1983.
38. Y.C. Ho and C.G. Cassandras. A new approach to the analysis of discrete event dynamic systems. *Automatica*, 19:149–167, 1983.
39. Y.C. Ho, A. Eyler, and D.T. Chien. A gradient technique for general buffer storage design in a serial production line. *International journal of production research*, 17:557–580, 1979.
40. Y.C. Ho and S. Li. Extensions of perturbation analysis of discrete event dynamic systems. *IEEE Trans. on Automatic Control*, AC-37:258–262, 1988.

41. Y.C. Ho, R. Sreenivas, and P. Vakili. Ordinal optimization of discrete event dynamic systems. *Journal of Discrete Event Dynamic Systems*, 2, 1:61–88, 1992.
42. J. E. Hopcroft and J. D. Ullman. *Introduction to Automata Theory, Languages, and Computation*. Addison-Wesley, 1979.
43. T. Kailath. *Linear Systems*. Prentice-Hall, 1980.
44. R. M. Karp. A characterization of the minimum cycle mean in a digraph. *Discrete Mathematics*, 23:309–311, 1978.
45. R. Kumar and V. K. Garg. *Modeling and Control of Logical Discrete Event Systems*. Kluwer Academic Publishers, 1995.
46. S. Lafortune and E. Chen. The infimal closed controllable superlanguage and its application in supervisory control. *IEEE Trans. Automatic Control*, 35(4):398–405, April 1990.
47. S. Lafortune and F. Lin. On tolerable and desirable behaviors in supervisory control of discrete event systems. *Discrete Event Dynamic Systems: Theory and Applications*, 1(1):61–92, May 1991.
48. F. Lin and W. M. Wonham. On observability of discrete-event systems. *Information Sciences*, 44:173–198, 1988.
49. Y.-J. Lin and N. D. Griffeth, Editors. Special Issue on Managing Feature Interactions in Telecommunications Systems. *IEEE Communications Magazine*, 31(8), August 1993.
50. A. R. Modarressi and R. A. Skoog. Signaling System No. 7: A tutorial. *IEEE Communications Magazine*, 28(7):19–35, July 1990.
51. S. Morioka and T. Yamada. Performance evaluation of marked graphs by linear programming. *International Journal of Systems Science*, 22:1541–1552, 1991.
52. T. Murata. Petri nets: Properties, analysis and applications. *Proceedings of the IEEE*, 77:541–580, 1989.
53. R. D. Nussbaum. Lattice isomorphisms and iterates of nonexpansive maps. *Nonlinear Analysis, Theory, Methods and Applications*, 22:945–970, 1994.
54. G.J. Olsder. Eigenvalues of dynamic min-max systems. *Journal of Discrete Event Dynamic Systems*, 1:177–207, 1991.
55. G.J. Olsder. Synchronized continuous flow systems. In S. Balemi, P. Kozàk, and R. Smedinga, editors, *Discrete Event Systems; Modelling and Control*, pages 113–124. Birkhäuser, Basel, 1993.
56. G.J. Olsder. On structural properties of min-max systems. In Guy Cohen and Jean-Pierre Quadrat, editors, *11th International Conference on Analysis and Optimization of Systems*, pages 237–246. Springer Verlag, 1994.
57. G.J. Olsder. On min-max-plus systems, nonexpansive mappings and periodic solutions. Technical report 95-09 of the Faculty of Technical Mathematics and Informatics, Delft University of Technology, 1995.
58. G.J. Olsder and R.E. de Vries. On an analogy of minimal realizations in conventional and discrete event dynamic systems. In *Algèbres exotiques et systèmes à évenements discrets*, pages 191–213. CRNS/CNET/INRIA, CNET, Issy-les-Moulineaux, France, 1987.
59. J.-E. Pin. Finite semigroups and recognizable languages: an introduction. Technical Report LITP 94.15, Laboratoire informatique théorique et programmation, Institut Blaise Pascal, 4, Place Jussieu, 75252 Paris, 1994.
60. P. J. Ramadge and W. M. Wonham. The control of discrete event systems. *Proc. IEEE*, 77(1):81–98, January 1989.
61. P.J. Ramadge and W.M. Wonham. Supervisory control of discrete event processes. In D. Hinrichsen and A. Isidori, editors, *Feedback Control of Linear and Nonlinear*

Systems, *Lecture Notes on Control and Information Sciences No. 39*, pages 202–214. Springer Verlag, 1982.
62. P.J. Ramadge and W.M. Wonham. Supervisory control of a class of discrete event processes. *SIAM J. Control and Optimization*, 25:206–230, 1987.
63. M.I. Reiman and A. Weiss. Sensitivity analysis for simulations via likelihood ratios. *Operations Research*, 37:285–289, 1989.
64. W. Reisig. *Petri nets*. Springer Verlag, 1985.
65. J.A.C. Resing, R.E. de Vries, M.S. Keane, G. Hooghiemstra, and G.J. Olsder. Asymptotic behavior of random discrete event systems. *Stochastic Processes and their Applications*, 36:195–216, 1990.
66. R. Rubinstein. *Monte Carlo optimization, simulation and sensitivity of queueing networks*. Wiley, 1986.
67. R. Y. Rubinstein and A. Shapiro. *Discrete Event Systems: Sensitivity Analysis and Stochastic Optimization by the Score Function Method*. Wiley, Chicester, 1993.
68. B. De Schutter and B. De Moor. The characteristic equation and minimal state space realization of SISO systems in the max algebra. In Guy Cohen and Jean-Pierre Quadrat, editors, *11th International Conference on Analysis and Optimization of Systems, Discrete Event Systems*, pages 273–282. Springer Verlag, 1994.
69. R. Sengupta and S. Lafortune. A deterministic optimal control theory for discrete event systems. In *Proc. 32nd IEEE Conf. on Decision and Control*, pages 1182–1187, San Antonio, TX, December 1993.
70. R. Suri and M. Zazanis. Perturbation analysis gives strongly consistent sensitivity estimates for the M/G/1 queue. *Management Science*, 34, 1:39–64, 1988.
71. J. G. Thistle. Logical aspects of control of discrete-event systems: A survey of tools and techniques. In *Proc. 11th International Conference on Analysis and Optimization of Systems 1994*, pages 3–15. Springer-Verlag, Lecture Notes in Control and Information Sciences, Vol. 199, 1994.
72. J. G. Thistle, H.-H. Hoang, and R. P. Malhamé. Modélisation et analyse du traitement des appels téléphoniques par l'approche de Ramadge-Wonham: Rapport préliminaire. Technical Report EPM/RT-93/15, École Polytechnique de Montréal, Montréal, Canada, August 1993.
73. J. G. Thistle, R. P. Malhamé, H.-H. Hoang, and S. Lafortune. Application of supervisory control theory to the development of intelligent network services. Technical report, École Polytechnique de Montréal, Montréal, Canada, 1994.
74. P. Vakili. A standard clock technique for efficient simulation. *Operations Research Letters*, 10:445–452, 1991.
75. P. Varaiya and A.B. Kurzhanski, editors. *Discrete Event Systems: Models and Applications*. Springer Verlag, 1988.
76. Y. Wardi and J.Q. Hu. Strong consistency of infinitesimal perturbation analysis for tandem queueing networks. *Journal of Discrete Event Dynamic Systems*, 1, 1:37–59, 1991.
77. W. M. Wonham. Notes on Control of Discrete-Event Systems for ECE 1636F/1637S. University of Toronto, September 1994.
78. W. M. Wonham and P. J. Ramadge. On the supremal controllable sublanguage of a given language. *SIAM J. Control and Optimization*, 25(3):637–659, May 1987.
79. W. M. Wonham and P. J. Ramadge. Modular supervisory control of discrete-event systems. *Math. Control Signals Systems*, 1(1):13–30, 1988.

Feedback Stabilization of Nonlinear Systems: Sufficient Conditions and Lyapunov and Input-output Techniques

Jean-Michel Coron[1], *Laurent Praly*[2] *and Andrew Teel*[3]

[1] Ecole Normale Supérieure de Cachan
Centre de Mathématiques et de Leurs Applications
Unité associée au CNRS - URA-1611
61 Avenue du Président Wilson
94235 Cachan Cedex
FRANCE
coron@cmla.ens-cachan.fr

[2] CAS Ecole des Mines
35 Rue Saint Honoré
77305 Fontainebleau Cedex
FRANCE
praly@cas.ensmp.fr

[3] Department of Electrical Engineering
University of Minnesota
4-174 EE/CS Building
200 Union St. SE
Minneapolis, Minnesota 55455
USA
teel@ee.umn.edu

1 Introduction.

This lecture is devoted to the survey of some recent results on feedback stabilization of nonlinear systems. This text can be seen as a prolongation of the overview written by E. Sontag in 1990 [83] in several directions where progress has been made. It consists of three parts:
The first part is devoted to sufficient conditions on the stabilization problem by means of discontinuous or time-varying state or output feedback.
In the second part, we present some techniques for explicitly designing these feedbacks by using Lyapunov's method. This introduces us with the notion of assignable Lyapunov function and leads us to concentrate our attention on systems having some special recurrent structure.
The third part presents some techniques for designing feedback based on $\mathcal{L}_\infty$ stability properties. This last section also addresses robustness through a small gain theorem.

We wish to thank Randy Freeman and Eduardo Sontag for the very helpful comments they made on some parts of this text while we were writing it.

2 Sufficient conditions for state or output feedback stabilization

2.1 Introduction.

It is a classical result, see e.g. [82] Theorem 7 p. 134, that any linear control system which is controllable can be asymptotically stabilized by means of continuous feedback laws. A natural question is if this result still holds for nonlinear control systems. In 1979 Sussmann has shown that the global version of this result does not hold for nonlinear control systems: in [89] he has given an example of a nonlinear analytic control system which is globally controllable but cannot be globally asymptotically stabilized by means of continuous feedback laws. In [4] Brockett has shown that the local version also does not hold. To get around the problem of impossibility to stabilize many controllable systems by means of continuous feedback laws two main strategies have been proposed
(i) Asymptotic stabilization by means of a discontinuous feedback law -see e.g. the pioneer work by H. Sussmann [89]-
(ii) Asymptotic stabilization by means of a continuous periodic time-varying feedback law -see the pioneer work by Sontag and Sussmann [86], [74], and Section 2.3 below.

In Section 2.2 we give a relation between these two strategies. In Section 2.3 we present results showing that, in many cases, controllability implies stabilizability by means of time-varying static feedback laws.

In many practical situations only part of the state – called the output – is measured and therefore state feedback cannot be implemented; only output feedback are allowed. It is well known, see e.g. [82] Section 6.2, that any linear control system which is controllable and observable can be asymptotically stabilized by means of dynamic continuous feedback laws. Again it is natural to look if this result can be extended to the nonlinear case. In the nonlinear case there are many possible definitions for observability. The weakest requirement for observability is that, given two different states, there exists a control $t \to u(t)$ which leads two outputs which are not identical. With this definition of observability, the nonlinear control system

$$\dot{x} = u \in \mathbb{R}, \, y = x^2 \in \mathbb{R} \tag{1}$$

where the state is x, the control u, and the output y is observable. This system is also clearly controllable and asymptotically stabilizable by means of (stationary) static feedback laws. But, see [14], this system cannot be asymptotically stabilized by means of stationary dynamic feedback laws. Again the introduction of time-varying feedback laws improve the situation; indeed control system (1) can be asymptotically stabilized by means of time-varying dynamic feedback laws. In Section 2.4 we present a result contained in [14] showing that many locally controllable and observable nonlinear control systems can be locally asymptotically stabilized by means of time-varying dynamic output feedback laws.

2 Sufficient conditions for state or output feedback stabilization

2.2 Discontinuous/continuous time-varying stabilizing feedback.

Throughout out all this survey, by (C) we denote the nonlinear control system

$$(C) : \dot{x} = f(x, u), \tag{2}$$

where $x \in \mathbb{R}^n$ is the state, $u \in \mathbb{R}^m$ is the control. We assume that $f \in C^\infty(\mathbb{R}^n \times \mathbb{R}^m; \mathbb{R}^n)$ and that

$$f(0, 0) = 0. \tag{3}$$

The goal of this section is to show a relation between stabilizability by means of discontinuous feedback laws and stabilizability by means of continuous time-varying stabilizing feedback laws.

Before stating this relation let us first recall the definition of asymptotically stable for a time-varying dynamic system – we should in fact say uniformly asymptotically stable –

Definition 1. Let X be in $C^0(\mathbb{R}^n \times \mathbb{R}; \mathbb{R})$. One says that 0 is locally asymptotically stable for $\dot{x} = X(x, t)$ if
(i) for all $\varepsilon > 0$, there exists $\eta > 0$ such that, for all $\tau \in \mathbb{R}$ and for all $t \geq \tau$,

$$(\dot{x} = X(x, t), |x(\tau)| < \eta) \Rightarrow |x(t)| < \varepsilon \tag{4}$$

and if
(ii) there exists $\delta > 0$ such that, for all $\varepsilon > 0$, there exists $M > 0$ such that, for all s in $\mathbb{R}$,

$$\dot{x} = X(x, t) \text{ and } |x(s)| < \delta \tag{5}$$

imply

$$|x(\tau)| < \varepsilon, \forall \tau > s + M. \tag{6}$$

If, moreover, for all $\delta > 0$, there exists $M > 0$ such that (5) implies (6) for all s in $\mathbb{R}$, one says that 0 is globally asymptotically stable for $\dot{x} = X(x, t)$.

Throughout all this paper, and in particular in (4) and (5), by $\dot{x} = X(x, t)$ we denote any maximal solution of this differential equation. Let us emphasize that, since the vector field X is only continuous, the Cauchy problem $\dot{x} = X(x, t), x(t_0) = x_0$, where t_0 and x_0 are given, may have many maximal solutions. Let us recall that Kurzweil in [46] has shown that, even for vector fields which are only continuous, asymptotic stability is equivalent to the existence of a Lyapunov function.

Let us now define "asymptotically stabilizable by means of a continuous periodic feedback law"

Definition 2. System (C) is locally (resp. globally) asymptotically stabilizable by means of a continuous periodic feedback law of period T if there exists $u \in C^0(\mathbb{R}^n \times \mathbb{R}; \mathbb{R}^m)$ satisfying

$$u(x, t+T) = u(x,t), \quad \forall (x,t) \in \mathbb{R}^n \times \mathbb{R}, \tag{7}$$

$$u(0, t) = 0, \quad \forall t \in \mathbb{R}, \tag{8}$$

such that, for the system $\dot{x} = f(x, u(x,t))$, 0 is a locally (resp. globally) asymptotically stable point.

We now need to specify the meaning of asymptotic stability for a system $\dot{x} = X(x)$ where X is a discontinuous vector field. Many definitions are possible but, following H. Hermes [31], it seems natural to adopt

Definition 3. Let $X \in L^\infty_{\text{loc}}(\mathbb{R}^n; \mathbb{R}^n)$. Then 0 is a locally asymptotically stable point of $\dot{x} = X(x)$ if (i) and (ii) of Definition 1 hold for any (maximal) solution in the Filippov sense of $\dot{x} = X(x)$. If, moreover, (ii) of Definition 1 holds for any $\delta > 0$ and any (maximal) solution in the Filippov sense of $\dot{x} = X(x)$, then 0 is a globally asymptotically stable point of $\dot{x} = X(x)$.

Let us recall that a solution in the Filippov sense of $\dot{x} = X(x)$ on an interval I is (see [F]) a locally absolutely continuous map from I into $\mathbb{R}^n$ such that

$$\dot{x}(t) \in F(x(t)) \quad \text{for almost all } t \in I \tag{9}$$

with

$$F(x) := \bigcap_{\epsilon > 0} \bigcap_{|N|=0} \overline{\text{conv}} X((x + \epsilon B) \setminus N), \tag{10}$$

where B is the unit ball of $\mathbb{R}^n$, and, for a set A, $|A|$ is the Lebesgue measure of A and $\overline{\text{conv}} A$ is the smaller closed convex set containing A. Of course, a maximal solution in the Filippov sense of $\dot{x} = X(x)$ is a solution x in the Filippov sense on some interval I such that there exists no solution in the Filippov sense defined on an interval which contains strictly I and which is equal to x on I.

Note that, if X is continuous,

$$F(y) = \{X(y)\} \tag{11}$$

and therefore in this case our definition of asymptotic stability coincide with the one given in Definition 1.

We now define "asymptotically stabilizable by means of a discontinuous feedback law".

Definition 4. System (C) is locally (resp. globally) asymptotically stabilizable by means of a discontinuous feedback law if there exists $u \in L^\infty_{\text{loc}}(\mathbb{R}^n; \mathbb{R}^m)$ such that

$$\text{Essential Sup } \{|u(x)|;\ |x| < \epsilon\} \to 0 \text{ as } \epsilon \to 0, \tag{12}$$

and 0 is a locally (resp. globally) asymptotically stable point of $\dot{x} = f(x, u(x))$.

2 Sufficient conditions for state or output feedback stabilization

The reason for considering in Definition 4 solutions in the Filippov sense is, as explained in [31], the following one : the feedback law $u(x(t))$ is determined after making a measurement of the state $x(t)$ at time t; of course this measurement gives only an approximation of $x(t)$: there is an "error" $e(t)$ between $x(t)$ and its measurement. A direct consequence of Lemma 3 in [31] is the following proposition which is proved in [19] -see also [31]-

Proposition 5. *Assume that f is locally Lipschitzian with respect to x. Let $x : [0,T] \to \mathbb{R}^n$ be a Filippov solution of $\dot{x} = f(x, u(x))$ where $u \in L^\infty_{loc}(\mathbb{R}^n; \mathbb{R}^m)$. Let ϵ be a positive real number. Then there exist $e \in L^\infty((0,T); \mathbb{R}^m)$ and an absolutely continuous function $y : [0,T] \to \mathbb{R}^n$ such that*

$$|e(t)| \leq \epsilon \text{ for all } t \text{ in } (0,T), \tag{13}$$

$$\dot{y}(t) = f(y(t), u(y(t) + e(t))) \text{ for almost all } t \text{ in } (0,T), \tag{14}$$

$$y(0) = x(0), \tag{15}$$

and

$$|y(t) - x(t)| \leq \epsilon \text{ for all } t \text{ in } [0,T]. \tag{16}$$

Proposition 5 justifies our definition of asymptotically stabilization by means of a discontinuous feedback law. With this definitions one has the following theorem, which is proved in [19],

Theorem 6. *Assume that $\dot{x} = f(x,u)$ can be locally (resp. globally) asymptotically stabilized by means of a discontinuous feedback law. Then, for any $T > 0$, $\dot{x} = f(x,u)$ can be locally (resp. globally) stabilized by means of a continuous time-varying feedback law of period T; if, moreover, $\dot{x} = f(x,u)$ is an affine system (i.e. $f(x,u) = f_0(x) + \sum_{i=1}^m u_i f_i(x)$), then $\dot{x} = f(x,u)$ can be locally (resp. globally) asymptotically stabilized by means of a continuous feedback law (independent of t : $u = u(x)$).*

Remark 7. There are completely controllable affine systems which are globally asymptotically stabilized by means of a continuous periodic time varying feedback law which cannot be locally asymptotically stabilized by means of a continuous feedback law (e.g. $\dot{x}_1 = u_1$, $\dot{x}_2 = u_2$, $\dot{x}_3 = x_1 u_2 - x_2 u_1$; see [4], [74], [9], or Section 2.3). By Theorem 6, these systems cannot be locally asymptotically stabilized by means of a discontinuous feedback law; see also [72] for the same conclusion but with a different approach. Note that this is not in contradiction with [89]: indeed our definition of asymptotic stability is different from the one used in [89] (see e.g. the definition of "steers M to p" in [89]; in particular we do not have any "exit rule" E on the singular set of u in our definition of asymptotic stability).

2.3 Time-varying feedback.

This section is divided in two subsections: the first subsection concerns nonlinear control system without drift, the second subsection concerns systems which may have a drift term.

Systems without drift. In this subsection we assume that

$$f(x, u) = \sum_{i=1}^{m} u_i f_i(x). \tag{17}$$

Let us denote by Lie$\{f_1, \ldots, f_m\}$ the Lie subalgebra generated by the vector fields $f_1, \ldots, f_m$. Then one has

Theorem 8. *Assume that, for all $x \in \mathbb{R}^n \setminus \{0\}$,*

$$\{h(x); h \in Lie\{f_1, \ldots, f_m\}\} = \mathbb{R}^n. \tag{18}$$

Then, for all $T > 0$, there exists u in $C^\infty(\mathbb{R}^n \times \mathbb{R}; \mathbb{R}^m)$ such that

$$u(0, t) = 0, \ \forall t \in \mathbb{R}, \tag{19}$$

$$u(x, t+T) = u(x, t), \ \forall x \in \mathbb{R}^n, \ \forall t \in \mathbb{R}, \tag{20}$$

and 0 is globally asymptotically stable for

$$\dot{x} = f(x, u(x,t)) = \sum_{i=1}^{m} u_i(x,t) f_i(x). \tag{21}$$

The proof of this theorem is given in [9]; it relies on a method, that we have called the "return method", which can also be used to prove controllability in some cases -see, e.g., [13], [15]- or obtain numerical techniques for the steering of arbitrary systems without drift, see [84].

General systems. Let us first point out that in [86] Sontag and Sussmann have proved that any one dimensional state nonlinear control system which is locally (resp. globally) controllable can be locally (resp. globally) asymptotically stabilized by means of time-varying static feedback laws. Let us also point out that it follows from Sussmann [89] that a result similar to Theorem 8 does not hold for systems with a drift term: more precisely there are analytic control systems (C) controls which are globally controllable for which there is no u in $C^0(\mathbb{R}^n \times \mathbb{R}; \mathbb{R}^m)$ for which 0 is globally asymptotically stable for $\dot{x} = f(x, u(x,t))$. In fact the proof of [89] requires uniqueness of the trajectories of $\dot{x} = f(x, u(x,t))$. But this can always been assumed; indeed it follows easily from Kurzweil's result [46] that, if there exists u in $C^0(\mathbb{R}^n \times \mathbb{R}; \mathbb{R}^m)$ such that 0 is globally asymptotically stable for $\dot{x} = f(x, u(x,t))$, then there exists $\bar{u}$ in $C^0(\mathbb{R}^n \times \mathbb{R}; \mathbb{R}^m) \cap C^\infty((\mathbb{R}^n \setminus \{0\}) \times \mathbb{R}; \mathbb{R}^m)$ such that 0 is globally asymptotically

2 Sufficient conditions for state or output feedback stabilization

stable for $\dot{x} = f(x, \bar{u}(x,t))$; for such a $\bar{u}$ one has uniqueness of the trajectories of $\dot{x} = f(x, \bar{u}(x,t))$. But we are going to see in this subsection that a local version of Theorem 8 holds for many control systems which are Small Time Locally Controllable (STLC).

Let us again introduce some definitions

Definition 9. The origin (of $\mathbb{R}^n$) is locally continuously reachable (for (C)) in small time if, for all positive real number T, there exist a positive real number ε and u in $C^0\left(\mathbb{R}^n; L^1\left((0,T); \mathbb{R}^m\right)\right)$ such that

$$\mathrm{Sup}\{|u(a)(t)|; t \in (0,T)\} \to 0 \text{ as } a \to 0,$$

$$(\dot{x} = f(x, u(x(0))(t)), |x(0)| < \varepsilon) \Rightarrow x(T) = 0.$$

Let us notice that, following a method due to M. Kawski [42] (see also [30]), we have proved in [10, Lemma 3.1 and Section 5] that "many" sufficient conditions for Small Time Locally Controllability imply that the origin is locally continuously reachable in small time. This is in particular the case for the Hermes condition [32] or [91] and its generalization due to H.J. Sussmann [92], Theorem 7.3 ; this is in fact also the case for the Bianchini and Stefani condition [3, Corollary p. 970], which extends [92, Theorem 7.3] .

Our next definition is

Definition 10. System (C) is locally stabilizable in small time by means of almost smooth periodic time-varying feedback laws if, for any positive real number T, there exist ε in $(0, +\infty)$ and u in $C^0(\mathbb{R}^n \times \mathbb{R}; \mathbb{R}^m)$ of class C^∞ on $(\mathbb{R}^n \setminus \{0\}) \times \mathbb{R}$ such that

$$u(0,t) = 0, \forall t \in \mathbb{R}, \tag{22}$$

$$u(x, t+T) = u(x,t), \forall t \in \mathbb{R}, \tag{23}$$

$$((\dot{x} = f(x, u(x,t)) \text{ and } x(s) = 0) \Rightarrow (x(t) = 0 \, \forall t \geq s)), \forall s \in \mathbb{R}, \tag{24}$$

$$((\dot{x} = f(x, u(x,t)) \text{ and } |x(s)| \leq \varepsilon) \Rightarrow$$
$$(x(t) = 0, \forall t \geq s + T)) \, \forall s \in \mathbb{R}. \tag{25}$$

Note that (23), (24), and (25) imply that 0 is locally asymptotically stable for $\dot{x} = f(x, u(x,t))$; see [12, Lemma 2.15] for a proof.

Definition 11. [11]. The strong jet accessibility subspace of (C) at $(\bar{x}, \bar{u}) \in \mathbb{R}^n \times \mathbb{R}^m$ is the subspace of $\mathbb{R}^n$, denoted by $a(\bar{x}, \bar{u})$, spanned by

$$\{h(\bar{x}); h \in \{\partial^{|\alpha|} f / \partial u^\alpha(\cdot, \bar{u}) \alpha \in \mathbb{N}^m, |\alpha| \geq 1 \cup \mathrm{Br}_2(f, \bar{u})\}\} \tag{26}$$

where $\mathrm{Br}_2(f, \bar{u})$ denotes the set of iterated Lie brackets of length at least 2 of vector fields in $\{\partial^{|\alpha|} f / \partial u^\alpha(\cdot, \bar{u}); \alpha \in \mathbb{N}^m\}$.

Remark 12. One easily checks that the usual strong accessibility subspace of (C) at $\bar{x}$ (see e.g. [93], p. 101) contains $a(\bar{x}, \bar{u})$ for all $\bar{u}$ in $\mathbb{R}^m$ and that, if f is analytic with respect to x and u or is a polynomial with respect to u, these inclusions are all equalities.

Our last definition before stating our main result is

Definition 13. System (C) satisfies the strong jet accessibility rank condition at $(\bar{x}, \bar{u})$ if

$$a(\bar{x}, \bar{u}) = \mathbb{R}^n. \tag{27}$$

Remark 14. It follows from Remark 12 that if (C) satisfies the strong jet accessibility rank condition at $(\bar{x}, \bar{u})$ then it satisfies the usual strong accessibility rank condition at $\bar{x}$ and the converse holds if f is analytic with respect to x and u or is a polynomial with respect to u.

Note that, if (C) is locally stabilizable in small time by means of almost smooth periodic time-varying feedback laws, then $0 \in \mathbb{R}^n$ is locally continuously reachable for (C). The main result of this subsection is that the converse holds if $n \notin \{2, 3\}$ and if (C) satisfies the strong jet accessibility rank condition at $(0, 0)$, i.e.

Theorem 15. *Assume that 0 is locally continuously reachable in small time, that (C) satisfies the strong jet accessibility rank condition at $(0, 0)$, and that*

$$n \notin \{2, 3\}. \tag{28}$$

Then (C) is locally stabilizable in small time by means of almost smooth periodic time-varying feedback laws.

This theorem is proved in [12] when $n \geq 4$ and in [16] when $n = 1$. Let us just sketch the proof of [12].

Let I be an interval of $\mathbb{R}$. By a trajectory of the control system (C) on I we mean $(\gamma, u) \in C^\infty(I; \mathbb{R}^n \times \mathbb{R}^m)$ satisfying $\dot{\gamma}(t) = f(\gamma(t), u(t))$ for all t in I. The linearized control system around (γ, u) is $\dot{\xi} = A(t)\xi + B(t)w$ where the state is $\xi \in \mathbb{R}^n$, the control is $w \in \mathbb{R}^m$, and $A(t) = \partial f/\partial x(\gamma(t), u(t)) \in \mathcal{L}(\mathbb{R}^\setminus; \mathbb{R}^\setminus)$, $B(t) = \partial f/\partial u(\gamma(t), u(t)) \in \mathcal{L}(\mathbb{R}^\updownarrow; \mathbb{R}^\setminus)$ for all t in I. We first introduce the following definition

Definition 16. *The trajectory (γ, u) is supple on $S \subset I$ if, for all s in S,*

$$\mathrm{Span}\{((d/dt) - A(t))^i B(t)\big|_{t=s} w \ ; \ w \in \mathbb{R}^m, i \geq 0\} = \mathbb{R}^n. \tag{29}$$

In (29) we use the classical convention $(d/dt - A(t))^0 B(t) = B(t)$. Let us recall that L. Silverman and H. Meadows have shown in [75] that (2.1) implies that the linearized control system around (γ, u) is controllable with impulsive

2 Sufficient conditions for state or output feedback stabilization

controls at time s (in the sense of [40] p. 614). Let T be a positive real number. For u in $C^0(\mathbb{R}^n \times [0,T]; \mathbb{R}^m)$ and a in $\mathbb{R}^n$, let $x(a, \cdot; u)$ be the maximal solution of $\partial x/\partial t = f(x, u(a,t))$, $x(a,0;u) = a$. Let, also, C^* be the set of $u \in C^0(\mathbb{R}^n \times [0,T]; \mathbb{R}^m)$ of class C^∞ on $(\mathbb{R}^n\setminus\{0\}) \times [0,T]$ and vanishing on $\{0\} \times [0,T]$. For simplicity, in this sketch of proof, we omit some details which are important to take care of the uniqueness property (24) (note that without (24) one does not have stability).

Step 1. Using (1.8), (1.9), and [10] or [11], one proves that there exist ϵ_1 in $(0, +\infty)$ and u_1 in C^*, vanishing on $\mathbb{R}^n \times \{T\}$, such that

$$|a| \leq \epsilon_1 \Rightarrow x(a,T;u_1) = 0, \tag{30}$$

$$0 < |a| \leq \epsilon_1 \Rightarrow (x(a,\cdot;u_1), u_1(a,\cdot)) \text{ is supple on } [0,T]. \tag{31}$$

Step 2. Let Γ be a closed submanifold of $\mathbb{R}^n\setminus\{0\}$ of dimension 1 such that $\Gamma \subset \{x \in \mathbb{R}^n; 0 < |x| < \epsilon_1\}$. Perturbing in a suitable way u_1 one obtains a map u_2 in C^*, vanishing on $\mathbb{R}^n \times \{T\}$, such that

$$|a| \leq \epsilon_1 \Rightarrow x(a,T;u_2) = 0, \tag{32}$$

$$0 < |a| \leq \epsilon_1 \Rightarrow (x(a,\cdot;u_2), u_2(a,\cdot)) \text{ is supple on } [0,T], \tag{33}$$

and

$$a \in \Gamma \to x(t,a;u_2) \text{ is an embedding of } \Gamma \text{ into } \mathbb{R}^n\setminus\{0\}, \forall t \in [0,T). \tag{34}$$

Here one uses the assumption $n \geq 4$ and one proceeds as in the classical proof of the Whitney embedding theorem (see e.g. [26] Chapter II, Section 5). Let us emphasize that this is only in this step that we use this assumption.

Step 3. From Step 2 one deduces the existence of u_3^* in C^*, vanishing on $\mathbb{R}^n \times \{T\}$, and of an open neighborhood $\mathcal{N}^*$ of Γ in $\mathbb{R}^n\setminus\{0\}$ such that

$$a \in \mathcal{N}^* \Rightarrow \S(\dashv, \mathcal{T}; \sqcap_\ni^*) = \prime, \tag{35}$$

$$a \in \mathcal{N}^* \to \S(\dashv, \sqcup; \sqcap_\ni^*) \text{ is an embedding of } \mathcal{N}^* \text{ into } \mathbb{R}\setminus\{\prime\}, \forall \sqcup \in [\prime, \mathcal{T}). \tag{36}$$

This embedding property allows to transform the open-loop control u_3^* into a feedback law u_3 on $\{(x(a,t;u_3),t); a \in \mathcal{N}, \sqcup \in [\prime, \mathcal{T})\}$. So - see in particular (2.7) and note that u_3^* vanishes on $\mathbb{R}^n \times \{T\}$ - there exist u_3 in C^* and an open neighborhood $\mathcal{N}$ of Γ in $\mathbb{R}^n\setminus\{0\}$ such that

$$(x(0) \in \mathcal{N} \text{ and } \Y = \{(\S, \sqcap_\ni(\S, \sqcup))\}) \Rightarrow (x(T) = 0). \tag{37}$$

One can also impose, for all τ in $[0,T]$,

$$(\dot{x} = f(x, u_3(x,t)) \text{ and } x(\tau) = 0) \Rightarrow (x(t) = 0 \quad \forall t \in [\tau, T]). \tag{38}$$

Step 4. In this last step one shows the existence of a closed submanifold of $\mathbb{R}^n\setminus\{0\}$ of dimension 1 included in the set $\{x \in \mathbb{R}^n; 0 < |x| < \epsilon_1\}$ such that for any neighborhood $\mathcal{N}$ of Γ in $\mathbb{R}^n\setminus\{0\}$ there exists u_4 in C^* such that, for some ϵ_4 in $(0, +\infty)$,

$$(\dot{x} = f(x, u_4(x,t)) \text{ and } |x(0)| < \epsilon_4) \Rightarrow (x(T) \in \mathcal{N} \cup \{\prime\}), \tag{39}$$

$$((\dot{x} = f(x, u_4(x,t)) \text{ and } x(\tau) = 0) \Rightarrow (x(t) = 0 \quad \forall t \in [\tau, T])) \forall \tau \in [0, T]. \tag{40}$$

Finally let $u : \mathbb{R}^n \times \mathbb{R} \to \mathbb{R}^m$ be equal to u_4 on $\mathbb{R}^n \times [0, T]$, $2T$-periodic with respect to time, and such that $u(x,t) = u_3(x, t-T)$ for all (x,t) in $\mathbb{R}^n \times (T, 2T)$. Then u vanishes on $\{0\} \times \mathbb{R}$, is continuous on $\mathbb{R}^n \times (\mathbb{R}\setminus\mathbb{Z}T)$, of class C^∞ on $(\mathbb{R}^n\setminus\{0\}) \times (\mathbb{R}\setminus\mathbb{Z}T)$, and satisfies

$$(\dot{x} = f(x, u(x,t)) \text{ and } |x(0)| < \epsilon_4) \Rightarrow (x(2T) = 0), \tag{41}$$

$$(\dot{x} = f(x, u(x,t)) \text{ and } x(\tau) = 0) \Rightarrow (x(t) = 0, \quad \forall t \geq \tau) \forall \tau \in \mathbb{R}, \tag{42}$$

which implies, see [12], that (25) holds, with $4T$ instead of T and $\epsilon > 0$ small enough, and that 0 is uniformly locally asymptotically stable for the system $\dot{x} = f(x, u(x, t))$. Since T is arbitrary, Theorem 15 is proved (modulo a problem of regularity of u at (x,t) in $\mathbb{R}^n \times \mathbb{Z}T$ that is fixed in [12]).

Remark 17. We conjecture that assumption (28) can be removed in Theorem 15.

2.4 Time-varying output feedback.

In this section only part of the state (called the output) is measured; let us denote by $(\tilde{C})$ we denote the control system

$$(\tilde{C}) : \dot{x} = f(x, u), \ y = h(x), \tag{43}$$

where $x \in \mathbb{R}^n$ is the state, $u \in \mathbb{R}^m$ is the control, and $y \in \mathbb{R}^p$ is the output. Again $f \in C^\infty(\mathbb{R}^n \times \mathbb{R}^m; \mathbb{R}^n)$ and satisfies (3); we also assume that $h \in C^\infty(\mathbb{R}^n; \mathbb{R}^p)$ and satisfies

$$h(0) = 0. \tag{44}$$

In order to state the main result of this section we first introduce some definitions

Definition 18. System $(\tilde{C})$ is said to be locally stabilizable in small time by means of continuous static periodic time-varying output feedback laws if, for any positive real number T, there exist ε in $(0, +\infty)$ and u in $C^0(\mathbb{R}^n \times \mathbb{R}; \mathbb{R}^m)$ such that (22), (23), (24), (25) hold and such that

$$u(x,t) = \bar{u}(h(x), t) \tag{45}$$

for some $\bar{u}$ in $C^0(\mathbb{R}^p \times \mathbb{R}; \mathbb{R}^n)$.

2 Sufficient conditions for state or output feedback stabilization

Our next definition concerns dynamic stabilizability.

Definition 19. System $(\tilde{C})$ is locally stabilizable in small time by means of continuous dynamic periodic time-varying state (resp. output) feedback laws if, for some integer $k \geq 0$, the control system

$$\dot{x} = f(x, u), \ \dot{z} = v, \ \tilde{h}(x, z) = (h(x), z), \tag{46}$$

where the state is $(x, z) \in \mathbb{R}^n \times \mathbb{R}^k$, the control $(u, v) \in \mathbb{R}^m \times \mathbb{R}^k$, and the output $\tilde{h}(x, z) \in \mathbb{R}^p \times \mathbb{R}^k$, is locally stabilizable in small time by means of continuous static periodic time-varying state (resp. output) feedback laws.

In the above definition, System (46) with $k = 0$ is, by convention, system $(\tilde{C})$. Let us also point out that it is proved in [10, Section 3], that if, for system $(\tilde{C})$, 0 is continuously reachable in small time then $(\tilde{C})$ is locally stabilizable in small time by means of continuous dynamic periodic time-varying state feedback; this also follows from Theorem 15 – but the proof given in [10, Section 3], which gives a weaker result, is much simpler than the proof of Theorem 15 –.

For our last definition one needs to introduce some notations. For α in $\mathbb{N}^m$ and $\bar{u}$ in $\mathbb{R}^m$, let $f_{\bar{u}}^\alpha$ in $C^\infty(\mathbb{R}^n; \mathbb{R}^n)$ be defined by

$$f_{\bar{u}}^\alpha(x) = \frac{\partial^{|\alpha|} f}{\partial u^\alpha}(x, \bar{u}) \ \forall x \in \mathbb{R}^n. \tag{47}$$

Let $\mathcal{O}(\tilde{C})$ be the subspace of $C^\infty(\mathbb{R}^n \times \mathbb{R}^m; \mathbb{R}^p)$ spanned by the maps ω such that, for some integer $r \geq 0$ -depending on ω- and for some sequence $\alpha_1, ..., \alpha_r$ of r multi-indices in $\mathbb{N}^m$, we have, for all $x \in \mathbb{R}^n$ and for all $u \in \mathbb{R}^m$,

$$\omega(x, u) = L_{f_u^{\alpha_1}} ... L_{f_u^{\alpha_r}} h(x), \tag{48}$$

where $L_{f_{\bar{u}}^{\alpha_i}}$ denotes Lie derivatives with respect to $f_{\bar{u}}^{\alpha_i}$ and where, by convention, if $r = 0$ the right hand side of (48) is $h(x)$. With these notations our last definition is

Definition 20. System $(\tilde{C})$ is locally Lie null-observable if there exists a positive real number $\bar{\varepsilon}$ such that
(i) for all a in $\mathbb{R}^n \setminus \{0\}$ such that $|a| < \bar{\varepsilon}$ there exists q in $\mathbb{N}$ such that

$$L_{f_0}^q h(a) \neq 0 \tag{49}$$

with $f_0(x) = f(x, 0)$ and the usual convention $L_{f_0}^0 h = h$,
(ii) for all $(a_1, a_2) \in (\mathbb{R}^n \setminus \{0\})^2$ with $a_1 \neq a_2, |a_1| < \bar{\varepsilon}$, and $|a_2| < \bar{\varepsilon}$, and for all u in $\mathbb{R}^m$ with $|u| < \bar{\varepsilon}$, there exists ω in $\mathcal{O}(\tilde{C})$ such that

$$\omega(a_1, u) \neq \omega(a_2, u). \tag{50}$$

Note that (i) implies the following property
(i)* for any $a \neq 0$ in $B_{\bar{\varepsilon}} := \{x \in \mathbb{R}^m, |x| < \bar{\varepsilon}\}$ there exists a positive real number τ such that

$$x(\tau) \text{ exists and } h(x(\tau)) \neq 0 \tag{51}$$

where $x(t)$ is defined by $\dot{x} = f(x,0), x(0) = a$. Moreover if f and g are analytic, (i)* implies (i). The reason of "null" in "null-observable" comes from condition (i) or (i)* : roughly speaking we want to be able to distinguish from 0 any a in $B_{\bar{\varepsilon}} \setminus \{0\}$ by using the control law which vanishes identically.
When f is affine with respect to u, i.e. $f(x,u) = f_0(x) + \sum_{i=1}^{m} u_i f_i(x)$ with $f_1, ..., f_m$ in $C^\infty(\mathbb{R}^n; \mathbb{R}^n)$, then a slightly simpler version of (ii) can be given. Let $\tilde{\mathcal{O}}(\tilde{C})$ be the observation space -see e.g. [29] or Remark 5.4.2 in [82]- i.e. the set of maps $\tilde{\omega}$ in $C^\infty(\mathbb{R}^n; \mathbb{R}^p)$ such that for some integer $r \geq 0$ - depending on $\tilde{\omega}$ - and for some sequence $i_1, ..., i_r$ of integers in $[0, m]$

$$\tilde{\omega}(x) = L_{f_{i_1}} ... L_{f_{i_r}} h(x), \quad \forall x \in \mathbb{R}^n, \tag{52}$$

with the convention that, if $r = 0$, the right hand side of (52) is $h(x)$. Then (ii) is equivalent to

$$((a_1, a_2) \in B_{\bar{\varepsilon}}^2, \tilde{\omega}(a_1) = \tilde{\omega}(a_2) \, \forall \tilde{\omega} \in \tilde{\mathcal{O}}(\tilde{C})) \Rightarrow (a_1 = a_2). \tag{53}$$

Finally let us remark that if f is a polynomial with respect to u or if f and g are analytic then (ii) is equivalent to
(ii)* for all $(a_1, a_2) \in \mathbb{R}^n \setminus \{0\}$ with $a_1 \neq a_2, |a_1| < \varepsilon$ and $|a_2| < \varepsilon$ there exists u in $\mathbb{R}^m$ and ω in $\mathcal{O}(\tilde{C})$ such that (50) holds.
Indeed in these cases the subspace of $\mathbb{R}^p$ spanned by $\omega(x,u); \omega \in \mathcal{O}(\tilde{C})$ does not depend on u: it is the observation space of $(\tilde{C})$ evaluated at x – as defined for example in [29] –.
With these definitions we have

Theorem 21. *Assume that the origin (of $\mathbb{R}^n$) is locally continuously reachable (for (C)) in small time. Assume that $(\tilde{C})$ is locally Lie null-observable. Then $(\tilde{C})$ is locally stabilizable in small time by means of continuous dynamic periodic time-varying output feedback laws.*

This theorem is proved in [14]. Let us just sketch the proof given in [14].
We assume that the assumptions of Theorem 21 are satisfied. Let T be a positive real number. Our proof of Theorem 21 is divided in three steps.

Step 1. Using the assumption that system (C) is locally Lie null-observable we prove, using [11], that there exist u^* in $C^\infty(\mathbb{R}^p \times [0,T]; \mathbb{R}^m)$ and a positive real number ε^* such that

$$u^*(y, T) = u^*(y, 0) = 0, \quad \forall y \in \mathbb{R}^p, \tag{54}$$

$$u^*(0, t) = 0, \quad \forall t \in [0, T], \tag{55}$$

2 Sufficient conditions for state or output feedback stabilization

and, for all (a_1, a_2) in $B_{\varepsilon^*}^2$, for all s in $(0, T)$,

$$(h_{a_1}^{(i)}(s) = h_{a_2}^{(i)}(s), \forall i \in \mathbb{N}) \Rightarrow (a_1 = a_2) \tag{56}$$

where $h_a(s) = h(x^*(a, s))$ with x^* defined by $\partial x^*/\partial t = f(x^*, u^*(h(x^*), t))$, $x^*(a, 0) = a$. Let us note that in [58] a similar u^* was considered, but it was taken depending only on time and so (55), which is important to get stability, was not satisfied - in general -. In this step we do not use any reachability property for (C).

Step 2. Let $q = 2n + 1$. In this step, using (56), we prove the existence of $(q+1)$ real numbers $0 < t_0 < t_1 ... < t_q < T$ such that the map $K : B_{\varepsilon^*} \to (\mathbb{R}^p)^q$ defined by

$$K(a) = \left(\int_{t_0}^{t_1} (s - t_0)(t_1 - s) h_a(s) ds, ..., \int_{t_0}^{t_q} (s - t_0)(t_q - s) h_a(s) ds \right) \tag{57}$$

is one-to-one and so, as we will see, there exists a map $\theta : (\mathbb{R}^p)^q \to \mathbb{R}^n$ such that

$$\theta \circ K(a) = x^*(a, T), \forall a \in B_{\varepsilon^*/2}. \tag{58}$$

Step 3. In this step we prove the existence of $\bar{u}$ in $C^0(\mathbb{R}^n \times [0, T]; \mathbb{R}^m)$ and $\bar{\varepsilon}$ in $(0, +\infty)$ such that

$$\bar{u} = 0 \text{ on } (\mathbb{R}^n \times \{0, T\}) \cup (\{0\} \times [0, T]), \tag{59}$$

$$(\dot{x} = f(x, \bar{u}(x(0), t)) \text{ and } |x(0)| < \bar{\varepsilon}) \Rightarrow (x(T) = 0). \tag{60}$$

Property (60) means that $\bar{u}$ is a "dead-beat" open-loop control. In this last step we use the reachability assumption on (C), but do not use the Lie null-observability assumption.

Using these three steps let us end the proof of Theorem 21. The dynamic extension of system (C) that we consider is

$$\dot{x} = f(x, u), \quad \dot{z} = v = (v_1, ..., v_q, v_{q+1}) \in \mathbb{R}^p \times ... \times \mathbb{R}^p \times \mathbb{R}^n \simeq \mathbb{R}^{pq+n}, \tag{61}$$

with $z_1 = (z_1, ..., z_q, z_{q+1}) \in \mathbb{R}^p \times ... \times \mathbb{R}^p \times \mathbb{R}^n \simeq \mathbb{R}^{pq+n}$. For this system the output is $\tilde{h}(x, z) = (h(x), z) \in \mathbb{R}^p \times \mathbb{R}^{pq+n}$. For $s \in \mathbb{R}$ let $s^+ = \max(s, 0)$ and let $\text{sgn}(s) = 1$ if $s > 0$, 0 if $s = 0$, -1 if $s < 0$. Finally, for r in $\mathbb{N}\backslash\{0\}$ and $b = (b_1, ..., b_r)$ in $\mathbb{R}^r$, let

$$b^{1/3} = (|b_1|^{1/3}\text{sgn}(b_1), ..., |b_r|^{1/3}\text{sgn}(b_r)). \tag{62}$$

We now define $u : \mathbb{R}^p \times \mathbb{R}^{pq+n} \times \mathbb{R} \to \mathbb{R}^m$ and $v : \mathbb{R}^p \times \mathbb{R}^{pq+n} \times \mathbb{R} \to \mathbb{R}^{pq+n}$ by requiring for (y, z) in $\mathbb{R}^p \times \mathbb{R}^{pq+n}$ and for all i in $[1, q]$

$$u(y, z, t) = u^*(y, t), \forall t \in [0, T], \tag{63}$$

$$v_i(y, z, t) = -t(t_0 - t)^+ z_i^{1/3} + (t - t_0)^+(t_i - t)^+ y, \forall t \in [0, T], \tag{64}$$

$$v_{q+1}(y,z,t) = -t(t_q-t)^+ z_{q+1}^{1/3} + 6\frac{(T-t)^+(t-t_q)^+}{(T-t_q)^3}\theta(z_1,...,z_q), \quad (65)$$

$$u(y,z,t) = \bar{u}(z_{q+1}, t-T), \quad \forall t \in [T, 2T], \quad (66)$$

$$v(y,z,t) = 0, \quad \forall t \in [T, 2T], \quad (67)$$

$$u(y,z,t) = u(y,z,t+2T), \quad \forall t \in \mathbb{R}, \quad (68)$$

$$v(y,z,t) = v(y,z,t+2T), \quad \forall t \in \mathbb{R}. \quad (69)$$

Roughly speaking the strategy is the following one
(i) During the interval of time $[0,T]$, one "excites" system (C) by means of $u^*(y,t)$ in order to be able to deduce from the observation during this interval of time what is the state at time T: at time T we have $z_{q+1} = x$
(ii) During the interval of time $[T,2T]$, z_{q+1} does not move and one uses the dead-beat open-loop $\bar{u}$ but transforms it into an output feedback by using in its argument z_q instead of the value of x at time T – this method has been used previously in the proof of Theorem 1.7 of [10] –.
In a context of adaptive control, a similar strategy has been used later on by Kreisselmeier and Lozano in [43].

One easily sees that u and v are continuous and vanishes on $\{(0,0)\} \times \mathbb{R}$. Let (x,z) be any maximal solution of the closed loop system

$$\dot{x} = f(x, u(\tilde{h}(x,z),t)), \quad \dot{z} = v(\tilde{h}(x,z),t); \quad (70)$$

then one easily checks that, if $|x(0)| + |z(0)|$ is small enough,

$$z_i(t_0) = 0, \quad \forall i \in [1,q], \quad (71)$$

$$(z_1(t),...,z_q(t)) = K(x(0)), \quad \forall t \in [t_q, T], \quad (72)$$

$$z_{q+1}(t_q) = 0, \quad (73)$$

$$z_{q+1}(T) = \theta \circ K(x(0)) = x(T), \quad (74)$$

$$x(t) = 0, \quad \forall t \in [2T, 3T], \quad (75)$$

$$z(2T+t_q) = 0. \quad (76)$$

Equalities (71) (resp. (73)) are proved by computing explicitly, for $i \in [1,q]$, z_i on $[0,t_0]$ (resp. z_{q+1} on $[0,t_q]$) and by seeing that this explicit solution reaches 0 before time t_0 (resp. t_q) and by pointing out that if, for some s in $[0,t_0]$ (resp.

2 Sufficient conditions for state or output feedback stabilization

$[0, t_q])$, $z_i(s) = 0$ (resp. $z_{q+1}(s) = 0$) then $z_i = 0$ on $[s, t_0]$ (resp. $z_{q+1} = 0$ on $[s, t_q]$)-note that $z_i \dot{z}_i \leq 0$ on $[0, t_0]$ (resp. $z_{q+1} \dot{z}_{q+1} \leq 0$ on $[0, t_q]$)-.

Moreover one has also, for all s in $\mathbb{R}$ and all $t \geq s$,

$$((x(s), z(s)) = (0,0)) \Rightarrow ((x(t), z(t)) = (0,0)). \tag{77}$$

Indeed, first note that without loss of generality we may assume $s \in [0, 2T]$ and $t \in [0, 2T]$. If $s \in [0, T]$, then, since u^* is of class C^∞ we get, using (55), that $x(t) = 0$, $\forall t \in [s, T]$ and then, using (44) and (64), get that, for all $i \in [1, q]$, $z_i \dot{z}_i \leq 0$ on $[s, T]$ and so z_i also vanishes on $[s, T]$; this, with (65) and $\theta(0) = 0$- see(57) and (58)-, implies that $z_{q+1} = 0$ also on $[s, T]$. Hence we may assume that $s \in [T, 2T]$. But, in this case, using (67), we get that $z = 0$ on $[s, 2T]$ and, from (59) and (66), we get that $x = 0$ also on $[s, 2T]$.

From (75), (76), and (77) we get – see Lemma 2.15 in [12] – the existence of ε in $(0, +\infty)$ such that, for any s in $\mathbb{R}$ and any maximal solution (x, z) of $\dot{x} = f(x, u(\tilde{h}(x, z), t))$, $\dot{z} = v(\tilde{h}(x, z), t)$, we have

$$(|x(s)| + |y(s)| \leq \varepsilon) \Rightarrow ((x(t), z(t)) = (0,0), \forall t \geq s + 5T). \tag{78}$$

Since T is arbitrary Theorem 21 is proved.

Remark 22. Concerning the proof, let us emphasize that we use an idea due to Lozano [53], Mazenc and Praly [58]: as in [53] and [58] we will first recover the state from the output. A related idea is also used in Section 3 of [10], where we first recover initial data from the state. Moreover as in [58] our proof relies on the existence -see [90] for analytic systems and [11] for C^∞ systems- of an output feedback which distinguishes every pair of distinct states. In [58] it is established that distinguishability with a universal time-varying control, global stabilizability by state feedback, and observability of blow-up are sufficient conditions for the existence of a time-varying dynamic (of infinite dimension and in a sense more general than the one considered in Definition 19) output feedback guaranteeing boundedness and convergence of all the solutions defined at time $t = 0$; the methods developed in [58] can be applied directly to our situation; in this case Theorem 21 gives two improvements: we get that 0 is asymptotically stable for the closed loop system, instead of only attractor for time 0, and our dynamic extension is of finite dimension, instead of infinite dimension.

If $(\tilde{C})$ is locally stabilizable in small time by means of continuous dynamic periodic time-varying output feedback laws, then the origin (of $\mathbb{R}^n$) is locally continuously reachable (for $(\tilde{C})$) in small time (use Lemma 3.5 in [14]) and, if moreover f and h are analytic, then $(\tilde{C})$ is locally Lie null-observable - see [14, Proposition 4.3].

Let us remark that it follows from our proof of Theorem 21 that it suffices to consider dynamic extension of dimension $n + (2n+1)p$, i.e. under the assumption of Theorem 21, System (46) with $k = n + (2n+1)p$ is locally stabilizable in small time by means of continuous static periodic time-varying output feedback laws. We conjecture that, as in the linear case, this result still holds for $k = n - 1$. Note that this conjecture is true if $n = 1$, i.e. we have

Proposition 23. *Assume that $n = 1$ and that the origin (of $\mathbb{R}$) is locally continuously reachable (for $(\tilde{C})$) in small time. Assume that $(\tilde{C})$ is locally Lie null-observable. Then $(\tilde{C})$ is locally stabilizable in small time by means of continuous periodic time-varying output feedback laws.*

Let us prove this theorem. Since $(\tilde{C})$ is locally Lie null-observable, there exist $i \in [1, m]$ and a positive integer l such that

$$h_i^{(l)}(0) \neq 0. \tag{79}$$

Hence there are two maps w_+ and w_- in $C^0(\mathbb{R}; \mathbb{R})$, of class C^∞ on $\mathbb{R}\setminus\{0\}$, and a positive real number ϵ_0 such that

$$w_+(h_i(x)) = x, \ \forall x \in [0, \epsilon_0], \tag{80}$$

$$w_-(h_i(x)) = x, \ \forall x \in [-\epsilon_0, 0]. \tag{81}$$

Modifying, if necessary, h outside a neighborhood of $0 \in \mathbb{R}$, we may assume, without loss of generality, that

$$h(x) \neq 0, \ \forall x \in \mathbb{R}\setminus\{0\}. \tag{82}$$

Similarly, without loss of generality, we may assume that

$$|f(x, u)| \leq 1, \ \forall (x, u) \in \mathbb{R} \times \mathbb{R}^m \tag{83}$$

Let T be a positive real number. One has the following lemma [16, Lemma 2.12]

Lemma 24. *There exist $\bar{u}_+$ and $\bar{u}_-$ in $C^\infty([0, T); \mathbb{R}^m) \cup C^0([0, T]; \mathbb{R}^m)$, vanishing for $t = T$, such that, if we denote by $\bar{x}_+$ and by $\bar{x}_-$ the solutions of*

$$\dot{\bar{x}}_+ = f(\bar{x}_+, \bar{u}_+(t)), \ \bar{x}_+(T) = 0 \tag{84}$$

$$\dot{\bar{x}}_- = f(\bar{x}_-, \bar{u}_-(t)), \ \bar{x}_-(T) = 0, \tag{85}$$

then

$$\bar{x}_+(t) > 0, \ \forall t \in [0, T), \tag{86}$$

$$\bar{x}_-(t) < 0, \ \forall t \in [0, T). \tag{87}$$

Straightforward arguments relying on partition of unity -proceed for example as in the proof of Lemma 2.11 of [12]- we get the existence of $\bar{u}$ in $C^0(\mathbb{R} \times \mathbb{R}; \mathbb{R}^m)$ of class C^∞ on $(\mathbb{R}\setminus\{0\}) \times \mathbb{R}$ satisfying (22) and (23) such that

$$\bar{u}(\bar{x}_+(t), t) = \bar{u}_+(t), \ \forall t \in [T/2, T], \tag{88}$$

$$\bar{u}(\bar{x}_-(t), t) = \bar{u}_-(t), \ \forall t \in [T/2, T], \tag{89}$$

2 Sufficient conditions for state or output feedback stabilization

$$\bar{u} \text{ vanishes on a neighborhood of } \{0\} \times [0,T] \text{ in } \mathbb{R} \times [0,T]. \tag{90}$$

$$\bar{u}(x,t) = 0 \ \forall t \geq T, \ \forall x \in \mathbb{R} \tag{91}$$

$$\bar{u}(x,t) = 0, \ \forall t \leq 0, \ \forall x \in \mathbb{R} \tag{92}$$

For y in $\mathbb{R}^p$, let us denote by y_i the i-th component of y. Finally, let us define $\tilde{u} : \mathbb{R}^p \times \mathbb{R} \to \mathbb{R}^m$ by

$$\tilde{u}(y,t) = \bar{u}(w_+(y_i),t), \ \forall (y,t) \in \mathbb{R}^p \times [0,T), \tag{93}$$

$$\tilde{u}(y,t) = \bar{u}(w_-(y_i),t), \ \forall (y,t) \in \mathbb{R}^p \times [T,2t), \tag{94}$$

$$\tilde{u}(y,t) = \tilde{u}(y,t+2T), \ \forall (y,t) \in \mathbb{R}^p \times \mathbb{R}. \tag{95}$$

Clearly $\tilde{u}$ is continuous on $\mathbb{R}^p \times \mathbb{R}$. Let us prove that this time-varying output feedback stabilizes $(\tilde{C})$ in finite time. Let $u : \mathbb{R} \times \mathbb{R} \to \mathbb{R}^m$ be defined by

$$u(x,t) = \tilde{u}(h(x),t), \ \forall (x,t) \in \mathbb{R} \times \mathbb{R}. \tag{96}$$

Then u is continuous on $\mathbb{R} \times \mathbb{R}$, of class C^∞ on $(\mathbb{R}\setminus\{0\}) \times \mathbb{R}$, is $2T$-periodic with respect to time and vanishes on $\{0\} \times \mathbb{R}$. Note, see in particular (90), that (24) holds. Let us point out that there exists $\tau \in [0,T)$ such that

$$\dot{\bar{x}}_+ = f(\bar{x}_+, u(\bar{x}_+,t)), \ \forall t \in [\tau, T] \tag{97}$$

$$\dot{\bar{x}}_-(t-T) = f(\bar{x}_-(t-T), u(\bar{x}_-(t-T),t)), \ \forall t \in [\tau+T, 2T]. \tag{98}$$

Let $x_+ : [0,T] \to \mathbb{R}$ be defined by

$$\dot{x}_+ = f(x_+, u(x_+,t)), \ \forall t \in [0,T], \tag{99}$$

$$x_+(t) = \bar{x}_+(t), \ \forall t \in [\tau.T]. \tag{100}$$

Then

$$x_+(t) > 0, \ \forall t \in [0,T), \tag{101}$$

$$x_+(T) = 0. \tag{102}$$

Moreover, for any solution of $\dot{x} = f(x, u(x,t))$, one has, for all $t \in [0,T]$,

$$x(0) \in [0, x_+(0)] \Rightarrow (x(t) \in [0, x(t)]). \tag{103}$$

So, for any solution of $\dot{x} = f(x, u(x,t))$, one has

$$x(0) \in [0, x_+(0)] \Rightarrow x(T) = 0. \tag{104}$$

Similarly, let $x_- : [0, 2T] \to \mathbb{R}$ be defined by

$$\dot{x}_- = f(x_-, u(x_-, t)), \ \forall t \in [0, 2T], \tag{105}$$

$$x_-(t) = \bar{x}_-(t - T), \ \forall t \in [T + \tau, 2T]. \tag{106}$$

Then $x_-(0) < 0$ and, for any solution of $\dot{x} = f(x, u(x, t))$, one has

$$x(0) \in [x_-(0), 0] \Rightarrow x(2T) = 0. \tag{107}$$

Hence, using (104), (107), and [12, Lemma 2.15], we get that for $\varepsilon > 0$ small enough

$$((\dot{x} = f(x, u(x, t)) \text{ and } |x(s)| \leq \varepsilon) \Rightarrow$$
$$(x(t) = 0 \ \forall t \geq s + 4T)) \ \forall s \in \mathbb{R}. \tag{108}$$

This ends the proof of Proposition 23.

Remark 25. There are linear control system which are controllable and observable which cannot be locally asymptotically stabilized by means of a continuous time-varying static feedback law. This is for example the case for the controllable and observable linear system, with $n = 2$, $m = 1$, and $p = 1$,

$$\dot{x}_1 = x_2, \ \dot{x}_2 = u, \ y = x_1 \tag{109}$$

Assume that this system can be locally asymptotically stabilized by means of a continuous time-varying static output feedback law $u : \mathbb{R} \times \mathbb{R} \to \mathbb{R}$. Hence there exist $r > 0$ and $\tau > 0$ such that, if $\dot{x}_1 = x_2, \dot{x}_2 = u(x_1, t)$,

$$x_1(0)^2 + x_2(0)^2 \leq r^2 \Rightarrow x_1(\tau)^2 + x_2(\tau)^2 \leq r^2/5 \tag{110}$$

Let $(u^n : n \in \mathbb{N})$ be a sequence of functions from $\mathbb{R}$ into $\mathbb{R}$ of class C^∞ which converges uniformly to u on each compact subset of $\mathbb{R} \times \mathbb{R}$. Then, for n large enough, we have, if $\dot{x}_1 = x_2, \dot{x}_2 = u^n(x_1, t)$,

$$x_1(0)^2 + x_2(0)^2 \leq r^2 \Rightarrow x_1(\tau)^2 + x_2(\tau)^2 \leq r^2/4. \tag{111}$$

But, since the time-varying vector field X on $\mathbb{R}^2$ defined by

$$X_1(x_1, x_2, t) = x_1, \ X_2(x_1, x_2, t) = u^n(x_1, t), \tag{112}$$

has a divergence equal to 0, the flow associated to X preserves area, which is in contradiction with (111).

3 Lyapunov design of stabilizing state or output feedback.

3.1 Introduction.

For the study of the (global) uniform asymptotic stability of $X = 0$, an equilibrium point in $\mathbb{R}^N$ of the time-varying dynamic system

$$\dot{X} = F(t, X) , \qquad (113)$$

it is well established that, a very efficient tool is provided by Lyapunov function theory. But it is also well known that finding an appropriate Lyapunov function is in general a very difficult task. Here, as in the previous part, instead of the uncontrolled system (113), we are concerned with the following controlled dynamic system

$$\dot{x} = f(x, u) \qquad (114)$$

where x is in $\mathbb{R}^n$, u is in $\mathbb{R}^m$, f is continuous on a neighborhood of $(0,0)$ and

$$f(0,0) = 0. \qquad (115)$$

We want to design an asymptotically stabilizing feedback law. Precisely, we want to find an integer p and two continuous functions φ and ϕ so that:
1. The control u is given by the dynamic system

$$\dot{\chi} = \varphi(t, h(x), \chi) , \qquad u = \phi(t, h(x), \chi) \qquad (116)$$

   with χ living in $\mathbb{R}^p$ and h is the imposed output function.
2. The point $(x = 0, \chi = 0)$ is a (globally) uniformly asymptotically stable equilibrium of the closed-loop system (114),(116).

The key difference between (113) and (114) is that the former is a given dynamic system whereas the latter is not completely defined. If, to solve the stabilizability problem for (114), we first find, by some way, the system (116) and then check that indeed we have stability, then we are back to the study of stability for a system like (113). But another approach consists in first finding a Lyapunov function and then complete the definition of (114) by choosing (116) so that this Lyapunov function will be appropriate. This approach is called Lyapunov design. This idea has been studied since at least the beginning of the sixties (see for example [27, 60]). To give a better grasp on the idea we wish to convey, we could say that in the first technique the problem is: *given a dynamic system, find a Lyapunov function*, whereas, in the second technique, the problem is: *given a Lyapunov function, find a dynamic system*. Of course this latter problem is not as simple as this, since part of the searched system is already given in the form of (114). This introduces implicitly a constraint on the Lyapunov function.

We shall be presenting various aspects of this Lyapunov design. But to limit the size of this presentation, we have chosen not to address two important topics:
1. *Lyapunov design of time-varying feedback laws for driftless systems.* There has been several publications on this topic, see [62, 17].

2. *Lyapunov design of output feedback laws.* This topic has received a lot of attention. Let us mention [101, 24] where results in the spirit of Theorem 27 below are presented. The case where the unmeasured components appear linearly has been exploited in [41, 55, 56, 64, 63, 25]. A class of systems where unmeasured components appear linearly is those linearly parameterized with unknown parameters. Then, we can have both unmeasured states and unknown parameters. Lyapunov design has a very long history in this topic and the literature is extremely rich. Let us mention two surveys [66, 45].

3.2 Assignable Lyapunov functions.

Using the analogy with the fact that poles can be assigned to a controllable linear system, we introduce the notion of assignable Lyapunov function

Definition 26. A function V is called an (resp. globally) assignable Lyapunov function for the system (114) if

1. it is in $C^1([0,+\infty) \times \mathcal{V};[0,+\infty))$ (resp. $C^1([0,+\infty) \times \mathbb{R}^n \times \mathbb{R}^p;[0,+\infty)))$, where $\mathcal{V}$ is a neighborhood of $(0,0)$ in $\mathbb{R}^n \times \mathbb{R}^r$,
2. it is (resp. radially unbounded) positive definite and decrescent[4],
3. there exist two continuous functions φ and ϕ such that the feedback law defined by (116) makes non positive the time derivative of V along all the trajectories issued from points in $\mathcal{V}$ and solutions of (114),(116), i.e., for all (t,x,χ) in $[0,+\infty) \times \mathcal{V}$ (resp. $[0,+\infty) \times \mathbb{R}^n \times \mathbb{R}^p$), we have

$$-W(x,\chi) \geq \tag{117}$$

$$\tfrac{\partial V}{\partial x}(t,x,\chi)\, f(x,\phi(t,h(x),\chi)) \;+\; \tfrac{\partial V}{\partial \chi}(t,x,\chi)\, \varphi(t,h(x),\chi) \;+\; \tfrac{\partial V}{\partial t}(t,x,\chi)$$

where W is a non negative continuous function. If W is in fact positive definite then V is called a strictly assignable Lyapunov function.

If we have a Lyapunov function which is assignable but not strictly assignable, we are guaranteed of having uniform stability but not uniform asymptotic stability. To prove the latter, we shall need to invoke an Invariance Theorem (see [28, Theorem 55.1] for instance) and for this, it will be more appropriate to restrict the functions φ, ϕ and V to be periodic in t if not time-invariant.

For the case where we restrict ourselves with time invariant feedback, as for uncontrolled dynamic system where we have equivalence between existence of Lyapunov functions and asymptotic stability of an equilibrium, Artstein has exhibited, in [1], a property in terms of a Lyapunov function which is equivalent to the existence of an asymptotically stabilizing feedback law[5] ϕ. This can be stated as

---
[4] See [28, p.194-195] for the definitions of these terms.
[5] In this context the dynamic extension χ is collected with the system state x so that the pair $[\dot{x}=f(x,u),u]$ represents in fact the pair $[(\dot{x}=f(x,u),\dot{\chi}=v),(u,v)]$. Of course this is possible only once the dimension p has been chosen.

3 Lyapunov design of stabilizing state or output feedback.

Theorem 27. *Let the set of admissible control be $\mathcal{U}$, a convex subset of $\mathbb{R}^m$. If the system (114) can be (resp. globally) asymptotically stabilized by means of a discontinuous feedback law (see Definition 4), then there exists a neighborhood $\mathcal{V}$ of 0 in $\mathbb{R}^n$ (resp. $\mathcal{V} = \mathbb{R}^n$) and a function V defined on $\mathcal{V}$ which is "a control Lyapunov function", i.e. it is positive definite (resp. radially unbounded), in $C^1(\mathcal{V};[0,+\infty))$ (resp. $C^1(\mathbb{R}^n;[0,+\infty))$) and such that*[6]

$$x \in \mathcal{V}\setminus\{0\} \quad \Longrightarrow \quad \exists u \in \mathcal{U} \;\; s.t. \;\; L_{f(x,u)}V(x) < 0, \qquad (118)$$

and the small control property holds. Namely, we have also that, for all $\varepsilon > 0$, there exists $\delta > 0$ such that

$$\{x \in \mathcal{V}, \; 0 < |x| < \delta\} \quad \Longrightarrow \quad \exists u \in \mathcal{U} \;\; s.t. \;\; \{|u| < \varepsilon, \; L_{f(x,u)}V(x) < 0\}. \qquad (119)$$

Conversely, if such neighborhood $\mathcal{V}$ (resp. $\mathcal{V} = \mathbb{R}^n$) and function V exist, then (114) can be (resp. globally) asymptotically stabilized by means of a discontinuous feedback law or a time-varying continuous feedback law with period T (see Definition 2) where T is an arbitrary strictly positive real number. If, moreover, f is affine in u, then (114) can be (resp. globally) asymptotically stabilized by means of a time-invariant continuous feedback law.

In fact, the notions of control Lyapunov function and of small control property – with u a relaxed control instead of a vector in $\mathcal{U}$ – are already present in the work of Sontag [77] where the problem of asymptotic controllability is addressed. In [1], Artstein studies stabilization but using relaxed controls instead of discontinuous or time-varying continuous controls, the latter being established by Coron and Rosier in [19].

With this theorem, we know that a strictly assignable Lyapunov function is a control Lyapunov function which satisfies the small control property.

Remark 28. A direct consequence of [81, Lemma 3.2], which has been exploited by Freeman and Kokotovic in [21] in the context of section 3.4, is the following: If V can be strictly assigned by a continuous feedback law ϕ, then there exists a positive positive definite function[7] r which can be as many times continuously

---

[6] For a "matrix field" indexed by u, $g(x,u) = (g_1(x,u),\ldots,g_m(x,u))$, for each u, we denote by $L_g(x,u)V(x)$ the row vector $(L_{g_1}V,\ldots,L_{g_m}V)$ where $L_{g_i}V$ is the derivative of V along the vector field g_i obtained by fixing u.

[7] A direct construction for r would be:
From [46, Remark p.74 and Theorem 7], there exists a C^∞ positive definite and proper function on the domain of attraction of $x = 0$ for the system $\dot{x} = f(x,\phi(x))$. This allows us to define two sequences of strictly positive real numbers

$$r_i^+ = \min\left\{R^+, \inf_{\{(x,u)|\, i+1 \leq V(x) \leq i+2,\, L_{f(x,u)}V(x) \geq 0\}} |u - \phi(x)|\right\},$$

$$r_i^- = \min\left\{R^-, \inf_{\{(x,u)|\, \frac{1}{i+2} \leq V(x) \leq \frac{1}{i+1},\, L_{f(x,u)}V(x) \geq 0\}} |u - \phi(x)|\right\}.$$

differentiable as we want and such that[8]

$$L_{f(x,u)}V(x) < 0 \quad \forall (x,u) \in \mathcal{V}\setminus\{0\} \times \overline{B}(\phi(x), r(x)). \tag{120}$$

Also, the continuity of ϕ and the positive definiteness of r imply that, conversely, there exists a continuous positive definite function $\overline{r}$ such that, for all x in $\mathcal{V}\setminus\{0\}$,

$$L_{f(x,\phi(x_m))}V(x) < 0 \quad \forall x_m \in \overline{B}(x, \overline{r}(x)). \tag{121}$$

This proves that state measurement is allowed. Indeed, if the actual state is x but its measurement is x_m, the Lyapunov function will still be decaying along the actual solution using the control $\phi(x_m)$ provided the norm of the measurement error $|x - x_m|$ is smaller than $\overline{r}(x)$. Unfortunately, typically $\overline{r}(x)$ tends to zero as x tends to 0 or to the boundary of $\mathcal{V}$. Freeman has displayed this problem with a counter-example in [20]. But in [22], Freeman and Kokotovic have exhibited a class of systems for which a feedback law can be constructed in such a way that the problem at infinity is rounded.

For the case where f is affine in u, Lin and Sontag have proposed an explicit expression for the feedback law ϕ which strictly assigns the Lyapunov function V (see [51] for more general sets $\mathcal{U}$)

Theorem 29 [79, 50]. *If the set of admissible control is*

$$\mathcal{U} = \left\{u \in \mathbb{R}^m \;\middle|\; |u| \leq \frac{1}{k}\right\} \tag{122}$$

where k is in $[0, +\infty]$ and if V is a C^1 (resp. global) control Lyapunov function satisfying the small control property with[9]

$$f(x, u) = a(x) + b(x) u \tag{123}$$

then a (resp. globally) stabilizing time invariant continuous feedback law is given by

$$\phi(x) = \begin{cases} -\dfrac{L_a V + \sqrt{(L_a V)^2 + |L_b V|^4}}{|L_b V|^2 \left(1 + k\sqrt{1 + |L_b V|^2}\right)} L_b V^\top & \text{if } L_b V \neq 0, \\ 0 & \text{if } L_b V = 0, \end{cases} \tag{124}$$

The feedback law ϕ given by this Theorem has the following property

$$L_b V(x) \neq 0 \quad \Longrightarrow \quad L_b V(x)\,\phi(x) < 0. \tag{125}$$

Such a control which also assigns strictly V is said to be sign optimal.

---

where R^+ and R^- are two strictly positive real numbers. Then the function $r(x)$ can be obtained by interpolation between the $\frac{1}{2}r_i$'s according to the time needed to reach the closest level sets of V to x, following the solution $\dot{X} = \frac{\partial V}{\partial x}(X)$, $X(0) = x$.

[8] $\overline{B}(x, \delta)$ denotes the closed ball with center x and radius δ.

[9] Here and in the following of this section, $a(x)$ is a vector and $b(x)$ is a matrix.

3 Lyapunov design of stabilizing state or output feedback.

Let us mention a first interesting application of Theorem 29: Assume that we know the feedback law ϕ only implicitly. Precisely assume the existence of a strictly assignable Lyapunov function V and of a C^1 function $\Phi : \mathcal{V} \times \mathcal{U} \to \mathbb{R}^m$ such that $\frac{\partial \Phi}{\partial u}(x, u)$ is invertible on a neighborhood of $(0,0)$,

$$\Phi(0,0) = 0 \tag{126}$$

and the solution

$$u = \phi(x) \tag{127}$$

of

$$\Phi(x, u) = 0 \tag{128}$$

satisfies (118) and (119). The properties of Φ implies that we can find explicitly[10] C^1 functions $\mathcal{F}$ satisfying, at least on a neighborhood of $(0, 0)$,

$$\left| \frac{\partial \mathcal{F}}{\partial u}(x, u) \right| \leq 1 - \varepsilon, \tag{129}$$

with ε in $(0, 1)$, and such that ϕ is also solution of

$$u = \mathcal{F}(x, u). \tag{130}$$

Our idea to get an explicit expression for the feedback law is to solve (130) "on-line". This will lead to a dynamic controller of the form (116)

$$\dot{\chi} = \varphi(x, \chi) \quad , \quad u = \chi \tag{131}$$

where φ is to be chosen. To design this function, we simply follow Theorem 27. We look for a positive definite function $U(x, \chi)$ such that (118) holds, i.e. for $(x, \chi) \neq 0$,

$$\frac{\partial U}{\partial \chi}(x, \chi) = 0 \tag{132}$$

implies

$$\frac{\partial U}{\partial x}(x, \chi) f(x, \chi) < 0. \tag{133}$$

In view of what we know, the following choice is appropriate

$$U(x, \chi) = V(x) + \tfrac{1}{2} |\chi - \mathcal{F}(x, \chi)|^2. \tag{134}$$

Also we can check that if f is C^1 and V is C^2 then the small control property (119) holds. It follows that (124) gives an explicit expression for φ. This technique has been used in [64].

---

[10] For instance, $\mathcal{F}(x, u) = u - \frac{\partial \Phi}{\partial u}(x, u)^{-1} \Phi(x, u)$

3.3 Robustness and Lyapunov redesign.

For uncontrolled dynamic systems, asymptotic stability implies total stability but the bound on the perturbations is fixed. In the case of controlled systems, we can take advantage of the freedom given by the control to increase the level of allowed perturbations in some "directions". Let us consider the following perturbation of (114) for the affine case (123)

$$\dot{x} = a(x) + b(x)u + c(x,d) \qquad (135)$$

where c is a continuous function and d represents an exogenous signal supposed to be in $L^\infty([0,+\infty);\mathbb{R}^q)$. We assume that we know a strictly assignable Lyapunov function V and that we have already implemented a corresponding sign optimal feedback law (see (125)) so that, for $x \neq 0$,

$$L_a V(x) < 0. \qquad (136)$$

If we do not modify this nominal feedback law, we are guaranteed that the state x of the system will be attracted in finite time and then remain in $\overline{B}(0,\delta)$, if d takes values in the set

$$\mathcal{D}_{\delta,V} = \left\{ d \in \mathbb{R}^q \mid x \notin \overline{B}(0,\delta) \Rightarrow L_{c(x,d)} V(x) < -L_a V(x) \right\}. \qquad (137)$$

This set is in general only a subset of the actual set of "admissible" perturbations d keeping the ball $\overline{B}(0,\delta)$ attractive and invariant. To get a better approximation of this set, we may have to reshape[11] the Lyapunov function V in order to minimize the ratio $\frac{L_{c(x,d)} V(x)}{-L_a V(x)}$. This shaping has been done for linear systems for example in the context of the so called quadratic stability when some a priori knowledge on the uncertainties is available (see [71, 8] for instance). For the nonlinear case, some aspects of this problem are addressed by Freeman and Kokotovic in [23]. Let us remark also that the technique used in section 3.4 solves this question for a particular class of systems.

For the time being, let us concentrate our attention on finding a feedback law u in order to increase $\mathcal{D}_{\delta,V}$. This problem has received a lot of attention. The surveys by Khalil [39, Chapter 5.5] and Corless [7]) give a good idea on the state of the art. To give the reader a flavor of the available results, let us state the following technical Lemma

Lemma 30 [68, Lemma 1]. *Assume we know a positive definite function r such that there exists a continuous function Δ on $\mathbb{R}^m$ satisfying*[12]

$$L_{c(x,d)} V(x) \leq -\frac{1}{1+r(x)} L_a V(x) + L_b V(x) \Delta(x,d). \qquad (138)$$

---
[11] modify the level sets of V.
[12] Note that this inequality is invariant under any transformation of V which does not change its level sets.

3 Lyapunov design of stabilizing state or output feedback.

Under this condition and (136), for any function μ in $C^0([0,+\infty);[0,+\infty))$, non decreasing and with $\mu(0) = 0$, there exists[13] m bounded functions $\theta_i : \mathcal{V} \to [0,1]$ which are continuous on $\mathcal{V}\backslash\{0\}$ and such that by choosing the ith component of the feedback law as

$$\phi(x)_i = -\theta_i(x)\operatorname{sign}(L_{b_i}(x))\,\mu(V(x)), \qquad (140)$$

we get

$$L_{a(x)+b(x)\phi(x)+c(x,d)}V(x) \leq \frac{r(x)}{2(1+r(x))}L_aV(x) - |L_bV(x)|\,[\mu(V(x)) - |\Delta(x,d)|]. \qquad (141)$$

The condition (138) can be interpreted as saying that the remainder of the "division" of $L_{c(x,d)}V(x)$ by $L_bV(x)$ is strictly smaller than $-L_aV(x)$. The case where $r = \infty$ has been considered for example by Qu in [69]. This Lemma, which is at the basis of the proof of Lemma 49 in the third part of this survey, shows in particular that, for each solution of (135),(140) which satisfies

$$|\Delta(x(t),d(t))| \in \mu([0,+\infty)) \qquad \forall t \in [0,+\infty), \qquad (142)$$

we have

$$\limsup_{t\to+\infty} V(x(t)) \leq \limsup_{t\to+\infty} \mu^{-1}\left(|\Delta(x(t),d(t))|\right), \qquad (143)$$

$$\limsup_{t\to+\infty} |\phi(x(t))| \leq \limsup_{t\to+\infty} |\Delta(x(t),d(t))|. \qquad (144)$$

It follows that by choosing μ appropriately, we can make the ball $\overline{B}(0,\delta)$ attractive.

Remark 31. By choosing the function μ as a continuous, strictly increasing function mapping $[0,+\infty)$ onto itself and such that[14]

$$\mu(V(x)) > \sup_{\{d|\,|d|\leq|x|\}}\{|\Delta(x,d)|\}, \qquad (145)$$

it follows from [88] that there exists a C^1 positive definite, (resp. radially unbounded) function which we still denote by V and continuous and strictly increasing functions σ and α which are 0 at 0 and with α onto $[0,+\infty)$ such that

$$L_{a(x)+b(x)\phi(x)+c(x,d)}V(x) \leq -\alpha(V(x)) + \sigma(|d|). \qquad (146)$$

---
[13] For example, we can take

$$\theta_i = \begin{cases} \operatorname{sat}\left(\dfrac{\sqrt{\left(\frac{r(x)L_aV(x)}{2m(1+r(x))}\right)^2 + 3(L_{b_i}V(x)\mu(V(x)))^2} + \frac{r(x)L_aV(x)}{2m(1+r(x))}}{|L_{b_i}V(x)|\mu(V(x))}\right) & \text{if } |L_{b_i}V(x)|\mu(V(x)) \neq 0, \\ 0 & \text{if } |L_{b_i}V(x)|\mu(V(x)) = 0, \end{cases} \qquad (139)$$

where $\operatorname{sat} : (-\infty,+\infty) \to [-1,1]$ is the standard saturation function.
[14] which is always possible since V is positive definite (and radially unbounded in the global case).

3.4 Adding Integrators.

The undisturbed case. We consider the system

$$\dot{x}_1 = f_1(x_1, u_1), \tag{147}$$

where x_1 is in $\mathbb{R}^{n_1}$, u_1 is in $\mathbb{R}^{n_2}$, f_1 is C^1 and we assume

Assumption A1: *We know a C^1 Lyapunov function V_1 which can be strictly assigned by the time-invariant continuous feedback law ϕ_1 to the system (147).*

Can we design a time-invariant asymptotically stabilizing continuous feedback law for the system

$$\begin{cases} \dot{x}_1 = f_1(x_1, x_2), \\ \dot{x}_2 = f_2(x_1, x_2, u_2), \end{cases} \tag{148}$$

with u_2 in $\mathbb{R}^m$? This problem is called adding one integrator. Its solution allows us to prove for example that systems admitting the following special recurrent structure, called feedback form

$$\begin{cases} \dot{x}_1 = f_1(x_1, x_2), \\ \dot{x}_2 = x_3 + f_2(x_1, x_2), \\ \quad \vdots \\ \dot{x}_n = u + f_n(x_1, \ldots, x_n). \end{cases} \tag{149}$$

are (resp. globally) asymptotically stabilizable by means of a time invariant continuous feedback law if the functions f_i's are C^{n-i} and $\dot{x} = f_1(x, u)$ is (resp. globally) asymptotically stabilizable by means of a time invariant C^{n-1} feedback law.

This problem of adding one integrator has received many answers and most of them can be obtained by following Theorem 27. For this, we first remark that if we can solve the asymptotic stabilization problem for

$$\begin{cases} \dot{x}_1 = f_1(x_1, x_2), \\ \dot{x}_2 = u, \end{cases} \tag{150}$$

then the problem is also solved for the system (148) if

Assumption A2: *There exists a continuous function $\mathcal{K}$ satisfying*

$$f_2(x_1, x_2, \mathcal{K}(x_1, x_2, u)) = u. \tag{151}$$

So we concentrate our attention on the system (150). We look for a positive definite function $V_2(x_1, x_2)$ such that (118) holds, i.e. for $(x_1, x_2) \neq 0$,

$$\frac{\partial V_2}{\partial x_2}(x_1, x_2) = 0 \quad \Longrightarrow \quad \frac{\partial V_2}{\partial x_1}(x_1, x_2) f_1(x_1, x_2) < 0. \tag{152}$$

3 Lyapunov design of stabilizing state or output feedback.

From Remark 28, we know the existence of a sufficiently smooth positive definite function r (resp. we choose $r \equiv 0$) such that it is sufficient to choose V_2 satisfying

$$\frac{\partial V_2}{\partial x_2}(x_1, x_2) = 0 \quad \Rightarrow \quad \left\{ x_2 \in \overline{B}(\phi_1(x_1), r(x_1)), \; \frac{\partial V_2}{\partial x_1}(x_1, x_2) = l(x_1)\frac{\partial V_1}{\partial x_1}(x_1) \right\} \tag{153}$$

where l is any positive definite function. We conclude that a function V_2 of the following form should be appropriate:

$$V_2(x_1, x_2) = k(V_1(x_1)) + \int_0^1 (x_2 - \phi_1(x_1))^\top \Theta(x_1, \phi_1(x_1) + s(x_2 - \phi_1(x_1))) \, ds \tag{154}$$

where k is any C^1 function, with $k(0) = 0$ and positive definite derivative, and the vector Θ is to be chosen such that V_2 is C^1, positive definite (resp. radially unbounded) and

$$\Theta(x_1, x_2) = 0 \quad \Longrightarrow \quad x_2 \in \overline{B}(\phi_1(x_1), r(x_1)). \tag{155}$$

For example, by taking

$$\Theta(x_1, x_2) = (x_2 - \phi_1(x_1)) \max\left\{ 1 - \frac{r(x_1)}{|x_2 - \phi_1(x_1)|}, 0 \right\}, \tag{156}$$

we get

$$V_2(x_1, x_2) = k(V_1(x_1)) + \frac{1}{2} \max\{|x_2 - \phi_1(x_1)| - r(x_1), 0\}^2. \tag{157}$$

Assumption A3: ϕ_1 is a C^1 function.

In this case, V_2, in (157), is a control Lyapunov function which satisfies the small control property, it is therefore a strictly assignable Lyapunov function. So we have

Theorem 32. *Under assumptions A1 to A3, the system (148) can be (resp. globally) asymptotically stabilized by means of a time-invariant continuous feedback law.*

Moreover, since the system (150) is affine in the control, Theorem 29 applies. But one can also check that an asymptotically stabilizing continuous feedback law is

$$\phi_2(x_1, x_2) = \tag{158}$$

$$-\left(k'(V_1(x_1)) \frac{\partial V_1}{\partial x_1}(x_1) \int_{\min\left\{1, \frac{r(x_1)}{|x_2-\phi_1(x_1)|}\right\}}^1 \frac{\partial f_1}{\partial x_2}(x_1, \phi_1(x_1) + s(x_2 - \phi(x_1))) \, ds \right)^\top$$

$$+ \theta(x_1, x_2) \left\{ \left[\frac{\partial \phi_1}{\partial x_1}(x_1) + \frac{x_2 - \phi_1(x_1)}{|x_2 - \phi_1(x_1)|} \frac{\partial r}{\partial x_1}(x_1) \right] f_1(x_1, x_2) - S(x_1, x_2) \right\}$$

where $\mathcal{S}$ on $\mathbb{R}^p$ and θ on $[0,1]$ are continuous functions satisfying[15]

$$|x_2 - \phi_1(x_1)| > r(x_1) \quad \Longrightarrow \quad \begin{cases} \theta(x_1, x_2) = 1, \\ (x_2 - \phi_1(x_1))^\top \mathcal{S}(x_1, x_2) > 0, \end{cases} \quad (159)$$

and

$$r \not\equiv 0, \ |x_2 - \phi_1(x_1)| = 0 \quad \Longrightarrow \quad \theta(x_1, x_2) = 0. \quad (160)$$

The Lyapunov function (157) and the feedback law (158), and therefore Theorem 34, have been obtained by Byrnes and Isidori [5] and Tsinias [100] (see also [82, Lemma 4.8.3]) for the case where the function $r \equiv 0$. The idea of introducing a non zero function r (see Remark 28) has been proposed by Freeman and Kokotovic in order to "flatten" the Lyapunov function in the neighborhood of the manifold $\{(x_1, x_2) | x_2 = \phi_1(x_1)\}$ which is a desirable property, this manifold not being a naturally invariant manifold of the closed loop system (see [21, 22] for more details). The interest of the functions k and Θ is in particular appreciated when ϕ_1 is not C^1 (see [65, 18]).

Assumption A1 can be weakened to the case where V_1 is assignable but not strictly

Assumption A1': *We know a C^1 Lyapunov function V_1 which can be assigned by the time-invariant feedback law ϕ_1 to the system (147) and such that $x_1 = 0$ is the only solution of*

$$\dot{x}_1 = f_1(x_1, \phi_1(x_1)) \quad , \quad \frac{\partial V_1}{\partial x_1}(x_1) f_1(x_1, \phi_1(x_1)) = 0. \quad (161)$$

Indeed, we remark that, by choosing $r \equiv 0$ in (157) the derivative of V_2 along the solutions of (148),(158) is zero if and only if $x_2 = \phi_1(x_1)$. So, we have

Theorem 33 [17, Lemma 1]. *Under assumptions A1', A2 and A3, the system (148) can be (resp. globally) asymptotically stabilized by means of a time-invariant continuous feedback law.*

The disturbed case. Let us now address the problem where a disturbance is present in the dynamics of the x_1 subsystem of (150), i.e.

$$\begin{cases} \dot{x}_1 = f_1(x_1, x_2) + c(x_1, d), \\ \dot{x}_2 = u_2, \end{cases} \quad (162)$$

where c is continuous, x_2 and u_2 are in $\mathbb{R}$ – to simplify the notations – and, as in section 3.3, d is the disturbing signal supposed to be in $L^\infty([0, +\infty); \mathbb{R}^q)$. We assume that we are in the context of Remarks 28 and 31, i.e.

---

[15] For example $\mathcal{S}(x_1, x_2) = \Theta(x_1, x_2)$ and $\theta(x_1, x_2) = \text{sat}\left(\frac{|x_2 - \psi_1(x_1)|}{r(x_1)}\right)$.

3 Lyapunov design of stabilizing state or output feedback.

We know a C^1 (resp. radially unbounded) positive definite and decrescent function V_1, a C^1 feedback law ϕ_1 and a C^1 positive definite function r (resp. $r \equiv 0$) such that

$$L_{f_1(x_1,u_1)+c(x_1,d)}V_1(x_1) \leq -\alpha_1(V(x_1)) + \sigma_1(|d|) \quad \forall u_1 \in \overline{B}(\phi_1(x_1), r(x_1)) \tag{163}$$

where σ_1 and α_1 are continuous and strictly increasing functions which are 0 at 0, with α onto $[0, +\infty)$.

Following the previous section, we shall try to assign the function V_2 defined in (157). For this, we let

$$u_2 = \phi_2(x_1, x_2) + u \tag{164}$$

with ϕ_2 given by (158). We get, along the solutions of (147),

$$\dot{V}_2 \leq -k'(V_1(x_1))\left[\alpha_1(V(x_1)) - \sigma_1(|d|)\right] - \Theta(x_1, x_2)\theta(x_1, x_2)\mathcal{S}(x_1, x_2) \tag{165}$$
$$+ \Theta(x_1, x_2)\left[u - \left(\frac{\partial \phi_1}{\partial x_1}(x_1) + \frac{x_2 - \phi_1(x_1)}{|x_2 - \phi_1(x_1)|}\frac{\overline{\partial r}}{\partial x_1}(x_1)\right)c(x_1,d)\right]$$

where Θ, θ and $\mathcal{S}$ satisfy (156), (159) and (160) and so in particular

$$\Theta(x_1, x_2)\,\theta(x_1, x_2) = \Theta(x_1, x_2). \tag{166}$$

Then, let us omit the arguments and choose the control

$$u = -\theta\, n_1 \left|\frac{\partial \phi_1}{\partial x_1} + \frac{x_2 - \phi_1}{|x_2 - \phi_1|}\frac{\partial r}{\partial x_1}\right| \gamma\left(\left|\frac{\partial \phi_1}{\partial x_1} + \frac{x_2 - \phi_1}{|x_2 - \phi_1|}\frac{\partial r}{\partial x_1}\right|\Theta\right), \tag{167}$$

where γ is an arbitrary continuous, strictly increasing function mapping $[0, +\infty)$ onto itself and extended on $\mathbb{R}$ by symmetry. By using the fact that θ is smaller than 1 and the following inequality,

$$x\, y \leq |x|\gamma_1(|x|) + |y|\gamma_1^{-1}(|y|), \tag{168}$$

we obtain finally an inequality similar to (141)

$$\dot{V}_2 \leq -[k'\alpha_1(V(x_1)) + \Theta\,\mathcal{S}] + k'\sigma_1(|d|) + n_1\,|c|\gamma^{-1}(|c|). \tag{169}$$

It remains several degrees of freedom in this inequality. They are determined by the choice of the feedback laws (157) and (167).

To appreciate the interest of this result, let us choose simply a bounded function for r and the identity function for k and γ and $\mathcal{S} = \Theta$. We introduce the following bounding functions

$$C_1(x_1) = \sup_{\{d\mid |d| \leq |x_1|\}} \left\{|c(x_1,d)|\, \gamma^{-1}\left(|c(x_1,d)|\right)\right\}, \tag{170}$$

$$C_d(d) = \sup_{\{x_1\mid |x_1| \leq |d|\}} \left\{|c(x_1,d)|\, \gamma^{-1}\left(|c(x_1,d)|\right)\right\}. \tag{171}$$

We get

$$\dot{V}_2 \leq -[\alpha_1(V(x_1)) - n_1 C_1(x_1)] - \Theta(x_1, x_2)^2 + [\sigma_1(|d|) + C_d(d)]. \tag{172}$$

Assume the feedback law ϕ_1 has been chosen such that the term between brackets is a positive definite (resp. radially unbounded function) of x_1. Then, since Θ, in (156), is radially unbounded in x_2 for all x_1, (172) allows us to get an inequality similar to (146). In particular, in this case, the solutions are bounded if the function d is bounded.

The technique presented here and many other improvements have appeared in the literature. We refer the reader in particular to [61, 21, 57, 70, 76].

3.5 Case of dissipative uncontrolled part.

A drawback of the technique of Lyapunov function assignment is the lack of information about the construction, in the general case, of assignable Lyapunov function. On the other hand many dynamic equations representing the dynamics of practical systems are obtained via a variational formulation. In such cases, the "total" energy function provides typically positive definite (resp. radially unbounded) function and the variational approach is based on the fact that without control this function is non increasing along the solutions. This latter property implies that we are almost but not exactly in the context of Theorem 27. To be exactly in that context, we would need a strict decrease along the solutions.

This new context has been the subject of many studies which started with the contributions of Jacobson [34, Theorems 2.5.1 and 2.5.2] and Jurdjevic and Quinn in [38].

To be more explicit let us consider the system

$$\dot{x} = a(x) + b(x, u) u \tag{173}$$

where x is in $\mathbb{R}^n$, u in $\mathbb{R}^m$, a, b are C^2 functions, $a(0) = 0$ and we denote $\widehat{b}(x) = b(x, 0)$. We introduce the following assumptions:

Assumption B1: *There exists a positive definite and radially unbounded C^2 function V so that, for all $x \in \mathbb{R}^n$,*

$$\frac{\partial V}{\partial x}(x) a(x) = -W(x) \leq 0. \tag{174}$$

Assumption B2: $x = 0$ *is the only solution of*

$$\dot{x} = a(x) \quad , \quad \frac{\partial V}{\partial x}(x) \widehat{b}(x) = 0 \quad , \quad \frac{\partial V}{\partial x}(x) \widehat{b}(x) = 0. \tag{175}$$

Assumption B1 expresses the fact that the uncontrolled system is dissipative. Assumption B2 is related to a controllability assumption. It is difficult to check it directly in practice but many sufficient conditions under which it holds have

3 Lyapunov design of stabilizing state or output feedback.

been proposed in the literature (see for instance [48, 49]). For example, following [38], it holds if

There exists an integer r such that

$$E_r = \{x : L_a^k L_{ad_a^i b} V(x) = 0, \forall k \leq r, i \leq n-1\} = \{0\}. \quad (176)$$

We have:

Theorem 34 [11, Corollary 1.6]. *Under Assumptions B1 and B2, for any $\bar{u}$ in $(0, +\infty]$, the origin can be made a globally asymptotically stable solution of the system (173) by means of a time-invariant continuous feedback law bounded by $\bar{u}$.*

This Theorem relies on the fact that the function V of Assumption B1 is an assignable Lyapunov function and the control is given by any continuous function ϕ – guaranteed to exist – satisfying
1. The function $|\phi(x)|$ is bounded by $\bar{u}$.
2. For all x, the scalar $\frac{\partial V}{\partial y}(y) b(x, \phi(x)) \phi(x)$ is non positive and zero if and only if the vector $\frac{\partial V}{\partial y}(y) b(x, 0)$ is zero.

3.6 Adding Integration.

The problem we address now is to design a feedback law for the system

$$\begin{cases} \dot{x} = h(y, u), \\ \dot{y} = f(y, u), \end{cases} \quad (177)$$

where f and h are C^1 and assuming that we know a time-invariant globally asymptotically stabilizing feedback law for the system

$$\dot{y} = f(y, u) \quad (178)$$

As opposed to (148), this time, we add state components which "integrate" functions of the other components. In particular this implies that the x-part of the system is weakly dissipative. This remark explains the strong links between, the results of this section with those of section 3.5.

The knowledge of a solution for this problem, called "adding one integration", allows us to deal with another recurrent structure, called feedforward form, where each state component acts on those following it, in the chain of integration starting from the control, i.e.[16]

$$\begin{cases} \dot{x}_n = f_n(x_1, \ldots, x_{n-1}, u)\,, \\ \quad \vdots \\ \dot{x}_2 = f_2(x_1, u)\,, \\ \dot{x}_1 = f_1(x_1, u)\,. \end{cases} \quad (179)$$

In particular, a repeated application of Theorem 35 stated below proves that, stabilizability of the system linearized at the origin being assumed, global asymptotic stabilizability by means of a C^1 feedback law holds if $\dot{x} = f_1(x, u)$ is globally asymptotically stabilizable by means of a C^1 feedback law with local exponential stability.

More generally than (177), we consider the following system

$$\begin{cases} \dot{x}_1 = h_0(x_1) + h_1(x_1, x_2, y)y + h_2(x_1, x_2, y, u)u \\ \dot{x}_2 = e_0(x_2) + e_1(x_1, x_2, y)y + e_2(x_1, x_2, y, u)u \\ \dot{y} = f_0(y) + f_1(x_1, x_2, y)y + f_2(x_1, x_2, y, u)u \end{cases} \quad (180)$$

where y is in $\mathbb{R}^n$, x_1 in $\mathbb{R}^{n_1}$, x_2 in $\mathbb{R}^{n_2}$, u in $\mathbb{R}^m$, all the functions are C^2, $h_0(0) = 0$, $e_0(0) = 0$, $f_0(0) = 0$ and we denote $\widehat{h}_2(x_1) = h_2(x_1, 0, 0)$. Mazenc and Praly have proposed in [59], to study this system under the following assumptions:

Assumption C1: *There exist three positive definite and radially unbounded C^2 functions Q, S and V so that*

$$\frac{\partial Q}{\partial x_1}(x_1)\, h_0(x_1) = -R(x_1) \leq 0 \quad \forall x_1\,, \quad (181)$$

$$\frac{\partial S}{\partial x_2}(x_2)\, e_0(x_2) = -T(x_2) < 0 \quad \forall x_2 \neq 0\,, \quad (182)$$

$$\frac{\partial V}{\partial y}(y)\, f_0(y) = -W(y) < 0 \quad \forall y \neq 0\,. \quad (183)$$

Assumption C2: $x_1 = 0$ *is the only solution of*

$$\dot{x}_1 = h_0(x_1)\,, \quad \frac{\partial Q}{\partial x_1}(x_1)\, h_2(x_1, 0, 0) = 0\,, \quad \frac{\partial Q}{\partial x_1}(x_1)\, h_0(x_1) = 0\,. \quad (184)$$

Assumption C3: *There exist positive Lipschitz continuous functions κ and ρ which are zero at zero and such that*

---

[16] Systems in the form (179) are generically not feedback linearizable. In particular this is the case when, controllability of the system linearized at the origin being assumed, $\frac{\partial^2 f_2}{\partial u^2}\frac{\partial f_1}{\partial u} - \frac{\partial f_2}{\partial u}\frac{\partial^2 f_1}{\partial u^2}$ is not identically equal to zero on an open neighborhood of the origin.

3 Lyapunov design of stabilizing state or output feedback.

$$\left|\frac{\partial Q}{\partial x_1}(x_1)\,h_1(x_1,x_2,y)\,y\right| + \left|\frac{\partial S}{\partial x_2}(x_2)\,e_1(x_1,x_2,y)\,y\right| \tag{185}$$

$$\leq \sqrt{\kappa(y)}\,(1+\rho(Q(x_1)+S(x_2)))\left[\sqrt{\kappa(y)}\,(1+\rho(Q(x_1)+S(x_2))) + \sqrt{T(x_2)}\right]$$

$$\frac{1}{1+\rho} \notin L^2([0,+\infty)), \tag{186}$$

$$\limsup_{y \longrightarrow 0}\left|\frac{\kappa(y)}{W(y)}\right| < +\infty. \tag{187}$$

and we have

$$\left|\frac{\partial V}{\partial y}(y)\,f_1(x_1,x_2,y)\,y\right| \leq \frac{1}{4}W(y). \tag{188}$$

Assumptions C1 and C2 are nothing but B1 and B2 of section 3.5. The new assumption here is C3. It concerns the coupling terms h_1, e_1, f_1. Clearly (188) implies the term f_1 cannot change the asymptotic stability of $y = 0$ whatever $x_1(t), x_2(t)$ are, as long as they are measurable and locally essentially bounded. Inequality (185) with (187) and (186) puts a restriction on the growth in y and (x_1, x_2) on h_1 and e_1. More specifically, (186) is a constraint at infinity for (x_1, x_2) whereas (187) is a constraint on a neighborhood of the origin for y. Unfortunately, with the smoothness of V and f_0, this latter constraint implies, for all x_1

$$\limsup_{y \longrightarrow 0} \frac{\left|\frac{\partial Q}{\partial x_1}(x_1)\,h_1(x_1,0,y)\right|}{|y|} < +\infty. \tag{189}$$

And therefore, we must have $h_1(x_1, 0, 0) = 0$. This shows that a preparatory step may be needed before trying to check if C3 holds. Indeed, in [59], it is shown that, under extra assumptions on the linearization at the origin of (180), the condition (187) is met after a change of variables. To go around this problem, Jankovic, Sepulchre and Kokotovic have proposed an alternative to Assumption C3 in [35]

Assumption C3': *There exist positive Lipschitz continuous functions κ and ρ and $\mathcal{H}$ which are zero at zero and such that (185),(186) hold,*

$$v \mapsto \max_{v \leq V(y) \leq 1}\left\{\frac{\kappa(y)}{W(y)}\right\} \in L^1((0,1];[0,+\infty)), \tag{190}$$

the set $\{(z, x_1, x_2, y)\,|\,z = \mathcal{H}(x_1, x_2, y)\}$ is a C^1 invariant manifold of[17]

$$\begin{cases} \dot{z} = -\frac{\partial Q}{\partial x_1}(x_1)\,h_1(x_1,x_2,y)\,y - \frac{\partial S}{\partial x_2}(x_2)\,e_1(x_1,x_2,y)\,y \\ \dot{x}_1 = h_0(x_1) + h_1(x_1,x_2,y)\,y \\ \dot{x}_2 = e_0(x_2) + e_1(x_1,x_2,y)\,y \\ \dot{y} = f_0(y) + f_1(x_1,x_2,y)\,y \end{cases} \tag{191}$$

---

[17] Sufficient condition for existence of this manifold are given in [35].

and (188) holds.

We have

Theorem 35 [59, 35]. *If Assumptions C1, C2 and C3 or C3' hold, then for any $\bar{u}$ in $(0, +\infty]$, the origin of the system (180) is globally asymptotically stabilizable by means of a time-invariant continuous feedback law bounded by $\bar{u}$.*

In the case of Assumption C3, an assignable Lyapunov function is

$$U(x_1, x_2, y) = l(Q(x_1) + S(x_2)) + k(V(y)) . \tag{192}$$

where l and k are any C^1, positive and radially unbounded functions with strictly positive derivative – guaranteed to exist – satisfying

$$\frac{1}{6} k'(V(y)) W(y) \geq \kappa(y) \quad , \quad l(r) = \int_0^r l'(s) ds \tag{193}$$

where l' is any function which does not belong to L^1 and satisfies

$$l'(0) = 1 \quad , \quad 0 < l' \leq \frac{1}{(1+\rho)^2} . \tag{194}$$

In the case of Assumption C3', an assignable Lyapunov function is

$$U(x_1, x_2, y) = Q(x_1) + S(x_2) + \mathcal{H}(x_1, x_2, y) + V(y). \tag{195}$$

In both case, the control is given by any continuous function ϕ – guaranteed to exist – satisfying
1. The function $|\phi(x)|$ is bounded by $\bar{u}$.
2. For all x, the scalar $B(x_1, x_2, y, u) \phi(x)$ is non positive and zero if and only if the vector $B(x_1, 0, 0, 0)$ is zero where, in the case of C3,

$$B(x_1, x_2, y, u) = \tag{196}$$

$$\left(k'(V(y)) \frac{\partial V}{\partial y}(y) , \, l'(Q(x_1) + S(x_2)) \frac{\partial Q}{\partial x_1}(x_1) , \, \frac{\partial S}{\partial x_2}(x_2) \right) \begin{pmatrix} f_2(x_1, x_2, y, u) \\ h_2(x_1, x_2, y, u) \\ e_2(x_1, x_2, y, u) \end{pmatrix} .$$

and, in the case of C3',

$$B(x_1, x_2, y, u) = \tag{197}$$

$$\left(\frac{\partial V}{\partial y}(y) + \frac{\partial \mathcal{H}}{\partial y}(x_1, x_2, y) , \, \frac{\partial Q}{\partial x_1}(x_1) + \frac{\partial \mathcal{H}}{\partial x_1}(x_1, x_2, y) , \, \frac{\partial S}{\partial x_2}(x_2) + \frac{\partial \mathcal{H}}{\partial x_2}(x_1, x_2, y) \right)$$

$$\times \begin{pmatrix} f_2(x_1, x_2, y, u) \\ h_2(x_1, x_2, y, u) \\ e_2(x_1, x_2, y, u) \end{pmatrix}$$

3 Lyapunov design of stabilizing state or output feedback.

The implementation of such feedback laws requires the explicit knowledge of the function V if not of $\mathcal{H}$. If this information is not available, a feedback law can still be designed as follows:
Let

$$\mathcal{B}(x_1, x_2, y) = \tag{198}$$

$$l'(Q(x_1) + S(x_2)) \left[\frac{\partial Q}{\partial x_1}(x_1) h_2(x_1, x_2, y, 0) + \frac{\partial S}{\partial x_2}(x_2) e_2(x_1, x_2, y, 0) \right].$$

Note that $\mathcal{B}$ does not depend on V or k but depends on l'. However l' can be determined, via (194), from the data of the (x_1, x_2)-subsystem only. Let also R and $\bar{u}$ be two arbitrary strictly positive real numbers and let us introduce two functions independent of V:

1. Let φ_R be a smooth non positive function onto $[0, 1]$ such that

$$\Psi_R(0) = 1 \ , \qquad \Psi_R(|y|^2) = 0 \quad \forall y : |y| \geq R \ . \tag{199}$$

2. Let $\psi_{R,\bar{u}}$ be a smooth function satisfying

$$\psi_{R,\bar{u}}(x_1, x_2) \geq \max \left\{ 1, \ \sup_{\substack{|u| \leq \bar{u} \\ |y| \leq R}} \left\{ \widehat{\psi}(x_1, x_2, y, u) \right\} \right\} \tag{200}$$

with

$$\widehat{\psi}(x_1, x_2, y, u) = \left| \frac{f_2(x_1, x_2, y, u) - f_2(x_1, x_2, y, 0)}{u} \right| \tag{201}$$

$$+ \left| \frac{\frac{\partial Q}{\partial x_1}(x_1) \left[h_2(x_1, x_2, y, u) - h_2(x_1, x_2, y, 0) \right]}{u} \right|$$

$$+ \left| \frac{\frac{\partial S}{\partial x_2}(x_2) \left[e_2(x_1, x_2, y, u) - e_2(x_1, x_2, y, 0) \right]}{u} \right|$$

We have:

Proposition 36. *Assume the system (180) satisfies Assumptions C1, C2 and C3. Under these conditions, if*[18]

$$\liminf_{y \to 0} \frac{W(y)}{\left| \frac{\partial V}{\partial y}(y) \right|^2} > 0 \ , \tag{202}$$

---
[18] This condition is satisfied if $y = 0$ is a locally exponentially stable equilibrium point of $\dot{y} = f_0(y)$.

then, for any $\overline{u}$ in $(0, +\infty)$, there exists a positive real number ζ^* in $(0, \overline{u}]$ so that the origin of the system (180) can be made a globally asymptotically stable solution by a state feedback bounded by $\overline{u}$ and of the form

$$u(x_1, x_2, y) = -\frac{\zeta \Psi_R(|y|^2) \mathcal{B}(x_1, x_2, y)}{\psi_{R,\overline{u}}(x_1, x_2) \left(1 + |\mathcal{B}(x_1, x_2, y)|\right) \left(1 + |f_2(x_1, x_2, y, 0)|^2\right)} \quad (203)$$

where ζ is any real number in $(0, \zeta^*]$ and $\mathcal{B}$ is defined in (198).

4 Results based on $\mathcal{L}_\infty$ input-output stability.

4.1 Introduction.

In this section of the paper, we focus on global asymptotic stability and stabilizability (GAS) for nonlinear systems using input-output methods. In particular, we will work with $\mathcal{L}_\infty$-type stability properties. This setting has recently proved to be a very useful domain in which to analyze nonlinear systems and it yields results very similar to Lyapunov stability. The section is intended to highlight a growing, and significant, group of research results on stabilization based on $\mathcal{L}_\infty$-type stability properties while providing pointers to more detailed information in the literature. It is not intended to be a comprehensive overview of the use of input-output methods for nonlinear stability. Consequently, passivity and other issues pertaining to $\mathcal{L}_2$ stability, for example, are not addressed. Finally, local versions of the results presented here are fairly straightforward and, thus, are not pursued.

The definition of global asymptotic stability for the equilibrium of an autonomous ordinary differential equation consists of three parts: 1) local ($\epsilon - \delta$) stability, 2) global boundedness, and 3) global convergence. Frequently, as in the previous section, this property is established by producing a positive definite, radially unbounded function with a negative definite derivative along the trajectories of the differential equation. In this part of the paper, instead, we will consider nonlinear systems decomposed into subsystems and we will use appropriate characterizations of the subsystems input-output behavior to establish the pieces which constitute the GAS property. (It turns out that there are many connections between the input-output properties we will use and corresponding Lyapunov function properties. The references [88, 85], [37] and [87] are useful supplements in this direction.) Our key tool for analyzing the interconnection of subsystems will be a version of the nonlinear small gain theorem, based on a condition made precise in [54] and [36]. In general, the small gain theorem is a very efficient tool for discussing robustness of the GAS property.

4.2 Clarifications for this section.

Throughout, we will assume that all vector fields are smooth enough so that solutions exist locally and are unique. We will use the so-called ∞-norm for vectors $x \in \mathbb{R}^n$, i.e. $|x| := \max_i |x_i|$. We will use the word 'signal' to refer to a time function that is measurable and locally essentially bounded. We will use the phrase 'bounded signal' for a signal that is essentially bounded. For a bounded signal d we will use

$$||d||_\infty := \sup_{t \in [0,\infty)} |d(t)| \quad , \quad ||d||_a := \limsup_{t \to \infty} |d(t)| \qquad (204)$$

where each supremum is understood to be an essential supremum.

4.3 Cascades as a starting point.

The interest in the GAS property for nonlinear systems decomposed into subsystems accelerated with the development of global, partially linear normal forms via the tools of geometric nonlinear control. See [33, 6] and the references therein. When applying these tools, a common structure that emerges after feedback is that of a cascade of subsystems :

$$\dot{x}_1 = f_1(x_1, x_2) \quad , \quad \dot{x}_2 = f_2(x_2). \tag{205}$$

For such a system to have the GAS property, the origins of the systems

$$\dot{x}_1 = f_1(x_1, 0) \quad , \quad \dot{x}_2 = f_2(x_2) \tag{206}$$

must be GAS. Moreover, for the system

$$\dot{x}_1 = f_1(x_1, u), \tag{207}$$

for each initial condition and each input u generated by the solutions of the autonomous x_2 subsystem in (206) according to $u(t) = x_2(t)$, the solution must converge to the origin. For nonlinear systems, this property is not guaranteed by the property that the x_1 subsystem in (206) is GAS. For example, as pointed out in [80], the scalar system

$$\dot{x} = -x + (x^2 + 1)u \tag{208}$$

is GAS when $u \equiv 0$, but with

$$u(t) = \frac{1}{\sqrt{2t+2}} \quad , \quad x(0) = \sqrt{2} \tag{209}$$

the resulting solution $x(t) = \sqrt{2t+2}$ is unbounded.

4.4 Addressing cascades via the ISS property.

In general, efficient methods for checking whether converging inputs produce converging states are not known. Instead, and also motivated by robustness issues to be discussed later, we will impose a stronger property which can be verified using Lyapunov arguments. In what follows, we will define and employ what is essentially the input-to-state stability (ISS) property introduced by Sontag in [78]. In our definition, we will call a function from the nonnegative real numbers to the nonnegative real numbers which is continuous, zero at zero and nondecreasing a **gain function**.

Consider the system

$$\dot{x} = f(x, d_1, d_2) \quad , \quad x(0) = x_\circ \tag{210}$$

where $x \in \mathbb{R}^n$, $d_1 \in \mathbb{R}^{m_1}$, and $d_2 \in \mathbb{R}^{m_2}$, along with an 'output' function $h(x, d_1, d_2)$. Let γ_0, γ_1 and γ_2 be gain functions. We will say that h satisfies an

4 Results based on $\mathcal{L}_\infty$ input-output stability.

a-$\mathcal{L}_\infty$ stability bound[19] with gain $(\gamma_0, \gamma_1, \gamma_2)$ if, for each $x_o \in \mathbb{R}^n$ and each pair of bounded signals $(d_1(\cdot), d_2(\cdot))$, the solution to (210) exists for all $t \geq 0$ and satisfies

$$||h(x, d_1, d_2)||_\infty \leq \max \left\{ \gamma_0(|x_o|), \quad \gamma_1(||d_1||_\infty), \quad \gamma_2(||d_2||_\infty) \right\}$$
$$||h(x, d_1, d_2)||_a \leq \max \left\{ \gamma_1(||d_1||_a), \quad \gamma_2(||d_2||_a) \right\}. \tag{211}$$

It will be convenient to think of the subscripts on the gain functions as channel numbers. We will always use channel 0 for the initial conditions. When we are working with the interconnection of two subsystem, each with a channel 1 gain for example, we will use γ_{11} to refer to the channel 1 gain for system 1 and γ_{12} to refer to the channel 1 gain for system 2.

There is a connection between this property when $h(x, d) := x$ and the existence of a Lyapunov-type function with a particular property for its derivative.

Fact 37 [78]. *If there exist a function V, globally invertible gain functions $\underline{\alpha}$ and $\overline{\alpha}$ and a gain function γ such that*

$$\underline{\alpha}(|x|) \leq V(x) \leq \overline{\alpha}(|x|) \tag{212}$$

and

$$|x| > \gamma(|u|) \implies \frac{\partial V}{\partial x}(x) f(x, u) < 0 \tag{213}$$

then the state x of the system $\dot{x} = f(x, u)$ satisfies an a-$\mathcal{L}_\infty$ stability bound with gain

$$(\gamma_0, \gamma_1) := \left(\underline{\alpha}^{-1} \circ \overline{\alpha}, \, \underline{\alpha}^{-1} \circ \overline{\alpha} \circ \gamma \right). \tag{214}$$

Remark 38. A converse of this result, for Sontag's ISS property, is reported on in [88] (see also [52] and [85]).

Now, returning to the cascade system in (205), if the state of the system

$$\dot{x}_1 = f_1(x_1, d) \tag{215}$$

satisfies an a-$\mathcal{L}_\infty$ stability bound then converging inputs produce converging states and, thus, the cascade is GAS. Thus, with this condition on the x_1 subsystem, the GAS control problem for the system

$$\dot{x}_1 = f_1(x_1, x_2), \quad \dot{x}_2 = f_2(x_2, u) \tag{216}$$

reduces to the GAS control problem for the x_2 subsystem.

---

[19] You may wish to read the 'a' in 'a-$\mathcal{L}_\infty$' as 'augmented' or 'asymptotic' or simply as the 'a' subscript in (204).

Adding integrators. When it is not the case that the state of the x_1 subsystem in (216) satisfies an a-$\mathcal{L}_\infty$ stability bound, certain control systems in the form (216) benefit from the following lemma:

Lemma 39 [81]. *If the origin of the system $\dot{x} = f(x,0)$ is GAS then there exists a smooth, globally invertible matrix-valued function[20] $\beta(x)$ such that the state of the system $\dot{x} = f(x, \beta(x)d)$ satisfies an a-$\mathcal{L}_\infty$ stability bound.*

Consider the GAS control problem for the system

$$\dot{x}_1 = f_1(x_1, x_2) \quad , \quad \dot{x}_2 = u \qquad (217)$$

where $x_1 \in \mathbb{R}^n$ and u and x_2 belong to $\mathbb{R}^m$. (The idea presented here easily generalizes to the case of adding chains of integrators.) Suppose that there exists a smooth function $k(x_1)$ so that the origin of the system

$$\dot{x}_1 = f_1(x_1, k(x_1)) \qquad (218)$$

is GAS. Using the above lemma, there exists $\beta(x_1)$, smooth and globally invertible, so that the state of the system

$$\dot{x}_1 = f_1(x_1, k(x_1) + \beta(x_1)d) \qquad (219)$$

satisfies an a-$\mathcal{L}_\infty$ stability bound. Define $\tilde{x}_2 := \beta^{-1}(x_1)[x_2 - k(x_1)]$, so that the original system (217) becomes

$$\begin{cases} \dot{x}_1 = f_1(x_1, k(x_1) + \beta(x_1)\tilde{x}_2) \\ \dot{\tilde{x}}_2 = \beta^{-1}(x_1)u + g(x_1, \tilde{x}_2) \end{cases} \qquad (220)$$

where the definition of g follows from differentiating $\tilde{x}_2$. Then choosing $u = \beta(x_1)[\alpha(\tilde{x}_2) - g(x_1, \tilde{x}_2)]$, where the origin of $\dot{x} = \alpha(x)$ is GAS, achieves the GAS property for the system (217).

4.5 GAS for feedback interconnections.

Imposing an a-$\mathcal{L}_\infty$ stability bound on the state of the x_1 subsystem in (205) is much stronger than is needed. However, it leads to natural robustness conditions for the GAS property made clear via a nonlinear small gain theorem. In particular, consider a nominally cascaded system which is perturbed in a way so that the x_1 subsystem feeds into the x_2 subsystem:

$$\begin{cases} \dot{x}_1 = f_1(x_1, x_2) \\ \dot{x}_2 = f_2(x_2, x_1). \end{cases} \qquad (221)$$

---

[20] We will have use later for the fact that this matrix-valued function can be chosen to be globally bounded and equal to the identity matrix on a neighborhood of the origin.

4 Results based on $\mathcal{L}_\infty$ input-output stability.

Suppose that for each subsystem its state satisfies an a-$\mathcal{L}_\infty$ stability bound. Let γ_{11} denote the channel 1 gain function for the x_1 subsystem and let γ_{12} denote the channel 1 gain function for the x_2 subsystem. We will say that the composition of these two functions (the channel 1 gains) is a **simple contraction** if $\gamma_{11}(\gamma_{12}(s)) < s$ (equivalently $\gamma_{12}(\gamma_{11}(s)) < s$) for all $s > 0$.

Theorem 40. *If the composition of the channel 1 gains is a simple contraction then the origin of (221) is GAS.*

Proof. Given a particular initial condition, let $[0, T)$ be the maximal interval of definition for the system (221). We will use x_τ to represent the truncation of the signal x at time τ. By causality, we have, for each $\tau \in [0, T)$,

$$\begin{aligned} ||x_{1_\tau}||_\infty &\leq \max\left\{ \gamma_{01}(|x_{1_\circ}|), \quad \gamma_{11}(||x_{2_\tau}||_\infty) \right\} \\ ||x_{2_\tau}||_\infty &\leq \max\left\{ \gamma_{02}(|x_{2_\circ}|), \quad \gamma_{12}(||x_{1_\tau}||_\infty) \right\}. \end{aligned} \quad (222)$$

Combining, we get

$$\begin{aligned} ||x_{1_\tau}||_\infty &\leq \max\left\{ \gamma_{01}(|x_{1_\circ}|), \quad \gamma_{11}\circ\gamma_{02}(|x_{2_\circ}|), \quad \gamma_{11}\circ\gamma_{12}(||x_{1_\tau}||_\infty) \right\} \\ ||x_{2_\tau}||_\infty &\leq \max\left\{ \gamma_{02}(|x_{2_\circ}|), \quad \gamma_{12}\circ\gamma_{01}(|x_{1_\circ}|), \quad \gamma_{12}\circ\gamma_{11}(||x_{2_\tau}||_\infty) \right\}. \end{aligned} \quad (223)$$

But, since the composition of γ_{11} and γ_{12} is a simple contraction, it follows that

$$\begin{aligned} ||x_{1_\tau}||_\infty &\leq \max\left\{ \gamma_{01}(|x_{1_\circ}|), \quad \gamma_{11}\circ\gamma_{02}(|x_{2_\circ}|) \right\} \\ ||x_{2_\tau}||_\infty &\leq \max\left\{ \gamma_{02}(|x_{2_\circ}|), \quad \gamma_{12}\circ\gamma_{01}(|x_{1_\circ}|) \right\}. \end{aligned} \quad (224)$$

Now, since the right hand sides are independent of τ, we have that x_1 and x_2 are uniformly bounded on $[0, T)$ which tells us that $T = \infty$ and that $||x_1||_a$ and $||x_2||_a$ are well-defined. So, we can also use

$$\begin{aligned} ||x_1||_a &\leq \gamma_{11}(||x_2||_a) \\ ||x_2||_a &\leq \gamma_{12}(||x_1||_a). \end{aligned} \quad (225)$$

Combining these inequalities and again using that the composition of γ_{11} and γ_{12} is a simple contraction, it follows that $||x_1||_a = ||x_2||_a = 0$. □

Remark 41. A similar result can be stated when the interconnection is made via generic output functions. Such a generalization is important for the result in the last part of section 4.7. This idea is also discussed in more detail in [36].

Remark 42. From the proof, it follows that if the composition of the gains is a contraction only for sufficiently small values of s then local asymptotic stability can be established. On the other hand, if the composition is a contraction only for sufficiently large values of s then global boundedness can be established. Indeed, suppose that $\gamma_1(\gamma_2(s)) < s$ for $s > s^*$ and we have $x \leq \max\{a, b, \gamma_1(\gamma_2(x))\}$. Then, we claim that $x \leq \max\{a, b, s^*\}$. If this is not the case then necessarily $x > s^*$ and $x \leq \gamma_1(\gamma_2(x))$ which contradicts the assumption.

Such situations are addressed in [96] and [36].

Remark 43. When there are multiple inputs, it is very common in the literature to work with an input-output bound which is expressed in terms of a summation rather than a maximum. In this case, the small gain condition needs to be stronger. The implication of the composition of γ_1 and γ_2 being a simple contraction is that, for each pair of positive real numbers b_1 and b_2, there exists a positive real number s^* such that the curve $\left(s, \max\{b_1, \gamma_1(s)\}\right)$ is below the curve $\left(\max\{b_2, \gamma_2(r)\}, r\right)$ for all $s > s^*$. Analogously for the summation case, we would want that, for each pair of positive real numbers b_1 and b_2, there exists a positive real number s^* such that the curve $(s, b_1 + \gamma_1(s))$ is below the curve $(b_2 + \gamma_2(r), r)$ for all $s > s^*$. But since b_1 and b_2 can be arbitrarily large, the curve $(s, b_1 + \gamma_1(s))$ can be shifted an arbitrarily large vertical distance from the curve $(s, \gamma_1(s))$ while the curve $(b_2 + \gamma_2(r), r)$ can be shifted an arbitrarily large horizontal distance from the curve $(\gamma_2(r), r)$. Thus, it is not enough for the curve $(s, \gamma_1(s))$ to simply be below the curve $(\gamma_2(r), r)$, i.e. for the composition to be simple contraction. Indeed what is required is that the distance between these two curves grows without bound. One way to characterize this is to require that there exists a globally invertible gain function ρ such that the composition of the functions $\gamma_1 + \rho$ and $\gamma_2 + \rho$ is a simple contraction. This is essentially the condition used in [54] and [36]. It is easy to see that this is sufficient by first using the fact that, for any globally invertible gain function ρ,

$$b + \gamma(a) \leq \max\left\{(\mathrm{Id} + \gamma \circ \rho^{-1})(b), (\gamma + \rho)(a)\right\} \tag{226}$$

and then using theorem 40. The inequality (226) follows from considering the two case: $b \leq \rho(a)$ and $a \leq \rho^{-1}(b)$.

These ideas are strongly connected to the ideas found in [73]. For the case where the gain functions are linear, there is no difference between a simple contraction and the stronger notion discussed above. Instead, both properties become that the product of the coefficients of the linear gains be less than one, i.e. the condition of the classical small gain theorem [102].

Example 1. As a simple example, consider the system

$$\begin{cases} \dot{x}_1 = -x_1 + \gamma_1(|x_2|) \\ \dot{x}_2 = -x_2 + \gamma_2(|x_1|) \end{cases} \tag{227}$$

where the functions γ_1 and γ_2 are gain functions. Using fact 37 and the function $V_i = x_i^2$, it follows that x_1 satisfies an a-$\mathcal{L}_\infty$ stability bound with channel 1 gain

4 Results based on $\mathcal{L}_\infty$ input-output stability.

γ_1 and x_2 satisfies an a-$\mathcal{L}_\infty$ stability bound with channel 1 gain γ_2. Therefore, if the composition of γ_1 and γ_2 is a simple contraction then the system is GAS.

It can be shown that if the composition is not a simple contraction then the origin is not GAS. For example, if there exists an $s^* > 0$ such that $\gamma_1(\gamma_2(s^*)) = s^*$ then $(s^*, \gamma_2(s^*))$ is a nonzero equilibrium. Otherwise, if $\gamma_1(\gamma_2(s)) > s$ for all $s > 0$ then the set

$$\left\{ (x_1, x_2) : \ 0 < x_2 \leq \gamma_2(|x_1|) \ , 0 < x_1 \leq \gamma_1(|x_2|) \right\}$$

is nonempty, positively invariant, and the function $x_1^2 + x_2^2$ is strictly increasing inside this set.

4.6 $\mathcal{L}_\infty$ stability for feedback interconnections.

For purposes of iteration, we may be interested in establishing an a-$\mathcal{L}_\infty$ stability bound for the composite state of the interconnection with respect to an external input rather than simply the GAS property. Consider the system

$$\begin{cases} \dot{x}_1 = f_1(x_1, x_2, d_1) \\ \dot{x}_2 = f_2(x_2, x_1, d_2) \end{cases} \tag{228}$$

and suppose that for each subsystem its state satisfies an a-$\mathcal{L}_\infty$ stability bound.

Theorem 44. *If the composition of the channel 1 gains is a simple contraction then the composite state $\binom{x_1}{x_2}$ for the system (228) satisfies an a-$\mathcal{L}_\infty$ stability bound.*

Remark 45. The proof of this result uses the same calculations as in the proof of theorem 40. In working through the proof, one can easily construct the gain functions for the closed loop system. They are simple combinations of the gain functions for the subsystems. In particular, with respect to $\binom{x_{1_\circ}}{x_{2_\circ}}$, d_1 and d_2, respectively, the gains are

$$\gamma_0(s) = \max \left\{ \gamma_{01}(s), \ \gamma_{11} \circ \gamma_{02}(s), \ \gamma_{02}(s), \ \gamma_{12} \circ \gamma_{01}(s) \right\}$$

$$\gamma_1(s) = \max \left\{ \gamma_{21}(s), \ \gamma_{12} \circ \gamma_{21}(s) \right\} \tag{229}$$

$$\gamma_2(s) = \max \left\{ \gamma_{11} \circ \gamma_{22}(s), \ \gamma_{22}(s) \right\}.$$

Remark 46. Similar to the comments in remark 42, if the composition of the gains is a contraction only for s sufficiently small then the a-$\mathcal{L}_\infty$ stability bound holds for sufficiently small initial conditions and inputs d_1 and d_2 with a sufficiently small $\mathcal{L}_\infty$-norm. Also, if the composition of the gains is a contraction only for s sufficiently large then the state of the closed loop system satisfies a modified a-$\mathcal{L}_\infty$ stability bound where a positive offset is included in the maximums on the right hand side of (211).

4.7 Robust control via gain assignment.

In this section we will state some important results on assigning closed loop gains for nonlinear control systems. These results will be used to solve certain robust stabilization problems.

Gain assignment for a scalar system. Consider the scalar system

$$\dot{x} = u + \phi(x, d_1) + d_2 =: f(u, x, d_1, d_2) \tag{230}$$

and suppose gain functions ρ_0 and ρ_1 are known so that

$$|\phi(x, d_1)| \leq \max\left\{\rho_0(|x|), \rho_1(|d_1|)\right\}. \tag{231}$$

Lemma 47 [36]. *Let γ_1 be a globally invertible gain function and suppose that $\rho_1 \circ \gamma_1^{-1}$ and ρ_0 are locally Lipschitz at the origin. Then there exists a smooth function $k(x)$ so that the state x of the closed loop system (230) with $u = k(x)$ satisfies an a-$\mathcal{L}_\infty$ stability bound with gain $(\mathrm{Id}, \gamma_1, \mathrm{Id})$.*

Remark 48. If $\rho_1 \circ \gamma_1^{-1}$ is not locally Lipschitz at the origin, for each strictly positive real number δ, one can always find a globally invertible gain function $\bar{\gamma}_1$ so that $\rho_1 \circ \bar{\gamma}_1^{-1}$ is locally Lipschitz at the origin and $\bar{\gamma}_1(s) \leq \max\{\gamma_1(s), \delta\}$. So, even without the extra condition on $\rho_1 \circ \gamma_1^{-1}$, one can still achieve the channel 1 gain γ_1 with a smooth control if one is willing to tolerate a small positive offset in the a-$\mathcal{L}_\infty$ stability bound. Also, if one is willing to settle for a control which is continuous and smooth everywhere except at the origin, the extra Lipschitz conditions are not needed.

Proof. Let $\alpha : \mathbb{R} \to \mathbb{R}$ be smooth, zero at zero, strictly increasing, odd and, with the definition $\tilde{\alpha}(s) := \alpha(s)$ for $s \geq 0$, satisfies

$$\max\left\{\rho_0(s), \rho_1 \circ \gamma_1^{-1}(s)\right\} \leq \tilde{\alpha}(s). \tag{232}$$

Such a smooth function exists from the local assumptions made for the functions on the left hand side. Then pick

$$u = -x - \alpha(x) =: k(x) \tag{233}$$

and consider the derivative of the function $V(x) = x^2$ along the solutions of the closed loop system. We have

$$\frac{\partial V}{\partial x}(x) f(k(x), x, d_1, d_2) \tag{234}$$

$$= 2x\left[-x - \alpha(x) + \phi(x, d_1) + d_2\right]$$

$$\leq 2|x|\left[-|x| - \tilde{\alpha}(|x|) + \max\left\{\rho_0(|x|), \rho_1(|d_1|)\right\} + |d_2|\right].$$

4 Results based on $\mathcal{L}_\infty$ input-output stability.

Suppose
$$|x| > \max\left\{\gamma_1(|d_1|), |d_2|\right\}. \tag{235}$$
Then, using (232), this implies
$$|x| > \max\left\{\tilde{\alpha}^{-1} \circ \rho_1(|d_1|), |d_2|\right\} \tag{236}$$
which in turn implies
$$\tilde{\alpha}(|x|) > \rho_1(|d_1|) \quad, \quad |x| > |d_2|. \tag{237}$$
Since we have from (232) that $\tilde{\alpha}(|x|) \geq \rho_0(|x|)$, it follows that (235) implies $\dot{V} < 0$. The lemma then follows from fact 37. $\square$

Adding perturbed integrators. (For more details see [36] and the related problem in [67].) Consider the system
$$\begin{cases} \dot{z} = f_1(z, x_1) \\ \dot{x}_1 = x_2 + \phi_1(z, x_1) \\ \dot{x}_2 = u + \phi_2(z, x_1, x_2) \end{cases} \tag{238}$$
where the state of the z subsystem satisfies an a-$\mathcal{L}_\infty$ stability bound with channel 1 gain γ_{11}, and where ϕ_1 and ϕ_2 are locally Lipschitz and vanish at the origin. In particular, consider the GAS control problem using feedback of x_1 and x_2 only. To solve this problem, we will use lemma 47 twice. We will assume the data of the problem is such that the local Lipschitz conditions in the lemma hold for each application. Otherwise, according to remarks 48 and 46, we can achieve a type of "practical GAS". Let γ_{12} be a gain function so that the composition of γ_{11} and γ_{12} is a simple contraction and let k be the solution to the gain assignment problem for γ_{12}. Now define $\zeta_2 = x_2 - k(x_1)$ so that we have
$$\begin{cases} \dot{z} = f_1(z, x_1) \\ \dot{x}_1 = k(x_1) + \phi_1(z, x_1) + \zeta_2 \\ \dot{\zeta}_2 = u + \tilde{\phi}_2(z, x_1, \zeta_2) \end{cases} \tag{239}$$
where the definition of $\tilde{\phi}$ follows from differentiating ζ_2. From the solution to the gain assignment problem, and according to remark 45, the state of the (z, x_1) subsystem satisfies an a-$\mathcal{L}_\infty$ stability bound with respect to ζ_2 with channel 1 gain
$$\tilde{\gamma}_{11}(s) = \max\left\{\gamma_{11}(s), s\right\}. \tag{240}$$
We then apply lemma 47 a second time, this time for the ζ_2 subsystem, for a gain function whose composition with $\tilde{\gamma}_{11}$ is a simple contraction. From theorem 44, this will give us that the state of the closed loop system satisfies an a-$\mathcal{L}_\infty$ stability bound with respect to initial conditions and an additive disturbance at the input. When this additive disturbance is zero, we have the GAS property.

Gain assignment for a general system. Consider the system

$$\dot{x} = f(x) + g(x)[u + d]. \tag{241}$$

Lemma 49 [68]. *Suppose the system (241) with $d = 0$ can be made GAS with smooth static state feedback. Let γ_x and γ_u be globally invertible gain functions. If the functions γ_x^{-1} and γ_u^{-1} are suitably smooth at the origin then there exists a smooth function $k(x)$ such that, for the closed loop system (241) with $u = k(x)$, the state x and the function $k(x)$ satisfy a-$\mathcal{L}_\infty$ stability bounds, the former with channel 1 gain γ_x and the latter with channel 1 gain $\mathrm{Id} + \gamma_u$.*

Remark 50. Remark 48 applies when the functions γ_x^{-1} and γ_u^{-1} aren't suitably smooth.

We use this result to discuss the GAS control problem for nonlinear systems affine in u where stable, unmodeled dynamics enter additively at the input. (For more details see [68]. Compare also with [44].) In particular we consider the GAS control problem for the system

$$\begin{cases} \dot{x}_1 = f_1(x_1) + g_1(x_1)[u + \phi(x_2, x_1, u)] \\ \dot{x}_2 = f_2(x_2, x_1, u) \end{cases} \tag{242}$$

using only feedback of x_1. We suppose that 1) the x_1 subsystem with $\phi \equiv 0$ can be made GAS with smooth static state feedback; 2) for the system

$$\dot{x}_2 = f_2(x_2, d_1, d_2), \tag{243}$$

the state x_2 and the function $\phi(x_2, d_1, d_2)$ satisfy a-$\mathcal{L}_\infty$ stability bounds, the function ϕ with channel 1 gain γ_{12} and channel 2 gain γ_{22}; and 3) there exist globally invertible gain functions γ_x and γ_u (with a suitably smooth inverses at the origin as required in lemma 49) so that the composition of γ_{12} and γ_x, as well as the composition of γ_{22} and $\mathrm{Id} + \gamma_u$, are simple contractions.

Apply lemma 49 to find a function $k(x_1)$ so that the closed loop system (242) with $u = k(x_1)$ satisfies: for each initial condition there is a maximal interval of definition $[0, T)$ and for each $\tau \in [0, T)$,

$$\begin{aligned}
\|\phi_\tau\|_\infty &\leq \max\left\{ \gamma_{0\phi}(|x_{2_o}|), \ \gamma_{12}(\|x_{1_\tau}\|_\infty), \ \gamma_{22}(\|k_\tau\|_\infty) \right\} \\
\|k_\tau\|_\infty &\leq \max\left\{ \gamma_{0k}(|x_{1_o}|), \ (\mathrm{Id} + \gamma_u)(\|\phi_\tau\|_\infty) \right\} \\
\|x_{1_\tau}\|_\infty &\leq \max\left\{ \gamma_{0x_1}(|x_{1_o}|), \ \gamma_x(\|\phi_\tau\|_\infty) \right\}
\end{aligned} \tag{244}$$

4 Results based on $\mathcal{L}_\infty$ input-output stability.

and, if $T = \infty$ and all signals are bounded,

$$||\phi||_a \leq \max\left\{ \gamma_{12}(||x_1||_a) , \gamma_{22}(||k||_a) \right\}$$
$$||k||_a \leq (\text{Id} + \gamma_u)(||\phi||_a) \tag{245}$$
$$||x_1||_a \leq \gamma_x(||\phi||_a) .$$

Using the same type of calculations as in the proof of theorem 40, and then combining with the a-$\mathcal{L}_\infty$ stability bound for the state x_2, it follows that all signals are defined on $[0, \infty)$, are bounded by a gain function of the initial conditions and converge to zero. Thus, the GAS property is established.

4.8 'Saturated' interconnections.

Having summarized several results for feedback interconnections, we draw attention back to cascades for a moment. We have worked with cascades where the state of the driven subsystem satisfied an a-$\mathcal{L}_\infty$ stability bound. However, we pointed out that this was stronger than was really needed. In fact, although for cascades it is somewhat awkward to think of it this way, we only need the second part of the a-$\mathcal{L}_\infty$ stability bound in (211) and, even then, we only need it for inputs that converge to zero. The reason that this is the case is that we know a priori, regardless of what the driven subsystem does, that the state of the autonomous system will converge to zero. This observation suggests a final class of interconnections that we will consider. This class will be such that the state of one subsystem is guaranteed to converge to a ball of a certain radius. This will be guaranteed using the second inequality of an a-$\mathcal{L}_\infty$ stability bound with a globally bounded gain function. (The state of the autonomous subsystem of a cascade can be thought of as satisfying such a bound with gain function identically zero.) The state of the second subsystem will also be assumed to satisfy the second inequality of an a-$\mathcal{L}_\infty$ stability bound but only for inputs that converge to a sufficiently small ball. For global boundedness, which is the second piece of the GAS property (cf. section 4.1), we will simply need that these two balls match. Global convergence will be guaranteed if the composition of the gain functions in the mentioned inequalities is a simple contraction. This is the third piece of the GAS property. Typically the remaining piece of the GAS property, namely the LAS property, can be check via the Jacobian linearization or a local version of theorem 40.

The description of the above type of interconnection may sound rather contrived. But, in fact, it has proved quite useful in the design and analysis of control laws for systems with saturation and/or with a type of feedforward structure similar to that discussed in the previous section. For references in this direction, see [99, 97, 98]. The same idea is used in [94] and [95] but without the same degree of formalism.

We now modify the notion of an a-$\mathcal{L}_\infty$ stability bound so that the first bound, the $\mathcal{L}_\infty$ bound, is removed while the second bound holds, but perhaps for a restricted class of inputs. Consider the system

$$\dot{x} = f(x, d_1, d_2) \quad, \qquad x(0) = x_\circ \tag{246}$$

where $x \in \mathbb{R}^n$, $d_1 \in \mathbb{R}^{m_1}$ and $d_2 \in \mathbb{R}^{m_2}$, along with an 'output' function $h(x, d_1, d_2)$. Let γ_1 and γ_2 be gain functions and $\Delta_1, \Delta_2 \in \mathbb{R}_{\geq 0} \cup \infty$. We will say that h satisfies an **asymptotic bound with gain** (γ_1, γ_2) **and restriction** (Δ_1, Δ_2) if, for each $x_\circ \in \mathbb{R}^n$ and each pair of signals $(d_1(\cdot), d_2(\cdot))$ satisfying

$$||d_1||_a \leq \Delta_1 \quad, \quad ||d_2||_a \leq \Delta_2 \tag{247}$$

the solution to (246) exists for all $t \geq 0$ and satisfies

$$||h(x, d_1, d_2)||_a \leq \max\left\{ \gamma_1(||d_1||_a)\,,\, \gamma_2(||d_2||_a) \right\}. \tag{248}$$

The arguments of the gain functions are not necessarily bounded since we are working here with signals which are not necessarily bounded a priori. Thus, we need the definition $\gamma(\infty) := \lim_{s \to \infty} \gamma(s)$. This quantity may also be infinite. We will refer to $\gamma(\infty)$ as the supremum of γ. Also, analogous to the labeling of gains, Δ_i will be referred to as the channel i restriction.

For an example of when this property holds, it was shown in [94] that, for matrices $A \in \mathbb{R}^{n \times n}$ and $B \in \mathbb{R}^{n \times m}$, if the pair of matrices (A, B) is stabilizable and the eigenvalues of A have nonpositive real part then for each strictly positive real number b there exists a smooth function $\alpha : \mathbb{R}^n \to \mathbb{R}^m$ and strictly positive real numbers (Δ, N) such that

$$|\alpha(x)| \leq b \qquad \forall x \in \mathbb{R}^n, \tag{249}$$

the state of the system

$$\dot{x} = Ax + B\alpha(x) + d \tag{250}$$

satisfies an asymptotic bound with channel 1 gain $N \cdot \text{Id}$ and channel 1 restriction Δ, and when $d \equiv 0$ the origin of the system (250) is GAS.

We will now consider the interconnection of subsystems in the form (228) where the state of each subsystem satisfies an asymptotic bound. Moreover, we will assume that for all initial conditions and signals (d_1, d_2) defined on $[0, \infty)$ there is no finite escape time.

Theorem 51. *Suppose the system (228) has no finite escape times. If*

1. *the channel 1 restriction for x_2 is ∞, i.e. $\Delta_{12} = \infty$,*
2. *the channel 1 restriction for x_1 is finite, i.e. $\Delta_{11} < \infty$,*
3. *the supremum of the channel 1 gain for x_2 is less than or equal to the channel 1 restriction for x_1, i.e. $\gamma_{12}(\infty) \leq \Delta_{11}$ and*
4. *the composition of the channel 1 gains is a simple contraction*

then the composite state $\begin{pmatrix} x_1 \\ x_2 \end{pmatrix}$ satisfies an asymptotic bound.

4 Results based on $\mathcal{L}_\infty$ input-output stability.

Proof. We will show that the composite state satisfies an asymptotic bound with gain (γ_1, γ_2) given in (229) and with restriction $(\Delta_{21}, \tilde{\Delta}_2)$ where $\tilde{\Delta}_2 \in (\mathbb{R}_{\geq 0} \cup \infty) \cap [0, \Delta_{22}]$ satisfies

$$\max\left\{ \gamma_{12}(\infty), \ \gamma_{22}(\tilde{\Delta}_2) \right\} \leq \Delta_{11}. \tag{251}$$

First, such a $\tilde{\Delta}_2$ exists since $\gamma_{12}(\infty) \leq \Delta_{11}$. Next, since $\tilde{\Delta}_2 \leq \Delta_{22}$ and since $\Delta_{12} = \infty$, $||d_2||_a \leq \tilde{\Delta}_2$ implies

$$||x_2||_a \leq \max\left\{ \gamma_{12}(||x_1||_a), \ \gamma_{22}(||d_2||_a) \right\} \leq \Delta_{11}. \tag{252}$$

This, together with $||d_1||_a \leq \Delta_{21}$ implies that

$$||x_1||_a \leq \max\left\{ \gamma_{11}(||x_2||_a), \ \gamma_{21}(||d_1||_a) \right\}. \tag{253}$$

Now, with the definition of γ_1 in (229), if $\gamma_{21}(||d_1||_a)$ is not finite then there is nothing to prove. Otherwise, both $||x_1||_a$ and $||x_2||_a$ are bounded and the inequalities (252) and (253) can be combined to arrive at the desired result. □

Applications. As a first application, consider the GAS control problem for the system

$$\begin{cases} \dot{x}_1 = Ax_1 + Bu \\ \dot{x}_2 = f(x_2, u) \end{cases} \tag{254}$$

where $u \in \mathbb{R}^m$. This may be a subproblem for the control of this system followed by chains of integrators. Suppose that the origin of the system

$$\dot{x}_2 = f(x_2, 0) \tag{255}$$

is GAS, the pair (A, B) is stabilizable and the eigenvalues of A have nonpositive real part. According to lemma 39, and its footnote, there exist a smooth, globally invertible, globally bounded matrix function $\beta(x_2)$, a gain function γ_2 and a strictly positive real number δ such that $|x_2| \leq \delta$ implies $\beta(x_2) = I_{m \times m}$ and such that the state of the system

$$\dot{x}_2 = f(x_2, \beta(x_2)v) \tag{256}$$

satisfies an a-$\mathcal{L}_\infty$ bound with channel 1 gain γ_2. Let $b > 0$ be such that

$$\gamma_2(b) \leq \delta. \tag{257}$$

Then, according to the result in [94] there exists a smooth function α and strictly positive real numbers N and Δ such that

$$|\alpha(x_1)| \leq b \qquad \forall x_1 \in \mathbb{R}^n, \tag{258}$$

the state of the system

$$\dot{x}_1 = Ax_1 + B(\alpha(x_1) + d) \tag{259}$$

satisfies an asymptotic bound with gain $N \cdot \mathrm{Id}$ and restriction Δ and the origin of the system (259) with $d \equiv 0$ is GAS. With these definitions, consider the control

$$u = \beta(x_2)[\alpha(x_1) + d_1] \tag{260}$$

yielding the closed loop system

$$\begin{cases} \dot{x}_1 = Ax_1 + B\alpha(x_1) + B\beta(x_2)d_1 + B(\beta(x_2) - 1)\alpha(x_1) \\ \dot{x}_2 = f(x_2, \beta(x_2)(\alpha(x_1) + d_1)). \end{cases} \tag{261}$$

We will show that the origin of the closed loop is GAS when $d_1 \equiv 0$ and satisfies an asymptotic bound with respect to d_1 otherwise. The signal d_1 may represent a decaying signal from an autonomous system if the original control problem was for the system (254) appended with integrators.

First, the closed loop (261) does not have finite escape times for any signal d_1 defined on $[0, \infty)$ since the x_2 subsystem satisfies an a-$\mathcal{L}_\infty$ stability bound with respect to $\alpha + d_1$ and α and β are globally bounded. Next, it can be shown that the state of the x_2 subsystem satisfies an asymptotic bound with gain $(\gamma_{21}, 0 \cdot \mathrm{Id})$ and with restriction $(\infty, 0)$ where $\gamma_{21}(\infty) \leq \delta$. Also, it can be shown that the state of the x_1 subsystem satisfies an asymptotic bound with respect to x_2 and d_1 with gain $(0 \cdot \mathrm{Id}, 0 \cdot \mathrm{Id})$ with restriction $(\delta, 0)$. The restrictions here are conservative but adequate for our needs. Thus, all of the conditions of the theorem are satisfied and the state of the closed loop system satisfies an asymptotic bound with respect to d_1 with restriction 0. This, in particular, gives us global boundedness and convergence when $d_1 \equiv 0$. Also in this case the LAS property holds since, near the origin, the closed loop system behaves like a cascade.

As a second application, consider a particular system in so-called feedforward form. Consider the GAS control problem for the system

$$\begin{cases} \dot{x}_1 = x_2 + x_3^2 \\ \dot{x}_2 = x_3 \\ \dot{x}_3 = u. \end{cases} \tag{262}$$

We will try the control

$$u = -x_2 - 2x_3 - \lambda \mathrm{sat}(\frac{x_1 + 2x_2 + x_3}{\lambda}) \tag{263}$$

where $0 < \lambda < 0.25$ and $\mathrm{sat}(s) = \mathrm{sgn}(s)\min\{|s|, 1\}$. The Jacobian linearization gives the LAS property. Also, there is no finite escape time since x_2 and x_3 are globally bounded. Now, for global boundedness and convergence of the full state using theorem 51, define $z_1 = x_1 + 2x_2 + x_3$. Then, the state of the (x_2, x_3) subsystem satisfies an a-$\mathcal{L}_\infty$ stability bound with respect to z_1 with gain $\gamma_{21}(s) =$

4 Results based on $\mathcal{L}_\infty$ input-output stability.

$\min\{2\lambda, 2s\}$ and also an asymptotic stability bound with this gain and restriction ∞. Also we have that

$$\dot{z}_1 = -\lambda\text{sat}(\frac{z_1}{\lambda}) + x_3^2. \tag{264}$$

So, the state of the z_1 subsystem satisfies an asymptotic bound with respect to x_3 with gain $\gamma_{11}(s) = 2\lambda s$ with restriction $\Delta_{11} = 2\lambda$. Since $\gamma_{21}(\infty) = \Delta_{11}$ and $\gamma_{21}(\gamma_{11})(s)) < s$ for all $s > 0$ (since $4\lambda < 1$), the GAS property follows from theorem 51.

Finally, several other applications of theorem 51, including stabilization of a general class of systems in so-called feedforward form, stabilization with rate saturation and time-delays, and stabilization of mechanical systems like the PVTOL, the ball and beam, and the inverted pendulum on a cart, are discussed in [99, 97, 98].

References

1. Z. Artstein, Stabilization with relaxed controls. Nonlinear Anal. TMA 7 (1983) 1163-1173.
2. A. Bacciotti, Local Stabilizability of non linear control systems. Series on Advances in Mathematics for Applied Sciences, Vol.8, World Scientific, 1992.
3. R.M. Bianchini and G. Stefani, Sufficient conditions for local controllability, Proc. 25th Conference on Decision and Control, Athens (1986) p. 967-970.
4. R.W. Brockett, Asymptotic stability and feedback stabilization, in: Differential Geometric Control Theory (R.W. Brockett, R.S. Millman and H.J. Sussmann Eds), Birkhäuser, Basel-Boston, 1983.
5. C. Byrnes, A. Isidori : New results and examples in nonlinear feedback stabilization. Systems & Control Letters. 12 (1989) 437-442
6. C.I. Byrnes and A. Isidori. Asymptotic stabilization of minimum phase nonlinear systems. IEEE Transactions on Automatic Control, 36(10):1122–1137, 1991.
7. M. Corless, Control of uncertain nonlinear systems, Journal of Dynamic Systems, Measurement, and Control, 115 (1993), pp. 362-372.
8. M. Corless, Robust stability and controller design with quadratic Lyapunov functions, in : Variable structure and Lyapunov control (A. Zinober Ed.), Springer-Verlag, 1993.
9. J.-M. Coron, Global asymptotic stabilization for controllable systems without drift, Math. Control Signals Systems, 5 (1992) p. 295-312.
10. J.-M. Coron, Links between local controllability and local continuous stabilization, in: IFAC Nonlinear Control Systems Design, Bordeaux, France, 1992, M. Fliess ed., p. 165-171.
11. J.-M. Coron, Linearized controlled systems and applications to smooth stabilization, SIAM J. Control and Optimization, 32 (1994) p. 358-386.
12. J.-M. Coron, Stabilization in finite time of locally controllable systems by means of continuous time-varying feedback laws, Preprint CMLA 1992, to appear in SIAM J. Control and Optimization 1995.
13. J.-M. Coron, Contrôlabilité exacte frontière de l'équation d'Euler des fluides parfaits incompressibles bidimensionnels, C. R. Acad. Sc. Paris, t. 317 (1993) p. 271-276.
14. J.-M. Coron, On the stabilization by an output feedback law of controllable and observable systems, Preprint, ENS de Cachan, Octobre 1993, to appear in Math. Control Signals and Systems.
15. J.-M. Coron, On the controllability of the 2-D incompressible perfect fluids, to appear in J. Math. Pures et Appliquées.
16. J.-M. Coron, Stabilizing time-varying feedback, IFAC Nonlinear Control Systems Design, Tahoe, USA, 1995.
17. J.-M. Coron and B. D'andréa-Novel, Smooth stabilizing time-varying control laws for a class of nonlinear systems. Application to mobile robots. IFAC NOLCOS'92 Symposium, Bordeaux, June 1992, 649-654.
18. J.-M. Coron and L. Praly, Adding an integrator for the stabilization problem. Systems & Control Letters 17 (1991) 89-104.
19. J.-M. Coron and Lionel Rosier, A relation between continuous time-varying and discontinuous feedback stabilization, J. Math. Systems, Estimations, and Control, 4 (1994) p. 67-84.

Bibliography

20. R. Freeman, Global internal stabilizability does not imply global external stabilizability for small sensor disturbances. Technical report CCEC, University of California, Santa Barbara. January 1995. Submitted for publication in IEEE Transactions on Automatic Control.
21. R. Freeman and P. Kokotovic, Design of 'softer' robust nonlinear control laws, Automatica, Vol. 29, No. 6,, pp. 1425-1437, 1993
22. R. Freeman and P. Kokotovic, Global robustness of nonlinear systems to state measurement, Proceedings of the 32nd IEEE Conference on Decision and Control, December 1993.
23. R. Freeman and P. Kokotovic, Inverse optimality in robust stabilization, Technical report CCEC, University of California, Santa Barbara. November 1994. To appear in SIAM J. Control and Optimization.
24. R. Freeman and P. Kokotovic, Robust control Lyapunov functions: the measurement feedback case. Proceeding of the 33rd IEEE conference on decision and control, pp. 3533-3538, December 1994.
25. R. Freeman and P. Kokotovic, Tracking controllers for systems linear in the unmeasured states, Technical report CCEC, July 1994. To appear in Automatica.
26. M. Golubitsky and V. Guillemin, Stable Mappings and their Singularities, Grad. Texts in Math. 14, Springer, New York-Heidelberg-Berlin, (1973).
27. L. Grayson, Design via Lyapunov's second method. Proceedings of the 4th Joint Conference on Automatic Control, pp. 589-598, 1963.
28. W. Hahn, Stability of Motion. Springer-Verlag, 1967
29. R. Hermann and A.J. Krener, Nonlinear controllability and observability, IEEE Transactions on Automatic Control, AC-22 (1977) p. 278-740.
30. H. Hermes, On the synthesis of a stabilizing feedback control via Lie algebraic methods, SIAM J. on Control and Optimization, 18 (1980) 352-361.
31. H. Hermes, Discontinuous vector fields and feedback control, in: Differential Equations and Dynamic Systems, (J.K. Hale and J.P. La Salle, Eds.), Academic Press, New-York and London, 1967.
32. H. Hermes, Control systems which generate decomposable Lie algebras, J. Differential Equations, 44 (1982) 166-187.
33. A. Isidori, Nonlinear Control Systems, Springer-Verlag, 1989.
34. D.H. Jacobson, Extensions of linear quadratic control, optimization and matrix theory. Academic Press, New York 1977.
35. M.J. Jankovic , R. Sepulchre , P. V. Kokotovic, Global stabilization of an enlarged class of cascade nonlinear systems. Technical report CCEC, University of California, Santa Barbara. February 1995. Submitted for presentation at the 34th Conference on Decision and Control.
36. Z.P. Jiang, A.R. Teel and L. Praly, Small-gain theorem for ISS systems and applications. Mathematics of Control, Signals, and Systems, 7(2):95–120, 1995.
37. Z.P. Ziang, I. Mareels and Y. Wang, A Lyapunov formulation of nonlinear small gain theorem for interconnected systems. In: Preprints of the 3rd IFAC Symposium on Nonlinear Control Systems Design, Tahoe City, CA, June 25-28, 1995, to appear.
38. Jurdjevic V. and J.P. Quinn, Controllability and stability. Journal of differential equations. vol. 4 (1978) pp. 381-389
39. H.K. Khalil, Nonlinear Systems. Macmillan Publishing Company, 1992.
40. T. Kailath, Linear Systems, Prentice Hall, Inc., London, 1980.
41. I. Kanellakopoulos, P. Kokotovic, A.S. Morse, A toolkit for nonlinear feedback design. Systems & Control Letters 18 (1992) 83-92

42. M. Kawski, High-order small time local controllability, in: Nonlinear Controllability and Optimal Control, (H.J. Sussmann ed.), Monographs and Textbooks in Pure and Applied Mathematics, 113, M. Dekker, Inc., New York, 1990, p. 431-467.
43. G. Kreisselmeier and R. Lozano, Adaptive control of continuous-time overmodeled plants, Preprint, Universities of Kassel and Compiègne, July 1994.
44. M. Krstic and P. Kokotovic, On extending the Praly-Jiang-Teel design to systems with nonlinear input unmodeled dynamics, Technical report CCEC 94-0211, February, 1994.
45. M. Krstic, I. Kanellakopoulos, P. Kokotovic, Nonlinear and adaptive control design. John Wiley & Sons, New York, 1995.
46. J. Kurzweil, On the inversion of Ljapunov's second theorem on stability of motion, Ann. Math. Soc. Trans. Ser.2, 24, 19-77, (1956)
47. V. Lakshmikantham, S. Leela, Differential and integral inequalities : Theory and applications. Volume 1 : Ordinary differential equations Academic Press 1969.
48. K.K. Lee, A. Arapostathis, Remarks on smooth feedback stabilization of nonlinear systems. Systems & Control Letters 10 (1988), 41-44.
49. W. Lin, Input saturation and global stabilization by output feedback for affine systems. Proceeding of the 33rd IEEE conference on decision and control. December 1994.
50. Y. Lin, E. Sontag, A universal formula for stabilization with bounded controls. Systems & Control Letters 16 (1991) 393-397
51. Y. Lin, E. Sontag, On control Lyapunov functions under input constraints, Proceedings of the 33rd Conference on Decision and Control, pp. 640-645, December 1994.
52. Y. Lin, E. Sontag, and Y. Wang, A smooth converse Lyapunov theorem for robust stability, SIAM J. Control and Optimization, to appear. preliminary version in: Recent results on Lyapunov-theoretic techniques for nonlinear stability, in: Proc. Amer. Automatic Control Conference, Baltimore, June 1994, pp. 1771–1775.
53. R. Lozano, Robust adaptive regulation without persistent excitation, IEEE Transactions on Automatic Control, AC-34 (1989) p. 1260-1267.
54. I.M.Y. Mareels and D.J. Hill, Monotone stability of nonlinear feedback systems. Journal of Mathematical Systems, Estimation and Control, 2(3):275–291, 1992.
55. R. Marino, P. Tomei, Global adaptive output feedback control of nonlinear systems, Part I: Linear parameterization, IEEE Transactions on Automatic Control, 38 (1993) 17-32.
56. R. Marino, P. Tomei, Global adaptive output feedback control of nonlinear systems, Part II: Nonlinear parameterization, IEEE Transactions on Automatic Control, 38 (1993) 33-48.
57. R. Marino, P. Tomei, Robust stabilization of feedback linearizable time-varying uncertain nonlinear systems. Automatica 29 (1), 181-189, 1993
58. F. Mazenc and L. Praly, Global stabilization for nonlinear systems, Preprint, Fontainebleau, January 1993.
59. F. Mazenc and L. Praly, Adding an integration and global asymptotic stabilization of feedforward systems, Accepted for publication in IEEE Transactions on Automatic Control. See also Proceedings 33rd IEEE Conference on Decision and Control, December 1994.
60. R. Monopoli, Synthesis techniques employing the direct method. IEEE Transactions on Automatic Control, Vol.10, pp. 369-370, 1965.
61. P. Myszkorowski, Practical stabilization of a class of uncertain nonlinear systems. Systems & Control Letters 18 (1992) 233-236.

62. J.-B. Pomet, Explicit design of time varying stabilizing control laws for a class of controllable systems without drift. Systems & Control Letters 18 (1992) 147-158
63. Pomet J.-B., R.M. Hirschorn, W.A. Cebuhar : Dynamic output feedback regulation for a class of nonlinear systems. Maths. Control Signals Systems (1993) 6: 106-124.
64. L. Praly, Lyapunov design of a dynamic output feedback for systems linear in their unmeasured state components. Proceedings of the IFAC Nonlinear Control Systems Design, Bordeaux, France, 1992.
65. Praly L., B. d'Andréa-Novel, J.-M. Coron, Lyapunov design of stabilizing controllers for cascaded systems. IEEE Transactions on Automatic Control, Vol.36, No.10, October 1991.
66. L. Praly, G. Bastin, J.-B. Pomet, Z.P. Jiang, Adaptive stabilization of non linear systems. In: Foundations of Adaptive Control (Kokotovic P Ed.) Lecture Notes in Control and Informations Sciences 160. Springer-Verlag, Berlin, 1991.
67. L. Praly and Z.P. Jiang, Stabilization by output feedback for systems with ISS inverse dynamics. Systems & Control Letters, 21, 19-33, 1993.
68. L. Praly and Y. Wang, Stabilization in spite of matched unmodelled dynamics and an equivalent definition of input-to-state stability. CAS report, September 1994, Submitted for publication in Mathematics of Control, Signal and Systems.
69. Z. Qu, Global stabilization of nonlinear systems with a class of unmatched uncertainties, Systems & Control Letters 18 (1992) 301-307.
70. Z. Qu, Robust control of nonlinear uncertain systems under generalized matching conditions. Automatica 29 (4), 985-998, 1993
71. M. Rotea, P. Khargonekar, Stabilization of uncertain systems with norm bounded uncertainty – a control Lyapunov function approach, SIAM J. Control and Optimization, Vol. 27, No. 6, pp. 1462-1476, November 1989.
72. E. P. Ryan, On Brockett's condition for smooth stabilizability and its necessity in a context of nonsmooth feedback, SIAM J. Control and Optimization, 32 (1994) p. 1597-1604.
73. M. Safonov, Stability and Robustness of Multivariable Feedback Systems. The MIT Press, Cambridge, MA, 1980.
74. C. Samson, Velocity and torque feedback control of a nonholonomic cart, Advanced Robot Control, Proceedings of the International Workshop on Nonlinear and Adaptive Control: Issues in Robotics, Grenoble, France, November 21-23, 1990 (C. Canudas de Wit, ed.), Lecture Notes in Control and Information Sciences, vol. 162, p. 125-151, Spinger-Verlag, Berlin Heidelberg New York, 1991.
75. L.M. Silverman and H.E. Meadows, Controllability and observability in time variable linear systems, SIAM J. on Control, 5 (1967) p. 64-73.
76. J.-J. Slotine and K. Hedrick, Robust input-output feedback linearization. Int. J. Control 57, 1133-1139, 1993
77. E. Sontag, A Lyapunov-like characterization of asymptotic controllability. SIAM J. Control and optimization, Vol. 21, No. 3, May 1983.
78. E. Sontag, Smooth stabilization implies coprime factorization, IEEE Transactions on Automatic Control, 34:435-443, 1989.
79. E. Sontag, A "universal" construction of Artstein's theorem on nonlinear stabilization. Systems and Control Letterrs 13 (1989) 117-123.
80. E. Sontag. Remarks on stabilization and input-to-state stability, Proceedings of the 28th Conference on Decision and Control, pages 1376-1378, December 1989.
81. E. Sontag. Further facts about input to state stabilization, IEEE Transactions on Automatic Control, 35(4):473-476, 1990.

82. E. Sontag, Mathematical control theory - Deterministic finite dimensional systems. Texts in Applied Mathematics 6, Springer-Verlag, 1990
83. E. Sontag, Feedback stabilization of nonlinear systems. in Robust control of Linear Systems and Nonlinear Control, M.A. Kaashoek, J.H. van Schuppen, A.C.M. Ran, Ed.. Birkhäuser, pages 61-81, 1990.
84. E. Sontag, Control of systems without drift via generic loops, IEEE Transactions on Automatic Control, July 1995.
85. E. Sontag, State-space and i/o stability for nonlinear systems, in: Feedback Control, Nonlinear Systems, and Complexity (B.A. Francis and A.R. Tannenbaum, eds.), Lecture Notes in Control and Information Sciences, Springer-Verlag, Berlin, 1995, pp. 215-235.
86. E. Sontag and H. Sussmann, Remarks on continuous feedback, IEEE CDC, Albuquerque, 2 (1980) p. 916-921.
87. E. Sontag and A.R. Teel, Changing supply functions in input/state stable systems. Submitted for publication in IEEE Transactions on Automatic Control.
88. E. Sontag and Y. Wang, On characterizations of the input-to-state stability property, Systems and Control Letters 24 (1995): 351-359.
89. H. Sussmann, Subanalytic sets and feedback control, J. Differential Equations, 31 (1979) p. 31-52
90. H. Sussmann, Single-input observability of continuous-time systems, Math. Systems Theory, 12 (1979) p. 371-393.
91. H. Sussmann, Lie brackets and local controllability : a sufficient condition for scalar-input systems, SIAM J. on Control and Optimization, 21 (1983) 686-713.
92. H. Sussmann, A general theorem on local controllability, SIAM J. on Control and Optimization, 25 (1987) 158-194.
93. H. Sussmann and V. Jurdjevic, Controllability of nonlinear systems, J. Differential Equations, 12 (1972) p. 95-116.
94. H. Sussmann, E. Sontag, and Y. Yang, A general result on the stabilization of linear systems using bounded controls, IEEE Transactions on Automatic Control, 39 (1994), pp.2411-2425.
95. A.R. Teel. Global stabilization and restricted tracking for multiple integrators with bounded controls. Systems and Control Letters, 18 (3):165–171, 1992.
96. A.R. Teel and L. Praly. Tools for semi-global stabilization by partial state and output feedback, SIAM J. of Control and Optimization, to appear. 1995
97. A.R. Teel, Additional stability results with bounded controls, Proceedings of 33rd IEEE Conference on Decision and Control, pp. 133-137, Orlando, FL, 1994.
98. A.R. Teel, Examples of stabilization using saturation: an input-output approach, In Preprints of the 3rd IFAC Symposium on Nonlinear Control Systems Design, Tahoe City, CA, June 25-28, 1995, to appear.
99. A.R. Teel, A nonlinear small gain theorem for the analysis of control systems with saturation, Accepted for publication in IEEE Transactions on Automatic Control.
100. J. Tsinias, Sufficient Lyapunov-like conditions for stabilization. Math. Control Signals Systems 2 (1989) 343-357
101. J. Tsinias and N. Kalouptsidis, Output feedback stabilization, IEEE Transactions on Automatic Control, Vol. 35, pp. 951-954, August 1990.
102. G. Zames, On the input-output stability of time-varying nonlinear feedback systems. Part I: Conditions using concepts of loop gain, conicity, and positivity. IEEE Transactions on Automatic Control, 11:228–238, 1966.

System-Theoretic Aspects of Dynamic Vision

Ruggero Frezza[1], *Pietro Perona*[3], *Giorgio Picci*[2], *and Stefano Soatto*[3]

[1] Department of Electronics and Computer Science, University of Padua, via Gradenigo 6/A, 35131 Padua, Italy
[2] Department of Electronics and Computer Science, University of Padua, via Gradenigo 6/A, 35131 Padua, Italy and LADSEB-CNR, Padua, Italy
[3] Department of Electrical Engineering, California Institute of Technology, Pasadena, Ca.

1. Introduction

Our aim in this minicourse is twofold. On one hand, we aim to show that the introduction of computer vision in control systems, i.e. "Vision in the Loop", raises exciting and yet unexplored problems in system theory. On the other hand, we explain how tools from control and estimation theory are, nowadays applied [3, 6, 7, 26, 27, 38, 54, 57, 63, 66] to "Dynamic Vision" problems with rather encouraging results in traditionally difficult applications, such as autonomous vehicle navigation [17, 18, 19], vision-based tracking and servo [9, 35, 36, 50], vision-based manipulation [5, 35, 36], docking [18, 42], vision-based planning [12], active sensing [67]. The course is therefore divided in two parts. In the first part, we shall pose two fundamental dynamic vision problems, namely 3–D motion and scene structure recovery, in a system theoretical framework. In the second part, we shall illustrate some relevant applications of dynamic vision to control systems.

What is *Dynamic Vision?*

A 3-dimensional scene is projected onto a 2-dimensional screen to form an *image*. In most applications of interest here the image materializes in the focal plane of a CCD camera. The image is then actually quantized in space in a finite number of discrete "pixels", the energy intensity at each pixel being read off numerically by some coupling device.

A number of factors which include: the geometry of the observed scene, the relative motion of the camera and of the objects in the scene, the properties of the light sources and of the camera, etc. interact in a complex fashion to generate an image. Computer Vision is the discipline concerned with the inverse problem of recovering information on some of the factors that generated a given image. Traditional computer vision is most of the times concerned with recovering 3-D information from a finite number (typically one or two) of images of the same scene.

A time-varying sequence of images carries information also on the dynamics of the scene, for example on the relative motion of the camera with respect to the scene and/or on the relative motion of various objects in the scene.

Dynamic Vision is the discipline that studies the inverse problem of recovering information on the scene from a *sequence* of images. While computer vision infers 3–D structure from an image which is a 2–D representation of the scene, dynamic vision reconstructs the time evolution of 3–D structure from a sequence of images. For this reason, dynamic vision has also been named in the literature *4–D Vision*. Dynamic vision exploits both the spatial and the temporal continuity of the scene for its reconstruction and interpretation. The general problem area may be seen as a chapter of nonlinear estimation and/or identification theory but this general classification is hardly of any help in practical solutions of the problem. It is only by exploiting the peculiar geometric and dynamic structure of the sequential inverse-projection problem of dynamic vision that useful and practically implementable solutions can be obtained.

In this spirit general tools from nonlinear estimation/identification theory start being exploited [14, 28, 60, 63] to solve dynamic vision problems and as the reliability and the performance of the algorithms improves, vision starts being acknowledged in the automatic control community as a powerful and versatile sensor to measure motion, position and to estimate the shape and structure of the environment.

2. Motion estimation via dynamic vision

Estimating the three-dimensional motion of an object and the scene structure from a sequence of projections is of paramount importance in a variety of applications in control and robotics, such as autonomous navigation, manipulation, servo, tracking, docking, planning, surveillance. Although "visual motion and structure estimation" is an old problem (the first formulations date back to the beginning of the century), only recently tools from nonlinear estimation theory have hinted at acceptable solutions. In words the problem can be defined as follows:

> Given a sequence of images taken from a moving camera, reconstruct the scene structure and the relative 3-D motion of the camera with respect to the environment (or scene).

Since our goal is that of posing the visual motion and scene structure estimation problem within a system-theoretical framework, we shall need to specify a "description" of the environment and of the motion of the viewer. We will restrict our attention to "static" scenes or, equivalently, to portions of a scene which are moving rigidly relative to the viewer.

In the last decade a variety of schemes has been proposed for recursively reconstructing structure for known motion [54], motion for known structure [7, 26, 27] or both structure and motion [3, 38, 57, 63, 66]. In general, given either the relative motion or the shape of the object being viewed, the other can be recovered easily, since the problem can be reduced to a linear estimation task. When neither the motion nor the shape of the scene are known, the problem of estimating both of them from visual information becomes a remarkably difficult one. Later in the course, we show that a crucial step in tackling such an estimation task consists in being able to *decouple* the estimation of motion from the estimation of structure. This decoupling has dramatic consequences also from the practical standpoint, since it allows integrating motion information in the presence of occlusions in the image plane, whereas previous structure *and* motion estimation schemes could integrate motion information only to the extent in which all initial feature points were still visible (as in [3, 39]). We shall present a framework for estimating rigid motion *independent* of the structure (shape) of the scene and viceversa. The estimates of motion, or of structure can later be fed, respectively, to any recursive "structure from known motion" or "motion from known structure" module [54, 57, 66].

The existing methods for motion and scene structure estimation may be classified, depending on the scene descriptors employed, as point-based, line-based, curve-based or model-based. In this course we will focus on the simplest case, when the scene is described by a number of point features in the Euclidean 3-D space. For line-based schemes see [28, 70, 76] and references therein. The curve-based approach has been addressed in [1, 10, 28, 69].

The point-based methods may be further classified in terms of the camera model in question. The simplest cases assume either parallel projection [59, 71, 72, 73] or ideal perspective projection (pinhole model, see [22]). More complicated camera models in terms of projective transformations allow parallel and perspective projection as a subcase [3, 24, 62, 68]. We will be mainly concerned with the classical pinhole model; however our schemes generalize to other camera representations and there are recursive schemes to identify the camera model along with visual motion (camera self-calibration, see [24, 62]). Other schemes recover projective, non-metric structure and motion independent on the camera parameters [21, 56, 59].

Motion reconstruction methods may be further classified in terms of the data processing technique as 2-frames schemes (see for example [45, 51, 74]), multiframe-batch methods [68, 73] or recursive algorithms.

Let us simplify the problem by assuming that the motion of the objects being viewed is *rigid* and has *constant velocity*. The "structure" of the scene is represented by a number of point-features whose coordinates in the ambient space are $\mathbf{x} \doteq [x_1 \ x_2 \ x_3]^T$; $v = [v_1 \ v_2 \ v_3]^T$ indicates the relative translational velocity of the object with respect to the viewer frame and $\omega := [\omega_1 \ \omega_2 \ \omega_3]^T$ is the rotational velocity vector also expressed in the viewer frame. The coordinates of the *projection* of a point feature $P = [x_1 \ x_2 \ x_3]^T$ onto the *image plane* (perpendicular to the x_3 axis) assumed at a conventional distance $f = 1$ from the origin, are

$$\begin{bmatrix} y_1 \\ y_2 \end{bmatrix} \doteq \begin{bmatrix} \frac{x_1}{x_3} \\ \frac{x_2}{x_3} \end{bmatrix},$$

so that we can write a nonlinear dynamical model having the position of the point in the ambient plane as the state, and the projection as the measured output:

$$\begin{cases} \frac{d}{dt} \begin{bmatrix} x_1 \\ x_2 \\ x_3 \end{bmatrix} = \begin{bmatrix} 0 & -\omega_3 & \omega_2 \\ \omega_3 & 0 & -\omega_1 \\ -\omega_2 & \omega_1 & 0 \end{bmatrix} \begin{bmatrix} x_1 \\ x_2 \\ x_3 \end{bmatrix} + \begin{bmatrix} v_1 \\ v_2 \\ v_3 \end{bmatrix} \\ \begin{bmatrix} y_1 \\ y_2 \end{bmatrix} \doteq \begin{bmatrix} \frac{x_1}{x_3} \\ \frac{x_2}{x_3} \end{bmatrix} \qquad x_3 \neq 0 \end{cases} \qquad (2.1)$$

which is in the form

$$\begin{cases} \frac{d}{dt}\mathbf{x} = f(\mathbf{x}) & \mathbf{x}(t_0) = \mathbf{x}_0 \in \mathbb{R}^n \\ y = h(\mathbf{x}) \end{cases} \qquad (2.2)$$

where

$$\begin{cases} n = 3 \\ \mathbf{x} = [x_1, x_2, x_3]^T \\ f(\mathbf{x}) = \Omega \mathbf{x} + \mathbf{v} \\ \Omega \doteq \begin{bmatrix} 0 & -\omega_3 & \omega_2 \\ \omega_3 & 0 & -\omega_1 \\ -\omega_2 & \omega_1 & 0 \end{bmatrix} \\ \mathbf{v} = \begin{bmatrix} v_1 \\ v_2 \\ v_3 \end{bmatrix} \\ \mathbf{y} = h(\mathbf{x}) = \frac{\mathbf{x}}{x_3} \qquad x_3 \neq 0. \end{cases} \qquad (2.3)$$

We call the above model the *standard model for structure and motion*. Estimating structure and motion is equivalent, respectively, to estimating the state and identifying the parameters of the above model.

3. Perspective system theory

B. Ghosh has posed the structure and motion reconstruction problem in a general system theoretical framework [28, 29]. He defines a new class of problems which he names *Perspective problems in system theory* the most basic of which are the observation of the initial condition and the identification of the parameters of a linear dynamical system whose output is given by a perspective observation function.

Consider a linear system

$$\dot{x} = Ax \qquad (3.1)$$

where A is a $n \times n$ constant matrix, x is a $n \times 1$ state vector given by

$$x = \begin{bmatrix} x_1 & x_2 & \dots & x_n \end{bmatrix}^T.$$

Assume $n > 1$ and let m be an integer such that $1 < m < n + 1$. Consider the perspective output function

$$Y : (\mathbb{R}^n - B) \mapsto \mathbb{RP}^{m-1}$$

defined as a linear map

$$Y : \begin{bmatrix} x_1 & x_2 & \ldots & x_n \end{bmatrix} \mapsto \{ y_1 \ \ldots \ y_m \} \quad (3.2)$$

where $\{y_1, \ldots, y_m\}$ are homogeneous coordinates in the projective space $\mathbb{RP}^{m-1}$ and

$$B = \{x \in \mathbb{R}^n : y_j = 0, j = 1, \ldots, m\}.$$

Any such linear map can be represented by a $m \times n$ matrix C, so as

$$C = \begin{bmatrix} C_1 \\ \vdots \\ C_m \end{bmatrix}, \quad y_j = C_j \begin{bmatrix} x_1 & x_2 & \ldots & x_n \end{bmatrix}^T,$$

for $j = 1, \ldots, m$.

A systematic study of perspective linear dynamical system such as (3.1-3.2) is initiated in [28]. Clearly, if the motion is known and constant the standard model for structure and motion defined above can be made into a perspective system by homogeneization, i.e. by rewriting (2.1) as

$$\dot{x} = \begin{bmatrix} \Omega & \mathbf{v} \\ 0 & 0 \end{bmatrix} x \quad (3.3)$$

where $x \doteq [x_1, x_2, x_3, x_4]^T \in \mathbb{R}^4$ may be interpreted as homogeneous coordinates in $\mathbb{RP}^3$ and the output map $Y : \mathbb{RP}^3 \to \mathbb{RP}^2$ is accordingly defined by

$$y = [I_3 \ 0]x$$

where I_3 is the 3×3 identity matrix.

A number of other dynamic vision problems, however, can be modelled by linear perspective dynamical systems and later we will show another example. One of the most remarkable results of the theory gives conditions for the observability of perspective systems. Dayawansa, Ghosh Martin and Wang [14] and later Ghosh and Rosenthal [30] proved the following theorem.

Theorem 3.1. *A perspective system is observable modulo a d-dimensional affine space iff*

$$\text{rank} \begin{bmatrix} (A - \lambda_0) \ldots (A - \lambda_d) \\ C \end{bmatrix} = n \quad (3.4)$$

over the field $\mathbb{C}$ of complex numbers, for any set of eigenvalues $\lambda_0, \ldots, \lambda_d$ of A.

For $d = 0$ the theorem gives conditions for observability *tout court* and reduces essentially to the well-known Hautus test in linear system theory.

The *structure observability* problem, i.e. determining the initial condtion $\mathbf{x}(0)$ of (2.1) from observations of $y(t)$, $t \in [0, T]$ can be discussed in the context of the above result for the homogenized system (3.3). It is immediate to check that the initial condition $x(0) = [x_1(0), x_2(0), x_3(0), 1]^T$ is observable (here $d = 0$) if and only if there is a nonzero translational velocity, i.e. $\mathbf{v} \neq 0$. In fact, for the homogenized system (3.3)

$$\text{rank}\left[\begin{bmatrix} \Omega & \mathbf{v} \\ 0 & 0 \\ I_3 & 0 \end{bmatrix} - \lambda \begin{bmatrix} I_2 & 0 \\ 0 & 1 \end{bmatrix}\right] \quad (3.5)$$

is equal to 4 for $\lambda = 0$ (which is an eigenvalue of the extended matrix) iff $\mathbf{v} \neq 0$.

Under some simplifying assumptions the motion recovery problem can also be posed in the framework of perspective system theory [28]. Consider the standard 3-D structure and motion model defined above. Assume that the camera is undergoing both a constant rotation at speed ω and a constant translation at speed (v_1, v_2, v_3) relative to the scene. Then, the dynamics of a point of the scene (x_1, x_2, x_3) relative to a frame fixed with the camera are described by

$$\begin{bmatrix} x_1 \\ x_2 \\ x_3 \end{bmatrix} = e^{\Omega t}\left(\begin{bmatrix} x_1(0) \\ x_2(0) \\ x_3(0) \end{bmatrix} + \begin{bmatrix} v_1 t \\ v_2 t \\ v_3 t \end{bmatrix}\right).$$

It is easy to see that $(x_1(t), x_2(t), x_3(t))$ can be obtained as the solution of the dynamical system

$$\begin{bmatrix} \dot{X} \\ \dot{Z} \end{bmatrix} = \begin{bmatrix} \Omega & I \\ 0 & \Omega \end{bmatrix} \begin{bmatrix} X \\ Z \end{bmatrix} \quad (3.6)$$

where

$$\begin{aligned} X(t) &= [x_1(t)\ x_2(t)\ x_3(t)]^T \\ Z(0) &= [v_1\ v_2\ v_3]^T. \end{aligned}$$

The observation map is defined by

$$(X\ Z) \mapsto \{y_1, y_2, y_3\} \quad (3.7)$$

where $\{y_1, y_2, y_3\}$ are homogeneous coordinates. The dynamical system (3.6) together with the observation (3.7) is a perspective system with $n = 6$ and $m = 3$. The observability criterion can be used to analyze the possibility of recoverying the 3-D coordinates of the point of the scene *and* the camera translational velocity. The recovery of the rotational velocity is, instead, a much more difficult identification problem.

4. Motion estimation independent of structure

In this section we present a framework for estimating rigid motion *independent* of the structure (shape) of the scene[1]. The estimates of motion can later be fed to any recursive "structure from known motion" module in order to estimate scene structure [54, 57, 66]. The movement of a rigid body in $\mathbb{R}^3$ can be described by

---

[1] The exposition which follows is a summary of results reported in [61]. For further details we refer the reader to the original reference.

a point in the Euclidean group of transformations, $g(t) \equiv (T(t), R(t)) \in SE(3)$, which act on points of $\mathbb{R}^3$ via[2]

$$\mathbf{X}(t+1) = R(t)(\mathbf{X}(t) - T(t)). \qquad (4.1)$$

We measure the *projection* of each feature point $\mathbf{X}^i := [X_1^i \; X_2^i \; X_3^i]^T, i = 1\ldots N$ on the image plane:

$$\begin{aligned}\pi : \mathbb{R}^3 &\rightarrow \mathbb{R}\mathbb{P}^2 \\ \mathbf{X}^i &\mapsto \mathbf{x}^i = [\frac{X_1^i}{X_3^i} \; \frac{X_2^i}{X_3^i} \; 1]^T \end{aligned} \qquad (4.2)$$

where the last expression describes the ideal perspective projection with unitary focal length. The space $SE(3)$ can be embedded in the matrix group $GL(4)$, and the matrix product used as group operation, via the homogeneous coordinates. Motion estimation can be formulated as the inversion/identification of the basic model defined by the rigid motion and the perspective constraints (4.2)(4.1). The inverse system, however, depends upon the (unknown) state $\mathbf{X}$, which needs to be observed. Since the model is driftless, both the state and the input of the model appear at the first level of Lie differentiation of the output $\mathbf{x}$, and therefore it is possible from the raw model to estimate either the motion, as a function of the structure, or viceversa, but not both. It is possible, however, to resort to dynamic extension and formulate the structure and motion estimation problem as the estimation of the state of an augmented model defined on $\mathbb{R}^{3N} \times SE(3)$. The trick is then to assume that the input to the system is white, zero-mean Gaussian noise, and use an EKF to perform the estimation. This approach has been pursued for example by [3]. However, the augmented model has a linearization which is not observable, in fact it is not even locally (weakly) observable [60].

This motivates the separation of structure from motion estimation and the introduction of the "essential" representation of rigid motion, which leads to a model which is is globally observable under general position conditions (i.e. when there does not exist a proper quadric surface in $\mathbb{R}^3$ that contains all the visible points and the path of the center of projection [61]), and hence better suited for the purpose of motion estimation. Motion can be estimated on the essential manifold independent on structure;

4.1 Representation of rigid motion via the "essential manifold"

A rigid motion may be represented as a point in the Lie group $SE(3)$, which can be embedded in the linear space $GL(4)$ (and hence exploit the matrix product as

---

[2] Here we want to describe the motion of the viewer with respect to the scene. We shall also change notations slightly from the standard model (2.1) of the previous section. In particular $R(t)$ represents the orientation of the (camera-fixed) reference frame at time t relative to the reference at time $t+1$, while $T(t)$ is the position of the origin of the reference at time $t+1$ in the reference of time t. This convention is chosen for consistency with the standard notation of the essential matrices [51].

composition rule) and is in local correspondence with $\mathbb{R}^6$ via the exponential coordinates and the isomorphism between the relative Lie algebra $se(3)$ and $\mathbb{R}^6$, as seen in the previous section. We now discuss an alternative matrix representation of rigid motion which is more "compact", in the sense that it can be embedded in a linear space of smaller dimensions. Such a representation is derived from the so-called "essential matrices" introduced by Longuet-Higgins [51].

Consider a point $g = (T, R) \in SE(3)$, then $T \wedge \; \in so(3)$ is a skew-symmetric matrix. Now define the space of *"essential matrices"* as

$$E \doteq \{SR \mid R \in SO(3) \, , \, S = (T \wedge) \in so(3)\} \subset \mathbb{R}^{3 \times 3}. \qquad (4.3)$$

Clearly the essential space does not inherit the group structure from the sum of matrices in $\mathbb{R}^{3 \times 3}$, since $\mathbf{Q}_1, \mathbf{Q}_2 \in E$ does not imply $\mathbf{Q}_1 + \mathbf{Q}_2 \in E$. One possible way of imposing the group structure is by forcing a group morphism with $SE(3)$, for which it is necessary to "unfold" T, R from $\mathbf{Q} = (T \wedge)R$, perform the group operation on $SE(3)$ and then collapse the result into $\mathbf{Q}$. We will see later in this section a way of unfolding an essential matrix into its rotation and translation components.

The essential space has many interesting geometrical properties: it is an algebraic variety [55] and a topological manifold of dimension 6. Later on we will provide a characterization of a local coordinate chart. The essential space may also be identified with $TSO(3)$, the tangent bundle of the rotation group, defined as $TSO(3) \doteq \cup_{R \in SO(3)} T_R SO(3)$. This proves that the essential space is indeed a differentiable manifold of dimension 6 [65].

The following theorem, due to Huang and Faugeras and reported by Maybank [55], gives a simple characterizing property of the essential space.

Theorem 4.1. *(Huang and Faugeras, 1989)*
Let $\mathbf{Q} = \mathbf{U}\Sigma\mathbf{V}^T$ be the Singular Value Decomposition (SVD) of a matrix in $\mathbb{R}^{3 \times 3}$. Then

$$\mathbf{Q} \in E \Leftrightarrow \Sigma = \Sigma_0 = \mathrm{diag}\{\lambda \; \lambda \; 0\} \mid \lambda \in \mathbb{R}^+.$$

Note that, since $\mathbf{Q} \doteq \mathbf{U}\Sigma\mathbf{V}^T \in E \Leftrightarrow \Sigma = \mathrm{diag}\{\lambda \; \lambda \; 0\}$, there is one degree of freedom in defining the the basis components of the subspaces $<\mathbf{V}_{.3}>^\perp$ and $<\mathbf{U}_{.3}>^\perp$, which corresponds to rotating the orthogonal bases $<\mathbf{V}_{.1}, \mathbf{V}_{.2}>$ and $<\mathbf{U}_{.1}, \mathbf{U}_{.2}>$ about their orthogonal complements. However, the effects cancel out in the multiplications when defining R and S.

Local coordinates of the essential manifold

For any given rigid motion $(T, R) \in SE(3)$, there exists an essential matrix $\mathbf{Q}$ defined by $\mathbf{Q} \doteq (T \wedge)R$. We are interested now in the inverse problem: *given an essential matrix $\mathbf{Q}$, can we extract its rotational and translational components? Is the correspondence $\mathbf{Q} \leftrightarrow (T, R)$ unique?*
Consider the map

4 Motion estimation independent of structure

$$\Phi : E \to \mathbb{R}^3 \times SO(3) \to \mathbb{R}^3 \times \mathbb{R}^3 \tag{4.4}$$

$$\mathbf{Q} \mapsto \begin{bmatrix} \pm\|\mathbf{Q}\|\mathbf{U}_{\cdot 3} \\ \mathbf{U}R_Z(\pm\frac{\pi}{2})\mathbf{V}^T \end{bmatrix} = \begin{bmatrix} T \\ e^{\omega \wedge} \end{bmatrix} \mapsto \begin{bmatrix} T \\ \omega \end{bmatrix}$$

where $\mathbf{U}, \mathbf{V}$ are defined by the Singular Value Decomposition (SVD) [31] of $\mathbf{Q} = \mathbf{U}\Sigma\mathbf{V}^T$; $\mathbf{U}_{\cdot 3}$ denotes the third column of $\mathbf{U}$ and $R_Z(\frac{\pi}{2})$ is a rotation of $\frac{\pi}{2}$ about the axis $[0\ 0\ 1]^T$. Note that the map Φ defines the local coordinates of the essential manifold modulo two signs; therefore, the map Φ associates to each element of the essential space four distinct points in local coordinates. This ambiguity may be resolved in the context of the visual motion estimation problem by imposing the *"positive depth constraint"*, which means that each visible point lies in front of the viewer [23, 33, 51, 52, 75]. In a case like this we will be able to identify a unique local coordinates homeomorphism. The inverse map is simply

$$\Phi^{-1} : \mathbb{R}^3 \times \mathbb{R}^3 \to E$$

$$\begin{bmatrix} T \\ \omega \end{bmatrix} \mapsto (T\wedge)e^{(\omega\wedge)}.$$

Projection onto the essential manifold

Theorem 4.1 suggests a simple "projection" of a generic 3×3 matrix onto the essential manifold: let us define

$$pr_{<E>} : \mathbb{R}^{3\times 3} \to E \tag{4.5}$$

$$M \mapsto \mathbf{U}\,\mathrm{diag}\{\lambda, \lambda, 0\}\,\mathbf{V}^T$$

where $\mathbf{U}, \mathbf{V}$ are defined by the SVD of $M = \mathbf{U}\,\mathrm{diag}\{\sigma_1, \sigma_2, \sigma_3\}\,\mathbf{V}^T$, and $\lambda \doteq \frac{\sigma_1 + \sigma_2}{2}$. It follows from the properties of the SVD [31] that $pr_{<E>}(M)$ minimizes the Frobenius distance from M to the essential manifold [33, 55].

As time progresses, the point $\mathbf{Q}(t)$, corresponding to the actual motion, describes a trajectory on E (and a corresponding one in local coordinates) according to

$$\mathbf{Q}(t+1) \doteq \mathbf{Q}(t) + n_{\mathbf{Q}}(t).$$

The last equation is indeed just a *definition* of the right-hand side, as we do not know $n_{\mathbf{Q}}(t)$. The identity of $n_{\mathbf{Q}}(t)$ will be unrevealed later. For now, we will consider the previous equation to be a discrete-time dynamical model for $\mathbf{Q}$ on the essential manifold, with $n_{\mathbf{Q}}$ as *unknown* input. If we accompany it with the essential constraint, we get

$$\begin{cases} \mathbf{Q}(t+1) \doteq \mathbf{Q}(t) + n_{\mathbf{Q}}(t) & \mathbf{Q} \in E \\ 0 = \chi_{\mathbf{y}'(t), \mathbf{y}(t)} \mathbf{Q}(t) \\ \mathbf{y}_i = \mathbf{x}_i + n_i & \forall\, i = 1\ldots N. \end{cases} \tag{4.6}$$

Now the visual motion estimation problem is characterized as the estimation of the state of the above model, which is defined on the essential manifold. It can be seen that the system is "linear" (both the state equation and the essential constraint are linear in **Q**). E, however, is not a linear space.

At this point we are ready to address the problem of recursively estimating motion from an image sequence. There are two approaches that may be derived naturally from the formulation introduced above.

The first approach we describe consists of composing the equations (4.6) with the local coordinate chart Φ, ending up with a *nonlinear* dynamical model for motion in $\mathbb{R}^5$. At this point we have to make some assumptions about motion: for gnerality we shall assume that we don't have a dynamical model and we will assume a statistical model. In particular, we will assume that motion is a *first order random walk in* $\mathbb{R}^5$ (see fig. 4.2 top). The problem then is to estimate the state of a nonlinear system on a linear space driven by white, zero-mean Gaussian noise.

In the second approach we change the model for motion: in particular we assume motion to be a *first order random walk in* $\mathbb{R}^9$ *projected onto the essential manifold*. We will see that this leads to a method for estimating motion that consists in solving at each step a *linear estimation* problem in the linear embedding space and then "projecting" the estimate onto the essential manifold

4.2 The essential estimator in local coordinates

The model defined by the rigid motion and the perspective constraints (4.2)(4.1) is a nonlinear dynamical model having the 3-D structure of the scene in the state. Estimating motion amounts to identifying the above model with the parameters T, R constrained on $SE(3)$. However, we do not know $\mathbf{X}_{i0}$, so that we end up with a mixed estimation/identification task which proves extremely difficult [60]. If we represent motion on the essential manifold, instead, it is possible to remove the 3-D structure of the scene from the model, ending up with a nonlinear and implicit dynamical model for the (measured) projective coordinates of the visible features, with motion as an unknown parameter constrained on E. This allows us to decouple the estimation of motion from the 3-D structure of the scene, which has many advantages, for it allows dealing with occlusions of feature points and to cross regions of motion-space which render the "natural" model unobservable [60].

Consider a rigid motion between two time instants. Given any transformed point $\mathbf{X}^i$, its coordinates in the reference frame at time t, the corresponding coordinates in the reference at $t+1$ and the translation vector are coplanar. The same holds for $\mathbf{x}^i$ in place of $\mathbf{X}^i$, since the two are parallel. For each visible point we can write the coplanarity constraint, for example in the reference frame at time t, as

$$\mathbf{x}^i(t+1)^T R(t) \left(T(t) \wedge \mathbf{x}^i(t)\right) = 0 \quad \forall i. \qquad (4.7)$$

As it turns out, the above constraint is not only a consequence of rigid motion, but also suffices to characterize it, once five or more such constraints are given [55, 51]. We measure directly the image plane coordinates $\mathbf{x}^i \in \mathbb{R}\mathrm{P}^2$ up to some noise, which we safely assume to be white, zero-mean and Gaussian:

4 Motion estimation independent of structure

$$y^i(t) = \mathbf{x}^i(t) + n^i(t) \qquad n^i \in \mathcal{N}(0, R_{n^i}) \tag{4.8}$$

The estimation of motion amounts, therefore, to identifying the following implicit dynamical model with parameters on the essential manifold

$$\begin{cases} \mathbf{x}^i(t+1)^T \mathbf{Q}(t) \mathbf{x}^i(t) = 0 & \mathbf{Q} \in E \\ y^i(t) = \mathbf{x}^i(t) + n^i(t) & \forall i = 1 \ldots N. \end{cases} \tag{4.9}$$

If we apply to the previous system the local coordinates homeomorphism Φ defined in eq. (4.4), we can write a corresponding estimation model in the local coordinates $\mathbb{R}^5$: let $\xi \doteq (v, \omega)$, then

$$\begin{cases} \xi(t+1) = \xi(t) + n_\xi(t) & \xi \in \mathbb{R}^5 \\ y^i(t)^T \Phi^{-1}(\xi(t)) y^i(t-1) = \tilde{n}_i(t) & \forall i \end{cases} \tag{4.10}$$

where $n_\xi(t)$ is the white noise that drives the random walk model, and $\tilde{n}^i(t)$ is an induced residual noise whose second order statistics can be characterized in terms of the variance of the measurement error $n^i(t)$ [61]. The map Φ introduced in eq. (4.4) defines the local coordinates of the essential space (of course after choosing the sign in the direction of translation and in the rotation angle of R_Z).

Note that if a dynamical model for motion is available, as for example when the camera is mounted on a moving vehicle or on a robot arm, we can substitute the random walk model with a dynamical model of the form $\xi^i(t+1) = f^i(\xi(t), n_\xi(t))$, where now n_ξ describes the state of the vehicle or robot arm.

The state of the model of eq. (4.10) is defined on a linear space and can now be estimated using a variation of the Extended Kalman Filter for implicit measurement constraints, which is derived in [61]. We summarize here the equations of the estimator. Write the coplanarity constraint (4.9) for N points in the form of a matrix equation $\chi \mathbf{Q} = 0$, where χ is a $N \times 9$ matrix and $\mathbf{Q} = \Phi^{-1}(\xi)$ is intended as a nine-dimensional column vector. Call $C \doteq \left(\frac{\partial \chi \Phi^{-1}}{\partial \xi}\right)$ and $D \doteq \left(\frac{\partial \chi \Phi^{-1}}{\partial \mathbf{x}}\right)$, R_α the variance of the process α, then we have

Prediction step:

$$\begin{aligned} \hat{\xi}(t+1|t) &= \hat{\xi}(t|t) \,;\, \hat{\xi}(0|0) = \xi_0 \\ P(t+1|t) &= P(t|t) + R_\xi \,;\, P(0|0) = P_0 \end{aligned}$$

Update step:

$$\begin{aligned} \hat{\xi}(t+1|t+1) &= \hat{\xi}(t+1|t) - \\ &\quad -L(t+1)\chi(t+1)\mathbf{Q}(\hat{\xi}(t+1|t)) \\ P(t+1|t+1) &= \Gamma(t+1) P(t+1|t) \Gamma^T(t+1) + \\ &\quad + L(t+1) R_{\tilde{n}}(t+1) L^T(t+1) \end{aligned}$$

Gain:
$$L(t+1) = P(t+1|t)C^T(t+1)\Lambda^{-1}(t+1)$$
$$\Lambda(t+1) = C(t+1)P(t+1|t)C^T(t+1) + R_{\tilde{n}}(t+1)$$
$$\Gamma(t+1) = I - L(t+1)C(t+1)$$

Innovation variance:
$$R_{\tilde{n}}(t+1) = D(t+1)R_n D^T(t+1)$$

Note that $P(t|t)$ is the variance of the motion estimation error which is used as variance of measurement error from subsequent modules of the structure from motion estimation scheme [66]. A similar formulation of the IEKF was used by Di Bernardo et al. [16]. Similar expressions were also used before in the literature on specific applications; the first instance to our knowledge was in the recursive computation of the Hough transform [13].

4.3 The "essential estimator" in the embedding space

The model (4.9) is *linear* in **Q**. However, it is defined on a state-space which is *not a linear space*. We could thus think of lifting the model to the (linear) embedding space $\mathbb{R}^9$, and at each step "project" the current estimate onto the manifold. In general, this could be a very bad idea, for we perform the update in a bigger space, and then impose the structure of the state manifold *a posteriori*. In this case, however, we can show that the only difference between the filter in local coordinates and the filter defined in the embedding space is *the model of motion* employed: in the first case it is a random walk on $\mathbb{R}^5$ *lifted to the essential manifold*, whereas in the second case it is a random walk in $\mathbb{R}^9$ *projected onto the manifold*. Therefore we reduce the comparison between the two schemes to a *modeling issue*, which can be assessed only a posteriori (see figures 4.1, 4.2).

We define the operation $\oplus$ as the projection onto the essential manifold of the sum of two essential matrices interpreted as elements of $\mathbb{R}^{3\times 3}$:

$$\mathbf{Q}_1 \oplus \mathbf{Q}_2 \doteq \mathrm{pr}_E(\mathbf{Q}_1 + \mathbf{Q}_2).$$

The filter defined in the embedding space is *linear*, and solves optimally (in the sense of least error variance) the prediction for the model

$$\begin{cases} \mathbf{Q}(t+1) = \mathbf{Q}(t) \oplus n_\mathbf{Q}(t) & \mathbf{Q}(t) \in E \\ y^i(t+1)^T \mathbf{Q}(t) y^i(t) = n^i(t) & \forall i = 1\ldots N \end{cases} \quad (4.11)$$

The solution of the estimation task is derived in [61] and can be summarized as follows:

Prediction step:
$$\hat{\mathbf{Q}}(t+1|t) = \hat{\mathbf{Q}}(t|t) \;;\; \hat{\mathbf{Q}}(0|0) = \mathbf{Q}_0$$
$$P(t+1|t) = P(t|t) + R_\mathbf{Q} \;;\; P(0|0) = P_0$$

4 Motion estimation independent of structure

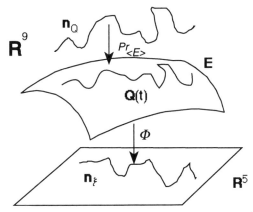

Fig. 4.1. *Model of motion as a random walk in $\mathbb{R}^5$ lifted to the manifold or as a random walk in $\mathbb{R}^9$ projected onto the manifold*

Update step:
$$\begin{aligned}
\hat{Q}(t+1|t+1) &= \hat{Q}(t+1|t) \oplus \\
&\oplus L(t+1)\chi(t+1)\hat{Q}(t+1|t) \\
P(t+1|t+1) &= \Gamma(t+1)P(t+1|t)\Gamma^T(t+1) + \\
&+ L(t+1)R_{\tilde{n}}(t+1)L^T(t+1)
\end{aligned}$$

Gain:
$$\begin{aligned}
L(t+1) &= -P(t+1|t)\chi(t+1)\Lambda^{-1}(t+1) \\
\Lambda(t+1) &= \chi(t+1)P(t+1|t)\chi(t+1) + \\
&+ R_{\tilde{n}}(t+1) \\
\Gamma(t+1) &= I - L(t+1)\chi(t+1) \\
R_{\tilde{n}}(t+1) &= D(t+1)R_n D^T(t+1)
\end{aligned}$$

We showed that the problem of estimating three-dimensional motion from a sequence of images can be naturally set in the framework of dynamic estimation and identification. Under the assumption of a static scene, the rigid motion constraint and the perspective projection map *define* in a natural way a nonlinear dynamical model, and estimating motion is equivalent to a mixed estimation/identification task.

Motivated by the structural limitations of the natural model (see [60]), we have proposed a new formulation for structure-independent motion estimation, based upon the representation of motion via the "essential matrices", introduced by Longuet-Higgins [51]. Motion estimation is equivalent to the identification of a nonlinear implicit model with parameters on the essential manifold. Other problems in dynamic vision may be cast as the identification of a nonlinear implicit model, as for example dynamic self-calibration, subspace motion factorization, partial motion reconstruction from weak perspective.

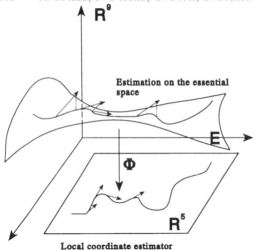

Fig. 4.2. *Estimation on the EssentialSpace*

5. Structure estimation

As it is well known, the scene structure estimation task is *nonlinear* due to the perspective nature of the measurements. One may ask whether there exists a smart choice of coordinates that simplifies the estimation task. In particular, since "linearity" is a coordinate-dependent notion, one may seek for a particular choice of coordinates such that the problem of estimating structure from motion becomes linear and spectrally assignable. Unfortunately, such a choice of coordinates does not exist, even if we allow for a nonlinear change of output coordinates or an embedding into a higher-dimensional state-space [64]. As a consequence of this result, some alternative estimators with nonlinear error dynamics which are proved to converge [46, 64] have been studied. This result also legitimates the use of local linearization-based techniques for estimating structure from known motion and visual information. In most of the cases, however, the true motion undergone by the viewer is unknown. We illustrate a dynamic estimator derived in [64] for scene structure which is independent of the motion of the viewer. The method consists in the identification of an Exterior Differential System with parameters on a sphere.

5.1 Local linearization–based structure from motion

Assuming that motion $(\mathbf{v}, \omega)$ is known, the model (2.2), (2.2) can be used to discuss the structure from motion problem. The model is not only locally weakly observable, but also its linearization about the current state is observable away from the center of projection. This is a very favorable situation for using local linearization-based observers, as for example the Extended Kalman Filter (EKF) [47]. This is essentially the approach taken in [54, 57, 66], and it has proven effective in most practical situation, when the *motion of the viewer is known*.

5.2 Alternative nonlinear observers, an adaptive observer

In the literature, alternative nonlinear observer schemes have been proposed; in particular, the adaptive observer proposed by [4] and further studied by [53] can be easily applied with some appropriate modifications to the structure-from-motion problem.

Consider for simplicity the planar case defined by the model (2.2) with $n = 2$ i.e.,

$$\begin{cases} n = 2 \\ f(\mathbf{x}) \doteq \Omega \mathbf{x} + \mathbf{v} \\ \Omega \doteq \begin{bmatrix} 0 & -\omega \\ \omega & 0 \end{bmatrix} \\ \mathbf{v} \doteq \begin{bmatrix} v_1 \\ v_2 \end{bmatrix} \\ h(\mathbf{x}) \doteq \frac{x_1}{x_2} \qquad x_2 \neq 0. \end{cases} \tag{5.1}$$

Note first that the derivative of the output y satisfies the differential equation

$$\dot{y} = -\omega(1 + y^2) + \frac{1}{\mathbf{x}_2}(v_1 - yv_2). \tag{5.2}$$

If $1/\mathbf{x}_2$ was an unknown constant parameter, (5.2) would be exactly in the form considered in [53]. It is easy to show using the Lyapunov function

$$V = k(y - \hat{y})^2 + k(\frac{1}{\mathbf{x}_2} - \frac{1}{\hat{\mathbf{x}}_2})^2, \tag{5.3}$$

where k is a positive constant, that the following scheme,

$$\begin{cases} \dot{\hat{y}} = -\omega(1 + y^2) + \frac{1}{\hat{\mathbf{x}}_2}(v_1 - yv_2) + k(y - \hat{y}) \\ \frac{d}{dt}\left(\frac{1}{\hat{\mathbf{x}}_2}\right) = (v_1 - yv_2)(y - \hat{y}) \end{cases} \tag{5.4}$$

is a globally asymptotically convergent observer since $\dot{V} = -2k^2(y - \hat{y})^2$. However, $1/\mathbf{x}_2$ is not a constant parameter, as it has its own dynamics

$$\frac{d}{dt}\left(\frac{1}{\mathbf{x}_2}\right) = -\omega y \frac{1}{\mathbf{x}_2} - v_2(\frac{1}{\mathbf{x}_2})^2 \tag{5.5}$$

and the convergence of (5.4) cannot be guaranteed by the above argument anymore. Consider then the following estimator, which is obtained from the previous one by adding the dynamics of the inverse depth

$$\begin{cases} \dot{\hat{y}} = -\omega(1 + y^2) + \frac{1}{\hat{\mathbf{x}}_2}(v_1 - yv_2) + k(y - \hat{y}) \\ \frac{d}{dt}\left(\frac{1}{\hat{\mathbf{x}}_2}\right) = (v_1 - yv_2)(y - \hat{y}) - \omega y \frac{1}{\hat{\mathbf{x}}_2} - v_2(\frac{1}{\hat{\mathbf{x}}_2})^2. \end{cases} \tag{5.6}$$

Some properties of the structure from motion problem come at hand in order to prove convergence. The output y is bounded by the dimensions of the image plane $|y| \leq m$ and the inverse of the depth satisfies $0 < 1/\mathbf{x}_2 \leq 1/f = 1$ since we assumed unitary focal length f. Then, under the following conditions

$$\begin{cases} \text{if } |\hat{y}(t)| > m \text{ then } \hat{y}(t^+) = \hat{y}(t)\frac{m}{|\hat{y}(t)|} \\ \text{if } \frac{1}{|\hat{x}_2(t)|} > 1 \text{ then } \hat{x}_2(t^+) = \frac{\hat{x}_2(t)}{|\hat{x}_2(t)|} \end{cases} \qquad (5.7)$$

the estimator described above can be easily shown to converge with a sufficiently high gain k, with the same Lyapunov function used previously. An observer in this form was already obtained by [46], who first came out with a provably convergent scheme for estimating structure from motion.

5.3 Motion–independent structure estimation

All we have said so far can be applied only when the relative motion between the scene and the viewer is *known* and has *constant velocity*. In many practical instances, however, this is not the case. As we have seen before the literature proposes a variety of motion estimation schemes which do not depend upon the structure of the scene either recursively from an image sequence as described in the previous sections [63] or from two views (see [22] for a review). However, errors in the reconstructed motion – if not treated properly – have a dramatic effect in the estimates of structure using the schemes described above.

In this section we describe a factorization method for recursively estimating structure *independent of motion* [37].

Consider the velocity vector of the image-plane projection $\mathbf{x}_i$ of a feature point $\mathbf{X}^i, i = 1\ldots N$ in the basic standard model (4.2):

$$\dot{\mathbf{x}}_i(t) = \left[\frac{1}{X_3(t)_i}\mathcal{A}(\mathbf{x}_i)\ \mathcal{B}(\mathbf{x}_i)\right]\begin{bmatrix}\mathbf{v}(t)\\ \omega(t)\end{bmatrix}, \qquad (5.8)$$

where

$$\mathcal{A}(\mathbf{x}_i) \doteq \begin{bmatrix} 1 & 0 & -(x_1)_i \\ 0 & 1 & -(x_2)_i \end{bmatrix}$$

$$\mathcal{B}(\mathbf{x}_i) = \begin{bmatrix} -(x_1)_i(x_2)_i & 1+(x_1)_i^2 & -(x_2)_i \\ -1-(x_2)_i^2 & (x_1)_i(x_2)_i & (x_1)_i \end{bmatrix}$$

depend only upon the measured function of the state $x_{1i} = (\frac{X_1}{X_3})_i$ and $x_{2i} = (\frac{X_2}{X_3})_i$. Assume we can observe exactly the velocities $\dot{\mathbf{x}}_i(t)$ of the N points moving on the image plane (this is called the *motion field*); we may then write

$$\dot{\mathbf{x}} = \mathcal{C}(\frac{1}{X_{31}},\ldots,\frac{1}{X_{3N}},\mathbf{x})[\mathbf{v}\ \omega]^T$$

where

$$\mathcal{C}(\frac{1}{X_{31}},\ldots,\frac{1}{X_{3N}},\mathbf{x}) \doteq \begin{bmatrix} \frac{1}{X_{31}}\mathcal{A}_1 & \mathcal{B}_1 \\ \vdots & \vdots \\ \frac{1}{X_{3N}}\mathcal{A}_N & \mathcal{B}_N \end{bmatrix}.$$

Assuming $\mathcal{C}$ of full rank, the motion parameters $\mathbf{v}, \omega$ can be computed (exactly if there is no observation noise) as

$$\begin{bmatrix} \mathbf{v} \\ \omega \end{bmatrix} = \mathcal{C}^\dagger \dot{\mathbf{x}}$$

where † indicates a left-inverse. Therefore, for $N > 3$, the motion field equations (5.8) specify the constraint

$$\dot{\mathbf{x}} = \mathcal{C}\mathcal{C}^\dagger \dot{\mathbf{x}} \Rightarrow \mathcal{C}^\perp(\frac{1}{X_{3\,1}}, \ldots, \frac{1}{X_{3\,N}}, \mathbf{x})\dot{\mathbf{x}} = 0,$$

where $\mathcal{C}^\perp \doteq I - \mathcal{C}\mathcal{C}^\dagger$.

Indeed, it is immediate to see that the problem of estimating the structure *independent of the motion of the viewer* can be rephrased as the problem of identifying the following Exterior Differential System [8] , embedded in $\mathbb{R}^N$:

$$\begin{cases} \mathcal{C}(Z, \mathbf{x})^\perp \dot{\mathbf{x}} = 0 & Z \in \mathbb{R}^N \\ \mathbf{y}_i = \mathbf{x}_i + n_i & \forall i = 1 \ldots N \end{cases}$$

where $Z \doteq [\frac{1}{X_{3\,1}} \ldots \frac{1}{X_{3\,N}}]^T$ is the inverse depth vector, $\mathbf{y}_i$ is the actual noisy measurement of $\mathbf{x}_i$, and n_i is a white, zero-mean Gaussian noise. We have implemented an identifier for the above model using the techniques in [63], and tested the scheme on synthetic image sequences.

6. Visual motion control

There are a variey of control problems where the goal of the controller is to keep the motion of the system "close" to the boundary of a surrounding environment or to drive the system so as to reach smoothly some specific desired position and orientation in space. Typical examples are docking of a space vehicle to a space station, airplane landing or terrain following, automatic car steering on roads, manipulation of objects by robot arms etc..

In these applications a large number of position and orientation parameters of the system with respect to the environment need to be monitored and controlled simultaneously. Use of traditional "local" sensors (sonar, radar, microwave sensors etc.) which provide only separate measurements of location from certain points of the environment give rise to complicated sensor fusion problems. Vision instead can provide integrated mesurements of the global postion and location of the vehicle with respect to the environment. For example, by using vision we can easily sense the misplacement (both in position and orientation) of a workpiece on a conveyor in a production line and provide feedback control to the robot arm so as to grasp the piece in the desired way or do the necessary operations on the workpiece correctly.

6.1 Vision in the Loop in autonomous systems

A general motivation for the use of vision comes from *autonomous systems*. These are becoming more and more important in modern technology: sytems capable of operating in unknown environment with only a bare minimum of human supervision are becoming essential in many applications like high depth offshore operations, space operations like docking and remote manipulation of space objects (telemanipulation in space is inhibited by the long time delays in communication), automatic car driving on highways, operations in hazardous environments etc..

Main tasks of an autonomous system are :

- Estimation and understanding of the geometry of the environment (which changes during the motion). This can in general be done only locally i.e. in a neighborhood of the current position of the vehicle.
- Reconstruction of the relative motion (in particular position and attitude) of the the moving system (or equivalently of the camera, assumed fixed with it) with respect to the environment.
- Path planning for obstacle avoidance and goal reaching. This must be done *on line* based on the current local reconstruction of the environment. Problems which arise are how to update the scene reconstruction recursively, how to do on line path planninng so as to update the planned trajectory in real time, etc.. This is essential for avoiding unknown obstacles.
- Tracking of the planned trajectory.

Vision has shown good applicability to autonomous systems. Successful applications to robotics and navigation of autonomous vehicles are well documented in the literature [35, 36, 34, 19, 18, 17].

Dickmanns [19, 18, 17]in particular has shown how one can take advantage of the *distributed sensor* capabilities of vision for estimation and tracking of an unknown contour. Since the future path to be followed by the vehicle is detected on the image plane at locations which are continuosly changing with time, a bank of local estimators (Kalman Filters) working in parallel is implemented. These filters process local information about the scene collected at particularly interesting spots of the image plane which can be changed automatically, tracking e.g. different areas containing the contour of interest.

Since vision has a capability of measuring directly on the image plane distances from a *predicted* trajectory to any point of the (visible) scene and in particular to the contour or boundary of obstacles etc., one could in principle implement very accurate *predictive control* strategies to do tracking and automatic obstacle avoidance. Most of this is however still to be explored, see [25].

6.2 Vision-based cartesian control of robotic arms

In the following sections we shall choose to discuss applications of vision to a particular application area, namely robotics. Most of the basic ideas however apply with only minor changes to other fields like space vehicles, autonomous cars etc.

Naturally the main purpose of vision in robotics is to provide real-time measurements of the relative location of the end-effector in the workspace. In the most common situation there is a workpiece of known shape and dimensions to be acted upon. A variety of local scene reconstruction algorithms can be based on the known geometry of the workpiece or of certain portions of the workspace. A preselected set of precisely known geometric features of the workpiece project on the image plane of the camera and the erlative image features are tracked during the motion of the system. This can be done by local tracking algorithms and with a relatively modest computational cost. A variety of position + orientation (pose) reconstruction algorithms exist which use the image feature location data read off on the image plane. Such algorithms are integrated in the control system of the robot arm to provide a direct task-space feedback signal to be used by the arm controller.

In order to appreciate the role of vision in robotics a brief discussion of the structure of a task-space control systems for robot arms will be given.

6.3 Joint-space control

Traditional robot-arm control is essentially open loop. This is because only *joint* position (and sometimes also velocity) measurements are available. Controller design typically consists of the following phases

1. Design of a suitable end-effector trajectory in the workspace i.e. path/motion planning. For this phase to be meaningful the geometry and location of the target object (workpiece) in the workspace must be known exactly and the relative pose of the end-effector at the starting instant must be known exactly.
2. Mapping of the planned path from the workspace to the joint (configuration) space to obtain a corresponding trajectory in terms of joint angles/displacements. An accurate knowledge of the kynematic model of the arm is required. Mapping of a task-space specified path to a corresponding joint-space path requires stepwise inversion of the jacobian map (or more complicated operations in case of redundant tasks) resulting in an interpolated trajectory matching the wanted positions only at discrete workspace locations.
3. Design of a controller that tracks the computed joint-space trajectory $t \to q^*(t)$. The trajectory is fed to a *joint controller* which generates the actuators command variables and eventually the joint torques necessary to track the command signal $q^*(t)$. Since only *joint* positions (and perhaps velocities) feedback is available to the controller the ability of the end-effector to follow the originally planned trajectory in the workspace depends on the accurate knowledge of the kynematical and dynamical model of the arm, on the absence of disturbances in workspace and on perfect positioning of the workpiece. In fact, no closed loop control is really exerted in the workspace.

6.4 Cartesian control

More sophisticated control actions are requires when the end effector must move in the workspace perhaps in contact with the environment, say slide on a surface while keeping a desired orientation and with a preassigned normal reaction force. It may then be necessary to add force or compliance sensors to the end effector. This introduces a form of (partial) workspace feedback and hence the joint space control philosophy described above, which is based only on models of the joint coordinates must be modified [48]. In particular, workspace coordinates must explicitely appear in the model of the arm used for control design. *Cartesian control* has been introduced primarily with this type of situation in mind [48] [2] [15][58].

Dynamic model in workspace coordinates

The dynamic model of a robot arm with external forces acting from the environment to the robot end-effector, can be written

$$M(q)\ddot{q} + C(q,\dot{q})\dot{q} + G(q) = u + J(q)'f \tag{6.1}$$

where q is the n-dimensional configuration (Joint angles and displacements) variable $M(q), C(q,\dot{q}), G(q)$ are the usual inertial, centrifugal and gravity terms, u is the n-dimensional control (joint torques and forces applied by the actuators), and $f \in \mathbb{R}^m$ is the vector of generalized linear reaction forces and angular momenta acting from the environment, expressed in the workspace (or base) frame of the system. We shall here only consider the case where $m = n$ i.e. non redundant tasks and, without loss of generality, take $n = 6$. Moreover in the following it will be assumed that *no kynematic singularities* are ever encountered. The square matrix $J(q)$ is the so-called *geometric Jacobian* (sometimes also called *basic* jacobian) of the robot, the matrix relating generalized end-effector velocity (translational and angular velocity) to joint velocities,

$$\begin{bmatrix} \dot{r} \\ \omega \end{bmatrix} = J(q)\dot{q} \tag{6.2}$$

Since the interaction between the robot and the environment occurs in workspace the model is most naturally described in the relative local coordinate frame. For this purpose introduce the kynematic equation of the arm, relating the position r and (say) the Euler angles triple ϕ of the end-effector to the joint coordinates,

$$x = \begin{bmatrix} r \\ \phi \end{bmatrix} = k(q) \tag{6.3}$$

Then the dynamical model of the arm in workspace local coordinates x is written

$$M_x(q)\ddot{x} + C_x(q,\dot{q})\dot{x} + G_x(q) = J_a(q)^{-T}u + f_a \tag{6.4}$$

where

$$M_x(q) = J_a(q)^{-T} M(q) J_a(q)^{-1}$$
$$C_x(q,\dot{q}) = J_a(q)^{-T} C(q,\dot{q}) J_a(q)^{-1} - M_x(q) \dot{J}_a(q) J_a(q)^{-1}$$
$$G_x(q) = J_a(q)^{-T} G(q)$$

and the generalized forces f_a and f are related by

$$J(q)^T f = J_a(q)^T f_a. \tag{6.5}$$

The *analytic jacobian* $J_a(q) := \frac{\partial k}{\partial q}$ can be computed from the geometric jacobian by a block diagonal transformation

$$J_a(q) = \begin{bmatrix} I & 0 \\ 0 & T(\phi) \end{bmatrix} J(q) \tag{6.6}$$

where the matrix $T(\phi)$ transforms the derivative of the Euler angles to the angular velocity of the end-effector, i.e. $\dot{\phi} = T(\phi)\omega$.

The functional dependence of the coefficient matrices in (6.4) is still on $q, \dot{q}$ and not on the new state variables $x, \dot{x}$. This substitution could in principle be performed by using the inverse kynematic relation of (6.3) but it is not essential and for control purposes it is actually more convenient to keep the model in its present form and exploit the explicit dependence on the joint variables for the design of feedback controllers based on both joint and workspace feedback.

6.5 Vision-based reconstruction of the relative pose

In this section we shall describe an algorithm which computes in real time the relative position and attitude (pose) x of the camera with respect to the workspace. The algorithm is based on the assumption that the target object is of a known shape and of accurately known geometry and dimensions. In particular, a suitable number of reference points has been selected having known mutual distances and are in a known position with respect to an object-fixed reference frame. (Note that this is exactly the situation in space docking and rendez-vous). These points are continuously tracked by the vision system during the phase of navigation and docking in workspace before the contact and manipulation phase, and their projections on the image plane of the camera are used as reference feature points for the computation of the relative pose of the camera with respect to the object.

We shall assume that the camera is ideal and the calibration parameters,in particular the camera focal length are known. In real applications this is usually far from true and camera calibration procedures are necessary.

We denote by Σ_c the camera reference frame with the origin in the camera optical center and let the z axis coincide with the camera optical axis. Let Σ_o denote a reference frame fixed with the target object. The camera is mounted on the manipulator in a fixed and known position with respect to the end effector, so that the transformation between the camera frame and any reference frame fixed with the end effector can be computed off line and is assumed to be known a priori. Therefore, without loss of generality we can identify the camera frame Σ_c

with the end effector frame. Also, since the typical workpieces to be acted upon by the manipulator are parts of (or generally tightly linked to) the workspace, we shall assume that the object reference frame is fixed in workspace.

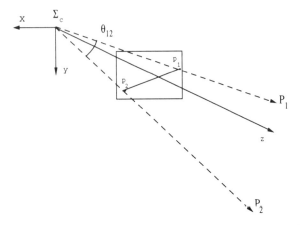

Fig. 6.1. *Relative pose reconstruction geometry*

Denote the vector of the Cartesian coordinates of a point $\mathbf{P}_k$ in the workspace, expressed with respect to the reference frames Σ_o or Σ_c by $\mathbf{a}_k$ or, respectively $\mathbf{p}_k$.

The relative position of the camera (or of the end effector) with respect to the object is defined by the translation vector $\overline{O_o O_c}$ joining the origin of the object frame O_o to the origin of the camera frame O_c. The relative orientation of the camera frame with respect to the object is defined by the rotation matrix $\mathbf{R}$ which transforms coordinates in the reference frame Σ_o into cordinates with respect to Σ_c (or, dually, rotates Σ_c into a reference frame with the same direction and orientation of Σ_o).

Let $\mathbf{t}$ denote the coordinate vector of the translation $\overline{O_o O_c}$ with respect to the object frame Σ_o and let $\bar{\mathbf{t}}$ denote the coordinates of $\overline{O_c O_o}$ with respect to Σ_c. Note that $\bar{\mathbf{t}} = -\mathbf{R}\,\mathbf{t}$.

It is immediate that the coordinate vectors with respect to Σ_o and Σ_c of a feature point P_k of the target object are related by the equation

$$\mathbf{p}_k = \mathbf{R}\,\mathbf{a}_k + \bar{\mathbf{t}}. \tag{6.7}$$

Clearly, if the coordinates $\mathbf{p}_k$, $\mathbf{a}_k$ $k = 1, 2, 3, \ldots$ of at least three not collinear points are known *both* with respect to the end effector frame and the object frame, in order to determine the relative position and the relative orientation $\mathbf{R}$ one may attempt to solve a system of linear equations of the type

$$\mathbf{p}_k - \mathbf{p}_o = \mathbf{R}(\mathbf{a}_k - \mathbf{a}_o) \qquad k = 1, 2, 3, \ldots \tag{6.8}$$

obtained by subtracting from each equation (6.7) a "reference" term $\mathbf{p}_o = \mathbf{R}\,\mathbf{a}_o + \bar{\mathbf{t}}$, where $\mathbf{p}_o$ and $\mathbf{a}_o$ are the coordinates of a suitable reference point. The general idea is to compute $\mathbf{R}$ from (6.8) and then compute $\bar{\mathbf{t}}$ from (6.7). In reality, the coordinates $\mathbf{p}_k$ are unknown and must be determined from the measurements on the image-plane of the camera. Hence, due to errors in the estimates of the coordinates $\mathbf{p}_k$, the matrix $\mathbf{R}$ which could in principle be computed from (6.8) will hardly be a rotation matrix, i.e. an orthogonal matrix of determinant $+1$. This is so because the representation of a rotation by a 3×3 matrix is vastly redundant. While a rotation is defined by three parameters, the 3×3 matrix $\mathbf{R}$ has nine free parameters. Furthermore, imposing the orthogonality constraint on the solution of (6.8) is not a simple matter. Hence, a more convenient and non-redundant representation of the relative orientation matrix is needed.

The classical idea from the photogrammetry literature around the end of fifties and in the vision engineering literature until rather recently, is to express the coordinates of the projections of the reference points on the image plane as functions of their (known) coordinates in the object-fixed reference frame and of the pose parameters, namely the translation vector and the rotation matrix defining the mutual location in space of one frame with respect to the other. One obtains in this way a system of nonlinear equations in the pose parameters which is usually solved by an iterative non-linear least squares algorithm. The difference in the various approaches lies on the different kinds of rotation coordinates employed.

For example Horn [43] parametrizes rotation directly by the 9 coefficients of the rotation matrix. In this way a large number of side constraints are necessary to guarantee that the solution will really be a rotation matrix. One should avoid if possible to work with constrained problems. This can be done to some extent by introducing Euler angles.

McClamroch and co-workers [41], use a slight variation of the three Euler angles to represent rotation. The resulting nonlinear equations are however still very complicated to solve.

The solution proposed by Hinsken [40] and further developed by Haralik and Shapiro, [32], uses unit quaternions to represent the rotation matrix which then becomes a function of four parameters with only one normalization constraint.

The best way to compute the rotation matrix both in terms of semplicity and computstional efficiency is probably Horn's method [44]. Using unit-quaternion representation the computation of the rotation matrix is reduced to the computation of the maximum eigenvalue of a 4×4 symmetric matrix. In fact Horn's algorithm has turned out to be a very reliable and accurate numerical method to compute the rotation matrix. Typical relative reconstruction errors in real experiments are less than one percent.

The pose reconstruction problem starting from projections on the image plane of points of a known object can be divided in three main steps, namely estimation of the coordinates of the reference points in the camera-fixed frame, computation of the rotation matrix and finally computation of the translation vector.

Estimation of the coordinates in the camera frame

Computing the coordinates $\mathbf{p}_i$ from the known projections $\mathbf{x}_i = [x_i, y_i]$ on the image plane, is just a matter of computing the lenghts, d_i, of the vectors $\overline{O_c P_i}$. Applying an elementary relation for triangles we can write the following equation

$$\|\overline{P_i P_j}\|^2 = \|\overline{O_c P_i}\|^2 + \|\overline{O_c P_j}\|^2 - 2\|\overline{O_c P_i}\| \|\overline{O_c P_j}\| \cos \theta_{ij}$$
$$= d_i^2 + d_j^2 - 2 d_i d_j \cos \theta_{ij},$$

where θ_{ij} denotes the angle between the vectors $\overline{O_c P_i}$ and $\overline{O_c P_j}$.

The cosines of the angles θ_{ij} can be obtained by normalizing to unit lenght the coordinate vectors $[x_i, y_i, f]^T = [\mathbf{x}_i, f]^T$ of the projected image features corresponding to the points $\mathbf{P}_i$. Denoting the normalized vectors by $\mathbf{v}_i$ we have

$$\cos \theta_{ij} = \mathbf{v}_i \cdot \mathbf{v}_j,$$

where $\cdot$ stands for scalar product of vectors. We end up with a system of equations in the unknown distances d_i, $i = 1, 2, \ldots, n$

$$d_i^2 + d_j^2 - 2 d_i d_j \cos \theta_{ij} = d_{ij}^2 \quad i, j = 1, 2, \ldots, n, \tag{6.9}$$

where

$$d_{ij} := \|\overline{P_i P_j}\| = \|\mathbf{p}_i - \mathbf{p}_j\| = \|\mathbf{a}_i - \mathbf{a}_j\|$$

are the *known* distances between the reference points of the object.

This is a system of quadratic equations in the d_i, $i = 1, 2, \ldots, n$, which is in general overdermined (for $n > 3$) and must be solved in the least squares sense. The solution can be computed by iterative linearization say by a Newton-Raphson method, which has very good local convergence properties, by applying least squares to each linearized problem in the iterative procedure.

With a good initialization two or threee iterations normally suffice.

Computing the rotation matrix.

After computing the coordinates $\mathbf{p}_i$ with respect to the camera frame, the second step of the algorithm is to estimate the rotation matrix $\mathbf{R}$. To this purpose we shall use a linear equation of the type (6.8) above, which we rewrite simply as

$$\mathbf{p}_i = \mathbf{R} \mathbf{a}_i,$$

where $\mathbf{R}$ is the rotation matrix from Σ_o to Σ_c. Later on we shall discuss how to choose the reference coordinates $\mathbf{p}_o$ and $\mathbf{a}_o$ in a convenient way.

Since the coordinate estimates $\mathbf{p}_i$ are noisy and in general we may have many more points than unknowns, it is necessary to solve the equation above in a mean-square sense. In this setting one looks for a rotation matrix $\mathbf{R}$ which minimizes the sum of squares of the errors, $\mathbf{e}_i = \mathbf{p}_i - \mathbf{R} \mathbf{a}_i$, corresponding to each measurement. This is a *constrained* minimization problem

$$\min_{\mathbf{R}} \sum_{i=1}^{n} \|\mathbf{e}_i\|^2,$$

with $\mathbf{R} \in SO(3)$, a rotation matrix.

This can easily be seen to be the same as

$$\max_{\mathbf{R}} \sum_{i=1}^{n} \mathbf{p}_i \cdot \mathbf{R}\mathbf{a}_i,$$

again with $\mathbf{R} \in SO(3)$.

Now, following Horn, [44] we express everything in terms of unit quaternions. We get

$$\sum_{i=1}^{n} \mathbf{p}_i \cdot \mathbf{R}\mathbf{a}_i = \sum_{i=1}^{n} (\overset{\circ}{q}\overset{\circ}{a}_i\overset{\circ}{q}^*) \cdot \overset{\circ}{p}_i = \sum_{i=1}^{n} (\overset{\circ}{q}\overset{\circ}{a}_i) \cdot (\overset{\circ}{p}_i\overset{\circ}{q}),$$

where $\overset{\circ}{q}$ denotes the unit quaternion corrisponding to the rotation $\mathbf{R}$, $\overset{\circ}{a}_i$ and $\overset{\circ}{p}_i$ are the quaternions corrisponding to $\mathbf{a}_i$ and $\mathbf{p}_i$ respectively and $\cdot$ stands for quaternion inner product.

Introducing coordinates, $\mathbf{a}_i = (a_{xi}, a_{yi}, a_{zi})^T$ and $\mathbf{p}_i = (p_{xi}, p_{yi}, p_{zi})^T$, the quaternion products can be written as matrix products in the following way

$$\overset{\circ}{q}\overset{\circ}{a}_i = \begin{bmatrix} 0 & -a_{xi} & -a_{yi} & -a_{zi} \\ a_{xi} & 0 & a_{zi} & -a_{yi} \\ a_{yi} & -a_{zi} & 0 & a_{xi} \\ a_{zi} & a_{yi} & -a_{xi} & 0 \end{bmatrix} \overset{\circ}{q} = \bar{\mathcal{A}}_i \overset{\circ}{q},$$

and

$$\overset{\circ}{p}_i\overset{\circ}{q} = \begin{bmatrix} 0 & -p_{xi} & -p_{yi} & -p_{zi} \\ p_{xi} & 0 & -p_{zi} & p_{yi} \\ p_{yi} & p_{zi} & 0 & -p_{xi} \\ p_{zi} & -p_{yi} & p_{xi} & 0 \end{bmatrix} \overset{\circ}{q} = \mathcal{P}_i \overset{\circ}{q}.$$

The matrices $\bar{\mathcal{A}}_i$ and $\mathcal{P}_i$ are skew-symmetric and orthogonal.

The sum to be minimized can now be rewritten in a more convenient matrix form as

$$\sum_{i=1}^{n} (\overset{\circ}{q}\overset{\circ}{a}_i) \cdot (\overset{\circ}{p}_i\overset{\circ}{q}) = \sum_{i=1}^{n} (\bar{\mathcal{A}}_i \overset{\circ}{q}) \cdot (\mathcal{P}_i \overset{\circ}{q}) = \sum_{i=1}^{n} \overset{\circ}{q}^T \bar{\mathcal{A}}_i^T \mathcal{P}_i \overset{\circ}{q}$$

$$= \overset{\circ}{q}^T (\sum_{i=1}^{n} \bar{\mathcal{A}}_i^T \mathcal{P}_i) \overset{\circ}{q} = \overset{\circ}{q}^T (\sum_{i=1}^{n} \mathbf{N}_i) \overset{\circ}{q} = \overset{\circ}{q}^T \mathbf{N} \overset{\circ}{q},$$

where $\mathbf{N}_i = \bar{\mathcal{A}}_i^T \mathcal{P}_i$ and $\mathbf{N} = \sum_{i=1}^{n} \mathbf{N}_i$. It is immediate that the $\mathbf{N}_i$'s are symmetric and hence $\mathbf{N}$ is also symmetric.

In conclusion, the best rotation matrix is found by solving the following classical constrained maximum problem for a quadratic form,

$$\max_{\overset{\circ}{q}} \overset{\circ}{q}^T \mathbf{N} \overset{\circ}{q} \quad \text{subjectto } \|\overset{\circ}{q}\| = 1.$$

Hence, see e.g. [44], the solution is just the maximum eigenvector $\mathbf{e}_m$, of $\mathbf{N}$

$$\overset{\circ}{q} = \mathbf{e}_m / \|\mathbf{e}_m\|,$$

normalized to unit norm. Thus estimation of the rotation matrix reduces to finding the maximum eigenvector of a symmetric matrix.

Computing the translation vector.

The third step of the algorithm is to compute the translation from the estimates of $\mathbf{p}_i$ and $\mathbf{R}$. To reduce noise one seeks a best solution in the mean square sense.
Recall from (6.7) that

$$\mathbf{p}_i = \mathbf{R}(\mathbf{a}_i - \mathbf{t}) = \mathbf{R}\mathbf{a}_i + \bar{\mathbf{t}},$$

where $\bar{\mathbf{t}} = -\mathbf{R}\mathbf{t}$ is the translation $\mathbf{t}$ expressed in the frame Σ_c. The $\bar{\mathbf{t}}$ minimizing $\sum_{i=1}^n \|\mathbf{e}_i\|^2$, where $\mathbf{e}_i$ is the error, $\mathbf{e}_i = \mathbf{p}_i - \mathbf{R}\mathbf{a}_i - \bar{\mathbf{t}}$, can be expressed very simply in terms of the averages, or "centroids" of the data

$$\hat{\mathbf{a}} = \frac{1}{n}\sum_{i=1}^n \mathbf{a}_i \text{ e } \hat{\mathbf{p}} = \frac{1}{n}\sum_{i=1}^n \mathbf{p}_i.$$

as,

$$\bar{\mathbf{t}} = \hat{\mathbf{p}} - \mathbf{R}\hat{\mathbf{a}}.$$

From this we can also compute the original translation vector $\mathbf{t}$ as $\mathbf{t} = \hat{\mathbf{a}} - \mathbf{R}^T\hat{\mathbf{p}}$.

6.6 Workspace control algorithms with vision feedback

For completeness we mention here some well-known ideas useful in the design of the vision-based controller.

In general the design of a nonlinear controller can be carried out in two main steps.

The first step is feedback linearization. Choose a preliminary feedback law of the type

$$u = J_a(q)^T [M_x(q)v + C_x(q,\dot{q})\dot{x} + G_x(q) - f_a] \quad (6.10)$$

to reduce the closed loop dynamics (in the task space) to bank of double integrators

$$\ddot{x} = v \quad (6.11)$$

where v is an external input still available for control.

In the second step, given a desired reference trajectory (a planned path) in workspace $x_d(t)$, select a desired *impedance model* of the form,

$$M_0(\ddot{x} - \ddot{x}_d) + D_0(\dot{x} - \dot{x}_d) + K_0(x - x_d) = f_a. \quad (6.12)$$

This model specifies the desired dynamical behaviour of the end effector. The matrices M_0, K_0 are positive definite (usually diagonal) and D_0 is positive semidefinite. If $f_a = 0$ no reaction forces are present from the environment and the system is traveling freely through the workspace. In this case the model (6.12) defines the desired dynamic tracking characteristics of the system for a prespecified reference trajectory x_d. When $f_a \neq 0$ (6.12) describes instead the mechanical impedance which the end-effector should ideally offer to balance the reaction forces exerted by the contact surface.

The desired mechanical impedance (model matching) is obtained by choosing v in (6.10) as

$$v = \ddot{x}_d + M_0^{-1}[D_0(\dot{x} - \dot{x}_d) + K_0(x - x_d) + f_a] \qquad (6.13)$$

Note that the implementation of (6.10) (6.13) requires in general feedback from the joint coordinates $(q, \dot{q})$, from the "cartesian" end-effector coordinates $(x, \dot{x})$ and on-line measurement of the contact force f_a.

The design of the mechanical parameters of the impedence model is discussed in many papers. See e.g. [15].

Important special cases of impedance control occur when the arm is moving slowly, say $\ddot{x}_d \approx 0$ and $\dot{q} \approx 0$, in which case (6.10) (6.13) particularize to the *PD controller*

$$u = G(q) + J_a(q)^T[K_0(x - x_d) + D_0(\dot{x} - \dot{x}_d)]. \qquad (6.14)$$

If the end effector travels without being subjected to contact forces ($f_a = 0$) and the motion does not cross singular configurations, this control law can be shown to be globally asymptotically stabilizing (i.e. tracking the desired trajectory x_d with zero asymptotic error).

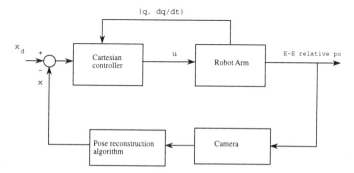

Fig. 6.2. *Vision based control in workspace*

7. Control on the image plane

Image plane control is based on the observation that the position and the orientation of a rigid body in 3-D space can be specified completely by assigning the coordinates $\mathbf{x}_i$, $i = 1, 2, 3$ on the image plane of the projections of three points $\mathbf{P}_i$ of the body assuming they are in generic position. In other words a local homeomorphism can be established between the 6-dimensional workspace and 3 copies of the image plane. This homeomorphism is in fact nothing else but the "pose reconstruction map"

$$\{\mathbf{x}_i | i = 1, 2, 3\} \mapsto x$$

that has been object of our study in section 6.5.

For this reason, in alternative to what we have described in the previous section, the control task in systems using vision as a sensor can be formulated directly on the image plane [35, 36, 34, 9, 20, 58].

This choice is very natural in certain applications, for example tracking, docking, navigation etc.. However, it results in methods that are intrinsically local, whereas there are applications in which one is required to track a globally prescribed path in the full configuration space. Furthermore, the control on the image plane exhibits some limitations due to the dependence of the controller on the structure (depth) of the observed scene.

In this section we briefly describe a simple tracking control on the image plane, and highlight its limitations. In the following section we propose to formulate the tracking problem in the configuration space in its *essential representation* (through the essential manifold). We anticipate the resulting control has more "global" features and does not depend on the structure of the observed scene. Instead, structure comes as a byproduct of the essential estimator once the control task has been accomplished.

Suppose our task is to maintain a given relative configuration between a platform and the scene. Such situation occurs in tracking the motion of a three dimensional object (of unknown shape and kinematics) or in maintaining a fixed pose with respect to a scene despite the action of disturbances on the platform (as for example in hovering or in underwater operation).

For simplicity we shall neglect here the dynamic constraints (inertia etc.) and assume to be able to control directly the translational and rotational velocity of the platform.

Consider the model defined by (4.1)(4.2), and the expression for the time derivative of the projected point $\mathbf{x}$ on the image plane (called "motion field" in section 5.3):

$$\dot{\mathbf{x}}^i(t) = \mathcal{J}(\mathbf{x}^i(t), X_3^i(t))u(t) \quad \forall i = 1 \ldots N \tag{7.1}$$

where

$$\mathcal{J}(x, X_3) \doteq \begin{bmatrix} \frac{1}{X_3} & 0 & -\frac{x_1}{X_3} & -x_1 x_2 & 1 + x_1^2 & -x_2 \\ 0 & \frac{1}{X_3} & -\frac{x_2}{X_3} & -(1 + x_2^2) & x_1 x_2 & x_1 \end{bmatrix} \tag{7.2}$$

and $\mathbf{x}^i$ indicates the image plane coordinates of the projection, while X_3^i denotes the third component of the space coordinate (depth) of each point. The vector $u(t) \doteq (v(t), \omega(t))$ is the control variable. Suppose the initial configuration of the points on the image plane is $\mathbf{x}^i(t_0|t_0) = \mathbf{x}_0^i$, and an exogenous agent acts by moving either the platform on which the camera is mounted or the target which the camera is looking at, producing a deformation of its image:

$$\mathbf{x}^i(t+1) = \mathbf{x}^i(t) + \tilde{\mathbf{x}}^i(t). \tag{7.3}$$

Suppose our goal is to keep the configuration of the observed points fixed at the value of the initial instant $\mathbf{x}_0^i$. At any time we can measure a noisy version of the instantaneous configuration modified by the external agent, and act with the control of the platform on which the camera is mounted. Using a first step approximation, one could write

$$\mathbf{x}^i(t+1) \cong \mathbf{x}^i(t) + \mathcal{J}(\mathbf{x}^i(t), X_3^i(t))u(t) \tag{7.4}$$

and use a minimum time controller:

$$u(t) \doteq \mathcal{J}^\dagger(\mathbf{x}^i(t), X_3^i(t))(\mathbf{x}_0^i - \mathbf{x}^i(t)), \tag{7.5}$$

where † denotes the pseudoinverse. Note that the control depends on the depth of each point of the scene $X_3^i(t)$. Such a strategy has been experimented by [35], who pioneered the control on the image plane. However, the expression of the deadbeat controller on the image plane depends on the inverse depth of each visible points, which needs to be "estimated" on line. This problem can be overcome by assuming that *the structure of the scene is known*, and therefore the inverse depth can be recovered linearly (the so-called "calibration" phase). Another alternative, which we do not pursue here, is the use of a stereo system.

If the structure (depth) of the scene is not known, we need to *estimate* it, unless the motion of the target is purely rotational about the center of the viewer's reference, in which case $\mathcal{J}$ does not depend on the depth. In order to estimate depth, we need non-zero *disparity* (also called visual parallax), which is the displacement of corresponding points across different images. When disparity is close to zero, the recovery of the depth is ill conditioned [66]. Therefore the image based controller, which depends on the depth, tries to drive the system towards a configuration of zero disparity, which does not allow to recover depth. As a result the controller either "drifts" or "swings".

7.1 Control on the essential manifold

Consider $\mathbf{Q}_0 \in E$ describing the relative configuration between the scene and the platform at the initial instant, and suppose we ask it to be constant despite the motion of the scene, encoded by an arbitrary $d(t) \in E$. We indicate with $\mathbf{Q}(t)$ the essential matrix describing the motion between the *initial* instant and the current time, which is therefore defined by the essential constraint $\mathbf{x}^i(t)^T \mathbf{Q}(t) \mathbf{x}_0^i \doteq 0$. Note that usually in the essential filter we consider $\mathbf{Q}(t)$ to be the instantaneous configuration with respect to the observer's reference at the previous time sample. The effect of the exogenous displacement (motion of the scene) $d(t)$ and the control action are described by the model

$$\begin{cases} \mathbf{Q}(t+1) = \mathbf{Q}(t) \oplus \Phi^{-1}(u(t)) \oplus d(t) \\ y^i(t)^T \mathbf{Q}(t) y(0) = n(t) \end{cases} \quad (7.6)$$

where $\oplus$ represents the composition operation in E corresponding to sum of the local coordinates and n describes the effect of measurement errors. In general we may want to specify the control task in terms of some *distance* defined on the essential space, $d_E(\mathbf{Q}_1, \mathbf{Q}_2)$, so that

$$e(t) \doteq d_E(\mathbf{Q}(t), \mathbf{Q}_d(t)) \quad (7.7)$$

satisfies a difference equation whose dynamics can be assigned by choice of the input.

Choice of a metric on the essential manifold

Since E can be interpreted as an alternative representation of $SE(3)$, any control strategy on the Euclidean group can be mapped onto the essential manifold. However, if we were able to formulate the control strategy directly on the essential manifold, the essential filter would then gives us a direct estimate of the full state which is optimal, independent of the structure and obtained linearly from the visual data [63].

The choice of a metric on the essential space is not a trivial issue, and we intend in this paper to hint at some possible choices. First of all any metric in the Euclidean space $SE(3)$ can be "mapped" onto the essential manifold by defining

$$d_E(\mathbf{Q}_1, \mathbf{Q}_2) \doteq d_{SE(3)}(\Psi^{-1} \circ \Phi(\mathbf{Q}_1), \Psi^{-1} \circ \Phi(\mathbf{Q}_2)) \quad (7.8)$$

where Ψ and Φ are local coordinatizations of $SE(3)$ and E respectively. An alternative (and equivalent) method is to set the metric directly in the local coordinates and then "lift" it to the manifold. It must be pointed out, however, that there is no natural (invariant) choice of a metric on the Euclidean group. Another possibility is to "project" a metric of the ambient space of the essential manifold, $\mathbb{R}^9$, by using the projection onto the manifold pr_E. It is unclear at the moment what the properties of such a metric may be. Note also that a possible way of generating a path between two points of the essential manifold, based on its interpretation as the tangent bundle of $SO(3)$, is to formulate a control that connects two points of $SO(3)$ with a given direction in the tangent plane. Such control strategies, called "dynamic interpolation" have been studied for Riemannian manifolds and Lie groups by [11, 49].

7.2 Structure–independent control on the essential manifold–[63]

In this section we consider a simple experiment: we want to formulate the control that drives the relative configuration to the desired one in minimum time, as we have done in section 7. for the control on the image plane. We do not make any assumption on the scene, and we want to develop a control strategy which is independent of depth, so that we do not have ill-conditioned controllers at unobservable configurations of the system.

The model described in eq. (4.9) gives an immediate expression for such a minimum time controller. Suppose we are only interested in maintaining the initial configuration, then $\mathbf{Q}(t_0) = 0$, and our control can be inferred from the "controlled random walk model" (7.6) and the relative Kalman filter equations in E,

$$\hat{\mathbf{Q}}(t+1|t) = \hat{\mathbf{Q}}(t|t) \oplus \Phi^{-1}(u(t)) \qquad (7.9)$$

$$\hat{\mathbf{Q}}(t|t) = \hat{\mathbf{Q}}(t|t-1) \oplus L(t)\mathbf{e}(t) \qquad (7.10)$$

$$\mathbf{x}^i(t+1)^T(\hat{\mathbf{Q}}(t+1|t) \oplus e_\mathbf{Q}(t+1))\mathbf{x}_0^i = 0 \qquad (7.11)$$

where $e_\mathbf{Q}(t)$ is the one-step prediction error of $\mathbf{Q}(t)$. Therefore, provided that our estimator is unbiased, the control

$$u(t) = -\Phi(\hat{\mathbf{Q}}(t|t-1)). \qquad (7.12)$$

gives a one-step correction which keeps the state about the goal $\mathbf{Q}(t_0) = 0$ instantaneously up to white, zero-mean noise.

8. Conclusions

The purpose of this paper is to stress the flexibility, robustness and accuracy reached by vision as a sensor, which could be considered in alternative to traditional accelerometers or range sensors in a number of applications in robotics. We have shown the image-plane control is not practical when the structure (depth) of the scene is not known, and proposed to perform visual motion control on the essential manifold, using the output of a causal motion estimator, called the "essential filter".

Acknowledgments

This work has been funded by the California Institute of Technology, a scholarship from the University of Padova, a fellowship from the "A. Gini" Foundation and grant ASI-RS-103 from the Italian Space Agency.

References

1. E. Arbogast and R. Mohr. An egomotion algorithm based on the tracking of arbitrary curves. *Proc. of the 2nd Europ. Conf. on Computer Vision*, 1992.
2. H. Asada and J.J. Slotine. *Robot Analysis and Control.* John Wiley, 1987.
3. A. Azarbayejani, B. Horowitz, and A. Pentland. Recursive estimation of structure and motion using relative orientation constraints. *Proc. CVPR*, New York, 1993.
4. G. Bastin and M. Gevers. Stable adaptive observers for nonlinear time-varying systems. *IEEE trans. Aut. Control*, 1988.
5. A. Blake, M. Taylor, and A. Cox. Grasping visual simmetry. *Proc. of the ICCV*, 1993.
6. T. Broida and R. Chellappa. Estimating the kinematics and structure of a rigid object from a sequence of monocular frames. *IEEE Trans. Pattern Anal. Mach. Intell.*, 1991.
7. T. Broida and R. Chellappa. Estimation of object motion parameters from noisy images. *IEEE Trans. Pattern Anal. Mach. Intell.*, Jan. 1986.
8. Bryant, Chern, Goldberg, and Goldsmith. *Exterior Differential Systems.* Mathematical Research Institute. Springer Verlag, 1992.
9. F. Chaumette and A. Santos. Tracking a moving object by visual servoing. *Proc. of the 12th IFAC World Congr. Vol. 9 pp. 409–414*, 1993.
10. R. Cipolla and A. Blake. Surface orientation and time to crash from image divergence and deformation. *Proc. of the European Conf. on Comp. Vision*, 1992.
11. P. Crouch and F. Silva Leite. The dynamic interpolation problem on riemannian manifolds, lie groups and symmetric spaces. *Technical report*, .
12. R. Curwen, A. Blake, and A. Zisserman. Real-time visual tracking for surveillance and path planning. *Proc. of the ECCV*, 1992.
13. Darmon. A recursive method to apply the hough transform to a set of moving objects. *Proc. IEEE, CH 1746 7/82*, 1982.
14. W. Dayawansa, B. Ghosh, C. Martin, and X. Wang. A necessary and sufficient condition for the perspective observability problem. *Systems and Control Letters*, in press, 1994.
15. C. Canudas de Wit et. al. *Theory of Robot control.* CNRS, France, 1992. Lect. notes of the Ecole d'ete' d'automatique de Grenoble, sect. 6, Sept. 7-11 1992, CNRS, Grenoble France.
16. E. Di-Bernardo, L. Toniutti, R. Frezza, and G. Picci. Stima del moto dell'osservatore e della struttura della scena mediante visione monoculare. *Tesi di Laurea–Università di Padova*, 1993.
17. E. D. Dickmanns and Th. Christians. Relative 3d-state estimation for autonomous visual guidance of road vehicles. In *Intelligent autonomous system 2 (IAS-2)*, Amsterdam, 11-14 December 1989.
18. E. D. Dickmanns and V. Graefe. Applications of dynamic monocular machine vision. *Machine Vision and Applications*, 1:241–261, 1988.
19. E. D. Dickmanns and V. Graefe. Dynamic monocular machine vision. *Machine Vision and Applications*, 1:223–240, 1988.
20. B. Espiau, F. Chaumette, and P. Rives. A new approach to visual servoing in robotics. *IEEE Trans. on Robotics and Autom.*, 8:313–326, 1992.
21. O. Faugeras. What can be seen in three dimensions with an uncalibrated stereo rig. *Proc. of the 2 ECCV*, 1992.
22. O. Faugeras. *Three dimensional vision, a geometric viewpoint.* MIT Press, 1993.
23. O. D. Faugeras and S. Maybank. Motion from point mathces: multiplicity of solutions. *Int. J. of Computer Vision*, 1990.
24. O.D. Faugeras, Q.T. Luong, and S.J. Maybank. Camera self-calibration: theory and experiments. *Proc. of the ECCV92, Vol. 588 of LNCS, Springer Verlag*, 1992.

25. R. Frezza and G. Picci. On line path following by recursive spline updating. In *Submitted to 34th Conf. on Decision and Control*, New Orleans, December 1995.
26. D.B. Gennery. Tracking known 3-dimensional object. In *Proc. AAAI 2nd Natl. Conf. Artif. Intell.*, pages 13–17, Pittsburg, PA, 1982.
27. D.B. Gennery. Visual tracking of known 3-dimensional object. *Int. J. of Computer Vision*, 7(3):243–270, 1992.
28. B. Ghosh, M. Jankovic, and Y. Wu. Perspective problems in systems theory and its application in machine vision. *Journal of Math. Systems, Est. and Control*, 1994.
29. B. K. Ghosh, E. P. Loucks, and M. Jankovic. An introduction to perspective observability and recursive identification problems in machine vision. In *Proc. of the 33rd Conf. on Decision and Control*, volume 4, pages 3229–3234, 1994.
30. B. K. Ghosh and J. Rosenthal. A generalized Popov Belevitch Hautus test of observability. *IEEE Trans. Autom. Control*, volume 40, pages 176–180, 1995.
31. G. Golub and C. Van Loan. *Matrix computations*. Johns Hopkins University Press, 2 edition, 1989.
32. R. M. Haralik and L. Shapiro. *Computer and Robot Vision*, volume 2. Addison-Wesley Publishing Company, Reading, MA, 1993.
33. R. Hartley. Estimation of relative camera positions for uncalibrated cameras. In *Proc. 2nd Europ. Conf. Comput. Vision, G. Sandini (Ed.), LNCS-Series Vol. 588*, Springer-Verlag, 1992.
34. K. Hashimoto and H. Kimura. Dynamic visual servoing with nonlinear model based control. In *Proc. of the 13th IFAC World Congress*, volume 9, pages 405–408, 1993.
35. K. Hasimoto, T. Kimoto, T. Ebine, and H. Kimura. Image-based dynamic visual servo for a hand-eye manipulator. In Kodama Kimura, editor, *Recent advances in mathematical theory of systems, control, networks, and signal processing II*, pages 609–614. Proceedings of the international symposium of MTNS, Mita Press, 1991.
36. K. Hasimoto, T. Kimoto, T. Ebine, and H. Kimura. Manipulator control with image-based visual servo. In *IEEE Intl' Conference on Robotics and Automation*, pages 2267–2272, 1991.
37. D. Heeger and A. Jepson. Subspace methods for recovering rigid motion i: algorithm and implementation. RBCV TR-90-35, University of Toronto – CS dept., November 1990. Revised July 1991.
38. J. Heel. Direct estimation of structure and motion from multiple frames. *AI Memo 1190, MIT AI Lab*, March 1990.
39. J. Heel. Temporal integration of 3-d surface reconstruction. *To appear on IEEE trans. PAMI, special issue on the interpretation of 3-D scenes.*, March 1991.
40. L. Hinsken. A singukarity-free algorithm for spatial orientation of bundles. *International Archives of Photogrammetry and Remote Sensing*, 27, 1988.
41. C.C. Ho and N.H. McClamroch. Autonomous spacecraft docking using a computer vision system. *Proc. of the 31st CDC Tucson AZ*, 1992.
42. C.C. Ho and N.H. McClamrock. Autonomous spacecraft docking using a computer vision systm. *Proc. of the 31st CDC – Tucson, AZ*, 1992.
43. B. Horn. *Robot Vision*. MIT press, 1986.
44. B.K.P. Horn. Closed-form solution of absolute orientation using unit quaternions. *Journ. of Optical Society of America A*, 4(4):629–642, 1987.
45. B.K.P. Horn. Relative orientation. *Int. J. of Computer Vision*, 4:59–78, 1990.
46. M. Jankovic and B. K. Ghosh. Visually guided ranging from observations of points, lines and curves via an identifier based nonlinear observer. –, June 30, 1993.
47. A.H. Jazwinski. *Stochastic Processes and Filtering Theory*. Academic Press, 1970.
48. O. Khatib. A unified approach for motion and force control of robot manipulators: the operational space formulation. *IEEE Journ. of Robotics and Autom.*, RA-3:43–53, 1987.
49. G. Heinzinger L. Noakes and B. Paden. Cubic splines on curved spaces. *IMA J. Math. Control and Information*, 1989.

50. M. Lei and B. K. Ghosh. A new nonlinear feedback controller for visually-guided robotic motion tracking. *Proc. of the ECC*, 1993.
51. H. C. Longuet-Higgins. A computer algorithm for reconstructing a scene from two projections. *Nature*, 293:133–135, 1981.
52. H.C. Longuet-Higgins. Configurations that defeat the eight-point algorithm. *Mental processes: studies in cognitive science, MIT press*, 1987.
53. R. Marino. Adaptive observers for single output nonlinear systems. *IEEE Trans. Aut. Control (35) 9*, 1990.
54. L. Matthies, R. Szelisky, and T. Kanade. Kalman filter-based algorithms for estimating depth from image sequences. *Int. J. of computer vision*, 1989.
55. S. Maybank. *Theory of reconstruction from image motion*, volume 28 of *Information Sciences*. Springer-Verlag, 1992.
56. R. Mohr. Projective reconstruction. *in Geometric Invariance in Computer Vision*, 1992.
57. J. Oliensis and J. Inigo-Thomas. Recursive multi-frame structure from motion incorporating motion error. *Proc. DARPA Image Understanding Workshop*, 1992.
58. C. Samson, M. Le Borgne, and B. Espiau. *Robot Control, The Task Function Approach*. Oxford Engineering science series, vol. 22, Clarendon Press, Oxford, 1991.
59. A. Shashua. Projective structure reconstruction. *AI Memo MIT AI Lab*, March 1993.
60. S. Soatto. Observability of rigid motion under perspective projection with application to visual motion estimation. *Technical Report CIT-CDS 94-001, California Institute of Technology*, 1994.
61. S. Soatto, R. Frezza, and P. Perona. Motion estimation via dynamic vision. *Submitted to the IEEE Trans. on Automatic Control*, 1994. Extended version in Technical Report CIT-CDS-94-004, California Institute of Technology.
62. S. Soatto, R. Frezza, and P. Perona. Recursive estimation of camera motion from uncalibrated image sequences. *Technical Report CIT-CDS 94-003, California Institute of Technology. In the proc. of the first IEEE conf. on Image Processing*, 1994.
63. S. Soatto, R. Frezza, and P. Perona. Recursive motion estimation on the essential manifold. In *Proc. 3rd Europ. Conf. Comput. Vision, J.-O. Eklundh (Ed.), LNCS-Series Vol. 800-801, Springer-Verlag*, pages 61–72, Stockholm, May 1994.
64. S. Soatto, R. Frezza, and P. Perona. Structure from visual motion as a nonlinear observation problem. *To appear in the Proc. of the NOLCOS Conference*, 1995.
65. S. Soatto and P. Perona. Structure-independent visual motion control on the essential manifold. In *Proc. of the IFAC Symposium on Robot Control (SYROCO), Capri, Italy*, Sept. 1994.
66. S. Soatto, P. Perona, R. Frezza, and G. Picci. Recursive motion and structure estimation with complete error characterization. In *Proc. IEEE Comput. Soc. Conf. Comput. Vision and Pattern Recogn.*, pages 428–433, New York, June 1993.
67. M. Swain and M. Stricker (editors). Promising directions in active vision. Technical Report T.R. CS 91-27, University of Chicago, November 1991. Written by the attendees of the NSF Active Vision Workshop – August 5-7 1991.
68. R. Szeliski. Recovering 3d shape and motion from image streams using nonlinear least squares. *J. visual communication and image representation*, 1994.
69. G. Taubin. Estimation of planar curves, surfaces and nonplanar space curves defined by implicit equations. *IEEE Trans. Pattern Anal. Mach. Intell.*, 1991.
70. C. J. Taylor and D. J. Kriegman. Structure and motion from line segments in multiple views. *Technical Report 9402, Yale University*, 1994.
71. C. Tomasi and T. Kanade. Shape and motion from image streams: a factorization method – 3. detection and tracking of point features. CMU-CS 91-132, School of CS – CMU, April 1991.

72. C. Tomasi and T. Kanade. Shape and motion from image streams: a factorization method – 2. point features in 3d motion. CMU-CS 91-105, School of CS – CMU, January 1991.
73. C. Tomasi and T. Kanade. Shape and motion from image streams: a factorization method – 1. planar motion. CMU-CS 90-166, School of CS – CMU, September 1990.
74. J. Weng, T. Huang, and N. Ahuja. Motion and structure from two perspective views: algorithms, error analysis and error estimation. *IEEE Trans. Pattern Anal. Mach. Intell.*, 11(5):451–476, 1989.
75. J. Weng, T.S. Huang, and N. Ahuja. Motion and structure from line correspondences: closed-form solution, uniqueness and optimization. *IEEE Trans. Pattern Anal. Mach. Intell.*, 14(3):318–336, 1992.
76. Z. Zhang and O. Faugeras. Estimation of displacement from two 3d frames obtained form stereo. *TR 1440 – INRIA*, 1991.

Numerical Algorithms for Subspace State Space System Identification *

Bart De Moor[1] *and Peter Van Overschee*[2]

[1] Associate Professor at the Department of Electrical Engineering of the Katholieke Universiteit Leuven, Kardinaal Mercierlaan 94, B-3001 Leuven, Belgium, tel: 32/ 16/ 321715, fax:32/ 16/ 321986, email: bart.demoor@esat.kuleuven.ac.be, and Senior Research Associate of the National Fund for Scientific Research (NFWO).
[2] Senior Research Assistant of the National Fund for Scientific Research (NFWO).

Abstract. We present the basic notions on subspace identification algorithms for linear systems. These methods first estimate a state sequence directly from input-output data, through an orthogonal or oblique projection of the row space of certain input-output block Hankel matrices into the row spaces of others. The state estimation procedure is then followed by a least squares problem which delivers the state space model. These algorithms can be elegantly implemented using well-known numerical linear algebra algorithms such as the LQ- and singular value decomposition.

1 Introduction

While at first sight, the class of linear time-invariant systems with lumped parameters, seems to be rather restricted, it turns out that the input-output behavior of many real-life industrial processes, for most practical purposes (such as simulation, prediction, monitoring or control system design), can be approximated very well by a linear time-invariant system. The problem of linear system identification, which is the problem of obtaining (approximate) linear system representations from measured data, is certainly not new. It has been studied now for more than 20 years in the mathematical engineering literature. Excellent books, such as [AE 71] [AW 84], [BJ 76], [Eyk 74], [Lju 87], [Nor 86], [SS 89] and several others, reveal the maturity of the field, which has culminated in the eighties into the development of so-called *prediction-error-methods* (PEM). These have been analyzed in much detail and applied with great success to many practical problems. Excellent software toolboxes, in which these algorithms (together with the subspace algorithms described in this work) have been implemented, such as the Matlab System Identification Toolbox [Lju 91c],

---

* The research reported on here was supported by Belgian Federal Government Interuniversity Attraction Poles IUAP-17 (Modeling and Control of Dynamic Systems), IUAP-50 (Automation in Design and Production), Flemish Government Concerted Action GOA-MIPS (Model-based Information Processing Systems) and the European Commission Human Capital and Mobility SIMONET (System Identification and Modeling Network).

or the Xmath Interactive System Identification Module [VODMAKB 94], are nowadays commercially available.

Yet, it can not be denied that with these (by now) classical PEM approaches, there are still some problems that are difficult to deal with in practical situations. It so turns out that a new breed of algorithms, called *subspace algorithms*, which have been developed the last couple of years, provide a meaningful alternative for PEM, with respect to the following problems:

Parametrization problems: In PEM the transfer matrix models are parametrized with a minimal number of parameters. This parametrization issue has triggered lots of research (e.g. [GW 74] [Gui 75] [Gui 81] [HD 88] [Kai 80] [Lue 67] [VOL 82]) or for so-called *overlapping* parametrizations (see e.g. [GeWe 82] [HD 88] and the references in these papers and books). However, determining the right parametrization (especially in the case of multi-output systems) starting from observed measured input-output data is certainly not trivial in practice!

Numerical robustness: Even when a parametrization has been fixed, the resulting model structure can be ill-conditioned, meaning that the identified parameters are extremely sensitive to perturbations. In subspace methods, the models are full state space models ('fully parametrized'), typically within a certain state space basis with good properties, such as e.g. the balanced realization, which guarantees insensitivity with respect to perturbations. The only variable to be decided upon is the number of states (the 'order' of the system), a choice which is guided by the singular values of certain matrices and/or trial-and-error based. This automatically leads to more user-friendly algorithmic implementations.

Algorithms: For almost all parametrizations, the identification problem translates into a *nonlinear numerical optimization problem*, with all its unpleasant side effects such as for instance convergence problems and local minima. Subspace algorithms are typically faster (when implemented correctly!) than Prediction Error Methods, because they are basically *non-iterative*, exploiting basic algorithms from numerical linear algebra, such as the LQ-decomposition (transpose of the QR-decomposition) and the singular value decomposition.

Nonzero initial conditions: In parametrized input-output models, the presence of a nonzero initial state requires additional terms in the parametrization. In subspace methods, the initial state is estimated (implicitly or explicitly). More specifically, subspace methods will also model those modes of the system that are observable but not necessarily controllable, including *unstable* ones.

This paper should be considered as a *first introduction* to subspace methods for system identification. We do not aim here for completeness nor complete mathematical rigor nor exhaustive comparison with other methods and algorithms that have appeared in the literature. Nor will we attempt to provide a complete list of variations on the themes proposed in this paper. We simply want to elaborate on the main insights that have been influential in developing subspace methods for linear system identification during the last couple of years.

This paper is organized as follows: In Section 2, we briefly describe the state space model that will be identified from input-output data. In Section 3, we comment on one of the main ideas behind subspace methods, which is to recover (implicitly or explicitly), the state directly from input-output data. It will be explained how, first of all, this is possible by using orthogonal or oblique projections using block Hankel matrices with data, and second, how first obtaining the state sequence *linearizes* the problem, and allows to obtain the state space model from a simple least squares solution. In Section 4, we introduce (past and future) block Hankel matrices with input-output data and also define the notations to be used throughout. In Section 5, we treat the geometrical operations to be used such as orthogonal and oblique projections, involving the data matrices. In Section 6, we explore a relationship between a certain orthogonal projection of 'future outputs' into the 'past inputs and outputs' and 'future inputs', and the state sequence as it would be generated by a bank of Kalman filters if the state space model and its initial conditions would be completely known. In Section 7, we explore the relationship between a certain oblique projection of 'future outputs', along 'future inputs', into 'past inputs and outputs' and the state sequence as it would be generated by a bank of Kalman filters, which is the same as the one from Section 6, but with different initial conditions. Here we show how the oblique projection contains important information about the model to be identified, such as its order and its observability matrix. In Section 8, we derive how the state space model can be found by solving a least squares problem using the states obtained in Section 6. In Section 9, we derive another algorithm which is more elegant in a certain way, but biased. It uses the state sequence obtained in Section 7. A robustified implementation and the role of the LQ-decomposition is discussed in Section 10. Conclusions and references to extensions and related work are summarized in Section 11.

2 State space models and problem formulation

We will restrict ourselves to discrete time, linear, time-invariant, state space models, which are described as[3]:

$$x_{k+1} = Ax_k + Bu_k + w_k, \qquad (1)$$
$$y_k = Cx_k + Du_k + v_k, \qquad (2)$$

with

$$\mathbf{E}[\begin{pmatrix} w_p \\ v_p \end{pmatrix} \begin{pmatrix} w_q^t & v_q^t \end{pmatrix}] = \begin{pmatrix} Q & S \\ S^T & R \end{pmatrix} \delta_{pq} \geq 0. \qquad (3)$$

The vectors $u_k \in \mathbb{R}^{m \times 1}$ and $y_k \in \mathbb{R}^{l \times 1}$ are the measurements at time instant k of respectively the m inputs and l outputs of the process. The vector x_k is the state vector of the process at discrete time instant k, $v_k \in \mathbb{R}^{l \times 1}$ and $w_k \in \mathbb{R}^{n \times 1}$

---
[3] **E** denotes the expected value operator and δ_{pq} the Kronecker delta.

are unobserved vector signals. v_k is called the measurement noise and w_k is called the process noise. It is assumed that they are zero mean, stationary white noise vector sequences. $A \in \mathbb{R}^{n \times n}$ is the system matrix, $B \in \mathbb{R}^{n \times m}$ is the input matrix, $C \in \mathbb{R}^{l \times n}$ is the output matrix while $D \in \mathbb{R}^{l \times m}$ is the direct feedthrough matrix. The matrices $Q \in \mathbb{R}^{n \times n}$, $S \in \mathbb{R}^{n \times l}$ and $R \in \mathbb{R}^{l \times l}$ are the covariance matrices of the noise sequences w_k and v_k. As explained in e.g. [Fau 76], the noise model is not unique: there is a set of 'equivalent' noise models that can be described using linear matrix inequalities or equivalently, a pair of algebraic Riccati equations (backward and forward)(see [Cai 88] [Fau 76] [VODM 93a] for details). In addition, we require the following orthogonality conditions to be true:

$$\mathbf{E}[\begin{pmatrix} x_k \\ u_k \end{pmatrix} (w_k^T \ v_k^T)] = 0 \ . \tag{4}$$

We are now ready to state the main problem treated here:

> Given a large number of measurements of the input $u_k \in \mathbb{R}^m$ and the output $y_k \in \mathbb{R}^l$ generated by the unknown system (1)-(2)-(3). Determine, using only the inputs u_k and outputs y_k, the order n of the unknown system, the system matrices $A \in \mathbb{R}^{n \times n}, B \in \mathbb{R}^{n \times m}, C \in \mathbb{R}^{l \times n}, D \in \mathbb{R}^{l \times m}$ up to within a similarity transformation and the matrices $Q \in \mathbb{R}^{n \times n}, S \in \mathbb{R}^{n \times l}, R \in \mathbb{R}^{l \times l}$ (or equivalent realizations in the sense of Faurre).

There are many questions that pop up. For instance: What is meant by a large number of measurements? (We'll show that we need an infinite number for consistency). Which noise model is being identified, as there are an infinite number of 'equivalent' ones? etc ....

3 The main idea: Re-discovering the state

One of the important conceptual ideas behind subspace algorithms is to re-introduce the concept of *the state of a dynamical system* within the system identification context. In contrast to 'classical' identification algorithms, subspace algorithms first estimate/calculate the state (sequence) (implicitly or explicitly), while next the (state space) model is determined. This difference between PEM methods and subspace methods is illustrated in Figure 1.

Why would one bother to first obtain the (Kalman filter) state sequence, directly from input-output data and only after that the state space model? The answer is threefold:

- When the state sequence x_k is available and since the inputs u_k and outputs y_k are available as well, it can be seen from the state space model (1)-(2) and the orthogonality conditions (4) that the state space matrices A, B, C and D could be obtained by solving a least squares problem. The covariance matrices Q, R and S could then be estimated as the covariance matrices

3 The main idea: Re-discovering the state

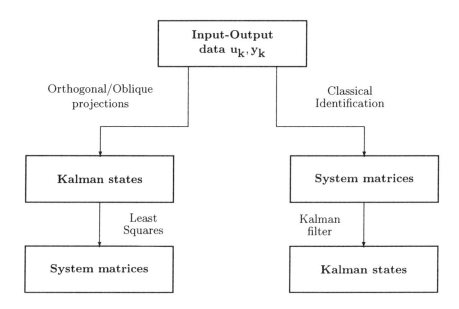

Fig. 1. System identification aims at constructing state space models from input-output data. The left hand side shows the subspace method approach: first the (Kalman filter) states are estimated directly from input-output data, then the system matrices can be obtained. The right hand side is the classical approach: first obtain the system matrices, then estimate the states.

of the least squares residuals. Therefore, it seems worthwhile to first try to obtain the state from input and output measurements, because in doing so, the identification problem is *linearized* in the sense that it is reduced to a simple least squares problem!

- Remains the problem of how to obtain the state sequence, which will be the main subject of Sections 6 and 7 below. The main conclusion will be that a certain Kalman filter state sequence can be calculated directly from input-output data, by doing one or two orthogonal or oblique projections with block Hankel matrices constructed from input-output data.

- It turns out that the state sequence can be calculated using some well-known tools from numerical linear algebra, such as the LQ-decomposition and the singular value decomposition, which ultimately leads to a numerically extremely efficient implementation.

4 The linear algebra tools: Block Hankel matrices with data

Block Hankel matrices with inputs and outputs are the basic linear algebra objects for subspace identification algorithms. Input block Hankel matrices are defined as:

$$U_{0|i-1} \stackrel{\text{def}}{=} \begin{pmatrix} u_0 & u_1 & u_2 & \ldots & u_{j-1} \\ u_1 & u_2 & u_3 & \ldots & u_j \\ \ldots & \ldots & \ldots & \ldots & \ldots \\ u_{i-1} & u_i & u_{i+1} & \ldots & u_{i+j-2} \end{pmatrix} \in \mathbb{R}^{mi \times j} .$$

The number of block rows (i) is a user-defined index which is large enough i.e. it should at least be larger than the maximum order of the system one wants to identify[4]. Most of the times, we will assume that $j \to \infty$ (which means in practice that j should be 'large'). Summarizing, we require that $n < i \ll j \to \infty$. We will also use the following input matrices:

$$U_p = U_{0|i-1} \ , \ U_f = U_{i|2i-1} \ , \ U_p^+ = U_{0|i} \ , \ U_f^- = U_{i+1|2i-1} \ .$$

Here, the subscript 'p' refers to 'past', 'f' refers to future and the superscripts '+' and '-' refer to adding or removing one more block row. Similar definitions hold for the block Hankel matrices with the output vectors, which will be denoted by Y_p, Y_f, Y_p^+ and Y_f^-. 'Double' block Hankel matrices with inputs and outputs, are defined as:

$$W_{0|i-1} = \begin{pmatrix} U_p \\ Y_p \end{pmatrix} = W_p \ , \ W_{0|i} = \begin{pmatrix} U_p^+ \\ Y_p^+ \end{pmatrix} = W_p^+ \ .$$

Subspace identification algorithms make extensive use of observability and controllability matrices and of their structure. The extended ($i > n$) observability matrix Γ_i (where the subscript i denotes the number of block rows) is defined as:

$$\Gamma_i \stackrel{\text{def}}{=} \begin{pmatrix} C \\ CA \\ CA^2 \\ \ldots \\ CA^{i-1} \end{pmatrix} \in \mathbb{R}^{li \times n} .$$

We assume the pair $\{A, C\}$ to be observable, which implies (see for instance [Kai 80]) that the rank of Γ_i is equal to n. The reversed extended controllability matrix Δ_i^d (where the subscript i denotes the number of block columns) is defined as:

$$\Delta_i^d \stackrel{\text{def}}{=} \begin{pmatrix} A^{i-1}B & A^{i-2}B & \ldots & AB & B \end{pmatrix} \in \mathbb{R}^{n \times mi} .$$

---

[4] Theoretically, i should be at least as large as the largest observability index of the observability matrix.

4 The linear algebra tools: Block Hankel matrices with data

Furthermore, we require the pair $\{A, [B\ ,\ Q^{1/2}]\}$ to be controllable. This implies that all modes are excited by either the external input u_k or the process noise w_k. The lower block triangular Toeplitz matrix H_i^d is defined as:

$$H_i^d \stackrel{\text{def}}{=} \begin{pmatrix} D & 0 & 0 & \ldots & 0 \\ CB & D & 0 & \ldots & 0 \\ CAB & CB & D & \ldots & 0 \\ \ldots & \ldots & \ldots & \ldots & \ldots \\ CA^{i-2}B & CA^{i-3}B & CA^{i-4}B & \ldots & D \end{pmatrix} \in \mathbb{R}^{li \times mi}.$$

The system (1)-(2) is split in a deterministic and a stochastic subsystem, by splitting the state (x_k) and output (y_k) in a deterministic ($\bullet^d$) and stochastic ($\bullet^s$) component:

$$x_k = x_k^d + x_k^s\ ,\ y_k = y_k^d + y_k^s\ .$$

The deterministic state (x_k^d) and output (y_k^d) follow from the deterministic subsystem, which describes the influence of the deterministic input (u_k) on the deterministic output while the stochastic state (x_k^s) and output (y_k^s) follow from the stochastic subsystem, which describes the influence of the noise sequences (w_k and v_k) on the stochastic output:

$$\begin{array}{ll} x_{k+1}^d = Ax_k^d + Bu_k\ , & x_{k+1}^s = Ax_k^s + w_k\ , \\ y_k^d = Cx_k^d + Du_k\ , & y_k^s = Cx_k^s + v_k\ . \end{array} \quad (5)$$

The controllable modes of $\{A, B\}$ can be either stable or unstable. The controllable modes of $\{A, Q^{1/2}\}$ are assumed to be stable. The deterministic inputs (u_k) and states (x_k^d) and the stochastic states (x_k^s) and outputs (y_k^s) are assumed to be quasi-stationary (as defined in [Lju 87, page 27]). Note that even though the deterministic subsystem can have unstable modes, the excitation (u_k) has to be chosen in such a way that the deterministic states and output are finite for all time. Also note that since the systems $\{A, B\}$ and $\{A, Q^{1/2}\}$ are not assumed to be controllable (only the concatenation of the deterministic and stochastic subsystem as a whole should be controllable), the deterministic and stochastic subsystem may have common as well as completely decoupled input-output dynamics. The state sequence is defined as:

$$X_i \stackrel{\text{def}}{=} \begin{pmatrix} x_i & x_{i+1} & \ldots & x_{i+j-2} & x_{i+j-1} \end{pmatrix} \in \mathbb{R}^{n \times j}.$$

The *deterministic* state sequence X_i^d and *stochastic* state sequence X_i^s are defined as:

$$X_i^d \stackrel{\text{def}}{=} \begin{pmatrix} x_i^d & x_{i+1}^d & \ldots & x_{i+j-2}^d & x_{i+j-1}^d \end{pmatrix} \in \mathbb{R}^{n \times j},$$

$$X_i^s \stackrel{\text{def}}{=} \begin{pmatrix} x_i^s & x_{i+1}^s & \ldots & x_{i+j-2}^s & x_{i+j-1}^s \end{pmatrix} \in \mathbb{R}^{n \times j}.$$

In a similar way, the past and future deterministic and stochastic state sequences are defined as $X_p^d = X_0^d$, $X_f^d = X_i^d$, , $X_p^s = X_0^s$, $X_f^s = X_i^s$. These are not the key state sequences for combined system identification. The introduction of the

more important Kalman filter state sequence is postponed to Section 6.
We also define the following covariance and cross covariance matrices (which exist due to the quasi stationarity of u_k and x_k^d, see above) [5]:

$$R^{uu} \stackrel{\text{def}}{=} \Phi_{[U_{0|2i-1}, U_{0|2i-1}]} = \begin{pmatrix} \Phi_{[U_p, U_p]} & \Phi_{[U_p, U_f]} \\ \Phi_{[U_f, U_p]} & \Phi_{[U_f, U_f]} \end{pmatrix} = \begin{pmatrix} R_p^{uu} & R_{pf}^{uu} \\ (R_{pf}^{uu})^T & R_f^{uu} \end{pmatrix},$$

$$S^{xu} \stackrel{\text{def}}{=} \Phi_{[X_0^d, U_{0|2i-1}]} = \begin{pmatrix} \Phi_{[X_p^d, U_p]} & \Phi_{[X_p^d, U_f]} \end{pmatrix} = \begin{pmatrix} S_p^{xu} & S_f^{xu} \end{pmatrix},$$

$$\Sigma^d \stackrel{\text{def}}{=} \Phi_{[X_p^d, X_p^d]}.$$

For the stochastic subsystem (5) we use the following notations. Defining the output covariance matrices as

$$\Lambda_i \stackrel{\text{def}}{=} \mathbf{E}[y_{k+i} y_k^T],$$

we find for

$$\Lambda_0 = \mathbf{E}[y_k . y_k^T] = \mathbf{E}[(Cx_k^s + v_k)(Cx_k^s + v_k)^T]$$
$$= C\mathbf{E}[x_k^s (x_k^s)^T]C^T + \mathbf{E}[v_k v_k^T] = C\Sigma^s C^T + R,$$

with an obvious definition for Σ^s. Defining

$$G \stackrel{\text{def}}{=} \mathbf{E}[x_{k+1}^s . y_k^T] = \mathbf{E}[(Ax_k^s + w_k).(Cx_k^s + v_k)^T]$$
$$= A\mathbf{E}[x_k^s (x_k^s)^T]C^T + \mathbf{E}[w_k v_k^T] = A\Sigma^s C^T + S,$$

we get (for $i = 1, 2, \ldots$)

$$\Lambda_i = CA^{i-1}G \quad \text{and} \quad \Lambda_{-i} = G^T (A^{i-1})^T C^T.$$

The reversed extended stochastic controllability matrix Δ_i^c (where the subscript i denotes the number of block columns and the superscript "c" stands for "covariance") is defined as

$$\Delta_i^c \stackrel{\text{def}}{=} \begin{pmatrix} A^{i-1}G & A^{i-2}G & \ldots & AG & G \end{pmatrix} \quad \in \mathbb{R}^{n \times li}.$$

---

[5] In subspace identification we typically assume that there are long time series of data available ($j \to \infty$), and that the data is ergodic. Due to ergodicity and the infinite number of data at our disposition, we can replace the expectation operator $\mathbf{E}$ (average over an infinite number of experiments) with the operator $\mathbf{E_j}$ applied to the sum of variables (average over one, infinitely long, experiment). For instance for the correlation $\mathbf{E}(ae^T)$ between two random variables a and e, on which we have j observations, a_k and e_k, $k = 1, \ldots, j$, we get $\mathbf{E}[ae^T] = \lim_{j\to\infty} [\frac{1}{j} \sum_{i=1}^j a_i e_i^T] = \mathbf{E_j}[\sum_{i=1}^j a_i e_i^T]$ with an obvious definition of $\mathbf{E_j}$: $\mathbf{E_j}[\bullet] \stackrel{\text{def}}{=} \lim_{j\to\infty} \frac{1}{j}[\bullet]$. We define the covariance $\Phi_{[A,B]}$ between two matrices $A \in \mathbb{R}^{p \times j}$ and $B \in \mathbb{R}^{q \times j}$ as $\Phi_{[A,B]} \stackrel{\text{def}}{=} \mathbf{E_j}[AB^T]$.

The block Toeplitz matrix L_i is constructed from the output covariance matrices as:

$$L_i \stackrel{\text{def}}{=} \begin{pmatrix} \Lambda_0 & \Lambda_{-1} & \Lambda_{-2} & \ldots & \Lambda_{1-i} \\ \Lambda_1 & \Lambda_0 & \Lambda_{-1} & \ldots & \Lambda_{2-i} \\ \Lambda_2 & \Lambda_1 & \Lambda_0 & \ldots & \Lambda_{3-i} \\ \ldots & \ldots & \ldots & \ldots & \ldots \\ \Lambda_{i-1} & \Lambda_{i-2} & \Lambda_{i-3} & \ldots & \Lambda_0 \end{pmatrix} \in \mathbb{R}^{li \times li} . \qquad (6)$$

5 The geometric tools: Orthogonal and Oblique Projections

In what follows, we will use the matrices $A \in \mathbb{R}^{p \times j}, B \in \mathbb{R}^{q \times j}$ and $C \in \mathbb{R}^{r \times j}$ as dummy variables, meaning that they are only 'local' variables in this Section, not to be confused with the system matrices A, B and C we have defined before. The elements of a row of one of the given matrices can be considered as the coordinates of a vector in the j-dimensional ambient space. The rows of each matrix A, B, C thus define a basis for a vector space in this ambient space. In Subsection 5.1 and 5.2 we define orthogonal and oblique projections, in which these row spaces are involved. It should be noted that these geometric operations can be easily implemented using an LQ decomposition, which is the subject of Section 10.

5.1 Orthogonal projections

Π_B denotes the operator that projects the row space of a matrix into the row space of the matrix $B \in \mathbb{R}^{q \times j}$ (where $\bullet^\dagger$ denotes the Moore-Penrose pseudo-inverse [6]) so that $\Pi_B \stackrel{\text{def}}{=} B^T.(BB^T)^\dagger.B$. A/B is shorthand for the projection of the row space of the matrix $A \in \mathbb{R}^{p \times j}$ into the row space of the matrix B so that $A/B \stackrel{\text{def}}{=} A.\Pi_B = AB^T.(BB^T)^\dagger.B$. The projection operator can be interpreted in the ambient j-dimensional space as indicated in Figure 2. Note that in the notation A/B the matrix B is printed bold face, which indicates that the result of the operation A/B lies in the row space of B.

$\Pi_{B^\perp}$ is the geometric operator that projects the row space of a matrix into the orthogonal complement of the row space of the matrix B, for which we have $A/B^\perp \stackrel{\text{def}}{=} A\Pi_{B^\perp}$, where $\Pi_{B^\perp} = I_j - \Pi_B$. Here $B^\perp$ is a matrix, the row space of which is the orthogonal complement to the row space of B. Once again these projections can be interpreted in the j-dimensional space as indicated in Figure 2. The combination of the projections Π_B and $\Pi_{B^\perp}$ decomposes a matrix A into two matrices for which the row spaces are orthogonal to each other as $A = A\Pi_B + A\Pi_{B^\perp}$. The projections also decompose the matrix A as a linear combination of the rows of B and those of $B^\perp$. With $L_B.B \stackrel{\text{def}}{=}$

---

[6] This generalized inverse could be replaced by less 'restricted' generalized inverses, but this will not be pursued here.

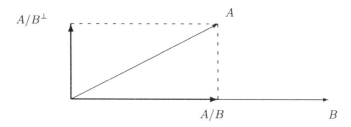

Fig. 2. Interpretation of the orthogonal projection in the j-dimensional space ($j = 2$ in this case). A/B is formed by projecting the row space of A orthogonally into the row space of B. $A/B^\perp$ on the other hand is formed by projecting the row space of A into the orthogonal complement of the row space of B.

$A\Pi_B$ and $L_{B^\perp} \cdot B^\perp \stackrel{\text{def}}{=} A\Pi_{B^\perp}$ we find $A = L_B \cdot B + L_{B^\perp} \cdot B^\perp$, which is indeed a decomposition of A as a linear combination of the rows of B and those of $B^\perp$.

5.2 Oblique projections

A matrix A can also be decomposed as a linear combination of two matrices B and C with *non-intersecting row spaces* and another matrix of which the row space is the orthogonal complement of the row spaces of B and C. This is illustrated in Figure 3. The matrix A is decomposed as a linear combination of the rows of B and C and of the rows of a third matrix orthogonal to B and C. This can be written as $A = L_B \cdot B + L_C \cdot C + L_{B^\perp, C^\perp} \begin{pmatrix} B \\ C \end{pmatrix}^\perp$. The row space of the matrix $L_C \cdot C$ is defined as the oblique projection of the row space of A along the row space of B in the row space of C as $A/_B C = L_C \cdot C$. The name *oblique* refers to the non-orthogonal projection direction. The oblique projection can also be interpreted through the following recipe: Project the row space of A orthogonally into the joint row space of B and C; and decompose the result along the row space of C. Mathematically, the orthogonal projection of the row space of A into the joint row space of B and C can be stated as:

$$A / \begin{pmatrix} C \\ B \end{pmatrix} = A \begin{pmatrix} C^T & B^T \end{pmatrix} \cdot \begin{pmatrix} CC^T & CB^T \\ BC^T & BB^T \end{pmatrix}^\dagger \cdot \begin{pmatrix} C \\ B \end{pmatrix}.$$

The oblique projection of the row space of $A \in \mathbb{R}^{p \times j}$ along the row space of $B \in \mathbb{R}^{q \times j}$ into the row space of $C \in \mathbb{R}^{r \times j}$ is then obtained as

$$A/_B C \stackrel{\text{def}}{=} A \begin{pmatrix} C^T & B^T \end{pmatrix} \cdot \left[\begin{pmatrix} CC^T & CB^T \\ BC^T & BB^T \end{pmatrix}^\dagger \right]_{\text{first } r \text{ columns}} \cdot C.$$

Another expression for the oblique projection is given by

$$A/_B C = [A/B^\perp] \cdot [C/B^\perp]^\dagger \cdot C. \tag{7}$$

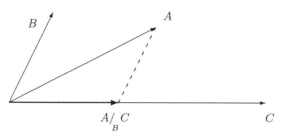

Fig. 3. Interpretation of the oblique projection in the j-dimensional space ($j = 2$ in this case). The oblique projection is formed by projecting the row space of A along the row space of B into the row space of C.

6 Kalman filter states and orthogonal projections

In this section, we first show that the Kalman filter state estimates can be obtained from certain linear combinations of the rows of the input-output block Hankel matrices. Next we show how this state sequence also shows up in the orthogonal projection of the row space of the future outputs into the row spaces of the past outputs and past and future inputs.

6.1 Explicit expressions for the Kalman filter state estimates

In the derivation of the subspace identification algorithms for combined deterministic - stochastic system identification, the Kalman filter plays a crucial role. In this Subsection, we introduce a closed form equation for the Kalman filter state estimate for the combined system. We also introduce a bank of Kalman filters generating a sequence of state estimates. We indicate the Kalman filter state estimate by a hat: $\hat{x}_k$. For the notations, we refer to Section 4.

Theorem 1. Kalman filter and explicit expressions for the state estimate

Let the following be given: An initial state estimate: $\hat{x}_0 = \bar{x}_0$, an initial estimate of the matrix: $P_0 = -\mathbf{E}[\bar{x}_0 . \bar{x}_0^T]$, the input and output measurements $u_0, y_0, \ldots, u_{k-1}, y_{k-1}$; Then the state estimate $\hat{x}_k$ from the Kalman filter, defined by the following recursive formulas:

$$\hat{x}_k = A\hat{x}_{k-1} + Bu_k + K_{k-1}(y_{k-1} - C\hat{x}_{k-1} - Du_k), \quad (8)$$

$$K_{k-1} = (G - AP_{k-1}C^T)(\Lambda_0 - CP_{k-1}C^T)^{-1}, \quad (9)$$

$$P_k = AP_{k-1}A^T + (G - AP_{k-1}C^T) \\ \times (\Lambda_0 - CP_{k-1}C^T)^{-1}(G - AP_{k-1}C^T)^T, \quad (10)$$

can be explicitly written as:

$$\hat{x}_k = \left(A^k - \Omega_k \Gamma_k | \Delta_k^d - \Omega_k H_k^d | \Omega_k \right) \begin{pmatrix} \overline{x}_0 \\ u_0 \\ \cdots \\ u_{k-1} \\ y_0 \\ \cdots \\ y_{k-1} \end{pmatrix}, \quad (11)$$

where:

$$\Omega_k \stackrel{def}{=} (\Delta_k^c - A^k P_0 \Gamma_k^T)(L_k - \Gamma_k P_0 \Gamma_k^T)^{-1}. \quad (12)$$

The explicit solution of the matrix P_k is equal to:

$$P_k = A^k P_0 (A^T)^k + (\Delta_k^c - A^k P_0 \Gamma_k^T)(L_k - \Gamma_k P_0 \Gamma_k^T)^{-1}(\Delta_k^c - A^k P_0 \Gamma_k^T)^T. \quad (13)$$

A proof can be found in [VODM 94a].
The covariance matrix of the state error $\widetilde{P}_k$ is given by:

$$\widetilde{P}_k = \mathbf{E}[\,(x_k - \hat{x}_k).(x_k - \hat{x}_k)^T\,] = \Sigma^s - P_k.$$

From this we conclude that the covariance of the initial error on the state estimate is given by $\widetilde{P}_0 = \Sigma^s - P_0$. This indirectly implies that P_0 should be negative definite (or smaller than Σ^s at least).

> The significance of Theorem 1 is that it indicates how the Kalman filter state estimate $\hat{x}_k$ can be written as a linear combination of the past inputs and output measurements $u_0, y_0, \ldots, u_{k-1}, y_{k-1}$ and of the initial state estimate $\overline{x}_0$ if the system matrices A, B, C, D, Q, S and R were known, which is not the case if only input-output data u_k and y_k are available.

We now define the state sequence

$$\widehat{X}_i = \left(\hat{x}_i \ \hat{x}_{i+1} \ \ldots \ \hat{x}_{i+j-1} \right)$$

$$= \left(A^k - \Omega_k \Gamma_k | \Delta_k^d - \Omega_k H_k^d | \Omega_k \right) \begin{pmatrix} \overline{X}_0 \\ U_p \\ Y_p \end{pmatrix}$$

$$= \left(A^k - \Omega_k \Gamma_k | (\Delta_k^d - \Omega_k H_k^d \ \Omega_k) \right) \begin{pmatrix} \overline{X}_0 \\ W_p \end{pmatrix}. \quad (14)$$

with $\overline{X}_0$ the sequence of initial states. This state sequence is generated by a bank of Kalman filters, working in parallel on each of the columns of the block Hankel matrix of past inputs and outputs, which is illustrated in Figure 4. The Kalman filters run in a *vertical* direction (over the columns). The state estimate is obtained from *partial* input-output information. Each vector in the sequence

6 Kalman filter states and orthogonal projections

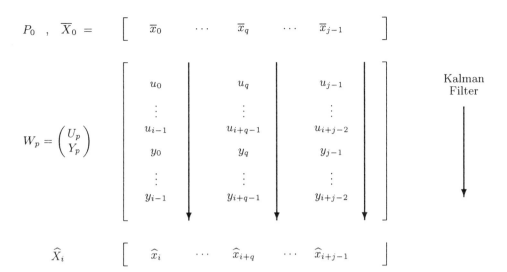

Fig. 4. Interpretation of the sequence $\widehat{X}_i$ as a sequence of Kalman filter state estimates based upon i measurements of u_k and y_k. When the system matrices A, B, C, D, Q, R, S would be known, the state $\widehat{x}_{i+q}$ could be determined from a Kalman filter as follows: Start the filter at time q, with an initial state estimate $\bar{x}_q$ and initial error covariance matrix $\Sigma^s - P_0$. Now iterate the Kalman filter over i time steps (the vertical arrow down). The Kalman filter will then return a state estimate $\widehat{x}_{i+q}$. This procedure could be repeated for each of the j columns, and thus we speak about a *bank* of Kalman filters. The major observation in subspace algorithms is that the system matrices A, B, C, D, Q, R, S do not have to be known to determine the state sequence $\widehat{X}_i$. It can be determined directly from input-output data. It is very important to realize that there is NO direct relation between the column vectors in $\widehat{X}_i$ (unless $i \to \infty$). The Kalman filter interpretation only 'works' in the vertical direction, for each column separately, and NOT in the horizontal direction, unless $i \to \infty$!

$\widehat{X}_i$ depends on its corresponding initial state in the sequence $\overline{X}_0$ and the initial covariance matrix P_0. In what follows we will encounter different Kalman filter sequences (in the sense of different initial states $\overline{X}_0$ and initial matrices P_0). Therefore we will denote the Kalman filter state sequence with initial state $\overline{X}_0$ and initial covariance matrix P_0 by $\widehat{X}_{i[\overline{X}_0, P_0]}$.

6.2 Kalman filter states and orthogonal projections

In this Subsection, we introduce the projection of the future outputs onto the past and future inputs and the past outputs. Through this projection, the Kalman filter states can be determined directly from the data. In the following Theorem, we show that there is a nice mathematical expression for the orthogonal projection of the row space of the future outputs Y_f onto the row space generated by the past and future inputs and the past outputs. This orthogonal projection is illustrated in Figure 5. Before we proceed, we need the following

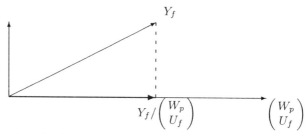

Fig. 5. The orthogonal projection of the row space of the future outputs into the row spaces of past outputs and past and future inputs.

technical definition [Lju 87]:

Definition 2. Persistency of excitation
The input sequence $u_k \in \mathbb{R}^m$ is persistently exciting of order $2i$ if the input covariance matrix $R^{uu} \stackrel{\text{def}}{=} \frac{1}{j}[U_{0|2i-1} U^T_{0|2i-1}]$ is of full row rank, which is $2mi$.

It is tedious though straightforward to prove the following Theorem which extracts Kalman filter state sequences directly from the input-output data.

Theorem 3. Kalman filter and orthogonal projection
Under the assumptions that:

1. *The deterministic input u_k is uncorrelated with the process noise w_k and the measurement noise v_k (see equation (4));*
2. *The input u_k is persistently exciting of order $2i$ (Definition 2) ;*
3. *The number of available data is large: $j \to \infty$;*
4. *The process noise w_k and the measurement noise v_k are not identically zero ;*

Then:

$$Z_i \stackrel{\text{def}}{=} Y_f / \begin{pmatrix} W_p \\ U_f \end{pmatrix} = \Gamma_i \widehat{X}_i + H_i^d U_f , \qquad (15)$$

with:

$$\widehat{X}_i \stackrel{\text{def}}{=} \widehat{X}_{i[\overline{X}_0, P_0]} , \qquad (16)$$

$$\overline{X}_0 = S^{xu}.(R^{uu})^{-1}.\begin{pmatrix} U_p \\ U_f \end{pmatrix} , \qquad (17)$$

$$P_0 = -[\Sigma^d - S^{xu}.(R^{uu})^{-1}.(S^{xu})^T] . \qquad (18)$$

A proof of this Theorem can be found in [VODM 94a].

6 Kalman filter states and orthogonal projections

> The importance of Theorem 3 is that it reveals one way in which the Kalman filter state sequence $\hat{X}_i$ (14) obtained from Theorem 1 relates directly to given input-output data. The projection matrix $\mathcal{Z}_i$ can be computed from the given data, without knowing the system matrices. $\mathcal{Z}_i$ could be considered as an optimal (least squares) prediction of the future output data $Y_{i|2i-1}$, given the past input and output data $U_{0|i-1}$ and $Y_{0|i-1}$ *and* the future input data $U_{i|2i-1}$ (see also Section 7.2).

Examining the formulas for the initial state sequence $\overline{X}_0$ (17) and the initial matrix P_0 (18) a little closer, leads to the following interesting formulas:

$$\overline{X}_0 = X_p^d / \begin{pmatrix} U_p \\ U_f \end{pmatrix}, \tag{19}$$

$$P_0 = -\Phi_{\left[X_p^d / \begin{pmatrix} U_p \\ U_f \end{pmatrix}^\perp , X_p^d / \begin{pmatrix} U_p \\ U_f \end{pmatrix}^\perp \right]}. \tag{20}$$

First note that the "real" initial state would be $X_p^d + X_p^s$. The stochastic part X_p^s is impossible to estimate and is thus set to zero. This explains why X_0^s does not appear in (19). The deterministic part X_p^d of the *real* initial state however enters (19). Note that the row space of $\overline{X}_0$ is a subspace of the combined row spaces of W_p and U_f (since $\mathcal{Z}_i$ is constructed by a projection on the combined row space). From (19) we can see that $\overline{X}_0$ is the best estimate of X_p^d lying in the row space of past and future *inputs*.

From formula (20) we find that the covariance $\tilde{P}_0$ of the *error* on the initial state estimate is given by:

$$\begin{aligned} \tilde{P}_0 &= \Sigma^s - P_0 \\ &= \Sigma^s + \Phi_{\left[X_p^d / \begin{pmatrix} U_p \\ U_f \end{pmatrix}^\perp , X_p^d / \begin{pmatrix} U_p \\ U_f \end{pmatrix}^\perp \right]} \\ &= \Phi_{[(X_p^s + X_p^d) - \overline{X}_0, (X_p^s + X_p^d) - \overline{X}_0]}, \end{aligned}$$

which is positive definite. The last equation indicates that the error on the initial state is given by the variance of the part of the *real* initial state $X_p^d + X_p^s$ that does not lie in the combined row space of U_p and U_f. Both equations (17) and (18) can thus be explained intuitively. The initial state estimate is the best estimate of the *real* initial state, lying in the combined row space of U_p and U_f. The initial state error covariance is the covariance of the difference between the *real* initial state and the estimated initial state $\overline{X}_0$. Finally note that when the inputs u_k are white noise the initial state $\overline{X}_0 = 0$, since in this case, there is no correlation between the *real* initial state $X_p^d + X_p^s$ and the inputs U_p and U_f.

7 Kalman filter states and oblique projections

In this Section, we show how an oblique projection with input-output block Hankel matrices forms one of the key observations in subspace system identification algorithms. The oblique projection which is of central interest here, is the projection of the row space of the future outputs Y_f, along the future input row space U_f into the past inputs and outputs row space W_p. This projection is illustrated in Figure 6.

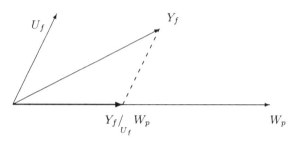

Fig. 6. Interpretation of the oblique projection in the j-dimensional space. The oblique projection is formed by projecting the row space of Y_f along the row space of U_f onto the row space of W_p.

7.1 The main theorem

The next Theorem allows to calculate a Kalman filter state sequence together with the column space of the extended observability matrix Γ_i directly from the input-output data, without any knowledge about the system matrices. The system matrices can then afterwards be extracted from the state sequence $\widehat{X}_i$ or from Γ_i. An overview of the general combined identification procedure is presented in Figure 7.

In what follows, $W_1 \in \mathbb{R}^{li \times li}$ and $W_2 \in \mathbb{R}^{j \times j}$ are user-defined weighting matrices, to which we will come back later.

Theorem 4. Kalman filter states and oblique projections
Under the assumptions that:

1. The deterministic input u_k is uncorrelated with the process noise w_k and measurement noise v_k (see (4)) ;
2. The input u_k is persistently exciting of order $2i$ (Definition 2) ;
3. The number of available data is large, so that $j \to \infty$;
4. The process noise w_k and the measurement noise v_k are not identically zero ;
5. The user-defined weighting matrices $W_1 \in \mathbb{R}^{li \times li}$ and $W_2 \in \mathbb{R}^{j \times j}$ are such that W_1 is of full rank and W_2 obeys (where W_p is the block Hankel matrix containing the past inputs and outputs): $\text{rank}(W_p) = \text{rank}(W_p.W_2)$;

7 Kalman filter states and oblique projections

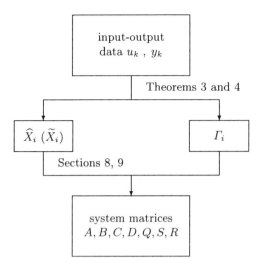

Fig. 7. An overview of the combined deterministic-stochastic subspace identification procedure. Through the Theorems 3 and 4 the state sequence $\widehat{X}_i$ ($\widetilde{X}_i$) and the extended observability matrix Γ_i are determined. The system matrices are then extracted using one of the algorithms described in Sections 8 and 9.

and with $\mathcal{O}_i$ defined as the oblique projection:

$$\mathcal{O}_i \stackrel{def}{=} Y_f /_{U_f} W_p , \qquad (21)$$

and the singular value decomposition:

$$W_1 \mathcal{O}_i W_2 = \begin{pmatrix} U_1 & U_2 \end{pmatrix} \begin{pmatrix} S_1 & 0 \\ 0 & 0 \end{pmatrix} \begin{pmatrix} V_1^T \\ V_2^T \end{pmatrix} = U_1 S_1 V_1^T , \qquad (22)$$

we have:

1. The matrix $\mathcal{O}_i$ is equal to the product of the extended observability matrix and a matrix $\widetilde{X}_i$ with Kalman filter state estimates:

$$\mathcal{O}_i = \Gamma_i . \widetilde{X}_i , \qquad (23)$$

with:

$$\widetilde{X}_i \stackrel{def}{=} \widehat{X}_{i[\widehat{X}_0, P_0]} ,$$
$$\widehat{X}_0 = X_p^d /_{U_f} U_p , \qquad (24)$$
$$P_0 = -[\Sigma^d - S^{xu}.(R^{uu})^{-1}.(S^{xu})^T] . \qquad (25)$$

2. The order of the system (1)-(2) is equal to the number of singular values in equation (22) different from zero ;

3. The extended observability matrix Γ_i can be chosen as:

$$\Gamma_i = W_1^{-1} U_1 S_1^{1/2}.T . \tag{26}$$

where T is an arbitrary similarity transformation.

4. The part of the state sequence $\tilde{X}_i$ that lies in the column space of W_2 can be recovered from:

$$\tilde{X}_i W_2 = T^{-1}.S_1^{1/2} V_1^T . \tag{27}$$

5. The state sequence $\tilde{X}_i$ can be recovered from:

$$\tilde{X}_i = \Gamma_i^\dagger.\mathcal{O}_i . \tag{28}$$

up to within an arbitrary similarity transformation T.

The proof of this Theorem can be found [VODM 94a].

This Theorem can be summarized algebraically as follows:

$$\text{rank}\,(Y_f/_{U_f} W_p) = n ,$$
$$\text{range}\,(Y_f/_{U_f} W_p) = \text{range}\,(\Gamma_i) ,$$
$$\text{range}\,((Y_f/_{U_f} W_p)^T) = \text{range}\,(\tilde{X}_i^T) .$$

Obviously, the singular value decomposition of the matrix $Y_f/_{U_f} W_p$ delivers these three results at once. It should be noted that the state sequence $\tilde{X}_i$ recovered through this Theorem differs from the state sequence $\hat{X}_i$ that was introduced in Theorem 3. The two sequences are different due to their different initial state $\hat{X}_0$. For Theorem 3, we have $\overline{X}_0 = X_p^d/\begin{pmatrix} U_p \\ U_f \end{pmatrix}$ and for Theorem 4 $\hat{X}_0 = X_p^d/_{U_f} U_p$. This difference in initial states will play a crucial role in the derivation of the algorithms. Finally note that even though their initial state sequence is different, both sequences $\hat{X}_i$ and $\tilde{X}_i$ are generated by a bank of Kalman filters.

7.2 Some intuition

In this Subsection we present some intuition behind the different projections of Theorem 3 and 4. More intuitive and rigorous explanations can be found in [VODM 93d]. The goal of an identification procedure is to find a model of which the input/output behavior approximates that of the system under consideration. This goal is classically solved by minimizing a "prediction error criterion" which expresses the "prediction performance" of the model on the given data

7 Kalman filter states and oblique projections

set. The minimizing solution is designated as the optimal model (see for instance [Lju 87]). In the framework of subspace identification, the identification goal is attained by solving two subsequent problems:

1. Optimal Prediction: As stated above we want to find a model that will predict the behavior of the process sufficiently accurately. This can be formulated as: predict the future outputs (Y_f) as accurately as possible, using all the information that can be obtained from the past (W_p), and using the knowledge of the inputs that will be presented to the system in the future (U_f).

Inspired by the linearity of the system, we propose to combine the past (W_p) and the future inputs (U_f) linearly to predict the future outputs (Y_f). We denote the linear combinations respectively with L_p and L_u. The quality of the prediction is measured in the Frobenius norm. Mathematically, the first part of the identification goal thus becomes:

$$\min_{\substack{L_p \in \mathbb{R}^{li \times (m+l)i} \\ L_u \in \mathbb{R}^{li \times mi}}} \| Y_f - \begin{pmatrix} L_p & L_u \end{pmatrix} \begin{pmatrix} W_p \\ U_f \end{pmatrix} \|_F^2 . \tag{29}$$

Obviously, this least squares problem leads to the orthogonal projection $\mathcal{Z}_i$ of Theorem 3 and hence provides exactly the desired linear combination of W_p and U_f as $\mathcal{Z}_i = \begin{pmatrix} L_p & L_u \end{pmatrix} \begin{pmatrix} W_p \\ U_f \end{pmatrix}$. The optimal linear combination of the past (W_p) to predict the future is the term $L_p W_p$, which is exactly equal to the oblique projection of Theorem 4, i.e. $\mathcal{O}_i = L_p . W_p$.

2. Low order model: Apart from the fact that the the model should predict the future, we also want its order to be as small as possible. We thus need to reduce the rank of $\mathcal{O}_i$. Since the rows of $\mathcal{O}_i$ span an li dimensional subspace in the j dimensional ambient space, we can introduce a complexity reduction by reducing the subspace dimension to n (the order of the system). Intuitively, this implies that we only have to remember n different directions of the past to predict the future. Mathematically, the second step can be formulated:

$$\min_{\mathcal{R} \in \mathbb{R}^{li \times j}} \| W_1 \left[\mathcal{O}_i - \mathcal{R} \right] W_2 \|_F^2 \text{ subject to rank}(\mathcal{R}) = n , \tag{30}$$

The user-defined weighting matrices W_1 and W_2 determine which part of the "information" of $\mathcal{O}_i$ is important to retain. A more rigorous interpretation of the weighting matrices is presented in [VODM 93d] [VODM 94d]. It is easy to show [VODM 93d] that $\mathcal{R}$ is determined through the singular value decomposition (22) as $\mathcal{R} = W_1^{-1} U_1 S_1 V_1^T W_2^\dagger$. Note that $\mathcal{R} = \mathcal{O}_i$ when all the assumptions of Theorem 4 are satisfied exactly (which is however never the case with real-life data). However, when $j \neq \infty$ or when the data generating system is not linear, the singular values of $W_1 \mathcal{O}_i W_2$ (22) are all different from zero. In that case, the row space of $\mathcal{O}_i$ is of dimension li, and the order has to be chosen equal to the number of "dominant" singular values. The complexity reduction step is then truly a reduction of the dimension of the row space of $\mathcal{O}_i$, and the weights W_1 and W_2 play an important role in determining which part of the original row space of $\mathcal{O}_i$ is retained.

7.3 Different weights, different algorithms

One might wonder about the effect of chosing the weights W_1 and W_2 in the singular value decomposition of Theorem 4. Without going into details here, it suffices to say that for certain appropriate choices of W_1 and W_2, one recovers some subspace algorithms that were recently proposed and more or less studied in the literature. Without going into detail, we present in the following table the acronyms and refererences of these algorithms and the weights to be plugged in into Theorem 4 to recover them

Acronym	Full name and Main ref.	W_1	W_2
N4SID	Numerical Algorithm for Subspace State Space System Identification [VODM 94a] [VOWL 93]	I_{li}	I_j
MOESP	Multivariable Output-Error State sPace [Ver 94]	I_{li}	$\Pi_{U_f^\perp}$
CVA	Canonical Variate Analysis [Lar 90]	$(Y_f/U_f^\perp)^T.(Y_f/U_f^\perp)$	$\Pi_{U_f^\perp}$

Our main contribution here is that we have unified these different approaches and put them on an equal footing in the sense that, in one way or another (i.e. implicitly or explicitly) they all recover the state sequence and the observability matrix. We refer to [VODM 93d] for proofs and details.

8 From the states to the state space model

What have we done so far? In Section 6 we have found a relation between the Kalman filter state sequence $\widehat{X}_i$ (14) and the orthogonal projection of the row space of Y_f (future outputs) into the row spaces of past and future inputs and past outputs. In this Section, we discuss an identification algorithm that uses the Kalman filter states as obtained from these orthogonal projections.

8.1 Shifting the state sequence

By the shifted state sequence $\widehat{X}_i$ we mean the state sequence (14) shifted to the right, i.e. starting with $\hat{x}_{i+1}$ and ending with $\hat{x}_{i+j}$. It turns out that this shifted sequence can be obtained from another orthogonal projection of shifted future outputs into shifted past inputs and outputs and future inputs. Why would we try to obtain this shifted state sequence? Because it will give us an elegant way to obtain the state space model A, B, C, D and an estimate of the noise covariance matrices Q, R and S, by just solving a linear least squares problem! This can seen from (1)-(2) and (4).
From Theorem 4, we find:

8 From the states to the state space model

- The order of the system from inspection of the singular values of equations (22).
- The extended observability matrix Γ_i from equation (26) and the matrix Γ_{i-1} as $\underline{\Gamma_i}$ [7].

The following side result of Theorem 3 can easily be proven. By shifting the separation line between "past" and "future" one block row downwards, we obtain the matrices W_p^+, U_f^- and Y_f^- (notations, see Section 4). A similar orthogonal projection as in Theorem 3 can now be performed with these matrices. This leads to the sequence $\mathcal{Z}_{i+1}$ and the Kalman filter states $\widehat{X}_{i+1}$:

$$\mathcal{Z}_{i+1} = Y_f^- / \begin{pmatrix} W_p^+ \\ U_f^- \end{pmatrix} = \Gamma_{i-1}\widehat{X}_{i+1} + H_{i-1}^d U_f^-, \tag{31}$$

$$\widehat{X}_{i+1} = \widehat{X}_{i+1[\overline{X}_0, P_0]}, \tag{32}$$

with $\overline{X}_0$ and P_0 given by equations (17)-(18). There is a very important observation to be made here:

> Corresponding columns of $\widehat{X}_i$ (16) and of $\widehat{X}_{i+1}$ (32) are consecutive state estimates of the <u>same</u> Kalman filters at two consecutive time instants, in the sense that the associated Kalman filters have the same initial state estimate $\overline{X}_0$ and the same initial error covariance P_0.

This statement is not trivial at all and is proven in [VODM 94a]! It implies that we can write with (8):

$$\widehat{X}_{i+1} = A\widehat{X}_i + BU_{i|i} + K_i(Y_{i|i} - C\widehat{X}_i - DU_{i|i}). \tag{33}$$

It is also true that:

$$Y_{i|i} = C\widehat{X}_i + DU_{i|i} + (Y_{i|i} - C\widehat{X}_i - DU_{i|i}). \tag{34}$$

It is now easy to prove (see [VODM 94a]) that the row space of $Y_{i|i} - C\widehat{X}_i - DU_{i|i}$ is orthogonal to the row spaces of W_p, U_f and $\widehat{X}_i$. This result can also be intuitively proven by noticing that the innovations $(Y_{i|i} - C\widehat{X}_i - DU_{i|i})$ of a Kalman filter are uncorrelated with the states $\widehat{X}_i$, the past inputs and outputs W_p and the future inputs U_f. We thus have:

$$\mathbf{E}_j(Y_{i|i} - C\widehat{X}_i - DU_{i|i}). \begin{pmatrix} W_p \\ U_f \\ \widehat{X}_i \end{pmatrix}^T = 0.$$

---

[7] A 'bar' under a symbol means that the matrix $\underline{\Gamma_i}$ is obtained by omitting the last block row of Γ_i.

This implies that (33)-(34) can be written as:

$$\begin{pmatrix} \widehat{X}_{i+1} \\ Y_{i|i} \end{pmatrix} = \begin{pmatrix} A & B \\ C & D \end{pmatrix} \begin{pmatrix} \widehat{X}_i \\ U_{i|i} \end{pmatrix} + \begin{pmatrix} \rho_w \\ \rho_v \end{pmatrix}, \qquad (35)$$

with obvious definitions for ρ_w and ρ_v as residual matrices, the row spaces of which are orthogonal to the row space of W_p, U_f and $\widehat{X}_i$. If we would now be able to compute the state sequences $\widehat{X}_i$ and $\widehat{X}_{i+1}$ from the input-output data, we could solve equation (35) in a least squares sense for the system matrices A, B, C, D. The matrices ρ_w and ρ_v would then contain the least squares residuals. Unfortunately, it is not possible to determine the state sequences $\widehat{X}_i$ and $\widehat{X}_{i+1}$ directly from the input-output data[8]. However, from (15) and (31) we can determine $\widehat{X}_i$ and $\widehat{X}_{i+1}$ as:

$$\widehat{X}_i = \Gamma_i^\dagger \cdot [\mathcal{Z}_i - H_i^d . U_f], \qquad (36)$$

$$\widehat{X}_{i+1} = \Gamma_{i-1}^\dagger \cdot [\mathcal{Z}_{i+1} - H_{i-1}^d . U_f^-]. \qquad (37)$$

In these formulas the only unknowns on the right hand side are H_i^d and H_{i-1}^d, since $\mathcal{Z}_i$ and $\mathcal{Z}_{i+1}$ can be determined as a projection of the input-output block Hankel matrices and Γ_i and Γ_{i-1} are determined through Theorem 4. Substitution of (36) and (37) into (35) leads to:

$$\begin{pmatrix} \Gamma_{i-1}^\dagger . \mathcal{Z}_{i+1} \\ Y_{i|i} \end{pmatrix} = \underbrace{\begin{pmatrix} A \\ C \end{pmatrix} . \Gamma_i^\dagger . \mathcal{Z}_i}_{\text{term 1}} + \underbrace{\mathcal{K}}_{\text{term 2}} . U_f + \underbrace{\begin{pmatrix} \rho_w \\ \rho_v \end{pmatrix}}_{\text{term 3}}, \qquad (38)$$

where we have defined:

$$\mathcal{K} \stackrel{\text{def}}{=} \begin{pmatrix} (B|\Gamma_{i-1}^\dagger . H_{i-1}^d) - A . \Gamma_i^\dagger . H_i^d \\ (D|0) - C . \Gamma_i^\dagger . H_i^d \end{pmatrix}. \qquad (39)$$

8.2 Determining the state space matrices

Observe that the matrices B and D appear linearly in the matrix $\mathcal{K}$. We can now solve equation (38) in a least squares sense for A, C and $\mathcal{K}$. Since the row spaces of ρ_w and ρ_v have been shown to be orthogonal to the row spaces of $\mathcal{Z}_i$ and U_f and since the least squares solution computes residuals that are orthogonal to the regressors, the least squares solution will compute asymptotically unbiased estimates of the system matrices (see [VODM 94a] for details). From (38) we find in this way (term by term):

term 1. The system matrices A and C.

---

[8] For all clarity, Theorem 4 determines the state sequence $\widetilde{X}_i$ directly from the input-output data. This sequence is different from $\widehat{X}_i$, in a sense that the initial conditions of the Kalman filter are different (see the discussion at the end of Subsection 7.1). We will use the sequence $\widetilde{X}_i$ in the second algorithm.

term 2. The matrix $\mathcal{K}$ from which B and D can be computed. This leads to a least squares problem for B and D, details of which can be found in [VODM 94a] [VO PhD]. Actually, one can show that

> The estimates of A, B, C and D are consistent as $j \to \infty$, provided that the system that generated the data belongs to the model class described by (1)-(2)-(4).

term 3. The covariances Q, S and R can be estimated from the residuals ρ_w and ρ_v as:

$$\begin{pmatrix} Q_i & S_i \\ S_i^T & R_i \end{pmatrix} \simeq \mathbf{E_j}[\begin{pmatrix} \rho_w \\ \rho_v \end{pmatrix} \cdot (\rho_w^T \; \rho_v^T)] \;.$$

This result is biased for finite i as indicated by the subscript i in the estimates Q_i, R_i and S_i. The reason is the fact that the bank of Kalman filters, which was discussed in Figure 4 is not in steady state for finite i (meaning that the Riccati difference equation (10) has not converged yet). As i grows larger, the approximation error grows smaller. For infinite i and j the stochastic system is determined asymptotically unbiased. However, as is obvious by construction, the matrix $\begin{pmatrix} Q_i & S_i \\ S_i^T & R_i \end{pmatrix}$ is guaranteed to be positive definite, which implies the following

> **Fact:** The noise model is guaranteed to be positive real, but its covariance matrix estimates Q_i, S_i and R_i are biased for finite i.

The steps of this first algorithm are summarized in Figure 8.

9 A biased subspace identification algorithm

The algorithm we have just presented, computes unbiased estimates of the system matrices A, B, C and D and biased ones for Q, R and S (unless $i \to \infty$). However, this algorithm is quite complicated (especially the extraction of B and D). One could wonder if there does not exist an algorithm in which the estimation of B and D does NOT require the solution of an additional least squares problem. In turns out that there is such an approach, which uses the the Kalman filter state sequence $\widetilde{X}_i$ (24) found in Section 7 from the oblique projection of the row space of the future outputs, along that of the future inputs, into the row space of past inputs and outputs. However, this algorithm calculates asymptotically biased solutions (see also [VODM 94a]). From a similar reasoning as in Theorem 4, we find that:

$$\mathcal{O}_{i+1} = Y_f^- /_{U_f^-} W_p^+ = \Gamma_{i-1} \cdot \widetilde{X}_{i+1} \;,$$

and we now have $\widetilde{X}_i$ and $\widetilde{X}_{i+1}$. The problem however is that this new Kalman filter sequence $\widetilde{X}_{i+1}$ has a different initial state as the sequence $\widetilde{X}_i$ and hence

Algorithm using the states:

1. Calculate the oblique and orthogonal projections:

$$\mathcal{O}_i = Y_f /_{U_f} W_p \,,$$

$$\mathcal{Z}_i = Y_f / \begin{pmatrix} W_p \\ U_f \end{pmatrix} \,,$$

$$\mathcal{Z}_{i+1} = Y_f^- / \begin{pmatrix} W_p^+ \\ U_f^- \end{pmatrix} \,.$$

2. Calculate the **SVD** of the weighted oblique projection $W_1 \mathcal{O}_i W_2 = USV^T$. where W_1 and W_2 are user-defined weighting matrices.
3. Determine the order by inspecting the singular values in S and partition the **SVD** accordingly to obtain U_1 and S_1.
4. Determine Γ_i and Γ_{i-1} as $\Gamma_i = W_1^{-1} U_1 S_1^{1/2}$ and $\Gamma_{i-1} = \underline{\Gamma_i}$.
5. Solve the set of linear equations for A, C and $\mathcal{K}$ (using least squares) which also gives residuals ρ_w and ρ_v:

$$\begin{pmatrix} \Gamma_{i-1}^\dagger . \mathcal{Z}_{i+1} \\ Y_{i|i} \end{pmatrix} = \begin{pmatrix} A \\ C \end{pmatrix} . \Gamma_i^\dagger . \mathcal{Z}_i + \mathcal{K} . U_f$$

$$+ \begin{pmatrix} \rho_w \\ \rho_v \end{pmatrix} \,.$$

6. Determine B and D from $\mathcal{K}, A, C, \Gamma_i, \Gamma_{i-1}$ as described in Subsection 8.2.
7. Determine Q_i, S_i and R_i from the residuals as:

$$\begin{pmatrix} Q_i & S_i \\ S_i^T & R_i \end{pmatrix} = \mathbf{E_j}[\begin{pmatrix} \rho_w \\ \rho_v \end{pmatrix} . \begin{pmatrix} \rho_w^T & \rho_v^T \end{pmatrix}] \,.$$

Fig. 8. A schematic overview of the first combined deterministic-stochastic identification algorithm. See Section 10 for implementation issues.

is not just a 'shifted' version of $\widetilde{X}_i$. Indeed, we have as an initial state for $\widetilde{X}_i$, $X_p^d /_{U_f} U_p$, while the initial state for $\widetilde{X}_{i+1}$ is $X_p^d /_{U_f^-} U_p^+$.

So, we can not write a formula similar to (35) with $\widehat{X}_i$ and $\widehat{X}_{i+1}$ replaced by $\widetilde{X}_i$ and $\widetilde{X}_{i+1}$!

It can be proven however (see [VODM 94a]) that the difference between $\widehat{X}_i$ and $\widetilde{X}_i$ is zero when at least one of the following conditions is satisfied:

1. $i \to \infty$. The difference between $\widehat{X}_i$ and $\widetilde{X}_i$ goes to zero at the same rate as the Riccati difference equation of the Kalman filter converges. This is intuitively clear, since by the time the Kalman filter is in steady state, the effect of the initial conditions has died out. In [VODM 94a] a rigorous proof

is given.
2. The system is purely deterministic.
3. The deterministic input u_k is white noise. In this case the deterministic state X_p^d is uncorrelated with U_p and U_f. This implies that the initial state sequences of $\widehat{X}_i, \widehat{X}_{i+1}, \widetilde{X}_i$ and $\widetilde{X}_{i+1}$ are all equal to zero (and are thus equal to each other).

When at least one of these conditions is satisfied, we can replace $\widehat{X}_i$ and $\widehat{X}_{i+1}$ in (35) with $\widetilde{X}_i$ and $\widetilde{X}_{i+1}$, and solve for A, B, C and D in a least squares sense, which will then give an asymptotically unbiased estimate. The details of this simple algorithm are summarized in Figure 9. If none of the above conditions is satisfied, this second algorithm will return biased estimates of the system matrices (since the states are replaced by approximations). In [VODM 94a] an expression for this bias is derived. Even though the algorithm is simpler than the first one we have presented, it turns out that it should be used with care in practice. For many practical measurements the input signal u_k is anything but white noise (steps, impulses, ...). This implies that the algorithm is biased, and this bias can be significant. Our experience with lots of practical data, is that, if computational complexity is no real problem, the first algorithm is to be preferred over the second one, but that the robust version that we will treat next, beats both of them.

10 Engineering a robust identification algorithm

The reader should be aware of the fact that the outlines of the algorithms presented so far only form the central backbone of a more sophisticated implementation. Details can be found in [VODM 94a]. In this Section, we will however not discuss these important alterations. Instead we will show in some detail how the oblique and orthogonal projections can be implemented in a numerically robust manner using the LQ-decomposition of the input-output block Hankel matrices.

10.1 An LQ-decomposition

The common factor in the implementation of all the subspace algorithms is the LQ-decomposition (see [GVL 89]) of the block Hankel matrix formed of the input and output measurements[9]:

$$\mathcal{H} \stackrel{\text{def}}{=} \frac{1}{\sqrt{j}} \underbrace{\begin{pmatrix} U_{0|2i-1} \\ Y_{0|2i-1} \end{pmatrix}}_{\in \mathbb{R}^{2(m+l)i \times j}} = L.Q^T ,$$

---
[9] The scalar $1/\sqrt{j}$ is used to be conform with the definition of $\mathbf{E_j}$.

with $Q \in \mathbb{R}^{j \times 2(m+l)i}$ orthonormal ($Q^T Q = I_{2(m+l)i}$) and $L \in \mathbb{R}^{2(m+l)i \times 2(m+l)i}$ lower triangular [10]. The LQ-decomposition is partitioned as follows:

$$\begin{pmatrix} U_{0|2i-1} \\ Y_{0|2i-1} \end{pmatrix} = \begin{array}{c} mi \\ mi \\ li \\ li \end{array} \begin{pmatrix} U_p \\ U_f \\ Y_p \\ Y_f \end{pmatrix} = \begin{array}{c} m(i+1) \\ m(i-1) \\ l(i+1) \\ l(i-1) \end{array} \begin{pmatrix} U_p^+ \\ U_f^- \\ Y_p^+ \\ Y_f^- \end{pmatrix} = \begin{array}{c} mi \\ m \\ m(i-1) \\ li \\ l \\ l(i-1) \end{array} \begin{pmatrix} U_{0|i-1} \\ U_{i|i} \\ U_{i+1|2i-1} \\ Y_{0|i-1} \\ Y_{i|i} \\ Y_{i+1|2i-1} \end{pmatrix}$$

$$= L\, Q^T = \begin{array}{c} mi \\ m \\ m(i-1) \\ li \\ l \\ l(i-1) \end{array} \begin{pmatrix} L_{11} & 0 & 0 & 0 & 0 & 0 \\ L_{21} & L_{22} & 0 & 0 & 0 & 0 \\ L_{31} & L_{32} & L_{33} & 0 & 0 & 0 \\ L_{41} & L_{42} & L_{43} & L_{44} & 0 & 0 \\ L_{51} & L_{52} & L_{53} & L_{54} & L_{55} & 0 \\ L_{61} & L_{62} & L_{63} & L_{64} & L_{65} & L_{66} \end{pmatrix} \begin{pmatrix} Q_1^T \\ Q_2^T \\ Q_3^T \\ Q_4^T \\ Q_5^T \\ Q_6^T \end{pmatrix}.$$

We will use the shorthand Matlab notation $L_{[4:6],[1:3]}$ for the submatrix of L consisting of *block* rows 4 to 6 and *block* columns 1 to 3, and the shorthand $Q_{3:4}^T$ in a similar way. For instance[11]:

$$L_{[4:5],[4:6]} = \begin{pmatrix} L_{44} & 0 & 0 \\ L_{54} & L_{55} & L_{56} \end{pmatrix} \text{, and also } L_{[1,4],[1,3]} = \begin{pmatrix} L_{11} & 0 \\ L_{41} & L_{43} \end{pmatrix}.$$

10.2 Expressions for the geometric operations

In this Subsection we give expressions for the geometric operations introduced in Section 5 in terms of the LQ-decomposition of the previous Subsection. Again, in this Section, A and B are *dummy* matrices (not to be confused with the model matrices).

Orthogonal projections Orthogonal projections can be easily expressed in function of the LQ-decomposition. We first treat the general case A/B, where A and B consist of any non-overlapping selection of rows of $\mathcal{H}$, which implies they can be expressed as linear combinations of the matrix Q^T as:

$$A = L_A Q^T,$$
$$B = L_B Q^T, \tag{40}$$

---

[10] Again, there might be some confusion in the notation here as Q is also one of the covariance matrices in the noise model. However, it will always be clear from the context which Q is actually meant. We stick to the notation Q here mainly because of a deeply rooted historical tradition in numerical analysis to call the numerical result of a Gram-Schmidt orthogonalization procedure applied to the rows of a matrix the LQ-decomposition.

[11] By convention, this Matlab subscripting, has priority over transposition $L_{[1,4],[1,3]}^T = \left[L_{[1,4],[1,3]}\right]^T$, $Q_{3:4}^T = [Q_{3:4}]^T$.

10 Engineering a robust identification algorithm

for certain matrices L_A and L_B (consisting of a non-overlapping selection of rows of L). We thus get:

$$A/B = \Phi_{[A,B]} \cdot \Phi^\dagger_{[B,B]} \cdot B = [L_A Q^T Q L_B^T] \cdot [L_B Q^T Q L_B^T]^\dagger \cdot L_B Q^T$$
$$= L_A L_B^T \cdot [L_B L_B^T]^\dagger \cdot L_B Q^T \ .$$

The important observation to be made here is *that the factor Q does not really play a role in these derivations*. One only needs to do certain operations (products and inverses) with certain submatrices of the L-factor. In many cases, this can be even further simplified. Consider for instance the expression for $\mathcal{Z}_i$ (see Equation (15)): Identifying this with (40), we have:

$$L_A = L_{[5:6],[1:6]} \ , \quad L_B = L_{[1:4],[1:6]} \ ,$$

which, when L_B is of full row rank, simplifies to

$$\mathcal{Z}_i = L_{[5:6],[1:6]} L^T_{[1:4],[1:6]} [L_{[1:4],[1:6]} L^T_{[1:4],[1:6]}]^{-1} L_{[1:4],[1:6]} Q^T$$
$$= L_{[5:6],[1:4]} L^T_{[1:4],[1:4]} [L_{[1:4],[1:4]} L^T_{[1:4],[1:4]}]^{-1} L_{[1:4],[1:4]} Q^T_{1:4}$$
$$= L_{[5:6],[1:4]} \cdot \underbrace{L^T_{[1:4],[1:4]} L^{-t}_{[1:4],[1:4]}}_{=I} \cdot \underbrace{L^{-1}_{[1:4],[1:4]} L_{[1:4],[1:4]}}_{=I} \cdot Q^T_{1:4}$$
$$= L_{[5:6],[1:4]} Q^T_{1:4} \ ,$$

which is a very simple expression for $\mathcal{Z}_i$ indeed! Similarly, we find

$$\mathcal{Z}_{i+1} = L_{[6],[1:5]} \, Q^T_{1:4} \ ,$$

which again, is very simple.

Oblique projections The oblique projection can also be written in function of the LQ decomposition. For instance with A, B and C being matrices the rows of which are a non-overlapping selection of the rows of $\mathcal{H}$ [12]:

$$A = L_A Q^T \ , \quad B = L_B Q^T \ , \quad C = R_C Q^T \ ,$$

where L_A, L_B and L_C are certain non-overlapping rows of the L-factor of $\mathcal{H}$, we find that

$$A/B^\perp = L_A[I - L_B^T[L_B L_B^T]^\dagger L_B].Q^T \ ,$$
$$C/B^\perp = L_C[I - L_B^T[L_B L_B^T]^\dagger L_B].Q^T \ ,$$

and with the orthogonal projection operator being idempotent ($\Pi_{B^\perp} \cdot \Pi_{B^\perp} = \Pi_{B^\perp}$), we find through formula (7) for the oblique projection:

$$A/_B C = A/B^\perp \cdot [C/B^\perp]^\dagger \cdot C$$
$$= L_A[I - L_B^T[L_B L_B^T]^\dagger L_B]L_C^T \times [L_C[I - L_B^T[L_B L_B^T]^\dagger L_B]L_C^T]^\dagger \cdot L_C Q \ .$$

---

[12] Again, A, B and C are *dummy* matrices of this section, not to be confused with the model matrices

For instance with:

$$A = Y_f = L_{[5:6],[1:6]}Q^T, B = U_f = L_{[2:3],[1:6]}Q^T, C = W_p = L_{[1:4],[1:6]}Q^T,$$

we could compute the oblique projection $Y_f/_{U_f} W_p$ using the above formula. Once again, observe that the matrix Q does not really 'participate' in these expressions. Even though this would lead to a valid expression for the oblique projection, there is a better way to calculate this quantity by first projecting Y_f into the row space of the past outputs and the past and future inputs, and then separating the effect of the future inputs U_f out of this projection (see also Section 5). This can be done as follows: With L_{U_p}, L_{U_f} and L_{Y_p} defined as:

$$\left(\underbrace{L_{U_p}}_{\in \mathbb{R}^{li \times mi}} \; \underbrace{L_{U_f}}_{\in \mathbb{R}^{li \times mi}} \; \underbrace{L_{Y_p}}_{\in \mathbb{R}^{li \times li}} \right) \stackrel{\text{def}}{=} L_{[5:6],[1:4]} L_{[1:4],[1:4]}^\dagger,$$

we get for the oblique projection:

$$Y_f/_{U_p} W_p = \left[L_{U_p}.L_{[1:1],[1:4]} + L_{Y_p}.L_{[4:4],[1:4]} \right] Q_{1:4}^T.$$

This computation is significantly faster than the previous one, since (when $L_{[1:4],[1:4]}$ is of full rank) the matrices L_{U_p}, L_{U_f} and L_{Y_p} can be computed using back-substitution (since $L_{[1:4],[1:4]}$ is a lower triangular matrix).

10.3 An implementation of the robust identification algorithm

To illustrate the simplicity of using the LQ-decomposition when implementing subspace identification algorithms, for completeness we present here one version of the subspace algorithms which has proven its usefulness on practical applications, without going into the rationales for the different steps. In this algorithm, we need an expression for the following projected oblique projection:

$$\mathcal{O}_i \Pi_{U_f^\perp} = \left[L_{U_p}.L_{[1:1],[1:4]} + L_{Y_p}.L_{[4:4],[1:4]} \right] Q_{1:4}^T$$

$$\times Q \left[I_{2(m+l)i} - \begin{pmatrix} L_{[2:3],[1:3]}^T \\ 0 \end{pmatrix} [L_{[2:3],[1:3]} L_{[2:3],[1:3]}^T]^\dagger \left(L_{[2:3],[1:3]} \; 0 \right) \right] Q^T.$$

With $\Pi = I_{2mi} - L_{[2:3],[1:3]}^T [L_{[2:3],[1:3]} L_{[2:3],[1:3]}^T]^{-1} L_{[2:3],[1:3]}$, this leads to:

$$\mathcal{O}_i \Pi_{U_f^\perp} = \left((L_{U_p} L_{[1:1],[1:3]} + L_{Y_p} L_{[4:4],[1:3]}).\Pi \, \big| \, L_{Y_p} L_{44} \right) Q_{1:4}^T.$$

The other steps of the implementation are straightforward (except maybe for the step where B and D are determined). The overall implementation is illustrated in Figure 10.

> Let us summarize that, by using the LQ-decomposition of the data matrix with input-output data, we achieve a drastic reduction in computational complexity. The Q factor, which has a dimension of $j \times 2(l+m)i$, never needs to be computed! It cancels out in all the expressions we need to obtain the state space model. Only the L-factor is required, the dimension of which is $2(m+l)i \times 2(m+l)i$, which is substantially smaller that Q. In brief, the LQ-decomposition, in which only L is computed, acts as an important data compression step. All quantities of interest can be calculated from L only.

11 Conclusions

In this paper, we have tackled the problem of multivariable system identification for multiple input multiple output, linear, combined deterministic-stochastic systems. Such models often provide good engineering models for real industrial plants, especially for design of model-based controllers. By combining insights, concepts and algorithms from system theory, (numerical) linear algebra and geometry, we have developed a new breed of system identification techniques, called subspace algorithms, that do not suffer from some shortcomings of 'classical' (prediction-error-method) identification approaches.

There are many extensions, special cases or properties that cannot be discussed in this paper due to space (and time) constraints. We restrict ourselves here to a brief literature survey: Some historical developments relating to the roots of subspace algorithms are surveyed in [DMVO 94], including realization theory [DMK 74a] [DKM 74b] [HK 66] [Kun 78] [Moo 81] [MR 76] [Sil 71] [ZM 74], deterministic system identification [Gop 69] [Bud 71] [LS 77] [DMo 88] [Wil 86] [DMV 87], stochastic realization theory [Aka 74] [Aka 75] [Aka 76] [Aok 87] [AK 90] [Cai 88] [DP 84] [DKP 85] [Fau 76] [VODM 93a]. Extensions are described for descriptor systems in [MDMRT 92], continuous time systems [MDMV 91], identification problems with known noise structure [MV 90], periodic systems [He 93] [Verr 94]) and frequency domain versions [McK 94b]. Further properties and interpretation can be found in [MDM 92] [JW 94] [VODM 93d] [MR 93] [OV 94] [VODM 94d] [LP 93] [LP 94]. Results on the statistical analysis can be found in [Vib 94] [DPS 94] [Lar 94] [VOWL 91] and in the context of array signal processing (which is much related to the system identification problem) [Ott 89] [OVSN 93] [OVK 92] [SROK 92] [Vib 89] [VO 91]. Software issues are treated in [AMKMVO 93] [AKVODMB 93] [VODMBAK 94] [VODMAKB 94], [Lju 91c]. Successful applications of subspace identifications methods are reported in vibrational analysis [AML 94], rapid thermal processing of VLSI silicium wafers [Gyu 93] [Cho 93] [CK 93], modelling of industrial processes [FVOMHL 94]

[VODM 93c] [ZVODML 94] and direction-of-arrival estimation for broad band sources [VP 93]. Relations to other identification methods, such as the Eigensystem Realization Algorithm (ERA, [Jua 85], [Jua 94]), Q-Markov covers [AnSk 88] [LSS 92] [Kin 88] [Ske 94], Observability Range Extraction [LS 91] [Liu 92] [LJM 94], etc... still need to be explored. Further research to speed up the calculations by exploitation of the (block) Hankel matrix structure using the notion of displacement rank, is reported in [CXK 94a] [CXK 94b] [CK 95] [Cho 93]. Another interesting idea is to apply subspace algorithms as a first initial guess to start up a nonlinear optimization problem, such as in [Chu 94] [Lju 91b] [Lju 91c].

Finally, there is one problem (which at first sight has to do with elegance but might also have more profound implications) which has bothered us for a long time. The algorithms we have presented here are *asymmetric* in two respects: One needs to make an a priori distinction between inputs and outputs, even if one wants to determine the states only from inputs and outputs (Indeed in the oblique projection it is Y_f that is projected along U_f onto W_p.) The second asymmetry has to do with 'time' and 'causality'. Indeed, it are the 'future' outputs Y_f that are projected onto the past inputs/outputs and the future inputs. Both of these asymmetries do not exist in the pure deterministic and stochastic cases (at least not in the first step where the state is estimated from inputs and outputs, without knowing the model). At present it remains an open problem whether such a double 'symmetric' approach, which in spirit would come very close to Willems' work [Wil 86], is possible.

12 Acknowledgments

We would like to thank Professors Gene Golub, Thomas Kailath, Robert Kosut and Stephen Boyd, all from Stanford University (California) for the many occasions at which we were their guest. That a real company is something completely different from our protected universitarian environment, was taught to us convincingly by Henk Aling and Alexandra Schmidt from Integrated Systems Incorporated, Santa Clara, California. The Swedish gang, consisting of Professor Lennart Ljung from Linköping University, professors Anders Lindquist, Björn Ottersten and Bo Wahlberg (Royal Institute of Technology, Stockholm) and professor Mats Viberg (Chalmers University, Göteborg) have shared with us lots of ideas and suggestions, many of which still need to be worked out. Professors Jan Maciejowski (Cambridge University, UK), Jan Willems en Jan Van Schuppen (Rijksuniversiteit Groningen) and Michel Gevers (UCL, Belgium) have contributed a lot in creating a stimulating European Research Network on System Identification.

References

[AML 94] Abrahamsson T., McKelvey T., Ljung L. *A study of some approaches to vibration data analysis.* Proc. of SYSID '94, Vol. **3**, 4-6 July, Copenhagen, Denmark, pp.289-294, 1994.

[Aka 74] Akaike H. *Stochastic theory of minimal realization.* IEEE Transactions on Automatic Control, **19**, pp. 667-674, 1974.

[Aka 75] Akaike H. *Markovian representation of stochastic processes by canonical variables.* Siam J. Control, Vol. **13**, no.1, pp. 162-173, 1975.

[Aka 76] Akaike H. *Canonical correlation analysis of time series and the use of an information criterion.* System Identification: Advances and case studies, R.K. Mehra and D.G. Lainiotis eds., New York, Academic Press, pp.27-96, 1976.

[AMKMVO 93] Aling H., Milletti U., Kosut R.L., Mesaros M.P., Van Overschee P., De Moor B. *An interactive system identification module for Xmath.* Proc. of the American Control Conference, June 2-4, San Francisco, USA, pp. 3071-3075, 1993.

[AKVODMB 93] Aling H., Kosut R., Van Overschee P., De Moor B., Boyd S. *Xmath Interactive System Identification Module, Part 1.* Integrated Systems Inc., Santa Clara, USA, 1993.

[AnSk 88] Anderson B.D.O., Skelton R.E. *The generation of all q-Markov covers.* IEEE Trans. on Circuits and Systems, 35, pp.375-384, 1988.

[Aok 87] Aoki M. *State space modeling of time series.* Springer Verlag, Berlin, 1987.

[AK 90] Arun K.S., Kung S.Y. *Balanced approximation of stochastic systems.* SIAM J. Matrix Analysis and Applications, **11**, pp. 42-68, 1990.

[AE 71] Åström K., Eykhoff P. *System identification - A survey.* Automatica, Vol. **7**, pp. 123-167, 1971.

[AW 84] Åström K., Wittenmark B. *Computer Controlled Systems: Theory and Design.* Prentice Hall, 1984.

[BJ 76] Box G.E., Jenkins G.M. *Time series analysis, forecasting and control.* Revised edition, Holden-Day series in time series analysis and digital processing, Holden-Day, Oackland, 1976.

[Bud 71] Budin M. *Minimal realization of discrete linear systems from input-output observations.* IEEE Transactions on Automatic Control, Vol. AC-**16**, no. 5, pp. 395-401, 1971.

[Cai 88] Caines P. *Linear Stochastic Systems.* Wiley Series in Probability and Mathematical Statistics, 1988.

[Cho 93] Cho Y.M. *Fast subspace based system identification: Theory and practice.* Ph.D. Thesis, Information Systems Lab, Stanford University, CA, USA, 1993.

[CK 93] Cho. Y.M., Kailath T. *Model Identification in Rapid Thermal Processing Systems.* IEEE Trans. on Semicond. Manuf., Vol. **6**, no. 3, 1993.

[CXK 94a] Cho Y.M., Xu G., Kailath T. *Fast Recursive Identification of State Space Models via Exploitation of Displacement Structure.* Automatica, Special Issue on Statistical Signal Processing and Control, Vol. **30**, no. 1, pp. 45-59, 1994.

[CXK 94b] Cho Y.M., Xu G., Kailath T. *Fast Identification of State Space Models via Exploitation of Displacement Structure.* IEEE Transactions on Automatic Control, Vol. AC-**39**, no. 10, 1994.

[CK 95] Cho Y.M., Kailath T. *Fast Subspace-Based System Identification : An Instrumental Variable Approach.* To appear in Automatica.

[Chu 94] Chun T.C. *Geometry of linear systems and identification.* PhD Thesis, Cambridge University, England, March 1994, 173 pp.

[DPS 94] Deistler M., Peternell K., Scherrer W. *Consistency and Relative Efficiency of Subspace Methods.* Proc. of SYSID '94, Vol. **2**, 4-6 July, Copenhagen, Denmark, pp.157-163, 1994.

[DMV 87] De Moor B., Vandewalle J. *A geometrical strategy for the identification of state space models of linear multivariable systems with singular value decomposi-*

tion. Proc. of the 3rd International Symposium on Applications of Multivariable System Techniques, April 13-15, Plymouth, UK, pp. 59-69, 1987.

[DMo 88] De Moor B. *Mathematical concepts and techniques for modeling of static and dynamic systems*. PhD thesis, Department of Electrical Engineering, Katholieke Universiteit Leuven, Belgium, 1988.

[DMo 94] De Moor B. *Numerical algorithms for state space subspace system identification*. Academia Analecta, Klasse der Wetenschappen, Koninklijke Akademie voor Wetenschappen, jaargang 55, no. 5, België, 1994.

[DMVO 94] De Moor B., Van Overschee P., *Graphical User Interface Software for System Identification*, Award winning paper of the *Siemens Award 1994*. ESAT-SISTA Report 94-06I, Department of Electrical Engineering, Katholieke Universiteit Leuven, 1994.

[DP 84] Desai U.B., Pal D. *A transformation approach to stochastic model reduction*. IEEE Transactions on Automatic Control, Vol. AC-**29**, no. 12, 1984.

[DKP 85] Desai U.B., Kirkpatrick R.D., Pal D. *A realization approach to stochastic model reduction*. International Journal of Control, Vol. **42**, no. 4, pp. 821-838, 1985.

[DMK 74a] Dickinson B., Morf M., Kailath T. *A minimal realization algorithm for matrix sequences*. IEEE Transactions on Automatic Control, Vol. AC-**19**, no. 1, pp. 31-38, 1974.

[DKM 74b] Dickinson B., Kailath T., Morf M. *Canonical Matrix fraction and state space descriptions for deterministic and stochastic linear systems*. IEEE Transactions on Automatic Control, Vol. AC-**19**, pp. 656-667, 1974.

[Eyk 74] Eykhoff P. *System identification*. Wiley, London, 1974.

[FVOMHL 94] Falkus H., Van Overschee P., Murad G., Hakvoort H., Ludlage J. *Advanced Identification and Robust Control of a Glass Tube Production Process*. Third Philips Conference on Applications of Systems and Control Theory (PACT), November 9-10, Doorwerth, The Netherlands, 1994.

[Fau 76] Faurre P. *Stochastic realization algorithms*. System Identification: Advances and case studies (Eds.) Mehra R., Lainiotis D., Academic Press, 1976.

[GeWe 82] Gevers M., Wertz V. *On the problem of structure selection for the identification of stationary stochastic processes*. 6th IFAC Symposium on Identification and System Parameter Estimation, Eds. G. Bekey and G. Saridis, Washington D.C., 1982, pp.387-392.

[GW 74] Glover K., Willems J. *Parametrizations of linear dynamical systems: canonical forms and identifiability*. IEEE Transactions on Automatic Control, Vol. AC-**19**, pp. 640-645, 1974.

[GVL 89] Golub G., Van Loan C. *Matrix computations.*, second edition, Johns Hopkins University Press, Baltimore, Maryland, 1989.

[Gop 69] Gopinath B. *On the identification of linear time-invariant systems from input-output data*. The Bell System Technical Journal, Vol. **48**, no. 5, pp. 1101-1113, 1969.

[Gui 75] Guidorzi R. *Canonical structures in the identification of multivariable systems*. Automatica, **11**, pp. 361-374, 1975.

[Gui 81] Guidorzi R. *Invariants and canonical forms for systems structural and parametric identification*. Automatica, **17**, pp. 177-133, 1981.

[Gyu 93] Gyugyi Paul. *Model-Based Control Applied to Rapid Thermal Processing*. Ph.D. Thesis Stanford University, CA, USA, 1993.

[HD 88] Hannan E.J., Deistler M. *The statistical theory of linear systems*. Wiley Series in Probability and Mathematical Statistics, John Wiley and Sons, New York,

1988.

[He 93] Hench J.J. *Identification of linear periodic state space models.* Inst. of Information Theory and Automation, Praag, 1993, to be published in the International Journal of Control.

[HK 66] Ho B.L., Kalman R.E. *Efficient construction of linear state variable models from input/output functions.* Regelungstechnik, **14**, pp. 545-548,1966.

[JW 94] Jansson M., Wahlberg B. *4SID Linear Regression.* Proc. 33rd IEEE Conference on Decision and Control, Lake Buena Vista, Florida, USA, 14-16 Dec., pp. 2858-2863, 1994.

[Jua 85] Juang J.N., Pappa R.S. *An eigensystem realization algorithm for model parameter identification and model reduction.* Journal of Guidance, Control and Dynamics, Vol. 8., no.5, Sept-Oct 1985, pp.620-627.

[Jua 94] Juang J.N. *Applied System Identification.* Prentice Hall, Englewood Cliffs, NJ, 1994.

[Kai 80] Kailath T. *Linear Systems.* Prentice Hall, Englewood Cliffs, New Jersey, 1980.

[Kin 88] King A.M., Desai U.B., Skelton R.E. *A generalized approach to Q-Markov Covariance Equivalent Realizations for Discrete Systems.* Automatica, 24, pp.507-515, 1988.

[Kun 78] Kung S.Y. *A new identification method and model reduction algorithm via singular value decomposition.* 12th Asilomar Conf. on Circuits, Systems and Comp., pp. 705-714, Asilomar, CA, 1978.

[Lar 83] Larimore W.E. *System identification, reduced order filtering and modeling via canonical variate analysis.* Proc. of the American Control Conference, San Francisco, USA, 1983.

[Lar 90] Larimore W.E. *Canonical variate analysis in identification, filtering and adaptive control.* Proc. 29th Conference on Decision and Control, Hawai, USA, pp. 596-604, 1990.

[Lar 94] Larimore W.E. *The Optimality of Canonical Variate Identification by Example,* Proc. of SYSID '94, Vol. **2**, 4-6 July, Copenhagen, Denmark, pp.151-156, 1994.

[LP 93] Lindquist A., Picci G. *On "Subspace Methods" Identification.* Proc. of MTNS, Regensburg, Germany, August 2-6, Vol. **2**, pp. 315-320, 1993.

[LP 94] Lindquist A., Picci G. *On "Subspace Methods" Identification and Stochastic Model reduction.* Proc. of SYSID '94, Vol. **2**, 4-6 July, Copenhagen, Denmark, pp. 397-404, 1994.

[LS 91] Liu K., Skelton R.E., *Identification and Control of NASA's ACES Structure.* Proceedings American Control Conference, Boston, MA, USA, pp. 3000-3006, 1991.

[Liu 92] Liu K., *Identification of Multi-Input and Multi-Output Systems by Observability Range Space Extraction.* Proc. 31st Conference on Decision and Control, Tucson, Arizona, USA, pp. 915-920, 1992.

[LSS 92] Liu K., Skelton R.E., Sharkey J.P., *Modeling Hubble Space Telescope Flight Data by Q-Markov Cover Identification.* Proc. of the American Control Conference, pp. 1961-1965, 1992.

[LJM 94] Liu K., Jacques R.N., Miller D.W. *Frequency Domain Structural System Identification by Observability Range Space Extraction.* Proc. of the American Control Conference, Baltimore, Maryland, Vol. **1**, pp. 107-111, 1994.

[LS 77] Liu R., Suen L.C. *Minimal dimension realization and identifiability of input-output sequences.* IEEE Transactions on Automatic Control, Vol. AC-**22**, pp. 227-232, 1977.

[Lju 87] Ljung L. *System identification - Theory for the User*. Prentice Hall, Englewood Cliffs, NJ, 1987.

[Lju 91b] Ljung. L. *A Simple Start-Up Procedure for Canonical Form State Space Identification, Based on Subspace Approximation*. 30th IEEE Conference on Decision and Control, Brighton, UK, pp. 1333-1336, 1991.

[Lju 91c] Ljung. L. *System Identification Toolbox For Use with Matlab*. The Mathworks Inc., Mass., U.S.A., 1991.

[Lue 67] Luenberger D.G. *Canonical forms for linear multivariable systems*. IEEE Transactions on Automatic Control, Vol. AC-**12:290**, 1967.

[McK 93] McKelvey T., *System identification using overparametrized state space models*. Dept. of EE, Linköping University, Sweden, 1993, Report LiTH-ISY-R-1454, 41 pp.

[McK 94a] McKelvey T., *On State-Space Models in System Identification*, Thesis no. 447, Department of Electrical Engineering, Linköping university, Sweden, 1994.

[McK 94b] McKelvey T. *An Efficient Frequency Domain State-Space Identification Algorithm*. Proc. 33rd IEEE Conference on Decision and Control, Lake Buena Vista, Florida, USA, 14-16 Dec., pp. 3359-3364, 1994.

[McK 94c] McKelvey T. *SSID - A MATLAB Toolbox for Multivariable State-Space Model Identification*. Dept. of EE, Linköping University, Linköping Sweden, 1994.

[MDMVV 89] Moonen M., De Moor B., Vandenberghe L., Vandewalle J. *On and off-line identification of linear state space models*. International Journal of Control, Vol. **49**, no. 1, pp. 219-232, 1989.

[MV 90] Moonen M., Vandewalle J. *A QSVD approach to on- and off-line state space identification*. International Journal of Control, Vol. **51**, no. 5, pp. 1133-1146, 1990.

[MDMV 91] Moonen M., De Moor B., Vandewalle J. *SVD-based subspace methods for multivariable continuous time system identification*. Identification of continuous-time systems, G.P. Rao, N.K. Sinha (Eds.), Kluwer Academic Publications, pp. 473-488, 1991.

[MDM 92] Moonen M., De Moor B. *Comments on 'State-space model identification with data correlation'*. International Journal of Control, Vol. **55**, no. 1, pp. 257-259, 1992.

[MDMRT 92] Moonen M., De Moor B., Ramos J., Tan S. *A subspace identification algorithm for descriptor systems*. Systems & Control Letters, Vol. **19**, pp. 47-52, 1992.

[MR 93] Moonen M, Ramos J. *A subspace algorithm for balanced state space system identification*. IEEE Transactions on Automatic Control, Vol. **38**, pp. 1727-1729, 1993

[Moo 81] Moore B.C. *Principal component analysis in linear systems: Controllability, Observability and Model Reduction*. IEEE Transactions on Automatic Control, Vol. AC-**26**, no. 1, pp 17-32, 1981.

[MR 76] Mullis C.T., Roberts R.A. *Synthesis of minimum round-off noise fixed point digital filters*. IEEE Transactions on Circuits and Systems, Vol. CAS-**23**, pp. 555-562, 1976.

[Nor 86] Norton J.P. *An introduction to identification*. Academic Press, London, 1986.

[Ott 89] Ottersten B. *Parametric subspace fitting methods for array signal processing*. Ph.D. Thesis, Information Systems Laboratory, Department of Electrical Engineering, Stanford University, CA, USA, 1989.

[OVK 92] Ottersten B., Viberg M., Kailath T., *Analysis of subspace fitting and ML techniques for parameter estimation from sensor array data*, IEEE Tr. on SP, Vol. **40**, no. 3, pp. 590-600, 1992.

[OVSN 93] Ottersten B., Viberg M., Stoica P., Nehorai A., *Exact and large sample maximum likelihood techniques for parameter estimation and detection in array processing*, in Radar array processing, S.Haykin, J.Litva, T.J.Sherpherd (Eds.), Springer Verlag, pp. 99-151, 1993.

[OV 94] Ottersten B., Viberg M., *A Subspace Based Instrumental Variable Method for State Space System Identification*. Proc. of SYSID '94, Vol. **2**, 4-6 July, Copenhagen, Denmark, pp.139-144, 1994.

[Pal 82] Pal D., *Balanced stochastic realization and model reduction*. Master's thesis, Washington State University, Electrical Engineering, 1982.

[Ske 94] Skelton R.E., Grigoriadis Karolus, Zhu Guoming. *An algorithm for iterative identification and control design using an impro ved Q-Markov cover.* SYSID '94, Copenhagen, pp.183-188.

[SS 89] Söderström T., Stoica P. *System Identification*. Prentice Hall International Series in Systems and Control Engineering, Prentice Hall, New York, 1989.

[Sil 71] Silverman L. *Realization of linear dynamical systems*. IEEE Transactions on Automatic Control, Vol. AC-**16**, pp. 554-567, 1971.

[SROK 92] Swindlehurst A., Roy R., Ottersten B., Kailath T. *System identification via weighted subspace fitting*. Proc. of the American Control Conference, pp. 2158-2163, 1992.

[VDS 93] Van Der Veen A., Deprettere E.F., Swindlehurst A.L. *Subspace-Based Signal Analysis Using Singular Value Decompositions*. Proceedings of the IEEE, Vol. **81**, no. 9, pp. 1277-1308, 1993.

[VOL 82] van Overbeek A.J.M., Ljung L. *On-line structure selection for multivariable state space models*. Automatica, **18**, no. 5, pp. 529-543, 1982.

[VODM 93a] Van Overschee P., De Moor B. *Subspace algorithms for the stochastic identification problem*. Automatica, Vol. **29**, no. 3, 1993, pp. 649-660.

[VODM 93c] Van Overschee P., De Moor B. *Subspace identification of a glass tube manufacturing process*. Proc. of the second European Control Conference, June 28-July 1, Groningen, The Netherlands, pp. 2338-2343, 1993.

[VODM 93d] Van Overschee P., De Moor B., *A Unifying Theorem for three Subspace System Identification Algorithms*. ESAT-SISTA Report 93-50I, Department of Electrical Engineering, Katholieke Universiteit Leuven, 1993. Accepted for *Automatica*.

[VODM 94a] Van Overschee P., De Moor B. *N4SID: Subspace Algorithms for the Identification of Combined Deterministic-Stochastic Systems*. Automatica, Special Issue on Statistical Signal Processing and Control, Vol. **30**, no. 1, pp. 75-93, 1994.

[VODMBAK 94] Van Overschee P., De Moor B., Boyd S., Aling H., Kosut R. *A fully interactive system identification module for Xmath (ISID)*, Proc. of SYSID '94, Vol. **4**, Copenhagen, Denmark, pp. 1, 1994.

[VODM 94d] Van Overschee P., De Moor P. *About the choice of State Space Bases in Combined Deterministic-Stochastic Subspace Identification*, ESAT-SISTA Report 94-24I, Department of Electrical Engineering, Katholieke Universiteit Leuven, 1994. Accepted for publication in *Automatica*.

[VODMAKB 94] Van Overschee P., De Moor B., Aling H., Kosut R., Boyd S. *Xmath Interactive System Identification Module, Part 2*. Integrated Systems Inc., Santa Clara, CA, USA, 1994.

[VO PhD] Van Overschee P., *Subspace identification: Theory - Implementation - Application*. PhD Thesis, Department of Electrical Engineering, Katholieke Universiteit Leuven, Belgium, February 1995.

[VP 93] Van Poucke F., Moonen M. *Direction finding of multiple wide-band emitters using state space modelling*. ESAT-SISTA TR 1993-30, Department of Electrical Engineering, Katholieke Universiteit Leuven, Belgium. To be published in IEEE Trans. on Signal Processing.

[Ver 91] Verhaegen M. *A novel non-iterative MIMO state space model identification technique*. Proc. 9th IFAC/IFORS Symp. on Identification and System Parameter Estimation, Budapest, Hungary, pp. 1453-1458, 1991.

[VD 92] Verhaegen M., Dewilde P. *Subspace model identification, Part I: The output-error state space model identification class of algorithms*. Int. J. Control, Vol. **56**, 1187-1210, 1992.; Part II: Analysis of the elementary output-error state space model identification algorithm, Vol. 56, no.5, pp.1211-1241, ; Part III: Analysis of the ordinary output error state space model identification algorithm, Vol. 58, no.3, pp.555-586, 1993.

[Ver 94] Verhaegen M. *Identification of the deterministic part of MIMO state space models given in innovations form from input-output data*, Automatica (Special Issue on Statistical Signal Processing and Control), Vol **30**, no. 1, pp. 61-74, 1994

[Verr 94] Verriest E.I., Kullstam J.A. *Realization of discrete-time periodic systems from input-output data*. Int. Report, Georgia Tech Lorraine, 1994, submitted for publication.

[Vib 89] Viberg M. *Subspace fitting concepts in sensor array processing*. Ph.D. thesis, Department of Electrical Engineering, Linköping University, Sweden, 1989.

[VO 91] Viberg M., Ottersten B. *Sensor array processing based on subspace fitting*. IEEE Transactions on Acoustics, Speech and Signal Processing, ASSP-**39**, 1991.

[VOWL 91] Viberg M., Ottersten B., Wahlberg B., Ljung L. *A Statistical Perspective on State-Space Modeling Using Subspace Methods*. Proc. of the 30th IEEE Conference on Decision and Control, Brighton, England, 11-13 Dec, Vol **2**, pp 1337-1342, 1991.

[VOWL 93] Viberg M., Ottersten B., Wahlberg B., Ljung L., *Performance of Subspace Based State Space System Identification Methods*. Proc. of the 12th IFAC World Congress, Sydney, Australia, 18-23 July, Vol **7**, pp 369-372, 1993.

[Vib 94] Viberg M. *Subspace Methods in System Identification*. Proc. of SYSID '94, Vol. **1**, 4-6 July, Copenhagen, Denmark, pp. 1-12, 1994.

[Wil 86] Willems J. *From time series to linear systems*. Automatica, Part I: Finite Dimensional Linear Time Invariant Systems, Vol. **22**, no. 5, pp. 561-580, 1986, Part II: Exact Modelling, Vol. **22**, no. 6, pp. 675-694, 1986, Part III: Approximate modelling, Vol. **23**, no. 1, pp. 87-115, 1987.

[ZM 74] Zeiger H., McEwen A. *Approximate linear realizations of given dimension via Ho's algorithm*. IEEE Transactions on Automatic Control, **19**, pp. 153, 1974.

[ZVODML 94] Zhu Y., Van Overschee P., De Moor B., Ljung L. *Comparison of three classes of identification methods*. Proc. of SYSID '94, Vol. **1**, 4-6 July, Copenhagen, Denmark, pp. 175-180, 1994.

> **Algorithm (simpler to estimate B and D)**
>
> 1. Calculate the oblique projections:
>
> $$\mathcal{O}_i = Y_f /_{U_f} W_p ,$$
> $$\mathcal{O}_{i+1} = Y_f^- /_{U_f^-} W_p^+ .$$
>
> 2. Calculate the **SVD** of the weighted oblique projection:
>
> $$W_1 \mathcal{O}_i W_2 = USV^T ,$$
>
>   where W_1 and W_2 are user-defined weighting matrices.
> 3. Determine the order by inspecting the singular values in S and partition the **SVD** accordingly to obtain U_1 and S_1.
> 4. Determine Γ_i and Γ_{i-1} as:
>
> $$\Gamma_i = W_1^{-1} U_1 S_1^{1/2} , \quad \Gamma_{i-1} = \underline{\Gamma_i} .$$
>
> 5. Determine the state sequences:
>
> $$\widetilde{X}_i = \Gamma_i^\dagger . \mathcal{O}_i ,$$
> $$\widetilde{X}_{i+1} = \Gamma_{i-1}^\dagger . \mathcal{O}_{i+1} .$$
>
> 6. Solve the set of linear equations for A, B, C and D in a linear least squares sense (which also delivers the residuals ρ_w and ρ_v):
>
> $$\begin{pmatrix} \widetilde{X}_{i+1} \\ Y_{i|i} \end{pmatrix} = \begin{pmatrix} A & B \\ C & D \end{pmatrix} \begin{pmatrix} \widetilde{X}_i \\ U_{i|i} \end{pmatrix} + \begin{pmatrix} \rho_w \\ \rho_v \end{pmatrix} .$$
>
> 7. Determine Q_i, S_i and R_i from the residuals as:
>
> $$\begin{pmatrix} Q_i & S_i \\ S_i^T & R_i \end{pmatrix} = \mathbf{E_j}[\begin{pmatrix} \rho_w \\ \rho_v \end{pmatrix} . \begin{pmatrix} \rho_w^T & \rho_v^T \end{pmatrix}] .$$

Fig. 9. A schematic overview of the second combined deterministic-stochastic identification algorithm. This algorithm computes asymptotically biased solutions, unless at least one of the three conditions discussed above is satisfied. See Section 10 for implementation issues. Note the 'symmetry' in this algorithm in the determination of the matrices A, B, C and D (Compare to the first algorithm) which gives a certain elegance to this algorithm, although in general it provides biased estimates.

Robust implementation of N4SID

1. Compute $L_{U_p} \in \mathbb{R}^{li \times mi}$ and $L_{Y_p} \in \mathbb{R}^{li \times li}$ from:

$$\begin{pmatrix} L_{U_p} & L_{U_f} & L_{Y_p} \end{pmatrix} = L_{[5:6],[1:4]} L^\dagger_{[1:4],[1:4]} ,$$

where the pseudo inverse can be substituted by a back-substitution when $L_{[1:4],[1:4]}$ is of full rank.

2. With:

$$\Pi = I_{2mi} - L^T_{[2:3],[1:3]} [L_{[2:3],[1:3]} L^T_{[2:3],[1:3]}]^{-1} L_{[2:3],[1:3]} .$$

Compute the **SVD** of:

$$\left((L_{U_p} L_{[1:1],[1:3]} + L_{Y_p} L_{[4:4],[1:3]}).\Pi \,\middle|\, L_{Y_p} L_{44} \right) .$$

3. Determine the order by inspection of the singular values in S and partition the **SVD** accordingly to obtain U_1 and S_1.

4. Determine $\Gamma_i = U_1 S_1^{1/2}$ and $\Gamma_{i-1} = \underline{\Gamma_i}$. Compute A and C as the first n columns of:

$$\begin{pmatrix} \Gamma^\dagger_{i-1} L_{[6:6],[1:5]} \\ L_{[5:5],[1:5]} \end{pmatrix} \begin{pmatrix} \Gamma^\dagger_i L_{[5:6],[1:5]} \\ L_{[2:3],[1:5]} \end{pmatrix}^\dagger .$$

Recompute Γ_i and Γ_{i-1} from A and C.

5. Solve for B and D from the least squares problem discussed and referred to in Subsection 8.2.

6. Form the matrices $\mathcal{P}$ and $\mathcal{Q}$:

$$\mathcal{P} = \begin{pmatrix} \Gamma^\dagger_{i-1} L_{[6:6],[1:5]} \\ L_{[5:5],[1:5]} \end{pmatrix} - \begin{pmatrix} A \\ C \end{pmatrix} \Gamma^\dagger_i L_{[5:6],[1:5]} ,$$

$$\mathcal{Q} = L_{[2:3],[1:5]} .$$

7. Determine the matrix $\mathcal{K}$ (39) from A, B, C, D, Γ_i and determine estimates of the covariance matrices Q, S and R as:

$$\begin{pmatrix} Q_i & S_i \\ S_i^T & R_i \end{pmatrix} = (\mathcal{P} - \mathcal{KQ})(\mathcal{P} - \mathcal{KQ})^T .$$

Fig. 10. Implementation of a robust deterministic-stochastic identification algorithm. For details the reader is referred to [VODM 94a] [VO PhD]. Notice that the 'long' dimension j has disappeared, which implies that the factor Q in the LQ-decomposition of the block Hankel matrix with input-output data is never needed explicitly and hence should not be computed.